AF520595

Climate Change and Agriculture:
Adaptation and Mitigation

Climate Change and Agriculture:
Adaptation and Mitigation

Editors :
RK Bhatt, U Burman, DK Painuli,
DV Singh, Ramavtar Sharma and SPS Tanwar

ICAR – Central Arid Zone Research Institute
Jodhpur-342003 (Rajasthan), INDIA

SATISH SERIAL PUBLISHING HOUSE
403, Express Tower, Commercial Complex
Azadpur, Delhi-110033 (India)
Phone : 011-27672852, Fax : 91-11-27672046
e-mail : info@satishserial.com, hkjain1975@yahoo.com
Website : www.satishserial.com

Published by :

SATISH SERIAL PUBLISHING HOUSE

403, Express Tower, Commercial Complex, Azadpur, Delhi-110033 (INDIA)
Phone : 011-27672852 Fax : 91-11-27672046
E-mail : info@satishserial.com, hkjain1975@yahoo.com

ISBN 978-93-85055-86-7

Composed, Designed & Printed in India

Preface

Climate change, earlier known as global warming, is now an accepted fact and its effects like increase in temperature, rise in sea level, extreme weather events like intense rainfall, droughts, floods, etc. are frequently being experienced in different parts of the world. The level of greenhouse gases (GHGs) has increased substantially compared to preindustrial era.

Forest degradation and deforestation are considered to contribute significantly to the rise of global GHGs. Agriculture and allied sector is also a contributor to GHGs emissions. The major sources of the GHGs in the agriculture sector are livestock enteric fermentation, soil tillage, inefficient manure and fertilizer management and burning of crop residues.

Agriculture sector not only contributes to GHGs, but is very sensitive to climate change. Crop yields are influenced by variations in climatic factors such as precipitation, air temperature, relative humidity, wind speed and the frequency and severity of extreme events like droughts, floods, hot wave, cold wave, windstorms and hailstorms.

Adverse impact on agriculture will pose serious risks to economy and livelihood security of agrarian class in particular and the country as a whole. Farmers have, so far, been coping with erratic climatic conditions by adjusting their farming practices. Scientific management practices have also been developed to reduce adverse impacts of natural climate variability. The projected climate change and associated variability, however, brings new risks that will require new adaptation and mitigation strategies.

Considering the multi-dimensional facets of climate change, the book has been compiled to address various relevant issues to sensitize the stakeholders about different aspects of climate change and their impacts on arid and semi-arid agriculture.

We are thankful to all the contributing authors of different chapters for their dedicated and timely contribution. We expect that the book will be useful to different stakeholders involved in arid and semiarid agriculture under changing climatic conditions.

Editors

Contents

Chapter 1

Changing Perceptions in Carbon Pool and Fluxes in Relation to Climate Change

PRAVEEN KUMAR*, UDAY BURMAN AND R.K. BHATT

ICAR - Central Arid Zone Research Institute, Jodhpur-342003, Rajasthan,

* E-mail: pkumar_bhatnagar@yahoo.com

Introduction

Over geologic history, an ingenious system has evolved, whereby atmospheric CO_2 is redistributed between vegetation, ocean and rocks through a range of pathways referred as the carbon cycle. These pathways have maintained CO_2 concentration in the atmosphere at around 163 ppm for centuries (Bolin *et al.*, 1979; Defense *et al.*, 1981). However, after industrialization, CO_2 concentration in the atmosphere has increased consistently. The rate of increase during 1980 to 1999 was 3.2 to 3.3 Pg-Cyr^{-1} (Prentice *et al.*, 2001). Present atmospheric concentration of CO_2 (396 ppm; WMO 2014) is further expected to increase to 600 ppm in 2100 and to 1300 ppm in 2200 and may level off at 1800 ppm in the year 2400 (Lee & Dodson, 1996; Harvey, 2000). CO_2 enrichment in the atmosphere and its consequent impact on global temperatures (5-6°C increase by 2100) were first speculated by Arrhenius (1896) and were echoed again much later by the US National Academy of Science in 1980 (Nierenberg *et al.*, 2010). Subsequently, IPCC predicted global temperatures to increase from 1 to 6.4°C (IPCC, 1991, 1996, 2001 and 2007). Nierenberg *et al.*, (2010) estimated temperature to increase between 1-3°C by 2100.But, other different scenarios of CO_2 concentration and consequent rise in atmospheric temperature have also been presented (Stern, 2007; Hansena *et al.*, 2012). Against these predictions of warming, stabilization of temperature recoded during 2000-2010 (Dessler, 2013; Dessler & Zelinka, 2015) and attributed to Pacific ocean currents (Meehl *et al.*, 2013) further add uncertainty to the extent temperature may rise in future.

During the mid-eighties, the implications of global warming were also perceived in terms of climate change, food security and disease incidences and since then this topic has received a widespread attention from scientists, governments and the people at large. Some interpretations of these changes predict a daunting scenario for agricultural production, increased food insecurity, poverty and infant mortality (Giorgi & Hewitson, 2001; Parry *et al.*, 2004; Kramer, 2007; Oluoko-Odingo 2009; Ramasamy & Moorthy, 2012; Dev & Sharma, 2010; Masters *et al.*, 2010; Ahmad *et al.*, 2011; Greg *et al.*, 2011).

But, of late, it is also being argued that present climate models are far from being perfect due to their inability to simulate (i) the amount and character of clouds in the atmosphere (ii) behavior of the oceans (iii) interaction between global geochemical cycles of N, P, S with C cycle during the model functions, and (iv) CO_2 fertilization in plants and many others (Paltridge & Calder, 2010; Paltridge, 2014; Zher, 2014). Reducing these uncertainties would still be a couple of decades away as a gap of about 50 years exists between an increase in greenhouse gases and changes in climate parameters. (Watson,http://lgmacweb.env.uea.ac.uk/ajw/Reprints Certainty_ uncertainty_ formatted.pdf <http://lgmacweb.env.uea.ac.uk/ajw/ %20Reprints%20 Certainty_uncertainty_formatted%20.pdf>,assessed on April 21, 2015). Therefore, the effects of climate change may not be universally similar (Lee, 2009). Thus, climate change debate now is moving away from arguments about the reality of global warming to understanding the implications of the same for us in the future in more and more local context; and also, whether it is better to adapt to or to mitigate the change. At this stage imperfection in assessment procedures make it difficult to finalize the path to reach the goal. Our actions will be different depending on whether the world will be 6 degrees warmer or one degree, in 2100. Both the economics and the ethics of these two outcomes would lead most people to favor mitigation in the former case, but adaptation in the latter. Scientific uncertainty has been a highly prominent aspect of climate change articles which cloud our perceptions of future climate. Climate change has huge economic consequences as well (Stern 2008; Weitzman, 2009) and the perfection of climatic models is still decades away. Therefore, Hansena *et al.* (2012) opted for an empirical approach to improve perceptions of climate change.

In this review, we have attempted to build the understanding of the intricacies involved in the prediction of global climate change by first putting some basic questions and then searching for related information in literature to arrive at an empirical conclusion.

Will CO_2 emission from fossil fuels continue to increase progressively?

Burning fossil fuel (including that from cement production) released on an average 5.4 Pg-C yr^{-1} during 1980 to 1989, 6.3 Pg-C yr^{-1} during 1990 to 1999 and 7.8 Pg-C yr^{-1} during 2000-2009 (IPCC 2013) and 9.5 Pg-C in 2011 (Peter *et al.*, 2013). This raises two issues (i) for how long will the fossil fuels would support the energy requirement of the society, and (ii) for how long emissions due to fossil fuels would maintain an upward trend?

The amounts of proven and potential fossil fuel reserves are uncertain and often debated, as complete information is neither provided by industry nor the State (IEA, 2008). Quantity of extractable fossil fuels, however,was estimated to be between 4000 and 6000 Pg-C (Goudriaan & Ketner, 1984). Of this British Petroleum estimated global proven oil reserves to be 1687 thousand million barrels (nearly equivalent to 54.8 Pg-C) in 2013 (British Petroleum, 2014a). Some believe that the energy industry is already close to the maximum amount of oil that can physically be extracted at low cost (Campbell &Laherre're 1998; Cavallo 2007). Others however disagree and consider that oil reserves and current production could be expanded by increased geographical exploration (Maugeri, 2004) or by increased investment in countries making up the Organization of the Petroleum Exporting Countries (Kerr, 2008; IEA, 2008). Reports of about 62% increase in proven oil reserves between 1993 and 2013 also support this view (British Petroleum, 2014a).

Coal deposits in 2013 were estimated to range from 892 to 1052 billion tonnes (nearly 841 Pg-C) which is equivalent to 134.5 years of global coal output (World Coal Association, 2013). Coal production has increased from 4677 million tons in 1990 to 7823 million tons in 2013 and about 40% of it is used to generate electricity (World Coal Association, 2013). Global coal consumption was expected to increase by 45 per cent between the reference year of 1995 and 2015. Nearly 20.3% of the global electricity is produced through coal, which accounts for 41.2% of the C emission (Raghuvanshi *et al.*, 2006). Present average global efficiency rate of coal-fired power plants are 33% and the use of coal is tagged with the release of pollutants(excluding CO_2) such as oxides of sulfur (SO_x) and nitrogen (NO_x), particulate and trace elements such as mercury. A range of advanced coal combustion technologies have been developed to improve the efficiency of coal-fired power generation with reduced emissions of CO_2, NO_x, SO_x and particulates. Improving efficiency levels increase the amount of energy that can be extracted from a single unit of coal and reduce the impact on climate. It is estimated the 1% improvement in the efficiency of a conventional pulverized coal combustion plant reduced CO_2 emissions by 2 to 3%. Enhancing the current average global efficiency rate of coal-fired power plants from 33% to 40% by deploying more advanced off-the-shelf technology or further improving it through high efficiency, low emission coal-fired power plants, could cut CO_2-C emissions by 2 Pg yr^{-1}, a value close to India's current annual CO_2 emission (http://www.worldcoal.org/coal-the-environment/coal-use-the-environment/ improving-efficiencies/ <http://www.worldcoal.org/coal-the-environment/coal-use-the-environment/%20improving-efficiencies/> assessed on April 25, 2015). Additionally, in last five year coal is being gradually substituted by natural gas.

Proved natural gas reserves rose from 157 trillion m^3 in 1995 to 185.7 trillion m^3 (http://knoema.com/smsfgud/world-reserves-of-fossil-fuels, British Petroleum, 2014b) and about 32% of the total estimated technically recoverable natural gas resources (proven and unproven reserves) are in shale formations (EIA, 2013). Globally, Shale gas is swapping of coal electric power generation (EPA, 2012, Credit Suisse2013) as combustion of natural gas produce approximately half the CO_2 that coal does. This shift is clearly perceptible in US and may significantly alter the CO_2 emission scenario in the world (EPA, 2012, Cohen 2013) in coming years.

Enhanced use of renewable energy is being adopted world over to decrease CO_2 emission. The theoretical potential of renewable energy in the world is estimated to be 143000000 Exajoules of which more than 7500 Exajoules is considered as the technical potential which is about 14 times higher than the total electricity produced in the world in 2011 (Johansson et al 2004; http://data.worldbank.org/indicator/EG.ELC.PROD.KH assessed on April 29, 2015). Between 2000 and 2011, installed global renewable electricity grew by 72% (from 748 GW to 1,285 GW), accounting for 22% of all global electricity generated in 2011. According to recent global energy projections, this momentum will continue. The majority of electric generating capacity additions through 2035 will come from natural gas and renewable energy sources rather than traditional fossil fuels. Overall, renewable capacity is expected to triple by 2035, supplying close to one-third of electricity globally (NREL, 2013).

Empirical Conclusion: There are enough fossil fuels to support future energy demand and by increasing their utilization efficiency or by shifting to low CO_2 emitting fossil fuels, CO_2 emission can actually be curtailed.

Whether capacity of ocean as CO_2 sink will decrease in future?

The ocean is a large reservoir of C (about 36000 Pg-C) and a significant sink for anthropogenic carbon dioxide (Siegenthaler& Sarmiento, 1993). IPCC (2014) estimated that out of the 555 Pg-C released in atmosphere from fossil fuel and changes in land use, between 1750 and 2011, 155 Pg-C was absorbed in the oceans. However, the absolute amount of CO_2 that oceans will be able to absorb continues to be a matter of debate as the amount of CO_2 that can be taken up by the oceans in a given year is a function of wind speed, air and water temperatures and concentration gradients which are difficult to determine it accurately (Raghuvanshi *et al.*, 2006). Consequently the literature estimates in literature vary from 7.4 to 1.0 Pg-CO_2 yr^{-1} (Energy facts sheet-greenhouse gases: http:// www.iclei.org/efacts/greengas. htm; Maier-Reimer *et al.*, 1996; Sarmiento *et al.*, 1998; IPCC, 2014).

CO_2 from the atmosphere enters into ocean mainly through chemical dissolution in water and forms H_2CO_3. The dissolution of CO_2 is governed by Henry's Law and increases with CO_2 concentration in the atmosphere (Tans *et al.*, 1990). The additional CO_2 absorbed by ocean waters is balanced by the dissolution of carbonates from the skeletal remains of $CaCO_3$ producing organisms. This equilibrium effectively buffers the pH of ocean waters around 8. The limited surface area of contact between ocean waters and atmosphere limits the rate of absorption. A part of absorbed CO_2 is also used by phytoplankton for photosynthesis. Marine NPP is estimated to range between 27 to 56 Pg-Cyr^{-1} (Berger, 1989; Knauer, 1993; Antonie *et al.*, 1996, Buitenhuis *et al.*, 2013a,b) and in terms of quantity contribute almost half to the photosynthesis that takes place on the Earth. But phytoplankton often escape attention due to their small size and ephemeral nature. Also, some marine phytoplankton release large amount of dissolved organic carbon (a component of NPP) in the ocean

(Baines & Pace, 1991) which make the estimations of NPP very difficult. Li *et al.* (1983) found that 25 to 90% of the photosynthetic biomass in the Pacific Ocean passes through 1ìm filter. Such Pico-plankton may account for up to 50% of ocean NPP (Stockner&Antia, 1986). Further, coastal zones account for 20% oceanic NPP which is about 10 per cent of global NPP while open oceans account for the rest of the 80 per cent NPP (Buitenhuis *et al.*, 2013a,b). More than 40% of the marine NPP is degraded to CO_2, NO_3 etc. in the surface waters and the remaining sinks below the euphotic zone to deep ocean (mean sinking rate of 350 m day^{-1}) but its degradation continues during sinking process also (Duarte &Cebrian 1996; Asper *et al.*, 1992; Sayles *et al.*, 1994; Honjo *et al.*, 1982). About 95% of the particulate organic matter (POM) decompose within 3000 m depth of the ocean and only a small quantity reach the sediment of the deep ocean where decomposition continue (Smith, 1992; Jahnke, 1996). The amount of NPP that sinks to the bottom is estimated to be 7.4 Pg-C yr^{-1}. Berner (1982) estimated the rate of incorporation of organic C in sediments to be 0.157 Pg-C yr^{-1}which suggested that about 98% of the POM sinking in the sea is decomposed.

Phytoplankton keeps the surface water of ocean unsaturated in CO_2 and that help in further absorption of CO_2. The annual uptake of CO_2 by the ocean is about a 92 Pg-C yr^{-1}. Nearly 90Pg-C$yr^{-1}$is annually released back from ocean to atmosphere due to decomposition or respiration by phytoplankton making the ocean a net sink of CO_2. Therefore, Laurmann (1979) hypothesized that given enough time ocean could theoretically take up nearly all of the CO_2 released from fossil fuel.

Increasing temperature may initiate physico-chemical changes like it could increase the kinetic energy of ocean waters, increase their circulation and bring more surface area of water in contact with the atmosphere which could enhance CO_2 uptake by the ocean. Increased difference in partial pressure of CO_2 between the atmosphere and ocean surface could also promote its absorption. But temperature rise of marine waters will also lower solubility of gases. Solubility of CO_2 could reduce by 6.25% for every 2.5^{0}C rise (Rivkin & Legendre, 2001) and reduction in solubility of O_2 could effect marine NPP. These changes could be followed by biological effects. Carbon retention of the oceans relies on the balance between primary production, microbial respiration (Karl & Trenberth, 2003) and degradation of carbon in aquatic systems (Azam *et al.*, 1983). Temperature also strongly influence microbiological processes such as production (Rivkin *et al.*, 1996), growth rate (White *et al.*, 1991), and growth efficiency (Giorgio & Cole, 1997; Rivkin & Legendre, 2001) but all are not effected to the same degree. However, three scenarios can be argued (i) when warming positively affects the rates of bacterial production and respiration, but respiration rates increase more abruptly than production rates then bacterial growth efficiency would decline with increasing temperatures, and a higher proportion of the carbon demand would be respired (Rivkin & Legendre, 2001) (ii) when warming affects production more than the respiration rates, bacterial growth efficiency would increase and a higher proportion of the carbon consumed by bacteria would be converted into biomass (Azam *et al.*, 1983), and (iii) if temperature affects bacterial production and respiration rates similarly but bacterial growth efficiency stays low, more carbon would be consumed by microbes but most of it would end up as CO_2 (Vázquez-Domínguez *et al.*, 2007). These differences may account for the observed differences in the ocean's future contribution in carbon cycle (Laws *et al.*, 2000; Raupach *et al.*, 2014). On an ecosystem scale, equal probability of occurrence of either scenario would further complicates the issue and makes it difficult to assess.

Thus predicting behaviour of ocean in future is difficult due to (i) highly variable nature of oceanic environment (ii) varying degree to which different physio-chemical and biological processer are impacted by oceanic environment (ii) difficulties in generalizing composition of biota (McGowan *et al.*, 1998; Timmermann *et al.*, 1999 Pomeroy &Wiebe, (2001); Goes *et al.*, 2005).

Empirical Conclusion: Assessing as to how CO_2 sinks capacity of oceans would change in future would depend on extent of warming and correctness of contribution of phyto-plankton, microorganisms and kinetics of ocean waters. Our present understanding of these processes is limited. Therefore it might be difficult to predict the change in behavior of ocean water as sink of CO_2.

Would net primary productivity (NPP) decrease with increasing CO_2 and temperature?

Plant biomass on earth is estimated to contain 560 Pg-C (Olson *et al.*, 1983) and terrestrial plants absorb 4.5 per cent of fossil-fuel emissions every year (Sendax, 2010). Therefore, if biospheric uptake is decreased, the carbondioxide concentration in the atmosphere will increase even faster. On global scale most estimates of NPP are in the range of 40 to 105 Pg-C yr^{-1} Pg-Cyr^{-1} (Esther *et al.,* 1982; Potter *et al.,* 1993; Field *et al.,* 1995; Melillo *et al.,* 1993). Terrestrial and marine ecosystem account for half of the NPP each but average rating of NPP at land surface is higher (426 g cm^{-2} yr^{-1}) than in the oceans (140 g cm^{-2} yr^{-1}). Tropical forests, savannas and crop lands account for 32, 30 and 14% of terrestrial NPP (Field *et al.,* 1998) with contribution of other biomes being <5%. Contribution of Tundra and desert biomes being <1%.

Primary production is governed by the effect of temperature, precipitation and radiation on photosynthesis (Berry & Bjorkman, 1980; Farquhar *et al.,* 1980; Tian *et al.,* 1998; Churkina *et al.,* 1999; Tao *et al.,* 2003a,b). Nemani *et al.* (2003) suggested that global warming has increased NPP between 1982-1999, especially in northern high latitude ecosystems. Potter *et al.* (2012) also found an increasing trend in global NPP in the Northern Hemisphere between 2000 to 2009 as warming alleviated heat limitations in high latitude ecosystems. Zhao & Running (2010) on the contrary, visualized a reduction in the global NPP (0.55 Pg C) due to increased frequency of large-scale droughts in the Southern Hemisphere which nullified the increased NPP over the Northern Hemisphere. These results were however disputed later (Watts, 2011). Song *et al.* (2013) using a dynamic land ecosystem model, projected an increase in GPP and NPP by 0.6 kg-C $m^{''2}$ $yr^{''1}$ and 0.2 kg-C $m^{''2}$ $yr^{''1}$, respectively, between 2000 to 2009 across the South-Eastern US. Bala *et al.* (2013) reported an increase of 3.9% per decade in terrestrial net NPP over India between 1982 to 2006. Assessing NPP either on the basis of inventory or satellite based approaches do not allow partitioning of the effects of climate and elevated CO_2 on NPP. On a global scale, changing climate in the absence of elevated CO_2, reduced terrestrial NPP while doubling CO_2 concentration under changing climatic condition increased global NPP by 25% (Cao & Woodward, 1998). Pan *et al.* (2006) analyzed the responses doubling of CO_2 from 355 to 710 ppm on NPP using three biogeochemistry models and reported increase in NPP by 5-11%. King *et al.* (1997) using a Geo-referenced model of ecosystem dynamics suggested that doubling of atmospheric CO_2 from 280 ppm to 560 ppm enhanced global NPP by 16.9%.

Sic *et al.* (2008) projected interaction of climate and atmospheric CO_2 to increase terrestrial NPP. However the extent may vary with emission scenarios, model structure and parameters used to simulate the physiological response of plants to climatic changes *viz.* photosynthesis and stomatal conductance (Amthor & Koch, 1995). Mechanistic scheme has been developed by Farquhar *et al.* (1980) to describe leaf-level photosynthesis response to CO_2. Ball *et al.* (1987) developed the Ball-Berry empirical model to describe the behavior of stomatal conductance to water vapor as a function of environmental conditions and net photosynthetic rate. These two schemes have been widely used in existing process-based models to describe plant responses to CO_2 increase. Plant physiological response to warm air which may carry more moisture is not clearly modelled (Pan *et al.* 2014).Reduced transpiration of land plants due to warming may also increase possibility of runoff, a consequence which is seldom studied.

Empirical Conclusion: NPP is expected to increase under increasing temperature and CO_2 scenario but the projection of quantum vary with emission scenarios, model structure and parameters used to simulate the physiological response of plants.

Would above ground C sink in forest decrease with increasing CO_2 and temperature and whether afforestation is a viable global mitigation strategy?

Forests are a major reservoir of terrestrial above ground C and contain an estimated 66 per cent of the same. Estimates of storage in plant matter range from the 500 Pg (Mahli, 2002) to 650 Pg in forest biomass alone (FAO, 2010). Globally, forest vegetation and soils contain about 1146 Pg-C, with approximately 37 per cent of this C in low-latitude forests, 14 percent in mid-latitudes, and 49 per cent at high latitudes (Dixon *et al.*, 1999). Most soil C i.e. 44 per cent is present in the forest and 12 per cent in agro-ecosystem (Dixon & Turner, 1991; Smith *et al.*, 1992, 1993; Dixon *et al.*, 1994). In addition to this, global forest ecosystem account for approximately 90 per cent of annual C flux between atmosphere and terrestrial soil C (Dixon *et al.*, 1994, Dixon, 1995).

Forestry has been proposed as a means to reduce impact of CO_2 related emissions, by either reducing sources or enhancing sinks (Schlesinger & Lichter, 2001). Application of optimum forest and agroforestry management practices in 500-800 mha in 12 to 15 key nations could sequester 1 to 2 Pg-C yr^{-1} using current infrastructure (Winjum *et al.*, 1992, 1993 a,b; Ford-Robertson 1997). Shiraishi (1997) reported that selective cutting of natural forest can increase carbon sink capacity. Makipaa *et al.* (1998) reported an increase in forest NPP by 20% in a mesic site and 44% in a dry site through the application of 600 to 700 kg N ha^{-1} over a period of 26 to 30 years. Smith *et al.* (1997) estimated that afforestation of 30% of present arable land in the European Union would increase soil carbon stocks by about 8% over a century and would substitute up to 0.030 Pg-C yr^{-1} of fossil fuel carbon if the wood were used as biofuel. Nilsson *et al.* (1995) estimated that only about 345 million hectares land across the world is available for a global afforestation program that is economically, politically, and technically feasible. Under such a programme maximum annual rate of carbon fixation (1.48 Pg-C yr^{-1}) could be achieved after 60 years and over a century 104 Pg-C could be sequestered, a value nearly similar to the CO_2 evolved in 15 to 20 years.

Increasing CO_2 concentrations in the atmosphere could further enhance $CO_{2-}C$ sequestered in forests by increasing their growth. But this growth is also coupled with an increase in soil respiration and thus adversely effect C sequestered in soil (Willey *et al.*, 2009). This suggests that forests may sequester less carbon than predicted in response to increased CO_2concentrations.

But in spite of increasing sequestration afforestation could influence global temperatures in an adverse way due to changes in albedo. The albedo (proportion of incident radiation reflected from a surface) of grassland is typically 0.24 to 0.27 compared to approximately 0.12 for pine plantations (Moore, 1976). This means that replacement of grassland by pine plantation results in greater short-wave absorption and this can contribute to increasing temperatures through increase in long-wave radiation emitted from the landscape. Long-wave radiation above radiata pine forest land exceeded that above the grassland by 15% and 25% during winter and summer respectively (Moore, 1976). Kirschbaum *et al.* (2011) estimated that the albedo effect would reduce the climate change mitigation potential of radiata pine plantations in New Zealand by approximately 10-20%. On a global scale, it has been estimated that up to 40% of the sequestration benefits of replacing grassland with monospecies forest might be offset by this effect (Gibbard *et al.*, 2005). Ball *et al.* (2007) went further to suggest that afforestation of boreal regions should result in a net increase in global temperature therefore tree planting in temperate regions should be neutral, while afforestation only in tropical zones could significantly mitigate climate change. However, Pongratz *et al.* (2011) pointed out that past deforestation occurred on productive land, less subject to snow than normal at any given latitude, and so afforestation of most temperate grasslands is deemed to be beneficial. Snow has a very high albedo, and covering of foliage over snowcan greatly impact on energy absorption in the landscape.

Agroforestry could also be considered as a means of conserving 10 Mha forest every year (Dixon 1995; Sanchez &Benites 1987 and Schroeder *et al.* 1993; Brown *et al.*, 1993; Dixon *et al.*, 1994). Although agroforestry systems are not primarily designed for carbon sequestration, there are many recent studies that substantiate the evidence that agroforestry systems can play a major role in storing carbon in aboveground biomass (Sathaye *et al.*, 2001, Montagini & Nair, 2004, Mutuo *et al.*, 2005, Verchot *et al.*, 2007, Verchot, 2014) and in soil and below ground biomass (Nair *et al.*, 2009).

However, sequestering C in the forest has its own limitations. The effect of net sink is considerably reduced if these forests are harvested and the wood is converted to short lived products, such as paper, which are burnt. Large quantities of forest products are disposed of in landfills annually and the C stored there may be recycled to atmosphere, however, Micales & Skog (1997) calculated that maximally only 30% of the carbon from paper and 0.3 per cent of the carbon from wood are ever emitted as landfill gas. The remaining carbon, approximately 28 Tg in 1993, remain in the landfill indefinitely. Some of this carbon may be removed during leachate treatment, but a large portion is permanently sequestered. Nilsson *et al.* (1995) estimated that on global scale forestry has a maximum potential of sequestering 0.34 Pg-C yr^{-1} which is substantially lower than the amount of carbon required to offset current emissions (3.8 Pg-C). There are two other mechanisms that make C stored in forest vulnerable in terms of long-term storage.

(i) **Deforestation:** Deforestation returns large volume of C to atmosphere by reducing the amount C stored in above ground biomass and increasing the oxidation of soil organic C. Lal & Logan (1995) reported much higher losses of CO_2 from soil in shifting cultivation (0.093 to 0.223 Pg-C yr^{-1}) after deforestation than other arable (0.063 Pg-C yr^{-1}) or pasture soil (0.092 Pg-C yr^{-1}). Houghton *et al.* (1983) attempted to estimate the changes in global biomass between 1860 and 1980. They suggested that world biomass has reduced by 110 Pg-C since 1860, equivalent to 13% of the biomass at 1860 level and also showed that in 1960 CO_2 emission due to deforestation was higher than from fossil fuel. Global deforestation during 80s was 15Mha annually, releasing 1.6 Pg-C annually to the atmosphere (Dixon *et al.*, 1994; Brown *et al.*, 1993). Global net CO_2 emissions from land use change are estimated at 1.4, 1.5 and 1.1 Pg-C yr^{-1} for 1980 (s), 1990 (s) and 2000 (s) respectively (Houghton *et al.*, 2012). (ii) **Forest fires:** Fires rapidly convert biomass to CO_2 and restart the process of ecosystem development. Globally loss of NPP by the fire is estimated between 2-5 Pg-C yr^{-1} (Crutzen & Andreae, 1990; Hao & Liu, 1994). But the estimates (Vitousek *et al.*, 1986; Dobrovolsky, 1994) suggesting a loss of 25 to 40 per cent NPP due to wild fire lead to a much higher value. Frequency and area of fire in recent years have increased due to human interventions. As a consequence boreal forest that may have once acted as a store of atmospheric CO_2 is now the potential net source of CO_2 due to higher frequency of fire (Kasischke *et al.*, 1995).

Empirical Conclusion: C sequestered in forest is likely to increase with temperature and increasing CO_2 in atmosphere. But afforestation can sequester only a fraction of the CO_2 evolved and its potential as a mitigation strategy is further limited by albedo considerations.

Would soil continue to hold as much organic carbon with increasing CO_2 and temperature and is SOC sequestration in soil a viable strategy?

Soil C is a major C pool and contains C both in organic form (SOC) and inorganic form (SIC). The amount of C stored in organic pool is almost double of inorganic pool. Soil organic carbon up to 1 m depth is considered biologically active and consists of soluble organic compound, amorphous organic compound and organo-mineral complexes. Estimates of the SOC pool to 1 m depth range vary widely from 710 to 2946 Pg-C but most values in literature center around 1500 Pg-C (Meentemeyer *et al.*, 1985, Bouwman, 1990; Batjes, 1996; Post *et al.*, 1982; Kasting, 1998; Robert, 2001; Jobbagy & Jackson, 2000; FAO/IIASA/ISRIC/ISS-CAS/JRC 2009; Hiederer & Köchy, 2011). Notable differences in global stock of SOC in estimates undertaken in the last decade have been mainly attributed to the values used for the bulk density of organic soils (Scharlemann *et al.*, 2014).

The density of the SOC pool to 1 m depth for different soil types ranges from 30 Pg-C to 800 Pg-C with a predominant range of 50 to 150 Pg-C. The size of SOC pool increases to 2456 Pg-C if depth of 2m is considered for its approximation (Batjes & Sombroek, 1997). Among different soil orders Histosols contain maximum and Vertisols minimum carbon (Lal *et al.*, 1998). The density of SOC varies widely among soils of different eco-regions. High values of 20 to 24 kg m^{-2} are recorded in polar, boreal and alpine region and minimum in arid regions (Lal, 2001a). Between a biome SOC density is higher in cool and moist than in warm and dry biomes. Within a biome SOC density increases with an increase in precipitation. Among different biomes, tropics with 384 to 403 Pg-C are the largest reservoir of the SOC pool to 1 m depth which is almost 26% of the global SOC pool (Lal, 2001a). Lal (2001b) calculated the ratio of C pool in the soil/biotic pools in different ecosystems. The ratio is about 0.41 in temperate deciduous forest, 0.65 in tropical forest, 1.67 in tropical forest, 1.67 in boreal forest, 6.0 in dry-land savannah and 6.33 in temperate grasslands and 31 in the tundra.

The SIC pool in the form of carbonates, can be significant in soils of arid and semi-arid region (Lal & Kimble, 2000; Scharpenseel *et al.*, 2000). The global estimate of SIC pool ranges from 695 to 748 Pg-C in 1 m depth. The total amount of SIC pool to 1 m depth is 203 to 218 Pg-C in soils of tropics as compared with 492 to 530 Pg-C in soils of other regions (Batjes, 1999). Soils of the tropics, therefore, contribute a 29% of the global SIC pool to 1 m depth. The SIC pool comprises of two components: lithogenic inorganic carbon (LIC) and pathogenic inorganic C (PIC). The latter comprises secondary carbonates that are formed through decomposition/dissolution of LIC or carbonate bearing minerals and re-precipitation of weathering products, or through precipitation of atmospheric CO_2 with Ca^{++} and Mg^{++} or as other salts in soils. The formation of secondary carbonates and leaching of carbonates and bicarbonates are important processes in arid and semiarid regions leading to sequestration of inorganic C in soil and aquatic ecosystem (Nordt *et al.*, 2000).

Amount of carbon stored in the soil is dependent on the balance between carbon inputs from leaf litter and root detritus etc. and its outputs from microbial respiration, decomposition, methanogenesis, erosion and leaching (Davidson & Janssens, 2006). Microbial respiration in soil alone accounts for 25% of naturally emitted CO_2 (Willey *et al.*, 2009). Thus, even small changes in decomposition may influence carbon stored in the soil over the decades (Davidson & Janssens, 2006). Most of the soil C is stored in wetlands, peat lands, and permafrost, where microbial decomposition is limited. Microbial methanogenesis is responsible for both natural and human-induced CH_4 emissions. A part of CH_4 produced is also converted to CO_2 by methanotrophic bacteria (Willey *et al.*, 2009). Microorgamisms like *N. eutropha* produce NO and N_2O in the process of oxidizing ammonium ions (Willey *et al.*, 2009).

Conventionally, accelerated soil erosion through water on sloppy lands and through the wind on flat and dry regions is an important mechanism of SOC loss. Because SOC is concentrated on the soil surface, even a relatively small amount of erosion can cause severe reduction in SOC pool. World soils contain about 1500 Pg-C in top 1 m depth of this about 0.38% is displaced annually in terrestrial ecosystem (Lal, 1995). About one tenth of the displaced C is transported to the oceans and the rest is redistributed in terrestrial ecosystems. Erosion disrupts soil aggregates and exposes previously protected SOC and renders it easily mineralizable. Carbon flux into the atmosphere from soil physically displaced by erosion from terrestrial ecosystem is estimated at 1.14 Pg-C yr^{-1}. Currently, however, this perception has changed (Lal, 2003; Govers *et al.*, 2013). Erosion from agricultural soils by water is estimated to mobilize Ca. 30 Pg-C (Quinton *et al.*, 2010). However, this mobilization of soil does not immediately lead to the release of carbon stored in SOC to the atmosphere. Most of the soil and associated carbon is re-deposited on land or in aquatic sediments where it is stored for considerable time periods (Stallard, 1998, Van Oost *et al.*, 2007). Furthermore, some of the SOC that is removed from eroding sites is replaced by so-called dynamic replacement (Harden *et al.*, 1999; Berhe *et al.*, 2008). Thus SOC mineralization rates in eroded soils during or just after erosion events appear to be limited (Wang *et al.*, 2010; Van Hemelryck *et al.*, 2011). Overall, soil erosion is therefore now estimated to be a small source of C (Van Oost *et al.*, 2007; Harden *et al.*, 2008). Reducing erosion will therefore not greatly contribute to an increase of SOC stocks.

Water soluble carbon is leached from the soil as dissolved organic carbon (DOC). This fraction is eventually transported into ground water and rivers. Leaching losses of SOC are minor, but not negligible (Moore, 1998). The fate of organic carbon transported into rivers and other aquatic systems is also not very well understood. There are two types of organic carbon in aquatic systems: the DOC and particulate organic carbon (POC). Ludwig *et al.* (1996) estimated that 0.38 Pg-C $yr^{-1}$is transported from the continents to the ocean every year, of which

0.1 Pg-C is inorganic and 0.62 Pg-C is organic C. One third of organic C flux (0.2 Pg-C) is DOC. Estimates of global flux of POC (size range of 0.5μ to 5 mm) range from 0.07 to 0.2 Pg-C yr^{-1} of which 30 -75per cent is labile and easily mineralized to CO_2 (Ittekkot & Laane, 1991).

Transport of SIC in groundwater may be an important factor affecting sequestration of C as carbonates. Some of the dissolved C or particulate SOC may also be transferred to inland water bodies. Inland water bodies are now recognized as important sources of CO_2 to the atmosphere. Raymond *et al.* (2013) reported a global evasion rate of 2.1 Pg C yr^{-1} which was higher than previous estimates and also suggested that about 70 per cent of the flux occurred over just 20 percent of the land surface. Paneer-Selvam *et al.* (2014) estimated flux of CO_2 in inland water in India and suggested average emissions of 0.022 Pg CO_2 yr^{-1}.

In this context, it is also pertinent to assess the concept of C sequestration in soil to offset the effect of increasing CO_2. Carbon sink capacity of soil, created by the historic land use change and soil degradation, are estimated at 78 Pg or 10-60 Mg ha^{-1} (Lal, 2008). Management of the terrestrial biosphere to conserve and sequester C may potentially reduce or offset 10 to 40% (0.7-2.8 Pg-C yr^{-1}) of current total greenhouse gas emission (Brown *et al.*, 1993; Dixon *et al.*, 1993a,b; Kauppi *et al.*, 1992; Lal & Bruce, 1999; Lal *et al.*, 1999; Lal, 1999). But this approach has some serious limitations with reference to quantification of SOC and complexity in its composition and its interaction (Schmidt *et al.*, 2011; Heimann & Reichstein, 2008). Changes in baseline against which effects are measured, limited understanding of the temperature and environmental constraints on decomposition of SOM especially its less decomposable fractions or substrate (Davidson & Janssens, 2006; Conant *et al.*, 2008) and limited understanding of the saturation concept of SOC (Angers *et al.*, 2011, Govers *et al.*, 2013) are a few reasons for these limitations. Thus, changes in global SOC stock over the next century varies in literature from an increase of 300 Pg C to a decrease of 50 Pg-C over the next century (Eglin *et al.*, 2011) due to inaccuracies in the assessment (Don *et al.*, 2007; Rumpel & Koegel-Knabner, 2011).

Empirical Conclusion: Our capability to account for microbial behavior in a model to monitor changes in SOC and SIC stability suffers from same impediments as in case of the marine microbial community discussed in earlier paragraphs. This limits our understanding of these processes on terrestrial scale.

Conclusion

It is evident from above, that predicting how increasing CO_2 and temperature scenario will unfold in the future, is very difficult. Most of the uncertainty in this regard emerges from the variations in (i) inventory datasets for model calibration, validation and process understanding and (ii) model structure (Zheng *et al.*, 2003, Lauenroth *et al.*, 2006; Jung *et al.*, 2007). The first point is highlighted by reported doubling of the carbon sink in Europe since 2003 (as indicated by atmospheric and ground-based measurements) attributed to additional process understanding (Schulze *et al.*, 2009). Further, differences in representations of the ecological processes in different models can also lead to huge variation. It happens because on one hand the interactions between components of terrestrial ecosystems are complicated and sometimes not even very well understood and on the other assumptions related to parameter-base and structure of numerical models are often over simplified. Often, quantum of gross carbon uptake within a model is strongly linked with the way and extent of interactions between different biogeochemical cycles (carbon, water, and nitrogen) in the model. Also ecosystem models often lack the structural aspects to capture below-ground processes (Ostle *et al.*, 2009). Magnani *et al.* (2007) observed that relationships between forest GPP and mean annual temperature are strongly related to a corresponding gradient of nitrogen availability. The sensitivity of a model to respond to meteorological conditions, therefore depend on accuracy of interaction of biogeochemical cycles (Jung *et al.*, 2007; Tian *et al.*, 2010, Tianand Lu, 2013).

As a consequence of limitations discussed above, the extent of variations in observed rainfall analysis often precludes perception of any clear impact of the increasing CO_2. Also initial perceptions of adverse effect of temperature on NPP/crop productivity may be masked by the fertilizer effect of increasing CO_2. Thus, on a larger scale the effect on agricultural productivity may not be as drastic as presented in the opening paragraph, however, some perturbations may still be felt at the local level. The entire range of interaction between soil, climate and ocean encompassing the effect of the global greenhouse effect on soil and vegetation deserve much greater attention and a number of unknown factors still thwart our ability to predict precisely the future changes and these uncertainties make conceptualization of alternative strategies even more difficult.

References

Ahmad, J., Dastgir, A. and Haseen, S. 2011. Impact of climate change on agriculture and food security in India. *International Journal of Agriculture, Environment and Biotechnology* 4:129-137.

Amthor, J.S. and Koch, G.W. 1995. Biota growth factor â: simulation of terrestrial ecosystem net primary production of elevated atmospheric CO_2. In *Carbon dioxide and terrestrial ecosystems* (Eds G. W. Koch and J. Roy), Viewed at Google Scholar.

Angers, D.A., Arrouays, D., Saby, N.P.A. and Walter, C. 2011. Estimating and mapping the carbon saturation deficit of French agricultural topsoils. *Soil Use and Management* 27:448-452.

Antoine, D., Andre, J.M. and Morel, A. 1996. Oceanic primary production 2. Estimation at global scale from satellite (coastal zone color scanner) chlorophyll. *Global Biogeochemical Cycles* 10: 570-572.

Arrhenius, S. 1896. On the influence of carbonic acid in the air upon the temperature of the ground. *Philosophical Magazine and Journal of Science* (Series 5) 41: 237-276.

Asper, V.L., Deuser,W.G., Knauer, G.A. and Lohrenz, S.E. 1992. Rapid coupling of sinking particle fluxes between surface and deep ocean waters. *Nature* 357: 670-672.

Azam, F., Fenchel, T., Field, J.G., Gray, J.S., Meyer-Reil, L. A. and Thingstad, F. 1983. The ecological role of water-column microbes in the sea. *Marine Ecology Progress Series*10: 257–263.

Baines, S.B. and Pace, M.L. 1991. The production of dissolved organic matter by phytoplankton and its importance of bacteria: Patterns across marine and fresh water systems. *Limnology Oceanography* 36:1078-1090.

Bala, G., Caldeira, K., Wickett, M., Phillips, T.J., Lobell, D̀. B., Delire, C. andMirin, A. 2007. Combined climate and carbon-cycle effects of large-scale deforestation. *Proceedings of the National Academy of Sciences* 104: 6550–6555.

Bala, G., Joshi, J., Chaturvedi, R. K., Gangamani, H. V., Hashimoto, H. and Nemani, R. 2013. Trends and Variability of AVHRR-Derived NPP in India. *Remote Sensing*. 5: 810-829; doi: 10.3390/rs5020810.

Ball, J.T., Woodrow, I.E. and Berry, J.A. 1987. A model predicting stomatal conductance and its contribution to the control of photosynthesis under different environmental conditions. In: *Progress in photosynthesis research. Proceedings of the VII International Congress on Photosynthesis*(Ed I Biggins Vol. 4.) . MartinusNijhoff Dordrecht, pp. 221–224.

Batjes, N. H. 1996. Total carbon and nitrogen in the soils of the world. *European Journal of Soil Science* 47:151- 163.

Batjes, N.H. 1999.*Management options for reducing CO2 concentration in the atmosphere by increasing carbon sequestration in the soil*. ISRIC Tech Pap. 30 ISRIC, Wageningen, Holland.

Batjes, N. H. andSombroek, W. G. 1997. Possibilities of C sequestration in tropical and subtropical soils. *Global Change Biology* 3:161-173.

Berger, W.H. 1989. Global maps of ocean productivity. In: *Productivity of the Oceans: Present and Past*, (Eds. Wiley WH Berger *et al.*), New York, pp. 429-455.

Bar, A. A., Harden, J.W., Torn, M.S. and Harte, J. 2008. Linking soil organic matter dynamics and erosion-induced terrestrial carbon sequestration at different landform positions. *Journal of Geophysical Research*, 113, G04039, doi: 10.1029/2008JG000751.

Berner, R. A. 1982. Burial of organic carbon and pyrite sulfur in the modern ocean: Its geochemical and environmental significance. *American Journal of Science* 282: 451-473.

Berry, J. and Bjorkman, O. 1980. Photosynthetic response and adaptation to temperature in higher- plants. *Annual Review of Plant Physiology* 31: 491–543.

Bolin, B., Degens, E., Camps,T. and Ketner, P. 1979.*The Global carbon cycle*, SCOPE Report 13, Chichester, New York, pp. 491.

Bouwman, A. F. 1990.*Soils and greenhouse effect*. John Wiley and Sons, Chichester, New York, p. 27.

British Petroleum. 2014a.*BP Statistical Review of World Energy June 2014*. http://www.bp.com/content/dam/bp/pdf/Energy-economics/statistical-review-2014/BP-statistical-review-of-world-energy-2014-full-report.pdf.

British Petroleum. 2014b.*Natural gas reserves*. http://www.bp.com/en/global/corporate/about-bp/energy- economics/ statistical-review-of-world-energy/review-by-energy-type/natural-gas/natural-gas-reserves.html.

Brown, S., Hall, C.A.S., Knabe, W., Raich, J. and Woomer, P. 1993. Tropical forests: their past, present and potential future role in terrestrial carbon budget. *Water, Air and Soil Pollution* 70:71-94.

Buitenhuis, E.T., Hashioka, T. and LeQuéré, C. 2013a. Combined constraints on global ocean primary production using observations and models. *Global Biogeochemicals Cycles* 27: 847-858. DOI: 10.1002/gbc. 20074.

Buitenhuis, E.T., Vogt, M., Moriarty, R., Bednarsek, N., Doney, S. C., Leblanc, K. , Le Quer´e´, C., Luo, Y.W., O'Brien. C., O'Brien, T., Peloquin, J., Schiebel, R. and Swan, C. 2013b. MAREDAT: towards a world atlas of marine ecosystem data. *Earth System Science Data* 5: 227–239, 2013 doi: 10.5194/essd-5-227-2013.

Campbell, C. J. and Laherrère, J. H. 1998. The End of Cheap Oil. *Scientific American* March 1998:78-83.

Cao, M. and Woodward, F. 1998. Net primary and ecosystem production and carbon stocks of terrestrial ecosystems and their responses to climate change. *Global Change Biology* 4: 185–198.

Cavallo, A. J. 2007. When the oil supply runs out. *Science* 316:980.

Churkina, G., Running, S.W. and Schloss, A.L. 1999. The participants of the Potsdam NPP model inter-comparison (1999): Comparing global models of terrestrial net primary productivity (NPP): the importance of water availability. *Global Change Biology* 5 : 46–55.

Cohen, A.K. 2013.*The Shale Gas Paradox: Assessing the Impacts of the Shale Gas Revolution on Electricity Markets and Climate Change* (PhD Thesis) Harvard College Cambridge, Massachusetts, 135p.

Conant, R.T., Steinweg, J.M., Haddix, M.L., Paul, E.A., Plante, A.F. and Six, J. 2008. Experimental warming shows that decomposition temperature sensitivity increases with soil organic matter recalcitrance. *Ecology* 89: 2384-2391.

Credit Suisse. 2013.*The Shale Revolution*. Credit Suisse Securities Research & Analytics. CREDIT SUISSE AG, Switzerland. https://doc.research-and-analytics.csfb.com/doc View? language=ENG&format=PDF& document _id = 1005321471 & source_id = em & extdocid = 1005321471_1_ENG_pdf & serialid = 0sD72ky5PVE69YdSxRIsei9L5gnDVssxzgAuLedlEiQ%3d assessed on July 13, 2015.

Crutzen, P.J. and Andreae, M.O. 1990. Biomass burning in the tropics: Impact on atmospheric chemistry and biogeochemical cycles. *Science* 250: 1669-1678.

Davidson, E.A. and Janssens, I.A. 2006. Temperature sensitivity of soil carbon decomposition and feedbacks to climate change. *Nature* 440:165-173.

Degens, E.T., Wong, H.K. and Kempe, S. 1981. Factors controlling global climate of past and future. In: *Some perspectives of the major biogeochemical cycles*, (Eds. G. E. Likens), John Wiley & Sons, Chichester, pp.3-24.

Dessler, A.E. 2013. Observations of Climate Feedbacks over 2000–10 and Comparisons to Climate Models. *Journal of Climate* 26: 333–342. doi: http://dx.doi.org/10.1175/JCLI-D-11-00640.1.

Dessler, A.E. and Zelinka, M. D. 2015. Climate feedbacks. In: *Encyclopedia of atmospheric sciences*(Eds. G. R. North, *et al* Second edition) 18–25, http://dx.doi.org/10.1016/B978-0-12-382225-3.00471-0.

Dev, S.M. and Sharma, A.N. 2010. Food security in India: Performance, challenges and policies. *Oxfam India Working Paper Series* 08. New Delhi 42p.

Dixon, R.K. 1995. Agroforestry systems: sources or sinks of greenhouse gases. *Agroforestry Systems* 31:99-116.

Dixon, R.K. and Turner, D.P. 1991. The global carbon cycle and climate change: response and feedback from below ground systems. *Environmental Pollution* 73:245-262.

Dixon, R.K., Andrasko, K.J.,Sussman, F.A., Lavinson, M.A., Trexler,C.A. and Vinson, T.S. 1993a. Forest sector carbon offset programs: near-term opportunities to reduce greenhouse gas emission to atmosphere. *Water Air and Soil Pollution*70:561-577.

Dixon, R.K., Brown, S., Houghton, R.A., Solomon, A.M., Trexler, M.C. and Wisniewski, J. 1994. Carbon pool and flux of global forest ecosystem. *Science* 263:185-190.

Dixon, R.K., Smith, J.B., Brown, S., Masera, O., Mata, L.J., Buksha, I.,Larocque, G.R. 1999. Simulations of forest system response and feedbacks to global change: experiences and results from the U.S. Country Studies Program. In: *Future directions in modeling net primary productivity in forest ecosystems* (Special issue). Proceedings of a symposium held at the joint meeting of the North American Chapter of the International Society for Ecological Modeling (ISEM) and the American Institute of Biological Sciences (AIBS) in Montreal, Canada, 5-6 August 1997, 122: 3, 289-305.

Dixon, R.K., Winjum, J.K. and Schroeder, P.E. 1993b. Conservation and sequestration of carbon: the potential of forest and agro forestry management practices. *Global Environmental Change* 2:159-173.

Dobrovolsky, V.V. 1994. *Biogeochemistry of the World's Land*. CRC Press, Boca Raton, Floridapp.293-317.

Don, A., Schumacher, J., Scherer-Lorenzen, M., Scholten, T. and Schulze, E.D. 2007. Spatial and vertical variation of soil carbon at two grassland sites:implications for measuring soil carbon stocks. *Geoderma* 141: 272-282.

Duarte, C.A. and Cebrian, J. 1996. The fate of marine autotrophic production. *Limnology and Oceanography* 41: 1758-1766.

Eglin, T., Ciais, P., Piao, S.L.,Barre, P., Bellassen, V., Cadule, P., Chenu, C., Gasser, T., Koven, C., Reichstein, M. and Smith, P. 2011. Historical and future perspectives of global soil carbon response to climate and land-use changes. *Tellus* 62B, 700–718.

EPA. 2012.*Inventory of US greenhouse gas emissions and sinks: 1990-2010* US Environmental Protection Agency. EPA 430-R-12-001. April 12, 2012.

EIA. 2013.*Technically recoverable shale oil and shale gas resources: an assessment of 137 shale formations in 41 countries outside the united states*. U.S. Energy Information Administration. Web. http://www.eia. gov/analysis/studies/worldshalegas/.

Esser, G., Aselmann, I. and Lieth, H. 1982.*Modelling the carbon reservoir in the system compartment* "litter" Mitteilungenausdem Geologisch-PalaeontologischenInstitut Der Universitaet Hamburg 52:39-58.

FAO/IIASA/ISRIC/ISS-CAS/JRC. 2009.*Harmonized World Soil Database* (version 1.1). FAO, Rome, Italy and IIASA, Luxembourg, Austria.

FAO. 2010.*Global Forest Resource Assessment*, Food and Agriculture Organization of the United Nations, Rome 378 pp.

Farquhar, G.D. von Caemmerer, S. and Berry, J.A. 1980. A biochemical model of photosynthetic CO_2 assimilation in leaves of C_3 species. *Planta* 149: 78–90.

Field, C.B., Behrenfeld, M.J., Randerson, J.T. and Falkowski, P. 1998. Primary production of the biosphere: integrating terrestrial and oceanic Components. *Science* 281:237-240.

Field, C.B., Randerson, J.T. and Malmstroem, C.M. 1995. Global net primary production:combining ecology and remote sensing. *Remote Sensing of Environment* 51: 74-88.

Ford-Robertson, J.B. 1997.*Carbon balance calculations for forest industries - a review*. New-Zealand Forestry 42:32-36.

Gibbard, S., Caldeira, K., Bala, G., Phillips, T.J. and Wickett, M. 2005. Climate effects of global land cover change. *Geophysical Research Letters* 32, L23705, doi: 10.1029/2005GL024550.

Giorgi, F. and Hewitson, B. 2001. Regional climate information - evaluation and projections. Chapter 10. In: IPCC (2001) *Climate Change 2001: The Scientific Basis.* Contribution of Working Group I to the Third Assessment Report of the Intergovernmental Panel on Climate Change. Cambridge and New York: Cambridge University Press. pp. 583–638.

Giorgio, P.A. del and Cole, J.J. 1997. Photosynthesis or planktonic respiration. *Nature* 388: 133.

Goes, J.I., Thoppil, P.G., Gomes,H. do R. and Fasullo, J.T. 2005. Warming of the Eurasian landmass is making the Arabian sea more productive. *Science* 308: 545-547. DOI: 10.1126/science.1106610.

Goudriaan, J. and Ketner, D. 1984. A simulation study for global carbon cycle including man's impact on biosphere. *Climate Change* 6:167-192.

Govers, G., Merckx, R., Van Oust, K. and Van Wesemael, B. 2013. Soil Organic Carbon Management for Global Benefits: A Discussion Paper, paper presented at the workshop '*Soil Organic Carbon Benefits: a Scoping Study*', 10th-12th September 2012, Nairobi. Workshop organized by the Scientific and Technical Advisory Panel of the Global Environmental Facility, p.53, https://www.researchgate.net/publication/236730954_Soil_organic_carbon_management_for_global_benefits_a_discussion_paper.

Greg, E.E., Anam, B.E., William, M.F. and Duru, E.J.C. 2011. Climate change, food security and agricultural productivity in African: Issues and policy directions. *International Journal of Humanities and Social Science* 1: 205-223.

Hansena, J., Satoa, M. and Ruedyb. R. 2012. Perception of climate change. *Proceedings of the National Academy of Sciences* 109: 14726-14727, E2415-E2423, doi:10.1073/pnas.1205276109.

Hao, W. M. and Liu, M. H. 1994. Spatial and temporal distribution of tropical biomass burning. *Global Biogeochemical Cycles* 8:495-503.

Harden, J.W., Berhe, A.A., Torn, M., Harte, J., Liu, S. and Stallard, R.F. 2008. Soil erosion: Data say C sink.*Science* 320:178-179.

Harden, J.W., Sharpe, J.M., Parton, W.J., Ojima, D.S., Fries, T.L., Huntington, T.G. and Dabney, S. M. 1999. Dynamic replacement and loss of soil carbon on eroding cropland. *Global Biogeochemical Cycles* 13:885-901.

Harvey, L. and Danney, D. 2000.*Global warming: the hard science*, Prentice Hall, Pearson Educ. Ltd., Harlow, UKp.336.

Heimann, M. and Reichstein, M. 2008. Terrestrial ecosystem carbon dynamics and climate feedbacks.*Nature* 451: 289-292 doi:10.1038/nature06591; Published online 16 January 2008.

Hiederer, R. and Köchy, M. 2012.*Global soil organic carbon estimates and the harmonized world soil database*, Luxembourg, p.79.

Hiederer, R. and Köchy, M. 2011.*Global soil organic carbon estimates and the harmonized World soil database*. EUR 25225 EN. Publications Office of the EU, Luxembourg.

Honjo, S., Manganini, S.J. and Cole, J.J. 1982. Sedimentation of biogenic matter in deep sea ocean. *Deep Sea Research* 29: 609-625.

Houghton, R.A., Hobbie, J.E., Melillo, J.M., Moore, B., Peterson, B.J., Shaver, G.R. and Woodwell, G.M. 1983. Changes in carbon content of terrestrial biota and soils between 1860 and 1980: net release of CO_2 to atmosphere. *Ecological Monographs*53:235-262.

Houghton, R.A., House, J.I., Pongratz, J., van der Werf, G.R., DeFries, R.S., Hansen, M.C., Le Quere´, C. and Ramankutty, N. 2012. Carbon emissions from land use and land-cover change. *Biogeosciences* 9: 5125–5142.

IEA. 2008.*World energy outlook 2008*. International Energy Agency, Paris. http://www.worldenergyoutlook.org/media/weowebsite/2008-1994/weo 2008.pdf.

IPCC. 1991.*Climate Change : The IPCC response strategies*, WMO, UNEP and IPCC 270p.

IPCC. 1996. *Climate Change 1995: The Science of Climate Change*. Report of Working Group I. J.T.Houghton, L.G.MeiraFilho, B.A. Callander, N.Harris, A.Kattenberg and K. Maskell (eds.) Cambridge University Press, New York, pp. 4.

IPCC. 2001.*Climate Change 2001: The Scientific Basis*. Contribution of Working Group I to the Third Assessment Report of the Intergovernmental Panel on Climate Change. Cambridge and New York: Cambridge University Press Cambridge, 83p.

IPCC. 2007. *Climate change 2007: the physical science basis*. In: Solomon, SD, *et al.* (Eds.), Contribution of Working Group I to the Fourth Assessment Report of the Intergovernmental Panel on Climate Change. Cambridge University Press, Cambridge, United Kingdom and New York, NY, USA, 996 p.

IPCC. 2014. *Climate change 2013: The physical science basis*, Cambridge University Press, New York. 1308p.

Ittekot, V. and Laane, R.W.P.M. 1991. Fate of riverine particulate organic matter. In: *Biogeochemistry of Major Rivers*, (Eds E.T. Degens, *et al.*) Wiley, New York, pp.233-243.

Jahnke, R.A . 1996. The global ocean flux of particulate organic matter: areal distribution and magnitude. *Global Biogeochemical Cycles* 10:71-78.

Jobbagy, E.G. and Jackson, R.B. 2000. The vertical distribution of soil organic carbon and its relation to climate and vegetation. *Ecological Applications* 10:423-436.

Johansson, T.B., McCormick, K., Neij, L. and Turkenburg, W. 2004. The Potentials of Renewable Energy. *Thematic Background Paper, International Conference on Renewable Energy*, Bonn p.36.

Jung, M., Vetter, M.,Herold, M., Churkina, G., Reichstein, M., Zaehle, S.,Ciais, P., Viovy, N., Bondeau, A., Chen, Y.,Trusilova, K.,Feser, F. and Heimann, M. 2007. Uncertainties of modeling gross primary productivity over Europe: A systematic study on the effects of using different drivers and terrestrial biosphere models. *Global Biogeochemical Cycles* 21: GB4021, doi:10.1029/2006GB002915, 2007.

Karl, T.R. and Trenberth, K.E. 2003. Modern Global Climate Change. *Science* 302: 1719, DOI: 10.1126/science.1090228.

Kasischke, E.S., Christensen, N.L. and Stocks, B.J. 1995. Fire, global warming and the carbon balance of boreal forests. *Ecological Applications* 5:437-451.

Kasting, J.F. 1998. The carbon cycle, climate and the long-term effects of fossil fuel burnings. *Consequences* p.4.

Kauppi, P.E., Mielikainen, K. and Kuusela, K. 1992. Biomass and carbon budget of European forests (1971-1990). *Science* 256:70-74.

Kerr, R.A. 2008. World oil crunch looming? *Science* 322:1178–1779.

King, W., Post, W.M. and Wullschleger, S.D. 1997. The potential response of terrestrial carbon storage to changes in climate and atmospheric CO_2. *Climate Change* 35: 199–227.

Kirschbaum, M.U.F., Whitehead, D., Dean, S.M., Beets, P.N., Shepherd, J.D.,Ausseil, A. 2011. Implications of albedo changes following afforestation on the benefits as carbon sinks. *Biogeosciences* 8:8563–8589.

Knauer, G.A. 1993. Productivity and new production of oceanic systems. In *Interactions of C, N, P and S Biogeochemical Cycles and Global Change*(Eds. R. Wollast, *et al*), Springer-Verlag, New York, pp. 211-231.

Lal, R. 1995. Global soil erosion by water and carbon dynamics. In *Soils and global change* (Eds R. Lal., *et al.)*, CRC Boca Raton, Florida, pp.131-141.

Lal, R. 1999. Soil management and restoration for C sequestration to mitigate the accelerated greenhouse effect. *Progress in Environmental Sciences*1:307-326.

Lal, R. 2001a. Soils and greenhouse effect. In *Soil Carbon Sequestration and Greenhouse Effect*, SSSA special Publication 57, Soil Science Society of America, Madison, pp.1-8.

Lal, R. 2001b. The Potential of carbon sequestration in forest ecosystem to mitigate the greenhouse effect. In : *Soil Carbon Sequestration and Greenhouse Effect*, SSSA special Publication 57, Soil Science Society of America, Madison, pp.137-154.

Lal, R. 2003. Soil erosion and the global carbon budget. *Environment International* 29: 437-450.

Lal, R. 2008. Carbon sequestration in soil. CAB Reviews: *Perspectives in Agriculture, Veterinary Science, Nutrition and Natural Resources* 2008 3, No. 030. Web. http:// tinread.usarb.md:8888/tinread/ fulltext/lal/ carbon_sequestration.pdf.

Lal, R. and Bruce, J.P. 1999. The potential of world cropland soil to sequester carbon and mitigate greenhouse effect. *Environmental Science & Policy* 2:177-185.

Lal, R., Hassan, H. M. and Dumanski, J. M. 1999. Desertification control to sequester C and mitigate the greenhouse effect. In *Carbon Sequestration in Soils : Science, Monitoring and Beyond* (Eds Rosenberg R.J. *et al.*), Battelle Press, Columbus, Ohio, pp. 83-107.

Lal, R. and Kimble, J.M. 2000. Pedogenic carbonates and global carbon cycle. In *Global Climate change and pedogenic carbonates* (Eds Lal, R. *et al.*), Lewis Publisher, Boca Raton, pp. 291-30.

Lal, R. and Logan, T.J. 1995. Agricultural activities and greenhouse gas emission from soil of tropics. In : *Soil Management Greenhouse Effect* (Eds Lal, R. *et al.*), CRC Press, Boca Raton, FL, pp. 293-307.

Lal, R., Henderlong, P. and Flowers, M. 1998. Forages and row cropping effects on soil organic carbon and nitrogen contents. In *Management of carbon sequestration in soil* (Eds Lal, R. *et al.*), CRC Boca Raton, pp. 365-379.

Lauenroth, W.K., Wade, A.A., Williamson, M.A.,Ross, B.E., Kumar, S. and Cariveau, D.P. 2006. Uncertainty in calculations of net primary production for grasslands. *Ecosystems* 9: 843–851.

Laurmann , J.A. 1979. Market penetration characteristics for energy production and atmospheric carbon dioxide growth. *Science* 205:896-898.

Laws, E., Falkowski, P.G., Smith, W.O., Ducklow, H. and McCarthy, J.J. 2000. Temperature effects on export production in the open ocean. *Global Biogeochemical Cycles* 14: 1231–1246.

Lee, J.L. and Dodson, R. 1996. Potential carbon sequestration by afforestation of pasture in south-central united states. *Agronomy Journal* 88:381-384.

Lee, H.L. 2009. The impact of climate change on global food supply and demand, food prices, and land use. *Paddy and Water Environment* 7: 321-331.

Li, W.K.W., Subba, Rao. D.V., Harrison, W.G., Smith, J.C., Cullen, J.J., Irwin, B., Platt, T. 1983 Autotrophic picoplankton in tropical ocean. *Science* 219: 292-295.

Ludwig, W., Probst, J.L. & Kempe, S. 1996. Predicting the oceanic input of organic carbon by continental erosion. *Global Biogeochemical Cycles* 10:23-41.

Magnani, F., Mencuccini, M., Borghetti, M.,Berbigier, P., Berninger, F., Delzon, S., Grelle, A., Har, P., Jarvis, P.G., Kolari, P., Kowalski, A.S., Lankreijer, H., Law, B.E., Lindroth, A., Loustau, D., Manca, G., Moncrieff, J.B., Rayment, M., Tedeschi, V., Valentini, R. and Grace, J. 2007. The human footprint in the carbon cycle of temperate and boreal forests. *Nature* 447: 848–850.

Mahli, Y. 2002. Carbon in the atmosphere and terrestrial biosphere in the 21st century. *Philosophical Transanction of Royal Society A* 1098: 1–20.

Maier-Reimer, E., Mikolajewicz, U. and Winguth, A. 1996. Future ocean uptake of CO2: interaction between ocean circulation and biology. *Climate Dynamics* 12: 711–721.

Makipaa, R., Karjalainen, T., Pussinen, A., Kukkola, M. and Paoletti, E. 1998. Effects of nitrogen fertilization on carbon accumulation in boreal forests: model computations compared with the results of long-term fertilization experiments. In: Stress factors and air pollution. Selected papers from the 17th International meeting for specialists in air pollution effects on forest ecosystems held in Florence, Italy, 14-19 September, 1996. *Chemosphere* (Special issue) 36: 1155-1160.

Masters, G., Baker, P. and Flood, J. 2010 Climate change and agricultural commodities. *CABI Working Paper 02*. Available atwww.cabi.org.

Maugeri, L. 2004. Oil: Never cry wolf. Why the petroleum age is far from over. *Science* 304:1114–1115.

McGowan, J.A., Cayan, D.R. and Dorman, L.M. 1998. Climate-Ocean variability and ecosystem response in the Northeast Pacific. *Science* 281: 210- 217.

Meehl, G.A., Hu, A., Arblaster, J.M., Fasullo, J. and Trenberth, K.E. 2013. Externally Forced and Internally Generated Decadal Climate Variability Associated with the Interdecadal Pacific Oscillation. *Journal of Climate*26:7298–7310 doi: http://dx.doi.org/10.1175/JCLI-D-12-00548.1.

Meentemeyer, V., Gardner, J. and Box, E.O. 1985. World patterns and amounts of detrital soil carbon. *Earth Surface Processes and Landforms*10: 557–567.

Melillo, J.M., McGuire, A.D., Kicklighter, D.W., Moore, B., Vorosmarty, C.J. and Schloss, A.L. 1993. Global climate change and terrestrial net production. *Nature* 363:234-240.

Micales, J.A. and Skog, K.E. 1997. The decomposition of forest products in landfills. *International Biodeterioration & Biodegradation* (Special issue: Biodegradation of wood) 39:145-158.

Montagnini, F. and Nair, P.K.R. 2004. Carbon sequestration: An underexploited environmental benefit of agroforestry systems. *Agroforestry Systems* 61: 281–295.

Moore, T.R. 1998. Dissolved organic carbon sources, sinks and fluxes and role in soil carbon cycle. In Soil processes and the carbon cycle (EdsLal, R. *et al.*), CRC Press, Boca Raton, pp. 281-291.

Moore, C.J. 1976. A comparative study of radiation balance above forest and grassland. *Quarterly Journal Of The Royal Meteorological Society* 102: 889–899.

Mutuo, P.K., Cadisch, G., Albrecht, Palm. C.A. and Verchot, L. 2005. Potential of agroforestry for carbon sequestration and mitigation of greenhouse gas emissions from soils in the tropics. *Nutrient Cycling in Agroecosystems* 71:43-54.

Nair, P.K.R., Kumar, B.M. and Nair, V.D. 2009. Agroforestry as a strategy for carbon sequestration. *Journal of Plant Nutrition and Soil Science* 172:10-23.

Nemani, R.R., Keeling, C.D., Hashimoto, H., Jolly, W.M., Piper, S.C., Tucker, C.J., Myneni, R.B. and Running, S.W. 2003. Climate-driven increases in global terrestrial net primary production from 1982 to 1999. *Science* New Series 300: 1560-1563.

Nierenberg, N., Tschinkel, W.R. and Tschinkel, V.J. 2010. Early climate change consensus at the national academy: the origins and making of changing climate. *Historical Studies in the Natural Sciences* 40: 318–349 http://www.ucpressjournals.com/reprintinfo. asp. doi:10.1525 hsns. 2010. 40.3.318.

Nilsson, S., Schopfhauser, W., Hoen, H.F. and Solberg, B. 1995. The carbon sequestration potential of a global afforestation program. *Climate Change* 30:267-293.

Nordt, L.C., Wilding, L. P. and Drees, L. R. 2000. Pedogenic carbonate transformation in leaching soil system: Implications for global C cycle. In *Global climate change and pedogenic carbonates*(Eds. Lal, R. et al.), Lewis Publisher, Boca Raton, pp.43-46.

NREL. 2013.*National Renewable Energy Laboratory*. Office of Energy Efficiency and Renewable Energy, operated by the Alliance for Sustainable Energy, LLC. http://www.nrel.gov/ docs/fy14osti/ 60894.pdf.

Olson, J. S., Watts, J.A. and Allison, L.J. 1983.*Carbon in live vegetation of major world ecosystems*. DOE/NBB-0037 National Technical Information Service, Springer-field, Virginia.

Oluoko-Odingo, AA (2009) Determinants of Poverty: Lessons from Kenya. *Geo J.* 74:311-331.

Ostle, N.J., Smith, P., Fisher, R., Woodward, F. I., Fisher, J.B., Smith, J.U., Galbraith, D., Levy, P., Meir, P., McNamara, N.P. and Bardgett, R. D. 2009. Integrating plant-soil interactions into global carbon cycle models. *Journal of Ecology* 97: 851–863.

Paltridge, G. W. and Calder, N. 2010.*The Climate Caper: Facts and Fallacies of Global Warming*. Taylor Trade Publishing, Lanham, USA.

Paltridge, G. W. 2014.*Climate Change's Inherent Uncertainties*.https:// quadrant.org.au/magazine/2014/01-02/fundamental-uncertainties-climate-change/ . Assessed on MAY 19,, 2015.

Pan, S., Tian, H., Dangal, S.R.S., Ouyang, Z., Tao, B., Ren, W., Lu, C. and Running, S. 2014. *Modeling and monitoring terrestrial primary production in a changing global environment: toward a multiscale synthesis of observation and simulation*. Hindawi publishing corporation advances in meteorology volume 2014, article id 965936, 17 pages http://dx.doi.org/ 10.1155/2014/965936.

Pan, Y., Birdsey, R., Hom, J., McCullough, K. and Clark, K. 2006. Improved estimates of net primary productivity from modis satellite data at regional and local scales. *Ecology Applications*16: 125–132.

Panneer-Selvam, B., Natchimuthu, S., Arunachalam, L. and Bastviken, D. 2014. Methane and carbon dioxide emissions from inland waters in India – implications for large scale greenhouse gas balances. *Global Change Biology* 20: 3397–3407. doi:10.1111/gcb.12575.

Peters, G.P., Andrew, R.M., Boden, T., Canadell, J.G., Ciais, P., Le Quéré, C., Marland, G., Raupach, M. R. and Wilson, C. 2013. The challenge to keep global warming below 2%C. *Nature Climate Change* 3: 4–6.

Pomeroy, L. R. and Wiebe, W. J. 2001. Temperature and substrates as interactive limiting factors for marine heterotrophic bacteria. *Aquatic Microbial Ecology* 23: 187–204.

Pongratz, J., Reick, C.H., Raddatz, T., Caldeira, K. and Claussen, M. 2011. Past land use decisions have increased mitigation potential of reforestation. *Geophysical Research Letter* doi: 10.1029/2011GL047848

Post ,W. M., Emanuel, W.R. Zinke, P.J. and Stangenberger, A.G. 1982. Soil carbon pools and world life zones. *Nature* 298:156-159.

Potter, C.S., Randerson, J.T., Field, C.B., Matson, P.A., Vitousek, P.M., Mooney, H.A. and Klooster, S.A. 1993. Terrestrial ecosystem production: A process model based on global satellite and surface data. *Global Biogeochemical Cycles* 7:811-841.

Potter, C., Klooster, S. and Genovese, V. 2012. Net primary production of terrestrial ecosystems from 2000 to 2009. *Climate Change* 115: 365–378.

Prentice, I.C., Farquhar, G.D., Fasham, M.J.R., Goulden, M.L., Heimann, M., Jaramillo, V.J., Kheshgi, H.S., Le Quéré, C., Scholes, R.J. and Wallace, D.W.R. 2001. The Carbon Cycle and Atmospheric Carbon Dioxide. In:, *Climate change: the scientific basis* (Eds Houghton JT, Ding Y, Griggs DJ, Noguer M, van der Linden PJ, Dai X, Maskell K, Johnson CA). Contribution of working group I to the third assessment report of the intergovernmental panel on climate change, Cambridge University Press, Cambridge, United Kingdom and New York, NY, USA, p.881.

Quinton, J.N., Govers, G., Van,Oust. K. and Bardgett, R.D. 2010. The impact of agricultural soil erosion on biogeochemical cycling. *Nature Geoscience* 3:311-314.

Raghuvanshi, S.P., Chandra, A. and Raghav, A. K. 2006. Carbon dioxide emissions from coal based power generation in India. *Energy Conversion and Manage* 47: 427–441.

Ramasamy J, Moorthy P. (2012) Managing food insecurity and poverty in India in the era of globalization. *International Journal of Multidisciplinary Research and Development* 2:411-421.

Raupach, M.R., Davis, S.J.,Peters, G.P.,Andrew, R.M., Canadell, J.G., Ciais, P., Friedlingstein, P., Jotzo, F., Vuuren, van. D.P. and Le Quéré, C. 2014. Sharing a quota on cumulative carbon emissions. *Nature Climate Change* 4: 873–879 doi:10.1038/nclimate2384.

Raymond, P.A., Hartmann, J., Lauerwald, R., Sobek, S., McDonald, C., Hoover, M., Butman, D., Striegl, R., Mayorga, E., Humborg, C., Kortelainen, P., Dürr, H., Meybeck, M., Ciais, P. and Guth, P. 2013. Global carbon dioxide emissions from inland waters. *Nature* 503: 355–359 doi:10.1038/ nature 12760.

Rivkin, R. B. and Legendre, L. 2001. Biogenic Carbon Cycling in the Upper Ocean: Effects of Microbial Respiration. *Science* 291: 2398-2400. DOI: 10.1126/science.291.5512.2398.

Rivkin, R. B., Anderson, M. R. and Lajzerowicz, C. 1996. Microbial processes in cold oceans. I. Relationship between temperature and bacterial growth rate. *Aquatic Microbial Ecology* 10: 243–254.

Rumpel, C. and Koegel-Knabner, I. 2011. Deep soil organic matter-a key but poorly understood component of terrestrial C cycle. *Plant and Soil* 338: 143-158.

Sanchez, P.A. and Benites, J.R. 1987. Low input cropping for acid soils of humid tropics. *Science* 238:1521-1527.

Sarmiento, J. L., Hughes, T.M.C., Stouffer, R. J. and Manabe, S. 1998. Simulated response of the ocean carbon cycle to anthropogenic climate warming. *Nature* 393, 245–249.

Sathaye , J., Makundi, W. R., Andrasko, K., Boer, R., Ravindranath, N.H., Sudha, P., Rao, S., Lasco, R., Pulhin, F., Masera, O., Ceron, A., Ordonez, J., Deying, X., Zhang, X. and Zuomin, S. 2001. Carbon mitigation potential and costs of forestry options in Brazil, China, India, Indonesia, Mexico, The Philippines and Tanzania, *Mitigation and Adaptation Strategies for Global Change* 6: 185-211.

Sayles, F.L., Martin, W.R. and Deuser, W.G. 1994. Response of benthic oxygen demand on particulate organic carbon supply in deep sea near Bermuda. *Nature* 371: 686-689.

Scharlemann, J. P. W., Tannerc, E.V.J., Hiedererd, R. and Kaposbe, V. 2014. Global soil carbon: understanding and managing the largest terrestrial carbon pool. *Carbon Manage* 5: 81-91, DOI: 10.4155/ cmt.13.77.

Scharpenseel, H.W., Mtimet, A. and Freytag, J. 2000. Soil inorganic carbon and global change, In: *Global climate change and pedogenic carbonates* (Eds. R. Lal, Kimble JM, Eswaran H, Stewart BA), CRC Press, Boca Raton, pp.27-42.

Schlesinger, W.H. and Lichter, J. 2001. Limited carbon storage in soil and litter of experimental forest plots under increased atmospheric CO2. *Nature* 411:466–469.

Schmidt, M.W.I., Torn, M.S., Abiven, S., Dittmar, T., Guggenberger, G., Janssens, I. A., Kleber, M., Kögel-Knabner, I., Lehmann, J., Manning, D.A.C., Nannipieri, P., Rasse, D.P., Weiner, S. and Trumbore, S.E. 2011. Persistence of soil organic matter as an ecosystem property. *Nature* 478:49–56 doi:10.1038/ nature10386.

Schroeder, P.E., Dixon, R.K. and Winjum, J.K. 1993.*Forest management and agroforestry to sequester and conserve atmospheric carbon dioxide*. Unasylva 44:52-60.

Schulze, E.D., Luyssaert S., Ciais P. and *et al.* 2009. Importance of methane and nitrous oxide for Europe's terrestrial greenhouse-gas balance. *Natural Geoscience* 2:842–850. doi:10.1038/ngeo686.

Sendax, A. 2010.*Are Plants Around the World Really Dying? A new study reveals that plant growth across the globe has decreased in the past 10 years, despite the observed increasing trend of the prior two decades.* How can this be? http:// www.popularmechanics.com/science/environment/a6098/are-plants-around-the-world-really-dying/9Assessed on June 21-06-2015.

Shiraishi, N. 1997.*Developing techniques for sustaining and expanding carbon sink capacity of natural forest*. Annual-Report-of-the-Hokkaido-Research-Center, Forestry and Forest Products Research Institute. 1996:49-51.

Siegenthaler, U. and Sarmiento, S. L. 1993. Atmospheric carbon dioxide and ocean. *Nature* 365:119-125.

Sitch, S., Huntingford, C., Gedney, N., Levy, P.E., Lomas, M., Piao, S.L., Betts, R., Ciais, P., Cox, P., Friedlingstein, P., Jones, C.D., Prentice, I.C. and Woodward, F. I. 2008. Evaluation of the terrestrial carbon cycle, future plant geography and climate-carbon cycle feedbacks using 5 Dynamic Global Vegetation Models (DGVMs). *Global Change Biology* 14: 1-25.

Smith, P., Powlson, D.S., Glendining, M. J. and Smith, J.O.U. 1997. Potential for carbon sequestration in European soils: preliminary estimates for five scenarios using results from long-term experiments. *Global Change Biology* 3:67-79.

Smith, T. M., Cramer, W.P., Dixon, R.K., Leemans, R., Neilson, R.P. and Solomon, A. M. 1993. The global terrestrial carbon cycle. *Water Air and Soil Pollution* 70:19-38.

Smith, T.M., Leemans, R. and Shugart, H. H. 1992. Sensitivity of terrestrial carbon storage to CO2 induced climate change: comparison of four scenarios based on general circulation models. *Climate Change* 21:367-384.

Smith, V.H. 1992. Effect of nitrogen : phosphorus supply ratios on nitrogen fixation in agricultural and pastoral ecosystem. *Biogeochemistry* 18:19-35.

Song, X.,Tian, H., Xu, X., Hui, D., Chen, G., Sommers, G., Marzen, L. and Liu, M. (2013). Projecting terrestrial carbon sequestration of the Southeastern United States in the 21st century. *Ecosphere* 4(7): 88. htp://dx. doi.org/10.1890/ES12-00398.1.

Stallard, R. F. 1998. Terrestrial sedimentation and the carbon cycle: Coupling weathering and erosion to carbon burial. *Global Biogeochemical Cycles* 12:231-257.

Stern, N. 2007. *The Economics of Climate Change: The Stern Review*. Cambridge University Press p.862.

Stern, N. 2008.*The Economics of Climate Change*. American Economic Review: Papers & Proceedings 2008, 98:1–37 http://www.aeaweb.org/articles.php?doi=10.1257/aer.98.2.1.

Stockner JG, Antia NJ (1986) Algal picoplankton from marine and fresh water ecosystem: A multidisciplinary perspective. *Canadian Journal of Fisheries and Aquatic Sciences* 43:2472-2503.

Tans, P.P., Fung, I.Y. and Takahashi, T. 1990. Observational constraints on global atmospheric CO2 budget. *Science* 247:1431-1438.

Tao, F.M., Yokozawa, Y. and Hayashi, E.L. 2003a. Changes in agricultural water demands and soil moisture in China over the last half-century and their effects on agricultural production. *Agricultural and Forest Meteorology* 118: 251–261.

Tao, F.M., Yokozawa, Y. and Hayashi, E, L. 2003b. Future climate change, the agricultural water cycle, and agricultural production in China. *Agriculture Ecosystems and Environment* 95: 203-215.

Tian, H., Chen, G., Liu, M., Zhang, C., Sun, G., Lu, C., Xu, X., Ren, W., Pan, S. and Chappelka, A. 2010. Model estimates of net primary productivity, evapotranspiration, and water use efficiency in the terrestrial ecosystems of the Southern United States during 1895–2007. *Forest Ecology and Management* 259: 1311–1327.

Tian, B. and Lu, H. 2013. Net greenhouse gas balance in response to nitrogen enrichment: perspectives from a coupled biogeochemical model. *Global Change Biology* 19:571–588.

Tian, H., Melillo, J. M., Kicklighter, D.W., McGuire, A.D., Helfrich, J.V.K. III, Moore, B. III and VoÈroÈsmar, C.J. 1998. Effect of interannual climate variability on carbon storage in Amazonian ecosystems. *Nature* 396: 664–667.

Timmermann, A., Oberhuber, J., Bacher, A., Esch, M., Latif, M. and Roeckner, E. 1999. Increased El Nin˜o frequency in a climate model forced by future greenhouse warming. *Nature* 398: 694 – 696.

Va´Zquez-Domi´nguez, E., Vaque,´D. and Gasol, J. M. 2007. Ocean warming enhances respiration and carbon demand of coastal microbial plankton. *Global Change Biology* 13: 1327–1334 doi: 10.1111/j.1365-2486.2007.01377.x.

Van, Hemelryck. H., Govers, G., Van,Oust. K. and Merckx, R. 2011. Evaluating the impact of soil redistribution on the *in-situ* mineralization of soil organic carbon. *Earth Surface Processes Landforms* 36:427-438.

Van,Oust. K., Quine, T.A., Govers, G., De, Gryze. S., Six, J., Harden, J.W., Ritchie, J.C., McCarty, G.W., Heckrath, G., Kosmas, C., Giraldez, J.V., da Silva, J.R.M. and Merckx, R. 2007. The impact of agricultural soil erosion on the global carbon cycle. *Science* 318:626-629.

Verchot, L.V., Noordwijk, M.V., Kandji, S., Tomich, T., Ong, C., Albrecht, A., Mackensen, J., Bantilan, C., Anupama, K.V. and Palm, C. 2007. Climate change: linking adaptation and mitigation through agroforestry. *Mitigation and Adaptation Strategies for Global Change* 12:901-918. 16.

Verchot, L. V. 2014.*Challenges and opportunities for mitigation in the agricultural sector*. Technical Paper , Center for International Forestry Research, Bogor, Indonesia, 14p.

Vitousek, P.M., Ehrlich, P.R., Ehrlich, A.H. and Matson, P. A. 1986. Human appropriation of the products of photosynthesis. *BioScience* 36:368-373.

Wang, Z., Govers, G., Steegen, A., Clymans, W., Van den, Putte, A., Langhans, C., Merckx, R. and Van,Oust. K. 2010. Catchment-scale carbon redistribution and delivery by water erosion in an intensively cultivated area. *Geomorphology* 124:65-74.

Watts, A. 2011.*Zhao and Running's plant productivity claim rebuked in Science: modeling errors, use of corrupted satellite data and statistically insignificant trends cited.* http://wattsupwiththat.com/2011/08/25/zhao-and-runnings-plant-productivity-claim-rebuked-in-science/. Assessed on May 28, 2015.

Weitzman, M.L. 2009. On modeling and interpreting the economics of catastrophic climate change. *Review of Economics and Statistics* 91(1): 1-19. doi:10.1162/rest.91.1.1.

White, P.A., Kalff, J., Rasmussen, J.B. and Gasol, J.M. 1991. The effect of temperature and algal biomass on bacterial production and specific growth-rate in fresh-water and marine habitats. *Microbial Ecology* 21: 99–118.

Willey, J.M., Sherwood, L.M. and Woolverton, C.J. 2009.*Prescott's Principles of Microbiology*. McGrawHill, New York, NY.

Winjum, J. K., Dixon, R.K. and Schroeder, P.E. 1992. Estimating the global potential of forest and agroforest management practices to sequester carbon. *Air Water and Soil Pollution* 64:213-228.

Winjum, J.K., Dixon, R.K. andSchroeder, P.E. 1993a. Forest management and carbon storage: an analysis of 12 key forest nations. *Air Water Soil Pollution* 70:239-257.

Winjum, J.K., Meganck, R.A. and Dixon, R.K. 1993b. Expanding global forest management: an easy first proposal. *Journal of Forestry* 91:38-42.

WMO. 2014. Greenhouse Gas Bulletin : *The State of Greenhouse Gases in the Atmosphere Based on Global Observations through 2013*, World Meteorological Organization, Atmospheric Environment Research Division, Research Department, Geneva, p.8.

World Coal Association. 2013. http://www.worldcoal.org/resources/coal-statistics.

Zhao, M. and Running, S.W. (2010). Drought-induced reduction in global terrestrial net primary production from 2000 through 2009. *Science* 329: 940–943. View at Publisher · View at Google Scholar · View at Scopus.

Zheng, D., Prince, S. and Wright, R. 2003. Terrestrial net primary production estimates for 0.5 grid cells from field observations – a contribution to global biogeochemical modeling. *Global Change Bi*ology 9: 46–64.

Zher, S.C. 2000. Public representations of scientific uncertainty about global climate change. *Public Understanding Science* 9: 85–103.

□□□

Chapter 2

Climate Variability Under Hot Arid Environment

D.V. SINGH

ICAR - Central Arid Zone Research Institute, Jodhpur-342003, Rajasthan,
E-mail: dvsingh.cazri@gmail.com

Introduction

Climate is always in a dynamic state due to the complex interactions of the solar, oceanic, terrestrial, atmospheric and biotic components of the earth. The condition of climatic elements at a particular place at least for a moment, is known as weather, while the climate means the average weather conditions and their variance, determined by observations made over a long period of 30 years or more. Agricultural sector is very sensitive to climate variability. Climate plays an important role in selection of agricultural enterprises, while weather conditions, governed by natural climatic variability, during a season determine their performance in the particular season. For example, average temperature and rainfall of a region mainly deternmines the suitable crops for that region, while temperature regime and rainfall during a season governs the actual crop yields. Better understanding of variability and change of climatic elements of a region is crucial for efficient management and utilization of resources. Arid agriculture is highly vulnerable due to wide intra- and inter-seasonal climate variability resulting in frequent water scarcity and temperature stress. Farmers have been modifying their existing practices or adopting new stratigies to cope with climate variability. Climate variability is natural, but recently observed climate change, attributed mainly to anthropoginic activities, has become threat to their livelihoods due to increasingly unpredictable, frequent and intense climate extremes such as droughts, floods, high temperatures and frosts.

Rainfall variability

A major portion of annual rainfall in India is received during the southwest monsoon season (June–September) also known as Indian Summer Monsoon Rainfall (ISMR). Regional rainfall has large scale year-to-year fluctuations. Indian agriculture continues to be a gamble of the vagaries of monsoon, rainfall being most critical because 85 m ha of India's 141 m ha net sown area is rainfed (Venkateswarlu, 2012). Aberrant behaviour of monsoon such as low and poor rainfall, its erratic distribution, delayed onset and prolonged dry spells during the cropping season often results in water stress conditions at different growth stages causing decline in productivity and sometimes crop failures. Monsoon is a large-scale seasonal wind system, blowing over vast areas of the globe, persistently in the same direction, only to be reversed with change of season. They affect the largest land masses. Asia is the only continent in the world having regular visits of monsoons of reversing and persistent wind regimes. Australia and Africa are also affected by monsoons, though not to the same extent as Southeast Asia. North America also shows monsoonal tendencies. But here monsoon wind is neither persistent, nor the seasonal wind shifts are so consistent. The phenomena of monsoon is explained by different theories, such as (a) differential heating of land and water, (b) seasonal shifting of the tropical and inter-tropical wind belts, (c) upper atmospheric wind movements and jet streams and (d) most recently linked to El Nino and La Nina effects. The variability in annual and seasonal rainfall is the most important characteristic of ISMR resulting in droughts in some regions and floods in others. The western arid part of Rajasthan has high rainfall variability compared to rest of the Rajasthan, as east and southern parts experience higher rainfall and less variability. However, in absolute terms rainfall variability is more in wetter parts (Rathore, 2005) since standard deviations of annual rainfall in southern and eastern regions (182 and 207 mm, respectively) are higher than that of the western region (118 mm).

Droughts

Drought is a normal, recurrent feature of climate and occurs in all climatic zones, although its characteristics vary significantly from one region to another. The onset and persistence of drought are gradual processes. However, its impact may be more disastrous than that of flood or other natural calamities. Successive droughts of two or more years are not uncommon which have a devastating effect on natural resources, availability of water, production of food and fodder. The impact of drought often accumulates slowly and may linger on for years after termination of the drought.

An extensive survey of the definition of droughts by World Meteorological Organisation (WMO) found that droughts are classified on the basis of: (i) rainfall, (ii) combinations of rainfall with temperature, humidity and or evaporation, (iii) soil moisture and crop parameter, (iv) climatic indices and estimates of evapotranspiration, and finally (v) the general definitions and statements. Based on the causative factors, drought can be classified into meteorological, hydrological and agricultural droughts (Sastri, 2011).

Meteorological drought

India Meteorological Department (IMD) defines drought as a situation when the seasonal rainfall over the area is less than 75% of its long-term average. It is further divided into 'moderate drought', if the rainfall deficit is between 26 and 50% and 'severe drought' when rainfall deficit exceeds 50%.

Hydrological drought

Prolonged meteorological drought can result in hydrological drought with marked decline in flow of streams and rivers, depletion of surface water and consequent drying up of reservoirs, lakes and fall in ground water table.

Agricultural drought

An agricultural drought occurs when soil moisture and rainfall are inadequate during the growing season to support a healthy crop growth till maturity, causing water stress and reduction in crop yields.

The severity of drought depends on the duration, intensity, frequency and geographical distribution of rainfall as well as on corresponding water demand of human, animals, crops and vegetation cover of a region. Seasonal temperature, wind speed, solar radiation, density of vegetation, soil moisture holding capacity and soil moisture balance in surface, sub-soil and ground water influence water demand during the drought years.

Drought impacts many sectors of the economy and its effects may be seen well beyond the area experiencing drought. Drought impacts may be direct or indirect. Reduced productivity of crops, rangelands and forest; increased fire hazard; reduced water levels; increased livestock and wildlife mortality rates; and damage to wildlife and fish habitat are some examples of direct impacts. The consequences of these impacts illustrate indirect impacts. Widespread crop failures lead to acute shortages of food and fodder that adversely affect human and livestock health and nutrition; scarcity of drinking water due to decline in water table and surface water resources; human and livestock migration are the major manifestation of droughts. The influence of climatic variability is high in arid ecosystem. Increased deforestation, over grazing of pasture land, indiscriminate mining and unplanned urbanization resulting in silting of water bodies, abuse of traditional water resources further aggravates the drought situation.

The Indian economy has been described as a 'gamble of monsoon'. There is large special and temporal variability in the monsoon rainfall. Consequently the Indian region experiences drought or flood in some part of the country or the other almost every year during the monsoon period. The coefficient of variability of SW monsoon rainfall over drought prone and chronically drought prone areas ranges from 15 to 40% and 41to 65%, respectively. Although no part of India is immune to drought, but the arid and dry semi-arid regions of the country are more vulnerable and experience droughts of varying frequency and intensity. Though all the El Nino years are not drought years, there appears a clear association between El Nino events and low monsoon rainfall (Table 1).

There is a high degree of correlation between food grain production and the monsoon rain pattern; production is higher in years when rainfall is in excess of normal and a dip in production in years of deficit rainfall. India witnessed 23 major droughts during 1877 to 2005. During most severe six droughts, rainfall deficit varied from 19 to 29%, whereas geographical area affected ranged from 49 to 83%.

Table 1. El Nino years during 1875-2004 and drought years in India (Shewale and Kumar, 2005)

S. No.	> 20% of country's area affected		> 10% of deficiency in ISMR	
	Drought year with EI Nino	Drought year without EI Nino	Drought year with EI Nino	Drought year without EI Nino
1	1877	1901	1877	1901
2	1891	1904	1899	1904
3	1899	1907	1905	1920
4	1905	1913	1911	1966
5	1911	1915	1918	1968
6	1918	1920	1941	1974
7	1925	1966	1951	1979
8	1939	1968	1965	
9	1941	1974	1972	
10	1951	1979	1982	
11	1965	1985	1987	
12	1972	2000	2002	
13	1982		2004	
14	1987			
15	2002			
Total Al Nino years 31				

The drought of 1987 was one of the worst in past century (Narain *et al.*, 2006). The monsoon rainfall was normal only in 14 out of 35 meteorological sub-divisions in the country. The overall deficiency in rainfall was 19% as compared to 26% in 1918 and 25% in 1972, which were the other severe drought years in the past century. During 1987 drought, agricultural operations were adversely affected in 43% (58.6 million ha) of cropped area in 263 districts of 15 states and six union territories. In the two worst affected states of Rajasthan and Gujarat, the rainfall was less than 50% of normal.

Food production, particularly in the state of Rajasthan, nose dives during the years of drought and aberrant weather. Severe droughts reduced the food grain production in western Rajasthan by 70% during 1987-88, 50% during 2002-03 and by 57% during 2004-05 compared to the production of previous good monsoon year. Such reduction in food grain production was also noticed at state as well as at country level during the drought years (Table 2).

Table 2. Food production in Rajasthan in drought years (Narain *et al.*, 2006)

Year	Food grain production (million tons)		
	Western Rajasthan	Rajasthan state	India
1964-65	1.95	5.46	97.92
1965-66	1.27 (35)*	4.01 (27)	78.75 (20)
1971-72	2.2	6.58	114.25
1972-73	1.50 (32)	5.44 (17)	104.17 (09)
1978-79	3	8.33	142
1979-80	3.59 (00)	5.49 (34)	118.44 (17)
1983-84	4.57	11.15	165.06
1987-88	1.38 (70)	6.03 (46)	153.00 (07)
1999-2K	4.42	14	230.52
2000-01	4.24 (04)	12.00 (14)	214.32 (07)
2001-02	6.88	17.13	232.33
2002-03	3.44 (50)	9.28 (46)	187.86 (19)
2003-04	6.52 (05)	10.90 (57)	212.05 (09)
2004-05	2.96 (57)	5.53 (68)	204.60 (12)

*Figures in parenthesis shows per cent decrease in food production due to drought over production of previous good monsoon year.

A study of the impact of consecutive droughts on the productivity of pearl millet in western Rajasthan indicated that yield during the drought years of 1984, 1985, 1986 and 1987 decreased by 36, 78, 69 and 89% respectively, compared to that recorded during 1983. The mean productivity in eight out of the 11 districts was less than 50%, the most affected districts being Jaisalmer, Bikaner, Barmer, Jodhpur and Jalore. In 2002, net sown area under *kharif* crops in Rajasthan decreased from 13 m ha of normal years to 6.9 m ha (47% decline) and that too withered subsequently. During *rabi* season in 2002, cropped area went down by 43% due to lesser availability of soil moisture and irrigation water.

Identification of drought prone areas

Different agencies and sources quote different figures of drought prone areas in India. The irrigation commission in 1972 identified 67 drought prone districts comprising of 326 *talukas* of eight states covering an area of 40.7 m ha. The national commission on agriculture added some more drought prone areas with slightly different criteria in 1976. The Government of India has identified drought prone districts for implementation of Desert Development Program (DDP) in arid districts and Drought Prone Area Program (DPAP) in semi-arid areas (Fig. 1).

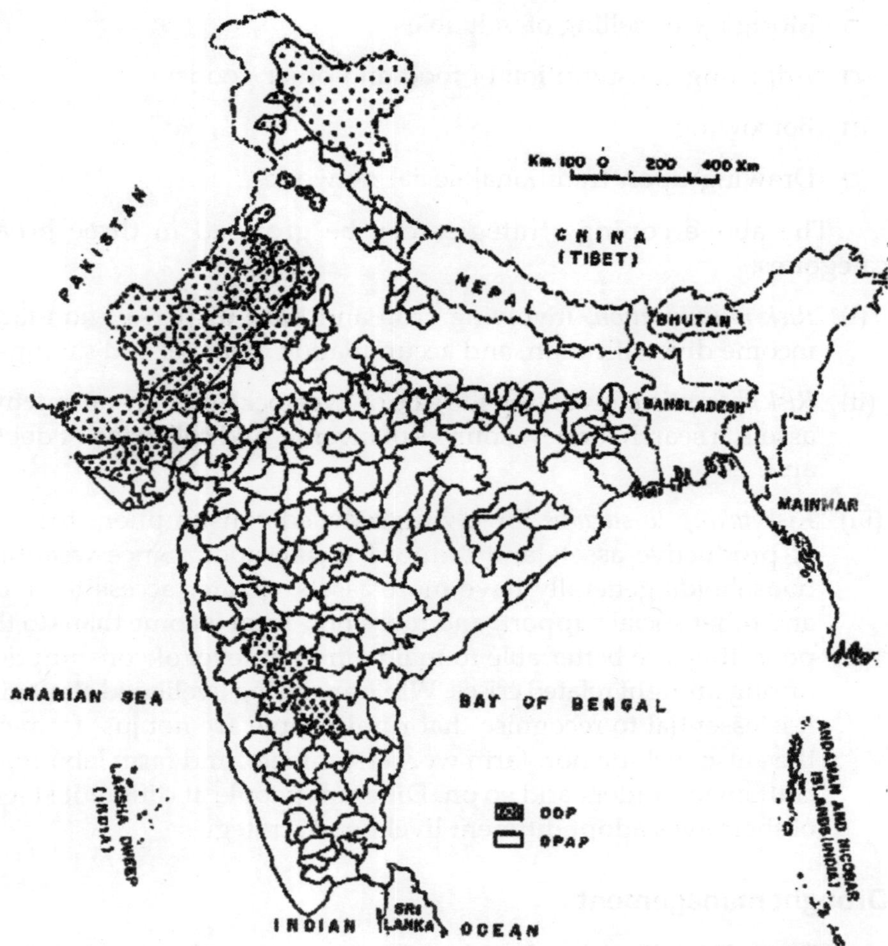

Fig. 1. Desert Development Program (DDP) and Drought Prone Areas Program (DPAP) districts in India.

Adaptation to drought and coping strategies of rural people

In drought prone areas of Rajasthan and other parts of India, the coping strategies adopted by households are relatively well known. They commonly involve the following:

- Adoption of integrated farming system involving a mix of crops, perennials and/or livestock
- Short- or long-term migration
- Earnings from non-agricultural enterprises

- Mortgage or selling of valuables
- Adjusting consumption of food and other goods
- Borrowing
- Drawing upon traditional social networks

The above coping strategies can be grouped in three broad categories:

(i) *Risk minimisation*, involving crop and herd dispersal, non-farm income diversification, and accumulation of assets and savings;

(ii) *Risk absorption*, involving the sale of livestock and non-productive assets, a search for new sources of income, and collection of debts; and

(iii) *Risk taking to survive*, involving reduced consumption, the sale of productive assets and reduced socialisation. Since wealthier households generally have more assets, greater access to credit and other social support, and more non-farm income than do the poor, they are better able to maintain their level of consumption during drought related crises. While analysing livelihood strategies it is essential to recognise that rural people are not just farmers; they also include non-farm workers, miners, and farm labourers, craftsman, traders and so on. Different people at different stages of their lives adopt different livelihood strategies.

Drought management

The traditional approach to drought management has been reactive, relying largely on crisis management. This approach has been ineffective because response is late, poorly coordinated, and poorly targeted to drought stricken groups or areas. In addition, drought response is post-impact and relief tends to reinforce existing resource management methods.

Drought management plans commonly have three major components:

- Monitoring and early warning
- Risk and impact assessment
- Mitigation and response

Monitoring and early warning

The overall goal of drought monitoring is to provide information that enables and persuades people and organisations to take action to maximise the probability of successful crop production and/or minimise the potential damage to established crops and other assets. In this regard, a reliable assessment of water availability and its outlook for near and long term is valuable information. In drought monitoring, data and information on each of the relevant indicators (e.g. precipitation, temperature, evapotranspiration, seasonal weather forecasts, soil moisture, stream flow, ground water, reservoir and lake level and snowpack) should be considered in the evaluation of water situation and outlook for the region.

There are a number of components that can be considered essential in the presentation of a comprehensive picture of droughts in a given region. These include information on:

- Timing of droughts
- Drought intensity
- Drought duration
- Spatial extent of a specific drought episode
- Analysis of the risk of the phenomenon and its likely effect on agricultural production.

There are a number of ways to provide information on the drought intensity:

a) Current rainfall data vis-a-vis long-term average rainfall

   This is the simplest means of presenting information on drought intensity and is used frequently around the world.

b) Presentation of current rainfall as a percentage of long-term average rainfall

   It is similar to above approach with the difference that rainfall data is presented as per cent of long term average.

c) Using different thresholds of current rainfall as a percentage of long-term average rainfall

   India Meteorological Department (IMD) use different thresholds of current rainfall as a percentage of long-term average (30 or more years rainfall) rainfall to delineate the drought intensity in different parts of the country. If the current rainfall in a given meteorological subdivision exceeds the Long Period Average (LPA) by 20%, the subdivision is deemed to have received excess rainfall. Threshold values of +19 to -19% of LPA are considered as normal, while current rainfall falling within -20 to - 59% of LPA would categorize a subdivision as 'deficient'. When the threshold value falls below - 60% of LPA, rainfall in a subdivision is considered 'scanty'. For example, in 1999, rainfall for India as a whole was 95.5% of the LPA rainfall, but seven out of the 35 meteorological subdivisions in the country received deficient rainfall i.e. 20 to 59% below the normal rainfall. In other words, nearly 8.1% of the country was affected by droughts in 1999. Rainfall in 2000 was 92% of the LPA and again seven meteorological subdivisions received deficient rainfall.

d) Use of drought indices to depict drought intensities

Numerous drought indices have been developed from known values of selected parameters to present a quantitative description of droughts. Some of the important drought indices are given below:

Standardized precipitation index (SPI)

The SPI, developed by McKee *et al.* (1993), requires precipitation data only and is widely used in India as well as in other countries. The SPI may be calculated on weekly, monthly or longer time scale. For calculation of SPI, gamma probability density function is fitted to precipitation data. The SPI is calculated as:

$$SPI = X - \bar{X} / c$$

Where, X = precipitation, $\bar{X}$ = mean precipitation of the location, and s = standard deviation

SPI value	Drought severity
-0.99 to –0.0	Mild drought
-1.0 to –1.49	Moderate drought
-1.50 to -1.99	Severe drought
< –2.0	Extreme drought

Palmer drought severity index (PDSI)

Palmer (1965) developed a soil moisture algorithm (a model), which uses precipitation, temperature data, available water content (AWC) of the soil, and climatic coefficients. The PDSI value is obtained based on water balance technique. PDSI varies between -6.0 to 6.0 indicating extreme drought conditions and adequate moisture conditions, respectively. The negative PDSI values show drought severity as shown below:

PDSI value	Drought severity
- 1.00 to –1.99	Mild drought
- 2.00 to –2.99	Moderate drought
- 3.00 to - 3.99	Severe drought
<= - 4.00	Extreme drought

Crop moisture index (CMI)

While PDSI is suitable as long-term drought measurement tool, CMI reflects short-term (weekly) moisture supply and it is suitable to identify agricultural droughts. Palmer (1968) developed it from procedures within the calculation of the PDSI. It is computed from mean temperature and total precipitation, as well as the CMI value of the previous week.

Severity of meteorological drought

IMD uses two measures; the first describes rainfall conditions (departures) while the second represents meteorological drought severity (Pai *et al.*, 2010). The precipitation is expressed on weekly, monthly or seasonal basis.

Departure of annual rainfall from normal (%)	Meteorological drought
0 or above	No drought
0 to –25	Mild drought
-26 to –50	Moderate drought
-50 or less	Severe drought

Aridity anomaly (I_a) as an index for demarcation of agricultural drought

IMD monitors agricultural drought from water balance technique. The monitoring is done during *kharif* for the country as a whole and during *rabi* for those areas which receives rainfall during post-monsoon. The departure of I_a from normal is expressed as a percentage. These values are then plotted on a map and analysed.

Drought duration

Information on drought duration depends not only on the onset of drought, but equally on when exactly the drought ends.

Spatial extent of a specific drought episode

The Spatial extent of specific drought episodes is best described using mapping tools.

Risk and impact assessment

Drought produces a complex web of impacts that not only ripple through many sectors of the economy but may be experienced well outside the affected region. To more clearly understand the impacts of drought, the phenomenon should not be viewed as merely a natural event. It is the result of interplay between a natural event and the demand placed on water and other natural resources by human-use systems. For example, societies can exacerbate the impacts of drought by placing demands on water and other natural resources that exceed the supply of those resources.

Drought mitigation

Mitigation is defined as short- and long-term actions, programs, or policies implemented in anticipation of drought or during the actual drought to reduce its adverse impacts. Drought related mitigation actions are, for the most part, different from those used for other natural hazards because of the insidious nature of the hazard and the non-structural characteristics of many of the impacts. In contrast to mitigation, response actions are those taken in an area that is experiencing drought and are intended to address impacts and expedite recovery of the affected area.

Floods

Floods in arid Rajasthan are rare in occurrence compared to drought which is a widespread phenomenon. However, floods create havoc by causing soil erosion, inundation of low lying areas, loss of life and property. On the other hand they also fill up the reservoirs, recharge the ground water and support the flora and fauna. Floods are mainly caused by heavy rainfall during the monsoon season.

The decade-wise flood frequency in western Rajasthan showed that floods occurred in the region with an average frequency of 2 to 3 out of 10 years. The decades 1911-1920, 1941-1950 and 1981-1990 were notable for severe floods. Though, wide-spread floods are not common in arid Rajasthan, such floods occurred in arid Rajasthan during 1908, 1917, 1944, 1975, 1983 and 1997. During 1917, the average rainfall of twelve arid districts exceeded by 161% and such high intensity flood has not been recorded in subsequent years. July 1979 flash flood in Luni basin was unprecedented. The rainfall recorded between 15 and 19 July, 1979 was up to 925 mm at Gorawar located in the Luni basin causing the severe flood. Pali (Luni basin area) experienced highest number of floods in eleven years, whereas, lowest number of floods occurred in Hanumangarh and Jhunjhunu (in four years each) districts.

The floods in Barmer and Jaisalmer districts during 2006 were due to high rainfall between 19 to 22 August with a total rainfall of 426 mm at Barmer, 392 mm at Ramsar, 383 mm at Gadraroad, 364 mm at Baitu, 311 mm at Sheo, 306 mm at Gida, 161 mm at Jaisalmer and 361 mm at Fatehgarh. The historical records at Barmer tehsil showed that such rainfall events occurred only thrice during the last century. During 2006 monsoon, two consecutive heavy rain-storms during 6-11 and 18-26 August caused excessive rainfall in Barmer, Jalor, Jaisalmer and Pali. The monsoon was erratic and uneven during 2006 and only three districts (Barmer, Jaisalmer and Jalor) received above normal rains.

Temperature variability

There are wide temporal and spatial variations in temperature in Rajasthan. Inter-annual, seasonal and diurnal temperature variations are particularly high in arid zone. Air temperatures sharply increase from April onwards and stand highest during May-June till pre-monsoon showers set in. The monsoon season (June-September) experiences a fall of about 3-5°C in temperature.

Temperature may reach up to 50°C in summer, while sub-zero temperature during winter is not uncommon. Extreme temperatures resulting in hot and cold wave conditions are frequently encountered in hot arid region. Extreme temperatures recorded in arid region show that the air temperature may rise up to 50˚C in summer months and go below -5˚C in winter season.

During summer season, westerly winds coming from sandy deserts of the Thar, Baluchistan and Iran, may bring dust storms that lower the temperatures for a brief period. High temperature, low relative humidity and high wind speed create heat wave condition (*loo*). For example, the heat wave conditions were recorded during May 2002 with air temperatures shooting above normal by 4-6°C at Jodhpur and Pali and 6-9°C above normal at Bikaner and Jaisalmer. Sometimes, the heat wave conditions continue or revive after a break of few days. The effect of heat wave on summer crops, vegetables and fodder crops is devastating. Many a times, farmers have to re-sow the crops due to burial of seeds or burning of young seedlings. Productivity of milch animals is also adversely affected by thermal stress.

Severe cold wave conditions may prevail during December and January in northwestern and northern parts of India due to passing western disturbances. Normally, cold waves occur in western Rajasthan for some days in the month of January and that too intermittently. Minimum temperature may dip below 4°C in sandy terrain of arid Rajasthan causing frost injury to sensitive *rabi* crops (Rao, 2009).

During 2005-06 winter season, the cold waves prevailed in the region for 13 to 15 days. The first cold spell of very severe nature started on 15 December, 2005 and continued for 12 days. The second spell of severe cold wave again set in from 4 January and continued for two weeks. The districts of Churu, Jhunjhunu, Ganganagar, Hanumangarh, Sikar, Jaisalmer and Bikaner were affected by intense cold wave conditions. Fatehpur (Sikar) recorded lowest minimum temperature of –5.5°C on 8 January, 2006. Churu recorded lowest night temperature of –3.4°C on 7 January, which was below normal by 7°C. Pilani also recorded –2.1°C on 8 January, which was 6°C below normal. In the extreme northwest, Ganganagar and Hanumangarh also experienced freezing temperatures (–0.8°C). On the western side, Bikaner recorded lowest of –2.8°C on 7 January. The minimum temperature remained as low as –2.0 to –5.5°C at Fatehpur continuously during 4 to 9 January 2006.

The cold wave of 2005-06 severely affected *rabi* crops like mustard, gram, castor; fruit plantations like *ber, anar* and vegetables like tomoto, brinjal. Due to the cold wave, castor crop was totally damaged in some regions. Early sown mustard crop, which was at grain filling stage, was also damaged in villages of Bimasar, Harsawa, Kayalnpur and Mandawa; In Churu and Jhunjhunu districts, there was about 80% damage in early and normal sown mustard crop while damage in late sown crop was about 60%. In irrigated gram, the newly developed branches turned red due to frost injury. Castor crop, which was at near maturity stage, was damaged to an extent of about 80%. Grain development and oil synthesis were stopped and loss in the oil percentage was very high. Bunches of bolls of castor broke down and drooped drown. Brinjal crop was totally damaged in Taranagar, Churu, Dungargarh and Sardarshahr areas.

Wind speed

The general wind direction in north-west India is southwesterly during monsoon season and northeasterly during winter season. In arid Rajasthan, the wind speed is usually low in winter months but it starts increasing from April onwards. Strong winds (15 to 18 kmph) are normally observed during June, however, the average wind speed during summer varies from 9 to 13 kmph. The wind velocity may reach as high as 60 - 80 kmph in the event of severe dust storms. These strong winds (Kharol *et al.*, 2013) occurring during May-June erode fertile soil layer particularly in drought years.

The number of hours wind speed exceeds a threshold value of 8 kmph at Jodhpur, which is considered useful for wind power generation prevails for about 4-6 hours in a day during mid-April to end-August. However, a long-term trend (1971-2010) of wind speed at Jodhpur indicates a reduction in wind speed (Singh *et al.*, 2011).

Wind erosion is a major problem in arid Rajasthan due to strong winds, sparse vegetation cover and poor soil structure. Jaisalmer, Bikaner and Barmer districts, which receive less than 250 mm of annual rainfall, are often prone to wind erosion. In these areas, shifting sand dunes cover fertile cultivated fields; block the roads and railway tracks. Wind action on drifting sand cause different types of dune formation. In western Rajasthan, moderate to severe wind erosion affected area is about 8.74 million ha. Soil loss due to wind erosion may be up to 273.7 kg ha^{-1} day^{-1} under sandy soil conditions in Bikaner district and 15.6 kg ha^{-1} day^{-1} under loamy sand in Jodhpur (Gupta, 1993).

The wind erosion can be checked by sand dune stabilization and shelterbelt plantation. Modification of wind regime and microclimate by shelterbelts and micro-crop shelterbelts provide favorable conditions for crop growth in the arid region. Tree shelterbelts of *Acacia tortilis* and *Cassia siamea* are suitable for arid regions. Three-rows of *Acacia tortilis, Cassia siamea* and *Prosopis juliflora* with *Albizzia lebbek* as the central row were found to be effective windbreaks (Kar *et al.*, 2009). Shelterbelts reduce wind speed and temperature, create favourable micro-climate and reduce soil moisture loss in the leeward side of shelterbelts. At least 14% higher soil moisture was recorded in the leeward side of shelterbelts compared to open fields (Gupta, 1993).

Relative humidity

Relative humidity in arid Rajasthan is often less than 30% during the summer months, but reaches up to 80% during monsoon season. It starts declining from October onwards with the retrieval of the monsoon (Choudhary *et al.*, 2013). The low humidity combined with strong wind regime leads to advection, causing higher evaporation losses.

Solar radiation

Solar radiation is high in arid Rajasthan throughout the year with a mean of 22 MJ m^{-2} day^{-1}. During winter months it varies from 15.1 to 17.7 MJ m^{-2} day^{-1}; while, in summer months, it ranges from 22.8 to 26.5 MJ m^{-2} day^{-1}. The daily duration of bright sunshine hours in the area remains above 10 h day^{-1} in May, 6.6 h day^{-1} in July-August and is above 8.8 h day^{-1} in the winter season (Rao, 2009). High solar radiation has great potential for small scale solar devices as well as solar power generation in the region.

Potential evapotranspiration

The annual potential evapotranspiration (PET) rate, estimated by modified Penman-Monteith formula, ranges from 1400 mm in eastern parts to more than 2000 mm $year^{-1}$ in the western parts of arid Rajasthan (Rao, 2009).

High values of temperature, solar radiation and wind speed, and low values of relative humidity result in high value of potential evapotranspiration demand. The daily PET ranges from 1.3 to 3.5 mm day^{-1} in winter months (December to February), 3.7 to 9.5 mm day^{-1}

in summer months (March to May) and 5.2 to 10.6 mm day^{-1} in monsoon months (June to September).

In future, the increase in temperature due to climate change is more likely to affect crop production in arid regions than in semi-arid or sub-humid regions of India since the prevailing temperatures are already high. Trend analysis has shown increase in temperature in past 40 years (Singh *et al.*, 2011). Awareness about the climate variability and change along with timely advice to farmers may help substantially in coping with weather vagaries. Monitoring of rainfall and temperature variations in conjunction with weather forecast may be required to provide timely weather-based information to the farmers. Crop losses due to weather vagaries can be reduced, while favourable weather conditions can be utilized fully, through weather monitoring, weather forecast and weather based agro-advisories for better crop management and efficient utilization of natural resources.

References

Choudhary, S.S., Garg, P.K. and Ghosh, S.K. 2013. Drought analysis using digital image processing & meteorological data. *International Journal of Advanced Remote Sensing and GIS* 2: 280-302.

Gupta, J.P. 1993. Wind erosion of soils in drought prone areas. In *Desertification in Thar, Sahara and Sahel Regions* (eds. A.K. Sen and Amal Kar). pp 91-105. Scientific Publishers, Jodhpur.

Kar, A., Moharana, P.C., Raina, P., Kumar, M., Soni, M.L., Santra, P., Ajai, Arya, A.S. and Dhinwa, P.S. 2009. Desertification and its control measures. In *Trends in Arid Zone Research in India* (eds. Amal Kar, B.K. Garg, M.P. Singh and S. Kathju). pp. 1-47. Central Arid Zone Research Institute, Jodhpur.

Kharol, S.K., Kaskaoutis, D.G., Badarinath, K.V.S., Sharma, A.R. and Singh, R.P. 2013. Influence of land use/land cover (LULC) changes on atmospheric dynamics over the arid region of Rajasthan state, India. *Journal of Arid Environments* 88: 90-101.

McKee, T.B., Doesken, N.J. and Kleist, J. 1993. The relationship of drought frequency and duration to time scales. Preprints, 8th *Conference on Applied Climatology*, Anaheim, California. pp. 179–184.

Narain, P., Rathore, L.S., Singh, R.S. and Rao, A.S. 2006. *Drought Assessment and Management in Arid Rajasthan*. 64p. Central Arid Zone Research Institute, Jodhpur and National Centre for Medium Range Weather Forecasting, Noida.

Pai, D.S., Sridhar, L., Guhathakurta, P. and Hatwar, H.R. 2010. *District-wise Drought Climatology of the Southwest Monsoon Season over India Based on Standardized Precipitation Index (SPI)*. 24p. India Meteorological Department, Pune.

Palmer, W.C. 1965. *Meteorological Drought*. Office of Climatology Research Paper 45. U.S. Weather Bureau, Washington, 58p.

Palmer, W.C. 1968. Keeping track of crop moisture conditions, nationwide: The new Crop Moisture Index. *Weatherwise* 21:156–161.

Rao, A.S. 2009. Climate variability and crop production in arid western Rajasthan. In *Trends in Arid Zone Research in India* (Eds. Amal Kar, B.K. Garg, M.P. Singh and S. Kathju). pp. 48-61. Central Arid Zone Research Institute, Jodhpur.

Rathore, M.S. 2005. State level analysis of drought policies and impacts in Rajasthan, India. IWMI, 40p. (*Working paper 93 : Drought Series Paper No. 6*), Colombo, Sri Lanka.

Sastri, A.S.R.A.S. 2011. Agricultural droughts: concepts, assessment and management. In *Agricultural Drought: Climate Change and Rainfed Agriculture* (Eds. Rao, V.U.M., Rao, A.V.M.S., Kumar, P.V., Desai, S., Saikia, U.S., Srivastava. N.N. and Ventateswarlu, B.), pp. 59-67. CRIDA, Hyderabad.

Shewale, M.P. and Kumar, S. 2005. *Climatological Features of Drought Incindences in India*. 22p. India Meteorological Department, Pune.

Singh, D.V., Purohit, R.S., Mertia, R.S., Laxminarayan, Poonia, S. and Kar, A. 2011. *Agrometeorological Data Handbook of Jodhpur (1971-2010)*. 110p. Central Arid Zone Research Institute, Jodhpur.

Venkateswarlu, B. 2012. Dryland agriculture: coping with climate change. *Annals of Arid Zone* 51: 1-9.

□□□

Chapter 3

Assessment and Monitoring of Drought and Desertification in Arid Rajasthan

P.C. MOHARANA

ICAR - Central Arid Zone Research Institute, Jodhpur, Rajasthan -342 003

E-mail: pcmoharana45@gmail.com

Introduction

Drought and desertification are the two very frequently used keywords by the researchers in the field of environment and land degradation. The way desertification has been equated with deserts or advancing of deserts, the common belief, still persists about drought that it enhances the possibility of desertification. Therefore, it is important to understand the rationale of this relationship and how such perception has been built upon over the period. During 1968-1974, Sahel region of West Africa (Southern margin of the Sahara to Ethiopia and Somalia) came under severe drought which affected 500 million people, resulted in widespread crop failure, livestock death, famine, large scale migration of the refugees out of their traditional grazing lands and villages to urban places. The incidence attracted worldwide attention in the form of land degradation. This was, perhaps for the first time, when researchers strongly thought of a possible cause and effect relationship between prolonged drought-climatic desiccation and land degradation. Since then, there have been many models explaining relationship like albedo drought model which suggested that over-gazing in Sahel led to depletion of surface plant cover and hence increase in the albedo which influenced regional rainfall pattern. The model however failed due to fact that overgrazing is not responsible for the prolonged nature of the Sahel drought and it is known that drought itself can lead to local overgrazing which can in turn accelerate soil erosion by wind and water. Glantz and Orlovsky (1983) indicated that density of the vegetative cover constitutes an important factor for desertification. According to them, as density decreases, the risks of wind erosion, water erosion and the adverse effect of increased solar radiation on bare soils are increased dramatically. The decreasing vegetation cover is also likely to enhance surface albedo (reflectivity), which is also considered to be a major contributor to desertification processes.

French forester Aubreville, (1949) used the word "Desertification" for the first time to an incident of replacement of African rainforest by secondary savanna and scrub as a result of tree clearing and associated with shifting cultivation. According to Mabbutt (1978) combination of climatic stress and landuse pressure (over grazing + over cropping) will result in land degradation. The incidence of Sahel drought acted as catalyst to hold first UN conference on "Combating Desertification" which was held at Nairobi on August 1977. One main outcome of this conference was the definition of desertification "as the spread of desert like conditions in arid or semi-arid areas up to 600 mm isohyet due to human influence or climatic change".

In general, there are two definitions of the term 'desertification' which need discussion for a good understanding of relationship between desertification, drought, land productivity etc. The first definition that was proposed by UNEP in 1991 suggested it as the land degradation in the arid, semi-arid and the dry sub-humid regions resulting mainly from adverse human impacts. From this definition, it is presumed that it is a state of natural resources like soil and vegetation, mainly resulting from human activity. For example, in a humid region, loss of vegetation due to tree felling exposes the land and accelerates soil erosion and loss of soil fertility, while in a dry region drought may compel people to cut trees making the land more vulnerable to erosion by rainfall. The second of the definitions adopted in Earth Summit at Rio, held in June 1992 (Anno, 1992) stated "Desertification is land degradation in arid, semi-arid, and dry sub-humid areas resulting from climatic variations and human activities".

The main difference between the two definitions is the inclusion of climate as a major cause in the Rio definition. However, it reduced the emphasis of human activities as being the primary responsible factor and linked the possible impact of climate change to desertification.

Drought

Drought can be defined as a recurring climatic event over land, characterized by below normal rainfall over a period of months to years in the pretext of arid region like western Rajasthan (Dai, 2011). The common perception about drought is that it is a regional weather phenomenon that occurs in all climatic zones. However, its intensity and frequency usually increases towards drier zones. In a country like India, where agriculture is the major source of livelihood and that by and large depends on the monsoon, drought often leads to shortage of food, fodder and water. Considering that the arid regions receive much less rainfall than the other regions on the earth and also less number of rainy days, the incidence of drought is also higher in this region. In Indian part of arid zone, which is more confined to western part of Rajasthan in 62% of country's arid lands, drought is a frequent phenomenon. Between 1901 and 2010, the region experienced droughts of moderate to severe intensities in last 57 years. In recent decades, the drought of 2002 was of disastrous proportion when annual rainfall was mere 127 mm (38% of the long-term average) and the crucial summer monsoon rainfall was 90 mm, against the normal of 280 mm (Saha *et al.*, 2012).

Drought is not easy to predict well in advance for most locations for simple reason that it depends how precisely we predict two variables, precipitation and temperature. One explanation for this is that droughts are a slow-onset, or "creeping," natural disaster, developing over months and years, and frequently they exist before it is realized that a drought is occurring (Albert et al., 2002). Secondly, a drought's severity varies by its precipitation deficit, spatial extent, and duration, making it difficult to compare one drought to another. Finally, the impacts of drought are based on the range of economic, environmental, and social resources within a region. Drought is perceived in different ways by different people. Its definition surrounds around specific objectives and type of drought. For example, to a hydrologist, it is a condition of less water availability in water bodies like streams, wells, tanks and reservoir; for an agronomist, it is soil moisture deficiency, crop yield reduction; and for a meteorologist, it is occurrence of less rainfall. However, to a common man, it is the scarcity of essential commodity (mainly food crops), drinking water, rise in price and unemployment. These otherwise represent the operational definitions (Table 1) of drought.

Table 1. Technical definition on types of drought (Anon, 2005)

Type of droughts	Definition	Database requirement
Meteorological	According to IMD (Indian Meteorological Department), for meteorological drought over an area, rainfall departure of -26 to -50 per cent from the normal is termed as 'moderate drought', more than -50 per cent is termed as 'severe drought' (Sikka, 2004)	Daily rainfall, temperature, humidity, wind velocity, pressure, evaporation.
Hydrological	This is an extension of meterological drought deciphered from water level in lake, reservoir and ground. Hydrological drought takes longer time than agricultural or meteorological droughts to manifest.	Surface area & volume, run-off, river flow measurement and infiltration for surface water. Water level fluctuation and aquifer characteristics for subsurface water availability.
Agricultural	It occurs when soil moisture and rainfall are inadequate during the growing season to support healthy crop growth to maturity and cause extreme crop stress. IMD calculates agricultural drought in terms of aridity index measured as [(AE-PE) x 100/PE] Where, AE is mean evapo-transpiration, and PE is potential evapo-transpiration. It is calculated on weekly basis. Accordingly, values < 26% indicate mild type drought, between 26 and 50 % is moderate type and values>50% represents severe drought situation.	Soil texture, soil fertility, soil moisture, crop type and area.

In India, more than 70 % population depend on agriculture, 68 % of net sown area (142.2 M ha) is drought prone and 50 % of drought prone is severe in nature (NRSC, 2010). At this scale, India experiences localized drought almost every year in some region or other. Major droughts that affected about one third of our country in post-Independence era, were reported during 1951, 1966-67, 1972, 1979,

1987-88 and 2002-03 (Subbaiha, 2004) and maximum (10 years) of drought occurred between 1901 and 1925 (Kulshreshta and Sikka 1989). Thus Indian agriculture continues to be affected by droughts. Arid and semi-arid regions of India are more prone to drought than the other climatic zones with a probability of drought being >20 %. Areas where more than 40 % of the rainfall recording years have experienced drought are classified as chronically drought-prone, while those experiencing drought for 20 to 40% years are termed as drought prone (CAZRI, 2005).The hot arid regions of our country (32.7 m ha) in western Rajasthan, Gujarat, Haryana and part of Karnataka and Andhra Pradesh are chronically drought prone.

Aridity

The word 'drought' has often been related to the term aridity. An aridity index (AI) is a numerical indicator of the degree of dryness of the climate at a given location. The UNEP (United Nations Environment Programme) has adopted the following indices defined as :

$$AI = P / PET$$

Where PET is the potential evapotranspiration and P is the average annual precipitation (UNEP, 1992). The PET and P are expressed in millimeters. Based upon this index, climatic boundaries (Table 2) for various categories of arid/humid areas are determined.

Table 2. Classification of climatic zones based on aridity index

Classification	Aridity Index	Global land area
Hyperarid	< 0.05	7.5%
Arid	Between 0.05 and 0.20	12.1%
Semi-arid	Between 0.20 and 0.50	17.7%
Dry subhumid	Between 0.50 and 0.65	9.9%

IMD monitors drought using water balance technique, which addresses agricultural drought.

Drought and desertification in western Rajasthan

As mentioned earlier, drought is a common phenomenon in arid Rajasthan and all the 12 districts of this region (Jaisalmer, Barmer, Jodhpur, Bikaner, Sri Ganganagar, Hanumangarh, Churu, Jhunjhunun, Sikar, Nagaur, Pali & Jalore) experience droughts of various intensities due to low and erratic nature of rainfall. Analysis of occurrences of drought indicate that this region has witnessed 6 major drought years between 1972 and 2002 (Rao & Roy, 2012). The percent deviation of rainfall from normal varies from maximum -69 in 2002 to minimum -12 during 2000 (Fig. 2). In century scale, the region experienced droughts in 49 to 60% of the years during 1901-2010 (Table 3). The year 2002 experienced the most severe drought. The isohyet distribution for western Rajasthan for 2002 is presented in Fig 1. This is mapped using GIS based on the occurrence of rainfall at various locations of western Rajasthan during the year (*Source:* Government of Rajasthan).

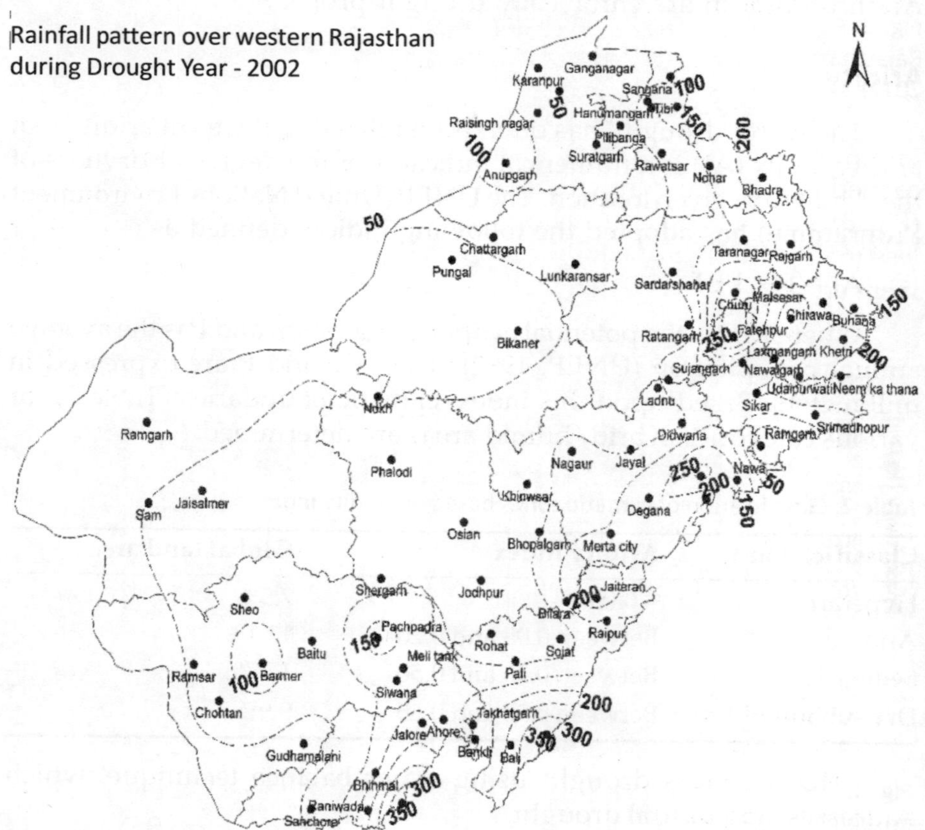

Fig. 1. Spatial distribution of mean annual rainfall of 2002

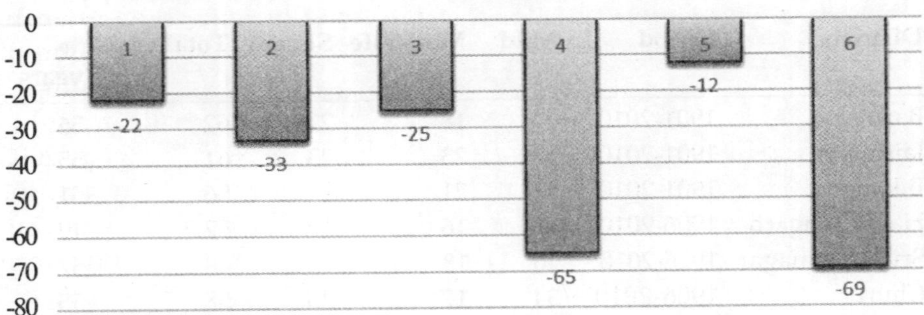

Fig. 2. Percent deviation of rainfall from normal during major droughts in western Rajasthan (*Source:* Weather variability and crop production in Arid Rajasthan, CAZRI, 2012).

An assessment of crop productivity for cereals of the region was carried out to see the impacts of drought for the succeeding periods (Fig. 3). It is clear that during 2002 and 1987, when the rainfall was very less (65% and 69% deviation from normal), the productivity of cereals were equally low.

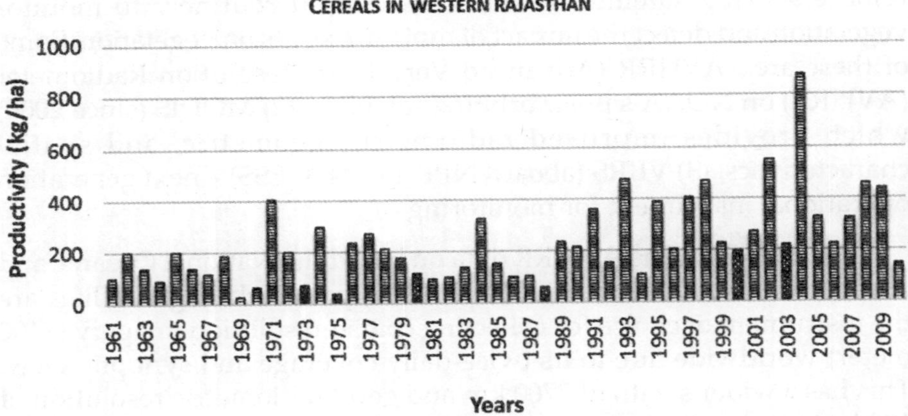

Fig. 3. Productivity of cereals in western Rajasthan (1961 – 2010) and during severe drought years (Agricultural Statistics, Rajasthan, Department of Economics and Statistics, 2010-11 and past years).

Table 3. Occurrence of types of drought in the districts of western Rajasthan (CAZRI, 2012)

Districts	Period	Mild	Moderate	Severe	Total years	% years
Barmer	1901-2010	22	18	22	62	56
Jaisalmer	1901-2010	15	23	23	61	55
Bikaner	1901-2010	22	21	13	56	51
Hanumangarh	1906-2010	17	16	19	52	49
Sri Ganganagar	1926-2010	20	13	15	48	57
Churu	1906-2010	34	17	05	58	55
Jhunjhunu	1901-2010	28	21	09	58	53
Sikar	1901-2010	23	21	15	59	54
Nagaur	1901-2010	29	23	14	66	60
Pali	1901-2010	18	30	11	59	54
Jalor	1901-2010	30	30	16	63	57

Application of Remote Sensing Technologies and Indices for monitoring and assessment of vegetation Cover

Over the last 20 years, several studies indicate how coarse resolution remote sensing satellite sensors can be used routinely to monitor vegetation and detect the impact of moisture stress on vegetation. Some of these are ; AVHRR (Advanced Very High Resolution Radiometer (AVHRR) on NOAA's polar orbiting satellites, (2) MODIS (since 2000) which provides improved radiometric, geometric, and spatial characteristics, (3) VIIRS (aboard NPP and NPOESS) – next generation operational instrument for monitoring.

AVHRR data: The AVHRR data on board the National Oceanic and Atmospheric Administration's (NOAA'S) polar-orbiting satellites are the instrument of choice for collecting coarse-resolution imagery (1100 meter) worldwide due to its twice-daily coverage and synoptic view. This has a wider swath of 2700 km and good radiometric resolution of 10 bits for assessing a large area (NRSC, 2010).

MODIS data: In December 1999, NASA launched the Terra spacecraft, under its Earth Observing System (EOS) program. A sensor called the Moderate-resolution Imaging Spectro-radiometer, or MODIS, was mounted on it that greatly improves scientists' ability to measure plant growth on a global scale. Briefly, MODIS provides spatial resolution up to 250 meter than AVHRR, while also matching AVHRR's almost-daily global cover and exceeding its spectral resolution. In other words, MODIS will provide images over a given pixel of land just as often as AVHRR, but in much finer detail and with measurements in a greater number of wavelengths using detectors that were specifically designed for measurements of land surface dynamics.

Indices

Enhanced Vegetation Index (EVI): The MODIS Science Team is preparing a new data product, called EVI that will improve upon the quality of the NDVI product. The EVI will take full advantage of MODIS' new, state-of-the-art measurement capabilities. While the EVI is calculated similarly to NDVI, it corrects for some distortions in the reflected light caused by the particles in the air as well as the ground cover below the vegetation. The EVI data product also does not become saturated as easily as the NDVI when viewing rainforests and other areas of the Earth with large amounts of chlorophyll.

Normalized Difference Vegetation Index (NDVI: It is derived from AVHRR data, has been extensively used for vegetation monitoring, crop yield, and drought assessment because of its simplicity in calculation and easy to interpret (Tucker and Chowdhary, 1987). Kaushalya *et al.*, 2013 used NOAA-AVHRR and MODIS time-series NDVI data products for assessment of vulnerability of Indian agriculture to rainfall variability, Piao Shilong et al. 2005 used NDVI based assessment to indicate decline in desertification in China in the past two decades.

The NDVI is a function of green leaf area and biomass. The values of NDVI is derived as (NIR - RED) / (NIR + RED), where NIR (Near Infra-Red) is the reflectance radiated in the near-infrared waveband and RED is the reflectance radiated in the visible red waveband of the satellite radiometer. Using the values of index which ranges from maximum 1 (higher vegetation) to -1 (lowest vegetation concentration), one can assess the condition in regard to start, growth and sequences in crop/vegetation growth. Higher NDVI indicates a greater level of photosynthetic activity (Tucker *et al.*, 1991). It has been demonstrated that multi-temporal NDVI derived from AVHRR data is useful for monitoring vegetation dynamics on a regional and continental scale. NDVI products can be generated from the data of following satellite sensor systems; MODIS NDVI of 250 m and 1000 m, SPOT VGT NDVI of 1000 m, NOAA AVHRR NDVI of 1000 m, IRS WiFS NDVI of 188 m and AWiFS NDVI of 56 m are widely used for drought monitoring (NRSC 2010). The last two sensors are Indian Remote Sensing data products.

Use of NDVI in the assessment of vegetation condition during 2002 drought year

Spatial distribution and variation of vegetation for monsoon, winter and summer months of 2002 for the entire western Rajasthan is presented in Fig. 4. The data was obtained from MODIS NDVI (https://earthdata.nasa.gov/earth-observation). The data indicates comparative vegetation covers during drought of 2002 in all the crop periods.

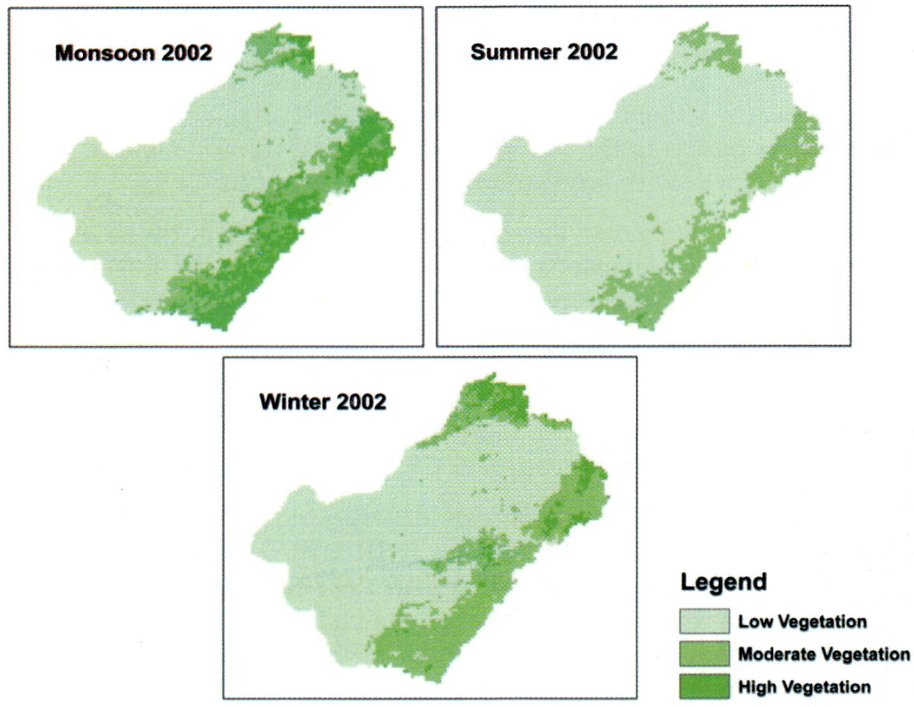

Fig. 4. Spatial variation in vegetation in western Rajasthan during different periods of 2002.

Spatial distribution of NDVI over Jaisalmer District in western Rajasthan during drought and normal years: a case study

This study was carried out using NASA MODIS CMG Monthly NDVI (Global) data, processed by NASA Goddard from the Terra sensor projected on a 0.05 degree (5.5 km resolution). IRS - AWiFs data were used as secondary data for checking vegetated area. Following methodology was used

- MODIS NDVI data (for the years 2000 to 2009 and for the monsoon period (July, August and September) were downloaded for visual observation of values.
- Conversion of MODIS NDVI data from 16 bit integer to 32 bit float to get the original values (+1 to -1)
- From this observation, we obtained maximum and minimum range of NDVI values (0-0.65).
- On the basis of this range, we have classified all the scenes into three categories (0.65/3) as 0-0.2 (Low), 0.2-0.4 (Moderate) and 0.4-0.7 (High) using ArcGIS spatial analysis tool.
- This followed statistics generation and area calculation for each class.

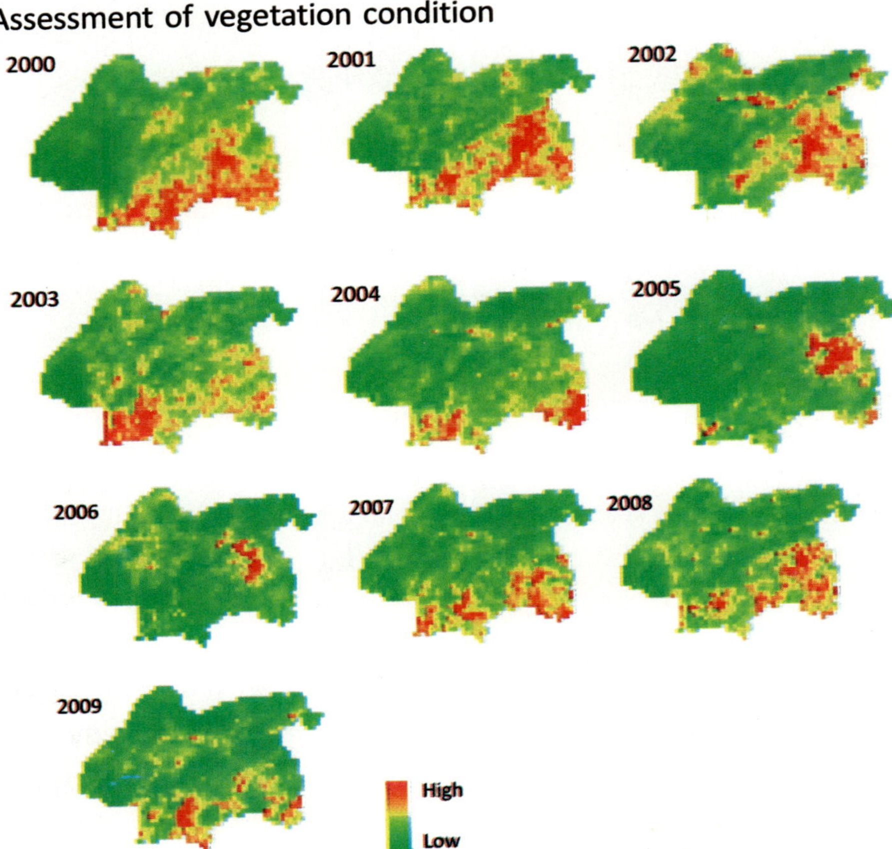

Fig. 5. NDVI derived from MODIS TERRA showing different levels of Greenness over Jaisalmer district from 2000 to 2009 (https://earthdata.nasa.gov/earth-observation). *Reclassification of NDVI values (categories)*

Time series analysis of NDVI values for the respective years (2000 to 2009) is presented in Fig. 5. The images present an over view of spatial changes in greenness over the region. Reclassification of this data into low, moderate and high (Fig. 6 & Table 4) using GIS indicates condition of vegetation and their spatial extent over the years. Area under high category was experienced during 2003 (4.8% area) and in 2001 and 2006 with <1% area. In the moderate situation, during not so good rainfall years, the areas remained at par with 10-12% while in the low categories, the percent area varied from 89 to 100 (2002), except in 2003 (36%) indicating very bad condition of vegetation.

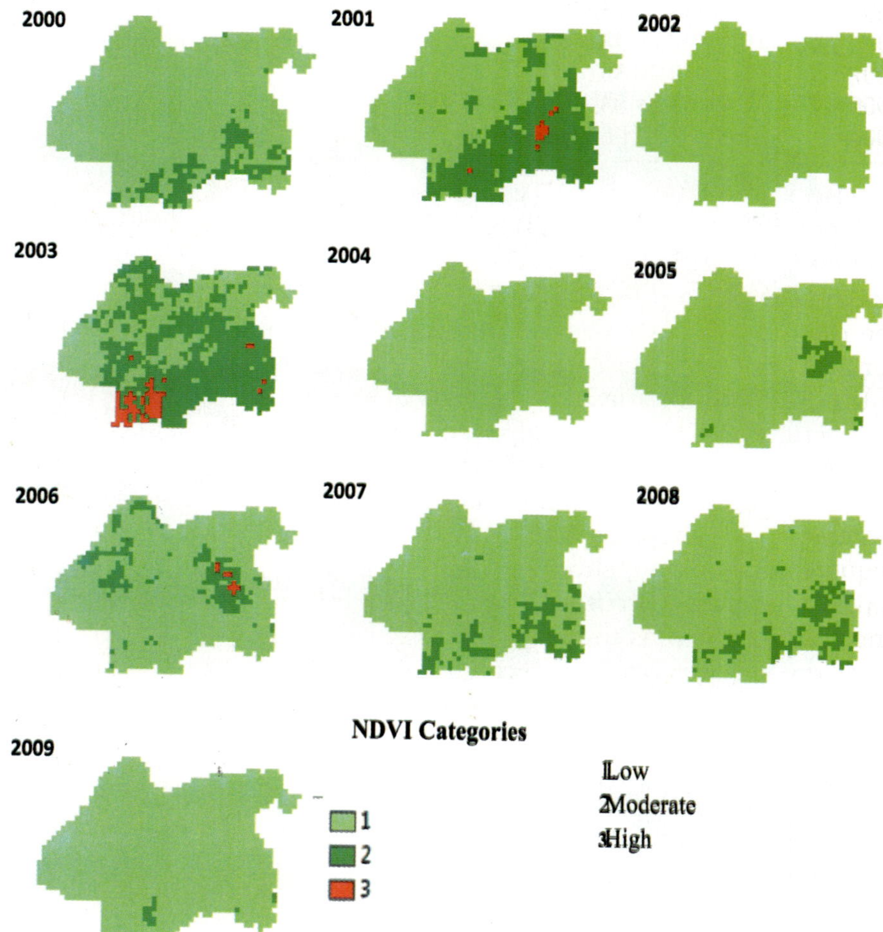

Fig. 6. Reclassification of data showing sequential changes in NDVI over the years (2002-2009) (Based on calculation using GIS).

Table 4. Reclassified values of NDVI over Jaisalmer district in western Rajasthan, 2000-2009 (Based on calculation using GIS)

Years	Area (km²) under different categories					
	Low (0-0.2)	Area (%)	Moderate (0.2-0.4)	Area (%)	High (0.4-0.7)	Area (%)
2000	34364.77	89.30	4116.57	10.70	0.00	0.00
2001	23182.96	60.25	14939.97	38.82	358.05	0.93
2002	38481.21	100.00	0.000	0.00	0.00	0.00
2003	15278.14	39.70	21349.60	55.48	1853.61	4.82
2004	38453.6	99.93	27.61	0.07	0.00	0.00
2005	36968.37	96.07	1512.84	3.93	0.00	0.00
2006	33293.27	86.52	4940.71	12.84	247.22	0.64
2007	33869.49	88.02	4611.95	11.98	0.00	0.00
2008	34014.08	88.39	4467.40	11.61	0.00	0.00
2009	38038.81	98.85	442.39	1.15	0.00	0.00

Reclassification of NDVI data showed that majority of areas correspond to low (0.02-0.04) NDVI values accounting for 60 to 100% area. Areas under high NDVI (0.4-0.7) is limited to < 1% area. Considering the upper limit of NDVI value (0.4 – 0.7) it can be said that the district has poor vegetal cover which is indicative of degraded conditions of vegetation. The year 2002 was a severe drought year and entire area was classified as "low NDVI". From the table 3, it is indicated that "High NDVI" values (depicted in red color) was observed only in the year 2001, 2003 and 2006. During all these three years, there were good rainfalls too. So rainfall may be one of the factors for better vegetation for these periods. In this region, the NDVI represents a sequence of cropland/shrub/grass/trees as per their area coverage. The rangelands which occur in about 47,000 km² area under sandy, rocky and saline habitat, is the major areas for grassland and shrubs. In the same time, these areas also receive less annual rainfall (150 – 200 mm) while temperature during summer ranges from 25^0 C - 45^0 C. Typical trees like *Prosopis cineraria, and Acacia tortilis;* shrubs (*Salvadora oleoidis and Caparis decidua*) and grasses (*Lasiurus sindicus, Eleusine compressa and Panicum turgidium*) occur dominantly even under heat and water stress conditions. Apart from vegetation cover, NDVI values also indicates state of other land use/land cover types. Mapping of land use /land cover shows that out of total cultivated lands, about 51.2% is under net sown area, 6.26 % is current fallow and 5.27% under other fallow category. Net irrigated area constitutes 13.0% of total mapped area, while area sown more than once is 10.0%. Other land uses contributing to vegetation cover are permanent pastures and other grazing lands (26.4%), notified forest area (2.2%) (Ram, 2009). Since majority of croplands are rainfed, the rainfall factor decides the vegetation health in entire region.

Time series analysis of Desertification in western Rajasthan

An image based interpretation on the state of desertification was carried out at CAZRI for western Rajasthan under a MoEF/ISRO (SAC) sponsored National level mapping (Desertification Status Map) using Indian Remote Sensing satellite images, IRS-P6-AWiFS of 2003-04 (Table 5). Systematic mapping of various land degradation units were carried out using a mapping schema under this project.

Table 5. Degradation types under different land use / land cover units in western Rajasthan (based on interpretation and mapping of Indian Remote Sensing satellite images of 2003-04)

Degradation process	Severity type	Rainfed cropland	Irrigated cropland	Scrubland	Sandy area	Barren land	Forest	Mine areas	Total area (km²)	% area
Wind erosion	1*	62868	27906	11640	17462	4675	0	0	124551	59.20
	2	12584	380	0	22046	0	0	0	35010	16.64
										75.84
Water erosion	1	1143	1350	926	416	0	0	0	3835	1.82
	2	39		64	0	0	0	0	103	0.05
										1.87
Salinity	1	1928	527	135	0	0	0	0	2590	1.23
	2	118	2	530	0	1311	0	0	1961	0.93
										2.16
Waterlogging	1	0	177	0	0	0	0	0	177	0.08
	2	0	6	0	0	0	0	0	6	0.00
										0.09
Vegetation deg	1	0	0	4655	0	0	1066	0	5721	2.72
	2	0	0	61	0	0	72	0	133	0.06
										2.78
Mining	1	0	0	0	0	0	0	111	111	0.05
	2	0	0	0	0	0	0	91	91	0.04
Rocky area	Rocky	0	0	0	0	0	0	0	2129	1.01
Total degraded area	176418	78680	30348	18011	39924	5986	1138	202	176418	**83.85**
Total mapped area (km²)	210399									100.00

* 1-Low 2-High

Salient features of state of land degradation in the region are described in the following paragraphs. The above database (Table 5) indicates 83.85% area of western Rajasthan is affected by various processes of desertification out of which, areas under wind erosion / deposition accounts for 75.84% indicating the vulnerability of western Rajasthan to aeolian activity. Water erosion (1.87%), vegetation degradation (2.78%), salinity/alkalinity (2.16%) and mining (0.09%) are the other processes. Since, this is derived from a 1:500,000 scale map, the database indicates the status of desertification at a regional level. Therefore, this database may not represent scenario of desertification at district or tehsil level. In the early 1990's, CAZRI carried out a satellite -based mapping at 1:1 M scale for western Rajasthan in which 1:250,000 scale FCCs of Landsat-TM images for 1989-1991 were visually interpreted. A set of field indicators was developed for gathering ground truth for the mapping. A publication on this mapping (Singh *et. al.*, 1992) indicated about 19.17 m ha (92% of 20.87 m ha mapped area) area was degraded out of which 76% area was affected by wind erosion, 15% area was by water erosion while waterlogging and salinity affected 6% area. Comparing these two time series databases, it was found that the differences may be due to two types of classification systems. The earlier one dealt with the processes that were defined either by sole agent (wind/water/salinity)/ impact or by combination of agents of land degradation (for example wind + vegetation, water + salinity). The data indicates wind erosion / deposition activity as the dominant process which can be observed by field manifestations (Fig. 7). The data also indicates how area under such process is decreasing over the years. Change in the areas affected by processes like water, salinity/alkalinity and vegetation degradation is not significant when mapped at 1:50,000 or 1:1 million scale. However, such land features can be observed when mapped at village level or watershed level. For example, mining as an important economic activity at village level (even though changes have been mapped in 0.02% area) is an important form of land degradation. This is not only a regular form of occupation but also it acts as a pool factor for migration of people during a drought year. Over the years, the methodology and interpretation however has seen drastic changes due to availability of precise information from large scale satellite images like IRS-LISS-IV or CARTOSAT DEM images. Here, in this map, there are only two categories; 1 & 2 representing low and high severity types. This is because of the fact that, at regional scale, there was very less scope for discriminating between low to moderate types, therefore, both were grouped under low.

Some useful websites for data availability;

- www.earthexplorere.usgs.gov.in
- http://www.bom.gov.au/climate/data-services/
- http://www.pancroma.com/data.html
- http://www.eumetsat.int/website/home/Data/DataDelivery/EUMETSATDataCentre/index.html
- http://www.ncdc.noaa.gov/data-access/quick-links#wind
- http://bhuvan.nrsc.gov.in/bhuvan/PDF
  NADAMS_July12_Report.pdf

Sand encroaching upon roads indicating severe wind erosion/deposition in Churu district

Sand depositional activity in high severity zone in the rangelands of Jaisalmer district

Gullied lands and land erosion due to run-off

Barren rocky/gravelly plains with shallow sand

Fig 7. Field photographs by the author showing features of desertification in western Rajasthan

Conclusion

Majority of developing countries of the world, and India in particular, very much depend on the occurrence of seasonal rainfall, and hence droughts are common features. Timely information related to rainfall and its spatial distribution is essential for agricultural activities in a large country like India. Besides conventional approaches to drought assessment and warning, satellite based information provides a quick input to understand and predict the state of climatic abnormalities and so as various types of drought. At global scale, information from a wide range of satellite systems is currently available which offer different dimension to agricultural drought assessment compared to conventional mechanisms. Drought is a common phenomenon in arid western part of Rajasthan where such incidences put considerable impact considering the demand of food and fodder for its 27.06 million (Census, 2011) human and a matching livestock population (30.17 million, 2012 census). For Rajasthan, the year 2002 was a century-scale drought year. At village level, the number of drought free years will be even less. At least half of 33,000 villages of Rajasthan are under the grip of drought almost every year. The region has also faced continuous droughts (for 3 to 6 years) during 1903-05, 1957-60 1966-71, 1984-87 and 1997-2000. Estimate shows that this region experienced droughts in 49 to 60% of the years during 1901-2010 in one or other districts (CAZRI, 2012). Studies have revealed that droughts do affect and increases the vulnerability of a region to desertification. For example, herds of animals would quickly remove the scanty grass cover that come up with meagre rainfall, thus aggravating the problems of soil erosion and desertification. Widespread crop failures lead to acute shortage of food and fodder. Dregne, 2000 thought that a better rainfall and land use practice may reverse the effect of drought. According to him, it is the bad practices by people during the drought period and still continuing the same even during the good rainfall period leads to aspects of desertification. Our assessment in western Rajasthan showed how good rainfall conditions after worst of drought years like 2000, 2002 even resulted in crop productivity coming to normal and also more than average. Therefore, incidences of droughts may not be a sufficient nor a necessary cause of desertification.

The way forward

Studies related to drought and desertification needs proper assessment of a situation which is both spatial and temporal. This can be assessed both by conventional survey and also by using remote sensing based information. Now, both coarse and moderate resolution satellite sensors as well as various drought indices products are available to routinely monitor vegetation and crops as indicator of drought severity. Similarly, desertification is also a case for long term observation and evaluation. But experiences show how conditions conducive to land degradation has a local dimension or begins from a smaller area like cadaster level plots. Therefore, it is suggested to begin assessment of land degradation at a cadastral / village level and upscale it later on at tehsil/district/state level to find out both natural and socio-economic reasons responsible for the situation arising out of climate change, drought and desertification.

Fig. 8 :

Low sand dunes or barchans in the crop field in the low severity zone in the canal command area of Hanumangarh district.

Sand spread on the irrigated fields indicating moderate wind activity in the canal command area in Suratgarh town.

References

Anonymous. 1992. Managing fragile ecosystems: Combating desertification and drought. In, Chapter 12, Agenda 21, *United Nations Conference on Environment and Development*, Rio de Janeiro. A/CONF.151/4 (Part II), United Nations, New York, pp. 46-66.

Anonymous. 2005. *Drought in western Rajasthan, impact, coping mechanism and management strategies* (Eds. Pratap Narain and Amal Kar), NAIP, Central Arid Zone Research Institute. Jodhpur. 104 p.

Aubreville, A. 1949. Climats, Forets et Desertification de l' Afrique Tropvcale. Societe d'Editions Geographiques, Maritimes et Coloniales. Paris, 255p.

Albert, J. Peters., Elizabeth, A. Walter-Shea., Lel, J.I., Andres, Vliia., Mlchael, Hayes. and Mark, D. Svoboda. 2002. Drought monitoring with NDVI-Based Standardized Vegetation Index, *Photogrammetric Engineering & Remote Sensing*. 68(1): 71-75.

Balak, Ram. 2009. Land use of arid lands in India, In *Trends in Arid Zone Research in India.* (Eds. Amal Kar, B.K.Garg, M.P.Singh and S.Kathju), Central Arid Zone Research Institute, Jodhpur.481 p.

Rao, A.S. and Roy, M. M. 2012. *Weather variability and crop production in Arid Rajasthan.* Central Arid Zone Research Institute, Jodhpur, India, 70 p.

CRIDA. 2007. Drought models. In *Agricultural drought: aspects of Micrometeorology, lecture notes of the 4th SERC School* (Eds. Y.S. Ramakrishna, G.G.S.N. Rao, P.S.N. Shastry and V.U.M. Rao), Hyderabad, India: Central Research Institute for Dryland Agriculture. 400p.

Dai, A. 2011. Drought under global warming. Wiley Interdisciplinary Review: *Climate change* 2(1): 45-65.

Dregne, H.E. 2000. Desertification: Problems and challenges. *Annals of Arid Zone* 39: 363-371.

Glantz, Michael, H. and Orlovsky. 1983. Nicolai. Desertification. *A review of the concept Desertification Control Bulletin* 9: 15-22

Kaushalya, R., Venkateshwarlu, B., Ramarao, C.A., Rao V.U.M., Raju, B.M.K., and others 2013. Assessing agricultural vulnerability to climate change using NDVI data products. In *Geospatial Technologies for natural resources management*. Central Research Institute for Dryland Agriculture, Hyderabad, pp 345-361.

Kulshreshta, S.M. and Sikka, D.R. 1989. Monsoons and droughts in India: Long term trends and policy choices. *National workshop on drought management*, New Delhi.

Mabbutt, J. 1978. Desertification in Australia. Report no. 4. Canberra: Water Research Foundation of Australia. 127 p.

NRSC 2010. *Remote sensing applications* (Eds. P.S.Roy, R.S.Dwivedi, D.Vijayan). National Remote Sensing Centre, Hyderabad, 377 p.

Piao, Shilong., Fang, Jingyun. and Liu, Hongyan. 2005. NDVI-Indicated decline in desertification in China in the past two decades. *Geophysical Research Letters*, 32: L06402.

Saha, D.K, Kar, Amal. and Roy, M.M. 2012. Indicators of drought vulnerability for assessing coping mechanism in arid western Rajasthan. *Annals of Arid Zone* 51(1): 1-9.

Subbaiah, A.R. 2004. State of the Indian Farmer - A Millennium study, *Natural Disaster Management,* 21, Department of Agriculture and Cooperation, Ministry of Agriculture, Government of India, New Delhi, 226 p.

Singh, S., Kar, A., Joshi, D.C., Kumar, S., Vats, P.C., Singh, N., Raina, P., Kolarkar, A.S. and Dhir, R.P. 1992. Desertification mapping in western Rajasthan. *Annals of Arid Zone* 31: 237-246.

Tucker, C. J. 1979. Red photographic infrared linear combinations for monitoring vegetation. *Remote Sensing Environment* 8:127-50

Tucker, C.J and Chowdhary, B.J. 1987. Satellite remote sensing of drought conditions. *Remote Sensing of Environment* 23: 243-251.

Tucker, C. J., Dregne, H.E. and Newcombe, W.W. 1991. Expansion and contraction of the Sahara Desert from 1980–1990. *Science* 253: 299–301.

UNEP. 1992. *World Atlas of Desertification* (Eds. N. Middleton and D.S.G. Thomas). Edward Arnold, London.

❑❑❑

Chapter 4

Climate Change and Land Use Pattern in the Thar Desert

MAHESH KUMAR GAUR

ICAR-Central Arid Zone Research Institute, Jodhpur-342003, Rajasthan
E-mail: maheshjeegaur@yahoo.com

Introduction

Human interference with the climate system is occurring, and climate change poses risks for human and natural systems of the Earth. Climate change is one of the major challenges to the human existence. It adds considerable stress to the societies and to the environment. From shifting weather patterns that threaten food production, to rising sea levels that increase the risk of catastrophic flooding, the impacts of climate change are global in scope and unprecedented in scale. The spatial differences in climate change are likely to imply a heterogeneous pattern of land use responses across the various eco-regions of the world.

Land use change is the result of multiple human-environment interactions operating across different scales ranging from global trade of food and energy to local management of land resources at the farm and landscape level. So, local-level land use and land cover changes are fundamental agents which influence global climate change and are significant forces. These impact biodiversity, water and radiation budgets, trace gas emissions, and ultimately, climate at all scales (Riebsame *et al.*, 1994). At local and regional scales, land cover change can have profound impacts on aquatic systems due to new land use practices that adversely affect water quality and sedimentation (Lowrance *et al.*, 1985). Such changes also modify the composition of plant communities through fragmentation, removal and introduction of species, alteration of nutrient and water pathways, and alteration of disturbance cycles (Ojima *et al.*, 1994).

Actually, land use and land cover changes are driven by: (1) *natural processes,* such as climate and atmospheric changes, wildfire, & pest infestation; (2) *direct anthropogenic activity,* such as deforestation, industrialization, urbanization and road-building; and (3) *indirect effects of human activity,* such as water extraction leading to lowering of the water table. Natural processes and anthropogenic activities can both improve or degrade the state of the land, so it is essential to distinguish beneficial from detrimental changes (Turner and Meyer, 1991).

Land use and land cover changes (LULC) occur at all scales, and remarkable changes at local scales can have dramatic, cumulative impacts at broader scales. Consequently, land use and land cover changes are not just of concern at local and regional levels (i.e., because of impacts on land management practices, soil fertility status, economic health and sustainability, and social processes), but globally as well. The challenge being faced by the policy-makers and scientists is that there is generally a lack of comprehensive and continuous data on the types and rates of land use and land cover changes, and even less systematic evidence on the causes and consequences of the changes. Lack of local and regional data of sufficient reliability and temporal and geographic detail disrupts any attempt towards fine-tuned assessments of the implications of such changes. Further, the impacts of land use and land cover change are critical to many government programs.

Climate change and climatic variability

Interactions between agriculture, climate and patterns of land use are complex. Agriculture is a major use of the land, and patterns of land use are shaped through climatic conditions. The characteristics of agriculture in arid western Rajasthan are primarily determined by climatic factors. Since agriculture remains a major component of the desert economy and the major activity of the majority of the inhabitants. So, climate is a major driver of the economy and of the livelihoods of the people, especially the marginalized rural poor for whom agriculture is extremely critical. The potential for agricultural production is set by many biophysical factors, including climate, soil quality, topography, latitude and altitude, but the main factor affecting to what degree potential production is achieved each year is the climate and most particularly the amount and distribution of rainfall.

'Climate change' refers to a change in the state of the climate that can be identified (e.g., using statistical tests) by changes in the mean

and/or the variability of its properties, and that persists for an extended period, typically for decades or longer. Climate change may be due to internal processes and/or external forces. Some external influences, such as changes in solar radiation, occur naturally and contribute to the total natural variability of the climate system. Other external changes, such as the change in composition of the atmosphere that began with the industrial revolution, are the result of human activity (IPCC, 2007).

Internal variability is present on all time scales. Atmospheric processes that generate internal variability are known to operate on time scales ranging from virtually instantaneous (e.g., condensation of water vapor in clouds) up to years (e.g., troposphere-stratosphere or inter-hemispheric exchange). Other components of the climate system, such as the ocean and the large ice sheets, tend to operate on longer time scales. These components produce internal variability of their own accord and also integrate variability from the rapidly varying atmosphere (Hasselmann, 1976). In addition, internal variability is produced by coupled interactions between components, such is the case with the El-Niño Southern Oscillation.

Actually, all major climate changes, including natural ones, are disruptive. The geologic record includes significant evidence for large-scale climate changes in Earth's past which led to extinction of many species, population migrations, and pronounced changes in the land surface and ocean circulation. The speed of the current climate change is faster than most of the past events, making it more difficult for human societies and the natural world to adapt.

Recent estimates of the increase in global average temperature since the end of the last ice age are 4 to 5°C (7 to 9°F). That change occurred over a period of about 7,000 years, starting 18,000 years ago. CO_2 has risen by 40% in just the past 200 years, contributing to alteration of the planet's energy budget that has so far warmed Earth by about 0.8°C (1.4°F). If the rise in CO_2 continues unchecked, warming of the same magnitude as the increase out of the ice age can be expected by the end of this century or soon after. This speed of warming is more than ten times that at the end of an ice age, the fastest known natural sustained change on a global scale (NAS, 2012).

Changing micro-climate: some reflections in the desert region

Despite universal fears about global warming, there are enough evidences of micro-climatic changes taking place in the Thar desert, that appears to be favorable to the human beings. Contrarily to the rising earth's temperature, the area is undergoing such changes which lead to mildness in its climate. Some of the changes are as follows –

1. Changing rainfall pattern
2. Natural vegetation: Growth, adaptation and succession
3. Land use: Changing pattern and agro-practices

Drought

The evidence of climate change is already all around us, with the occurrence of even more intense weather events, droughts, heat waves, floods and sea level rise. It is forecast that, as the Earth warms, deserts will expand and droughts will intensify causing demographic shifts even as the world's population burgeons. Drought is a normal, recurrent feature of climate and occurs in all climatic zones, although its characteristics vary significantly from one region to another. The onset and persistence of drought are gradual processes; its impact may be far more disastrous than that of flood or other natural calamities. Successive droughts for two or more years have a devastating effect on natural resources, availability of water, production of food and fodder. The impact of drought often accumulates slowly and may linger on for years after termination of the event. World Meteorological Organisation (WMO) has classified droughts on the basis of: (i) rainfall, (ii) combinations of rainfall with temperature, humidity and or evaporation, (iii) soil moisture and crop parameter, (iv) climatic indices and estimates of evapo-transpiration, and finally (v) the general definitions and statements.

The severity of drought depends on the duration, frequency, intensity and geographical distribution of rainfall as well as on corresponding water demand of human, animal, crops and vegetation cover of a region. Seasonal temperature, wind velocity, sunshine, density of vegetation and moisture retaining capacity of soil and soil moisture balance in surface, sub-soil and ground water influence water demand during the drought years.

The drought of 1987 was one of the worst in past century. The monsoon rainfall was normal only in 14 out of 35 meteorological sub-divisions in the country. The overall deficiency in rainfall was 19% as compared to 26% in 1918 and 25% in 1972 being worst years. Agricultural operations were adversely affected in 43% (58.6 million ha) of cropped area in 263 districts in 15 States and 6 Union Territories.

In the two worst affected states of Rajasthan and Gujarat, the rainfall was less than 50% from normal. In these states, the drought of 1987 was the third or fourth in succession resulting in distress to an unprecedented level. Nearly 54,000 villages faced acute drinking water problem.

Observing rainfall data of last two decades, it is noticed that this area is getting more rainfall than the eastern part of the state. The magnitude of rainfall in these desert districts (western area) is increasing whereas it is lowering down in the eastern districts (opposite area). It is clear while the expected and actual averages of rainfall for the four districts for last two decades are compared. Except the years when rainfall was low throughout country, there is increase in rainfall in the desert area, especially Barmer and Jaisalmer districts and thus the actual/observed average is almost 6 cm higher than the expected average in Barmer, and almost 6.4 cm higher in Jaisalmer (Bhardwaj and Gupta, 2014).

Food production particularly in the state of Rajasthan acutely declines during the years of drought and aberrant weather. Severe drought reduced the food grain production in western Rajasthan by 70 per cent during 1987-88, 50 per cent during 2002-03 and again by 57 per cent during 2004-05 against previous good monsoon year. Such reduction in food grain production was also noticed at state as well as at country level during drought years.

Temperature regime

Poonia and Rao (2013) has showed that air temperatures by the end of 21st century are likely to increase by 3.3°C at Bikaner, 3.4°C at Jaisalmer, 2.9°C at Jodhpur and 2.5°C at Pali. Similarly, the annual rainfall is likely to increase by 100 mm at Bikaner, 124 mm at Jaisalmer, -40 mm at Jodhpur and 21 mm at Pali. The spatial and temporal variation in potential evapotranspiration requirement of Thar region ranged from 2.1 mm/day to 12.2 mm/day and on anannual basis between 1500 mm to 2220 mm. Further, the impact of projected air temperature upto a riseof 4°C by end of 21st century increases in evapotranspiration requirements by 9 to 23% during monsoon period and 13-47% during winter period and such increased demand in water due to global warming will reduce the water and feed resources of Thar region. The studies for Jodhpur region showed that the changes in rainfall and air temperatures were not alarming, but the increase in human population (by 400%) and livestock (by 127%) during the twentieth century resulted in a major shift in land use pattern and put tremendous pressure on surface and ground water resources (Rao,1996; Rao and Miyazaki, 1997). The desertification process may continue due to increase in biological activity as a result of over-grazing and loss of vegetation cover with consequent more radiant energy loss and reduction in convective activity (Sikka, 1997). On a smaller scale, land use changes influence local temperatures.

Desert fauna suffers casuality during severe drought years, reducing population but multiplies during consecutive good rainfall years when adequate feed is available. Thar region experienced severe drought during 1918, 1987, 2002 and 2009, were most severe, when rainfall departure from the normal was -81, -65 and -70 per cent, respectively. In 2009, a rainfall deficiency of 40% from its normal rainfall caused drought affecting desert fauna due to scarcity for feed and drinking water. Drought followed by high temperatures touching 45-49°C during late summer period of June, 2010 resulted in causality of chinkaras and black bucks in Barmer, Churu and Jodhpur districts of Thar region (Poonia and Rao, 2013).

Floods in desert areas

Kawas

Desert district of Barmer of Rajasthan received heavy, widespread and continuous rain for nearly one week particularly during 20 to 23 August, 2006, amounting to nearly 650 mm of rainfall in 4 days (average annual rain 277 mm). Because of such heavy rains all water bodies got filled and ephemeral streams drained heavy run-off to the lower areas. As a result, the settlements, which were in lower areas (inter-dunal flats & depressions) got submerged resulting in heavy damage to human, livestock, property and infrastructure, particularly Kawas / Sarka Par and Malwa and six smaller villages (Fig. 1). Besides these, in all, more than 80 villages with 11,000 families and 61,000 people were affected due to floods.

Causalities were severe in Kawas and Malwa villages as the flood occurred in the night resulting in 107 person's deaths, 45,000 animals died and more than 2000 small water structures destroyed (Fig. 2).

Water spread area was analyzed using data of 24th August 2006 and 5th and 15th September, 2006. Total water spread in entire Barmer district was 244.51 sq. km, on 24th August, 2006 which reduced to 171.91 sq. km on 5th September, 2006 (RRSC, 2006).

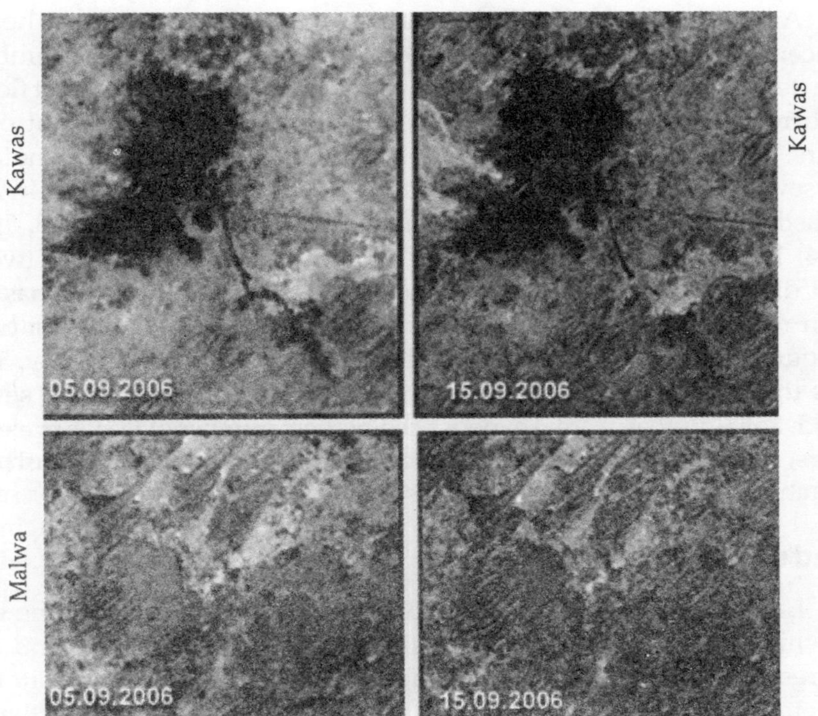

Fig. 1. Pre- and post- flood situation in Kawas & Malwa villages (*Source:* RRSC, 2006)

Field Photographs of 24th August, 2006

Fig. 2. Field Photographs of 24th August, 2006 (*Source:* RRSC, 2006)

As such, floods in Rajasthan are mainly caused by heavy concentrated rainfall during the monsoon season (July to September). The decade-wise flood frequency in western Rajasthan showed that floods occurred in the region with an average frequency of 2 to 3 out of 10 years. The decades 1911-1920, 1941-1950 and 1981-1990 were notable for severe floods. Though, wide-spread floods are not common in arid Rajasthan, such floods occurred in arid Rajasthan during 1908, 1917, 1944, 1975, 1983 and 1997. During 1917, the average rainfall of twelve arid districts exceeded by 161% and such high intensity flood has not been recorded in subsequent years. July 1979 flash flood in Luni basin was unprecedented. The rainfall recorded between 15 and 19 July, 1979 was upto 925 mm at Gorawar located in the Luni basin causing severe flood. Pali (Lunibasin area) experienced highest number of floods in eleven years, whereas, lowest number of floods occurred in Hanumangarh and Jhunjhunun (in four years each) districts.

Land use and its pattern

Land use term usually defined more strictly and refers to the way in which, and the purposes for which, humans employ the land and its resources (William, 2000). Land cover refers to the habitat or vegetation type present, such as forest and agriculture area. Land-use and land-cover change (LULC) also known as land change is a term for the human modification of Earth's terrestrial surface. It is widely accepted that LULC have an important effect on both the functioning of the Earth's systems as a whole (Lambin *et al.*, 2001) and the majority of ecosystems (Hansen *et al.*, 2001; MEA, 2005; Fischlin *et al.*, 2007). This change is based on the purposes of need, which is not necessarily only making the change in land cover but also change in intensity and management (Verburg *et al.*, 2000).

Land use and land cover change has been recognized as an important driver of environmental change on all spatial and temporal scales (Adepoju *et al.*, 2006), as well as emerging as a key environmental issue and on a regional scale is one of the major research endeavors in global change studies. These changes encompass the greatest environmental concerns of human populations today, including climate change, biodiversity loss and the pollution of water, soils and air. Land use change is largely driven by the decision of the people and population growth, declining household farm size and income.

Land use change models are tools to support the analysis of the causes and consequences of land use dynamics. Land use and land cover change (LULC) has been recognized as an important driver of environmental change on all spatial and temporal scales. Land Change Modeler for ecological sustainability is integrated software developed by IDRISI Selva for analyzing land cover changes. Land cover change model tools support the analysis of the land use changes. Use of such model also gives a better understanding of the functions of the land use systems and the support needed for planning and policy making. Such models can also predict the possible future change and use of the land cover under different scenario (Costanza and Ruth, 1998, Labs, 2009, Ahmed and Ahmed, 2012).

The land cover change variables that need to be mapped include: (1) general land cover type; (2) landscape biophysical properties; and (3) landscape pattern. The major emphasis is placed on mapping the general land cover types and changes for each sample block. The biophysical variables (i.e., vegetated fraction, bare fraction, and shadow fraction), developed using spectral unmixing methods, is used to understand landscape condition, such as successional or other gradual land cover transitions. Also, landscape pattern metrics, describing the number, size, shape, and spatial relationship of land cover patches (where patch is a contiguous set of pixels of the same land cover type). These are generated from the general land cover data for each block. These permit the analysis of the spatial dimension of land cover change, and also contribute to the assessment of consequences of land cover change over a period of time.

Table 1 illustrates that out of the 208,751 km^2 total geographical area of western arid Rajasthan the total reporting area is 208,228km^2. Out of this the total cultivated area comes to 66.56% or 13.86 mha. This includes 52.67% net sown area, 6.29% current fallow, 7.6% other fallow lands. The total cropped area comes to 64.15% and double cropped area 11.48%. Area under other land use categories includes 17.69% culturable waste, 0.02% miscellaneous tree crops & groves, 4.9% permanent pasture and other grazing lands, 4.58% barren and uncultivable land, 4.9% land put to non-agricultural uses and 2.95% forest as per land use statistics 2007-08. Through digital interpretation of IRS P6 LISS-III data from 2004 to 2009, CAZRI has prepared land use maps of the 12 arid districts on 1:250,000 scale using ArcGIS 9.2 version software. These maps show the spatial distribution of agricultural lands based on the intensity of cultivation. It is noticeable that Churu district occupies highest (88.4%) rainfed agricultural area while Jaisalmer has least one (21.1%). Of the gross irrigated area through canals 45.7% is shared by Ganganagar, 34.7% by Hanumangarh, 12.4% by Bikaner and 7.2% by Jaisalmer districts respectively. Through ground water irrigation Jhunjhunun district has 42.36% and Sikar 32.15%. Churu, Jaisalmer and Barmer have lowest concentration of 3.75%, 5.01% and 6.1% respectively (Ram and Chauhan, 2009). As per land use statistics 2007-08 the region has 13.69% (2.85 mha) net irrigated area and 20.1% (4.18 mha) gross irrigated area. There are 237,490 irrigation wells and 108,342 tube wells. Hanumangarh and Nagaur district have highest concentration of tube wells. Due to ground water degradation, about 91,438 wells have gone dry. The district wise details are given in table-2. Out of the total gross irrigated area, 45.56% (1,903,548 ha) is contributed by canals. The intensity of cultivation varies from 67% in Jodhpur to 92% in Churu district. Cropping intensity varies from 106% in Barmer to 153% in Jhunjhunun district. The irrigation intensity varies from 106% in Pali to 182% in Hanumangarh district. The percentage of net irrigated area to net sown area varies from 4.02% in Churu to 70.61% in Ganganagar district. The level of agricultural development (Varghese and Mordia, 2002) varies from 24.65% in Barmer district to 46.88% in Ganganagar district (Fig. 3).

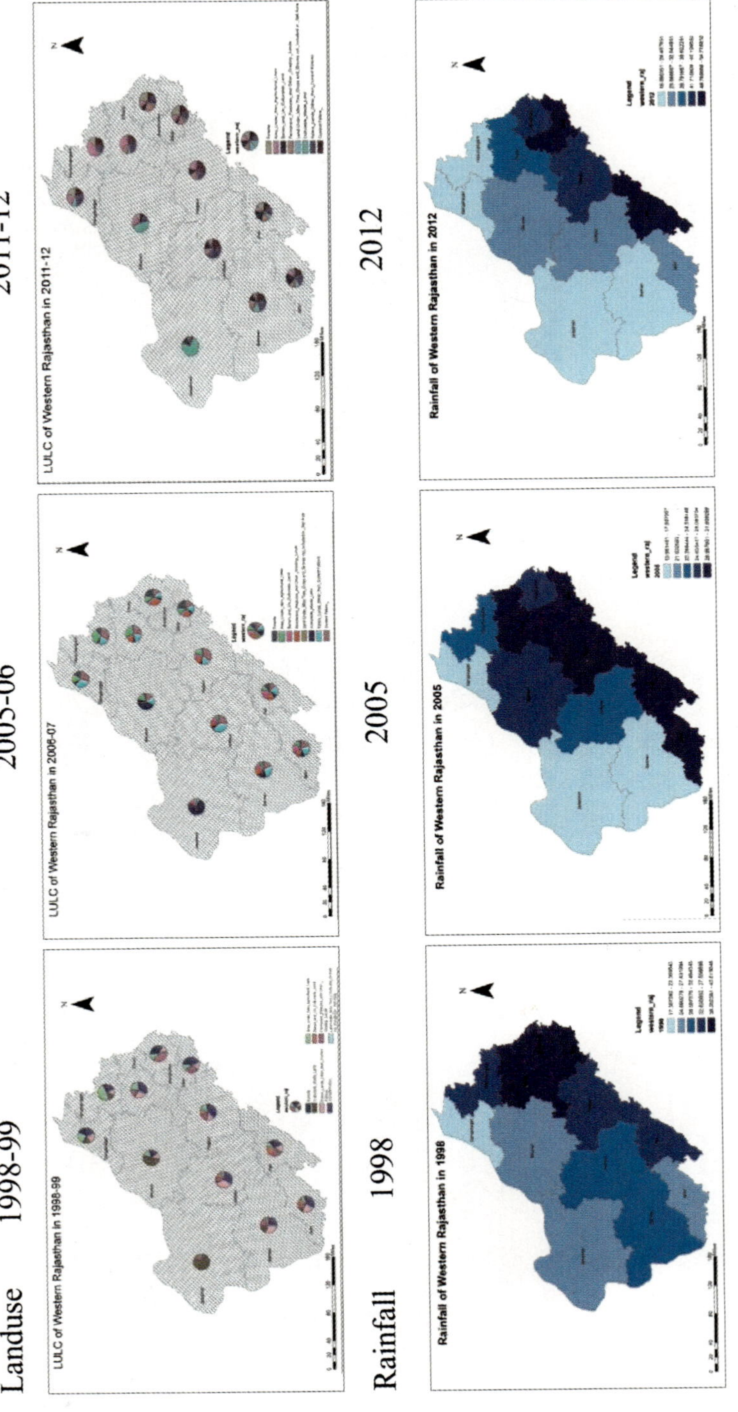
Landuse
1998-99
2005-06
2011-12
LULC of Western Rajasthan in 1998-99
LULC of Western Rajasthan in 2006-07
LULC of Western Rajasthan in 2011-12
Rainfall
1998
2005
2012
Rainfall of Western Rajasthan in 1998
Rainfall of Western Rajasthan in 2005
Rainfall of Western Rajasthan in 2012

contd...

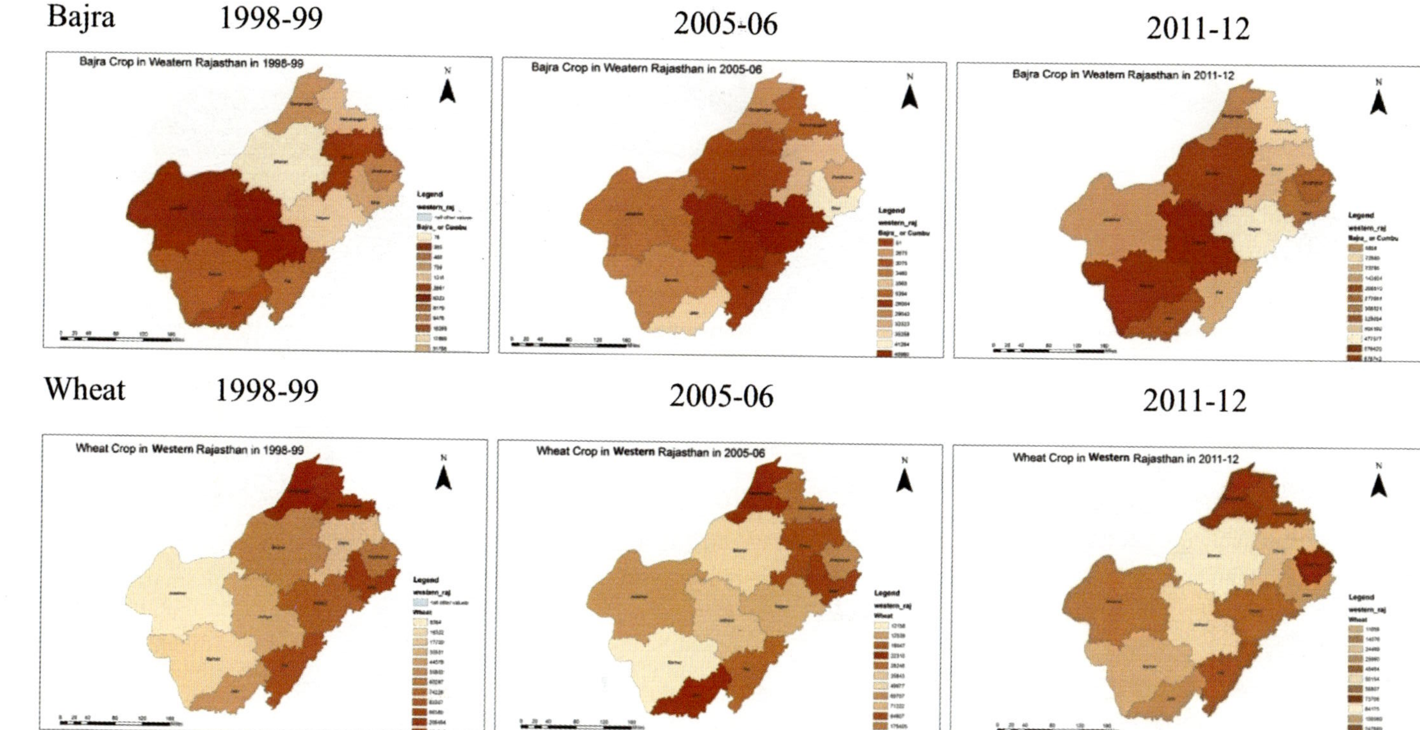

Fig. 3. Relation between land use-rainfall with area under *rabi* and *kharif* crops

Table1. Land uses in arid western Rajasthan* (1960-61 to 2012-13)

Year	Area under Forest (%)	Area under non-agriculture use (%)	Area under barren and uncultivated land (%)	Area under permanent pasture & grazing (%)	Area under misc. trees & groves (%)	Area under culturable waste land (%)	Area under old fallow (%)	Area under current fallow (%)	Area under net area sown (%)	Total cropped area (%)
1960-61	0.91	2.91	12.35	3.54	0.00	22.71	12.06	7.69	40.04	40.68
1961-62	0.89	2.95	12.38	3.57	0.00	22.38	10.93	7.19	42.02	43.08
1962-63	0.77	3.16	12.47	3.66	0.00	21.65	10.89	7.62	42.04	42.86
1963-64	0.83	3.22	12.41	3.54	0.00	22.52	11.42	8.09	40.22	40.96
1964-65	0.85	3.23	12.42	3.57	0.01	21.86	10.24	5.76	44.39	45.32
1965-66	0.89	3.16	12.31	3.75	0.05	22.25	8.82	7.92	43.20	44.08
1966-67	0.92	3.25	12.18	3.80	0.02	22.31	8.03	7.26	44.43	45.56
1967-68	0.92	3.20	12.19	3.86	0.01	21.79	7.85	6.63	45.89	48.40
1968-69	1.01	3.26	12.07	3.91	0.02	22.18	9.31	12.41	38.23	37.44
1969-70	1.03	3.27	12.09	3.94	0.01	22.82	12.54	10.53	36.19	37.42
1970-71	1.02	3.25	12.09	3.88	0.01	21.90	9.43	5.36	45.48	47.29
1971-72	1.02	3.27	12.07	3.88	0.01	22.21	7.36	6.62	45.99	47.47
1972-73	1.05	3.29	12.05	3.90	0.01	21.44	7.82	7.27	45.58	47.19
1973-74	1.23	3.29	11.80	3.95	0.01	20.53	7.03	5.16	49.38	52.02
1974-75	1.42	3.38	11.67	4.03	0.01	20.91	7.70	13.14	40.19	42.44
1975-76	1.69	3.31	5.80	4.07	0.01	25.54	8.67	6.78	46.13	49.74
1976-77	1.83	3.39	5.72	4.07	0.02	25.46	8.32	7.30	45.95	48.27
1977-78	1.88	3.39	5.35	4.10	0.01	25.50	8.07	6.98	46.90	49.73
1978-79	1.98	3.44	5.18	4.16	0.01	27.03	7.65	6.90	48.13	51.34

contd...

1979-80	2.05	3.55	5.20	4.22	0.01	24.79	8.31	11.12	42.98	45.92
1980-81	2.04	3.57	5.22	4.22	0.06	24.70	7.49	6.92	48.00	52.89
1981-82	2.00	3.52	5.48	4.25	0.09	23.95	7.39	7.13	48.43	55.56
1982-83	2.14	3.58	5.37	4.27	0.18	23.86	7.09	7.59	48.15	52.26
1983-84	2.19	3.62	5.33	4.33	0.32	21.81	6.82	7.04	50.76	54.94
1984-85	2.26	3.48	5.40	4.36	0.07	23.32	7.45	9.50	46.34	49.47
1985-86	2.35	3.49	5.40	4.34	0.10	23.13	8.53	7.45	47.43	51.31
1986-87	2.42	3.64	5.92	4.26	0.06	22.02	8.37	7.45	48.09	51.75
1987-88	2.45	4.30	5.39	4.25	0.09	23.10	11.49	20.91	30.26	32.37
1988-89	2.56	4.31	5.34	4.20	0.07	22.06	8.48	5.25	49.98	54.99
1989-90	2.58	4.34	5.31	4.21	0.05	21.78	7.51	7.84	48.63	52.61
1990-91	2.58	4.39	5.30	4.17	0.06	21.63	7.01	6.87	50.24	55.32
1991-92	2.60	4.40	5.23	4.22	0.04	21.62	8.15	9.36	46.62	45.44
1992-93	2.34	4.41	5.22	4.20	0.03	20.72	6.61	5.53	52.91	58.82
1993-94	2.68	4.39	5.23	4.21	0.02	20.34	7.13	8.32	49.90	50.77
1994-95	2.70	4.39	5.24	4.17	0.02	19.88	6.37	5.96	53.50	59.81
1995-96	2.70	4.43	5.25	4.15	0.02	19.54	6.88	7.55	51.70	58.03
1996-97	2.71	4.44	5.27	4.15	0.02	19.38	7.26	7.05	51.92	59.49
1997-98	2.73	4.53	5.21	4.12	0.02	19.31	7.21	6.10	52.96	65.40
1998-98	2.76	4.54	5.21	4.11	0.02	19.54	8.59	9.06	48.38	60.85
1999-00	2.78	4.58	5.21	4.10	0.02	19.17	9.46	10.64	46.22	49.08
2000-01	2.81	4.64	5.19	4.07	0.02	18.76	8.84	8.28	49.58	56.90
2001-02	2.86	4.67	5.16	4.06	0.01	17.97	8.08	6.38	53.02	61.78
2002-03	2.87	4.70	5.16	4.06	0.01	18.56	11.98	26.87	27.99	33.44

contd...

2003-04	2.89	4.71	5.13	4.06	0.02	17.25	8.61	4.48	55.06	64.49
2004-05	2.90	4.74	5.15	4.06	0.02	17.61	7.80	9.53	50.41	59.77
2005-06	2.92	4.95	4.94	4.07	0.04	17.67	8.37	7.47	51.79	62.00
2006-07	3.00	4.96	4.94	4.06	0.05	17.80	8.35	7.18	51.87	62.51
2007-08	3.10	5.01	4.93	4.05	0.02	17.69	7.75	6.38	53.31	86.15
2008-09	3.09	5.57	4.38	4.04	0.03	16.56	7.49	5.87	55.21	67.08
2009-10	3.10	5.60	4.37	4.04	0.03	17.24	7.20	8.24	52.40	62.08
2012-13	2.37	4.94	4.65	3.91	0.05	15.87	7.26	7.55	53.39	66.98

* area in %

Table 2. Trend in area and production of principal crops in western Rajasthan

Crop	Area (000 ha)			Production (000 tons)		
	1970-71	2009-10	Change (%)	1970-71	2008-09	Change (%)
Rice	13	33	153.8	23	12	-47.8
Jowar	229	255	11.4	93	89	-4.3
Bajra	4179	3851	-7.8	1966	2486	26.4
Wheat	405	880	117.3	525	2523	380.6
Barley	112	113	0.9	127	528	315.7
Kharif pulses	1614	2093	29.7	432	667	54.4
Gram	679	666	-1.9	438	744	69.9
Groundnut	20	207	935.0	4	329	8125.0
Sesame	262	248	-5.3	69	63	-8.7
Mustard	35	731	1988.6	42	1117	2559.5
Cotton	97	378	289.7	133	637	378.9

Source: Agricultural Statistics, Rajasthan 2009-10

The production of rainfed crops is entirely dependent on the availability of moisture and other climatic conditions and as such the production fluctuates each year. Therefore, during moderate and severe drought years the production goes down to very low level in spite the cropped area is more.

Pressure on underground water resources

Drought and famines are very frequent in the western Rajasthan. As a result, there has been scarcity of water for domestic as well as for livestock and agriculture purposes. After one short spell of monsoon, total or partial failure of rain is becoming quite often. It has been observed that during drought and famine years, number of wells and tube wells increases to meet demand of (i) domestic and livestock; and (ii) agriculture requirement. This is evident from the worst drought of 1987-89 in the country. For example, in Ganganagar district, prior to this drought, there were 2247 wells used for irrigation purpose whereas in 1987-88 and 1988-89 period, dependency on well irrigation increased due to less supply of canal water. As a result, number of wells increased to 4596 and 6409, respectively (Table 3).

Table 3. District-wise density of wells/1,000 km² in western Rajasthan, 1956-2000

Districts	Density of wells (all types)/1000 km²					
	1956-57	**1961-62**	**1971-72**	**1981-82**	**1999-2000**	**1999-2000 (tubewells only)**
Barmer	64	137	154	300	653	25
Jaisalmer	1	9	12	54	12	1
Bikaner	19	28	43	-	45	27
Churu	2	5	18	125	345	48
Ganganagar	2	-	2	66	0	566
Hanumangarh	-	-	-	-	55	1310
Jalore	768	994	1106	2272	5281	76
Jodhpur	188	259	300	432	850	475
Jhunjhunu	765	838	940	2602	6957	0
Nagaur	427	440	508	883	2602	267
Pali	1964	2129	2092	2620	4007	98
Sikar	2145	1771	1919	2984	6936	21

Source: State and Central Ground WaterBoard reports for various years, Government of Rajasthan and Government of India, Jaipur; Chatterji (1993).

All arid districts had shown appreciable increase in the number of wells. Whereas, parched areas like, Bikaner and Jaisalmer, showed decline in the number of wells during the same period due to lowering of water table and deterioration in water quality. In Pali district, area under gross irrigation by tank in 1986-87 was 27,152 ha which declined to 55 ha in 1987-88. Gross tank irrigated area and gross canal irrigated area in arid Rajasthan were 87 ha and 922,137 ha in 1987-88 and 30,210ha and 1157,424 ha in 1986-87, respectively. Whereas, gross irrigated area from well increased to 720,628 ha from 702,998 ha during same period. On the contrary, area under total cropped declined from 10,802,510 ha to 6,758,009 ha, thereby putting more pressure on the underground water resources.

The districts which have been severely affected by water level depletion are Jalore, Jhunjhunu, Jodhpur, Nagaur, Pali, and Sirohi. The prime reason for water table depletion is excessive withdrawal of ground water to meet out ever growing demand for agriculture, domestic and industrial purposes. Thus, resulting in shrinking of ground water resources due to which wells are getting dried and physico-chemical quality of ground water is getting deteriorated. It is nothing but tragedy of the worst order that when both rainfall and ground water are scarce commodity in the state and whatever and wherever available, it is under constant threat of being either polluted by industries or is being mismanaged in place and time to inflict unbearable damage on the fragile desert ecosystems.

Changes in the scenario

The changing environmental situation in the desert area certainly needs attention. The climate system of Thar desert, exhibits multi-decadal variability. It reveals the existence of an approximate 100-year variability timescale. Monsoon rainfall in the Thar desert is also at a decreasing rate (with some inter-annual exceptions), with indications of increase in rainfall during June and decrease in rainfall during subsequent months. The changes are not only in rainfall but they are visible in flora, fauna, moisture content in soil, changing economic scenario and fast growth of tertiary occupations in the area. Significant agro-biodiversity has already been lost from many production systems leaving them impoverished, vulnerable, dependent on external inputs and increasingly unsustainable. However, much of the desert's agro-biodiversity is still being used by traditional agricultural communities that depend on agro-biodiversity for their livelihoods. In this role they act as custodians of a diversity of crops, forages, livestock, agroforestry products, and the other plant, animal and microbial species found in and around their production areas that are managed and maintained to provide food, fuel, medicine and many other products necessary to their wellbeing. Drought followed by high temperatures touching 45-49°C during summer period of June 2010 has resulted causality in *chinkaras* and black bucks in Barmer, Churu and Jodhpur districts of Thar region. Further, agriculture, which was hardly being practiced in the area earlier, is now on expansion. This area is now booming with agro-producers. Cumin, *isabgol*, mustard, etc. are important *rabi* crops. Exploitation of reserves of the natural oil and gas, in Barmer is changing entire scenario of the region. Jodhpur, Barmer, Jaisalmer (with distinct yellow stone) are leading supplier of masonry stone. Barmer and Jodhpur are emerging as new textile industry centers, leather & wooden goods, and dry vegetables. In the last decade and more, scenario have changed drastically, owing to globalization and its localization. The economy, the landscape and the society have witnessed an intense reconfiguration due to excellent transport infrastructure. It has brought in dramatic effects on the villages, its culture, and economy. Newer technologies have come into play as a great resource for the rural communities for the communication.

Livelihood opportunities for those who have the resources have expanded with communication and infrastructure, improving the market and linkages prominently. The rural economy of the Thar is still largely pastoral, dependent on animal husbandry and localised production of handicrafts. However, their operations and market linkages have improved exponentially. Two major developments, MGNREGA and local self-governance (*Panchayti Raj*), have complicated the dynamics simultaneously, and are proving to be a definite game-changer in this fluid social setup, through empowering the local. However, it has invigorated a pattern of erratic and unequal modernization.

Conclusion

Land-use change is significantly related to climate change in the arid zone of Rajasthan as both a causal factor and a major way in which the effects of climate change are expressed. As a causal factor, land use influences the flux of mass and energy, and as land-cover patterns change, these fluxes are radically altered. Projected climate alterations will generate changes in land-cover patterns at a variety of temporal and spatial scales, although human uses of the land are expected to override many effects. A micro-level study will distinguish between the effects of external influences and internal climate variability. It requires careful comparison between observed changes and those that are expected to result from external forcing. These expectations are based on physical understanding of the climate system of the arid zone. Physical understanding is based on physical principles. So, for scientists, this complexity not only provides a major scientific challenge, but also an opportunity to supply appropriate information to policy makers and farmers. According to Verburg and Lesschen (2006), land use land cover change science can provide:

a) Insight into the driving factors and underlying processes that cause and modify land use change processes; and

b) Projections of plausible future land use change trajectories and land use patterns.

Actually, studies have pointed out that rising temperatures in the Thar Desert may ease the introduction of tropical crops in the region. Further, higher temperatures are expected to reduce the profitability of most crops, and farmers would be compelled to convert cropland to degraded pasture areas. Deforestation pressure would not be restricted to the available forests for fuel and fodder but is expected to go beyond them. Forest area reductions would be observed in all agro-eco-regions and food demand will also influence future patterns of land use in this desert region. But, it is also expected that per capita availability of water would decrease whereas demand by different sectors, particularly agriculture and infrastructure, would increase. Also, frequency and intensity of extreme climate events like dust storms, drought, floods, etc. may increase with the time. It would, therefore, lead to decrease in crop yields due to increasing temperature. Irrespective of rainfed or irrigated, both areas are vulnerable to climate change and extreme events.

As a matter of fact, rural areas are predominantly occupied by small, low-income farmers that have little capacity to adapt to climate change. Decreased agricultural productivity due to climate change may have significant socioeconomic impacts on the small and marginal farmers in this arid region.

References

Adepoju, M.O., Millington, A.C. and Tansey, K.T. 2006. Land use/land cover change detection in Metroploitian Lagos (Nigeria): 1984-2000. *AASPRS 2006 Annual Conference*, Reno Nevada, May 1-5, 2006, Maryland: American Society for Photogrammtery and Remote Sensing.

Ahmed, B. and Ahmed, R. 2012. Modeling urban land cover growth dynamics using multi-temporal satellite images: A case study of Dhaka, Bangladesh. *ISPRS International Journal of Geoinformation* 1: 3-31; doi:10.3390/ijgi1010003.

Bhardwaj, V.K. and Gupta, M. 2014. Micro-climatic changes in Thar Desert: Development and challenges. Paper presented in 9th *Himalayan Policy Research International Conference*, Madison, USA.

Chatterji, P.C. 1993. *Status of Ground Water in Rajasthan – Retrospect and Prospect.* Vol.III, 670p. Institute of Development Studies.

Costanza, R. and Ruth, M. 1998. Using dynamic modeling to scope environmental problems and build consensus. *Environmental Management* 22: 183-195.

Fischlin, A., Midgley, G.F., Price, J.T., Leemans, R., Gopal, B., Turley, C., Rounsevell, M.D.A., Dube, O.P., Tarazona, J. andVelichko, A.A. 2007. Ecosystems, their properties, goods and services. In *Climate Change 2007: Impacts, Adaptation and Vulnerability* (Eds. M.L. Parry, O.F. Canziani, J.P. Palutikof, P.J. Vander

Linden, and C.E. Hanson), pp. 211-272. Cambridge University Press, Cambridge.

Hansen, A.J., Neilson, R.P., Dale, V.H., Flather, C.H., Iverson, L.R., Currie, D.J., Shafer, S., Cook, R. and Bartlein, P.J. 2001. Global Change in Forests: Responses of Species, Communities and Biomes. *BioScience* 51: 765–779.

Hasselmann, K. 1976. Stochastic climate models. Part 1.Theory.*Tellus* 28: 473–485. http://www.clarklabs.org/applications/upload/Land-Change-Modeler-IDRISI-Focus-Paper-pdf.

IPCC. 2007.*IPCC Fourth Assessment Report: Climate Change.*

Labs, C. 2009. *The Land Change Modeler for Ecological Sustainability,IDRISI Focus Paper,* Worcester, MA: Clark University.

Lambin, E.F., Baulies, X., Bockstael, N., Fischer, G., Krug, T., Leemans, R., Moran, E.F., Rindfuss, R.R., Sato, Y., Skole, D., Turner, B.L. II, Vogel, C. 1999. Land-use and land-cover change (LULC): Implementation strategy. *IGBP Report No. 48, IHDP Report No. 10.* IGBP, Stockholm.

Lowrance, R.R., Leonard, R. and Sheridan, J. 1985. Managing riparian ecosystems to control non-point pollution. *Journal of Soil and Water Conservation* 40: 87-91.

MEA (Millennium Ecosystem Assessment) 2005. *Ecosystems and Human Well-Beings: Biodiversity Synthesis.* World Resources Institute, Washington DC.

NAS. 2012. *Climate Change Evidence & Causes: An overview* from the Royal Society and the US National Academy of Sciences.

Ojima, D.S., Galvin, K.A. and Turner, B.L. II. 1994. The global impact of land-use change. *BioScience:* 44: 300-304.

Poonia, S. and Rao, AS. 2013. Climate change and its impact on Thar desert ecosystem. *Journal of Agricultural Physics* 13: 71-79.

Ram, B. and Chauhan, J.S. 2009. Cartographic assessment of land use changes in arid Rajasthan using remote sensing and GIS. *Indian Cartographer* 29: 123-135.

Rao, A.S. 1996. Climatic changes in the irrigated tracts of Indira Gandhi canal region of arid western Rajasthan, India. *Annals of Arid Zone* 38: 111-116.

Rao, AS and Miyazaki, T. 1997. Climate changes and other causative factors influencing desertification in Osian (Jodhpur) region of the Indian arid zone.*Journal of Arid Land Studies* 1:1-11.

Riebsame, W.E., Parton, W.J., Galvin, K.A., Burke, I.C., Bohren, L., Young, R. and Knop, E. 1994. Integrated modeling of land use and cover change. *BioScience* 44: 350-356.

RRSSC. 2006. *Analysis of Flood in Barmer District*. RRSSC.

Sikka, D.R. 1997. Desert climate and its dynamics. *Current Science* 72: 35-46.

Turner, B.L. II and Meyer, W.B. 1991. Land use and land cover in global environmental change: Considerations for study. *International Social Science Journal* 130: 669-677.

Varghese, K.A. and Mordia, A. 2002. Classification of districts of Rajasthan according to level of development using multivariate indicators. http://www.nabard.org/databank/rajasthan.

Verburg, P.H. and Lesschen, J.P. 2006. Agriculture, climate and future land use patterns: potential for a simulation-based exploration. In *Agriculture and Climate Beyond 2015* (Eds. F. Brouwer and B.A. McCarl), pp. 5-32. Springer, the Netherlands.

Verburg, P.H., Chen, Y., Soepboer, W. and Veldkamp, A. 2000. GIS-based modeling of human environment interactions for natural resource management: Applications in Asia. In *Proceedings of 4th International Conference on Integrating GIS Environmental Modeling: Problems, Prospects and Research Needs*. pp.1-18. Banff, Alberta, Canada,.

William, N. 2000. Agricultural and small watershed hydrology: watershed characteristic. http://www.egr.msu.edu/~northco2/WshedChar.html.

□□□

Chapter 5

Multidimensional Role of Women for Livelihood Sustainability

PRATIBHA TEWARI

ICAR -Central Arid Zone Research Institute, Jodhpur-342003, Rajasthan

E- mail: pratibha.tiwari@icar.gov.in, pratibhatewari1@gmail.com

Introduction

Women play significant role in adaptation to climate change. Presently climate change is experienced by delayed rains, dry spell, higher intensity of rain fall and increasing temperature which are causing losses in crop yields, growing food insecurity and increase in food prices. Our understanding of the level of vulnerability of people to climate change and consequent global warming require more research and knowledge sharing. Gender is one critical dimension of all stakeholders involved and directly affected by climate change. Knowledge of gender, social caste system, agricultural wealth and climatic vagaries related to arid zones is fundamental to understanding the livelihood sustainability and priorities in the context of adaptation and mitigation to climate change. Climate change equally affects women but we have ignored to evaluate the multidimensional role played by women in formalizing adaptation strategies to climate change. Rural women besides performing household work have been traditionally predominantly engaged in farming, plantation, animal husbandry and dairy activities. Women in history are credited with initiating domestication of food crop plants and even today they are acknowledged for their efficient seed selection ability. Women's status in farming is adjudged as secondary to men but it is beyond doubt that women's multidimensional role in livelihood sustainability needs recognition and appreciation *par excellence*. The difference in male and female roles lies in mono tasking and multitasking respectively. While men typically play their role sequentially, focusing on a single productive role, women must play their roles simultaneously balancing the demand of each with in limited time availability. Systematized biases and discrimination determine the roles that men and women must play in society in order to generate sustainable livelihoods.

Climate challenge of twenty first century is ubiquitous and affects all people in all countries. However, the world's poor people are on the front line. India is home to 343.5 million destitute people, the poorest of the poor nearly constituting 28.5% of the population. Over all in South Asia over 420 million people are destitute (HDR, 2014). Destitute are the poorest of the poor people, 98% of them have substantially very less to cook and cooking is done with wood fuel, dung or straw, 83% destitute do not have any flooring in their household, it is either dirt, sand or dung floor. They do not have improved sanitation facility (forced to open defecation). Poor's carbon footprint on environment are minimum. As per various reports all greenhouse gases (CO_2, Methane, Nitrous Oxide, Hydro Fluorocarbon, Sulphur Hexa Fluoride) contributes to global warming and expresses in equivalent to CO_2. In developed countries like America the carbon footprint is average 20 ton/person/year, United Europe 12 ton/person/year, China 4 ton/person/year, but India has only 1 ton/person/year.

Human impact on the climate system is sparsely understood to the level that A) increasing greenhouse gas concentrations in the atmosphere and observed warming has substantial correlations; B) it has impacted global water cycle, reduction in glacier size, global mean sea level rise and changes in some climate extremes. Indiscriminate use of fossil fuels and other anthropogenic and abiotic factors have elevated levels of carbon dioxide. This represents a substantial multi century climate change commitment created by past, present and future emission of CO_2 (IPCC 2014). Climate change is a matter of justice. The richest countries caused the problem but it is the world's poorest who are suffering from its effects. The vast gap in resources available to rich and poor can be viewed is the deepest injustice of our time. The failure of resource fairness makes it impossible for the poor to lead decent lives. Climate change provides an opportunity to bridge the gap between rich and poor by exercising the principle of equality. Climate affects people equally throughout the Earth, it highlights our true interdependence and must lead to a new and respectful paradigm of sustainable development. It is an urgent need to scale up and transfer green technologies and to support low carbon climate resilient strategies for sustainable livelihood to mitigate climate change.

Gender justice In relation to climate justice

Gender refers not to male and female but to masculine and feminine, the roles and responsibility society ascribes to men and women. Gender is a central organizing factor in societies, and it can significantly affect the processes of production consumption and distribution. Rural women make up majority of the world's poor; they have the world's lowest levels of schooling and the highest rates of illiteracy. Perception of gender is deeply rooted, vary widely both within and between cultures and change over time. But in all cultures gender determines power and resources for men and women. The goals of developmental interventions therefore revolve around gender equality, as gender constrains women to an unequal position in society in comparison to men. Terry Geraldine (2009) while analyzing issues on climate and gender presented an overview to the extent that no climate justice is possible without gender justice. She states there are two main policy issues related to human induced climate change. The first is mitigation or reducing greenhouse gas emissions, the second is adaptation or preparing for the impact of climate change. There are powerful arguments for addressing gender issues in both these policy areas. Due to paucity of hard evidence in the form of data it is often difficult to arrive at existing gender inequalities in climate related matters. If gender is mentioned at all as a climate change issue it is usually with reference to the particular vulnerability of poor women. The connections between climate change and gender and development issues is that climate change is not happening in a vacuum, but rather in the context of other risks, including economic liberalization, globalization, conflict, unpredictable government policies, risks to health that threaten more of the women. There are arguments that poor women faces many gender specific barriers that limit their ability to cope with and adapt to a changing climate, these must be removed in the interest of both gender equity and adaptation efficiency. As men and women are vulnerable to the effects of climate change in different ways their views on risks associated with climate change also vary. So far, hardly there exists any study to validate these facts, but small representative studies are conducted in remote areas in Africa. This is an area that warrants further research, with a view to ensuring that adaptation policies take into account women's priorities as well as that of men.

Recent involvement of social scientists in climate change

IPCC in its second assessment report produced by working group III focuses on the socio economic aspects of the climate change. It consists of some eleven chapters on scope, decision making under uncertainty, equity issues, applicability of cost and benefit assessments etc. to address climate change (IPCC working group III). Involvement of social scientists in climate change field has shifted focus of natural sciences in the direction of people oriented approach. Engagements of social scientists in social, climate and gender justice have paved way for positive adaptation strategies and mitigation plans. Energy use and associated emissions of greenhouse gases are not abstract concepts. They are aspects of human interdependence, when a person switches on a light in Europe or an air conditioning unit in America, the act is linked to global climate system to some of the world's most vulnerable people, to small scale farmers eking out a living in Ethiopia, to slum dwellers in Manila and to people living in Ganga river's delta (Table 1). They are also linked to future generation, not only to their own but also to the children and grandchildren of people across the world. Sustainable human development in different countries can be indicated by global human indicators.

Global human development indicators

Table 1. Sustainable Human Progress

Indicators	Very high human develop-ment countries	High human develop-ment countries	Medium human develop-ment countries	Low human develop-ment countries	India	World
HDI, Value	0.89	0.735	0.614	0.493	0.586	0.702
Life expectancy at birth (Year)	80.2	74.5	67.9	59.4	66.4	70.8
Mean years of Schooling (Years)	11.7	8.1	5.5	4.2	4.4	7.7
Expected years of schooling (Years)	16.3	13.4	11.7	9.0	11.7	12.2
Gross material income per capita 2011 (ppp $)	40046	12231	5960	2904	5150	13723
Inequality adjust HD1 (Value)	0.780	0.590	0.457	0.332	0.418	0.541

contd...

Over all loss (%)	12.3	19.7	25.6	32.6	28.6	22.9
Inequality in life expectancy (%)	4.9	10.7	21.9	35.0	25.0	17.3
Inequality in education(%)	8.7	17.4	35.1	38.2	42.1	27.0
Inequality in Income(%)	22.4	29.9	18.6	23.9	16.1	24.1
Gender inequality index value	0.197	0.315	0.513	0.587	0.563	0.451
Share of seats in parliament (% held by women)	26.7	18.8	17.5	20.0	10.9	21.1
Population with at least some secondary education percentages female > 25 years	86.1	60.2	34.2	14.9	26.6	54.2
Population with at least some secondary education percentages male > 25 years	87.7	69.1	51.4	29.6	50.4	64.2
Labour force participation rates (female) >15 years	52.3	57.1	38.7	55.7	28.8	50.6
Labour force participation rates (male) >15 years	69.0	77.1	80.1	78.4	80.9	76.7
Ratio of female to male HDI (GDI)	0.975	0.946	0.875	0.834	0.828	0.920

Given the evidence about the implications of dangerous climate change for poverty and future catastrophic risks, it would be a denial of morality to disregard the responsibilities that come with the ecological interdependence that is driving climate change (HDR 2007-08, 2014).

Women's pivotal role in climate change

Although women are more vulnerable to the impacts of climate change they cope up by playing an active role in adapting to its impacts to secure food and a livelihood for their households. Women have greater knowledge of indigenous local plant varieties their nutritional and medicinal values. As seed keepers women often possess knowledge of a variety of genetic resources to adapt to varying climatic conditions such as resistance to drought or pests. New agricultural technologies primarily by pass women farmers despite their knowledge.

A livelihood comprises of the capabilities, assets (material and social resources) and activities required for a means of living. It is sustainable when it can cope up with and recover from stresses or shocks and maintain or enhance its capabilities and assets both now and in future while not undermining the natural resource base. (Chambers and Convey 1992). Primary functions of livelihood generation, nurturing and sustainability rest in natural and climatic setting in which an individual is residing. Minority of population may have individual stressors, shocks, addiction or personal setbacks that prevent them from sustainable livelihood generation. In way it is believed that livelihoods of majority of poor people is supported by environment. Man has substantial role in creation and degradation of environment, by choice he aspires for an environment that has fullest reserve of natural resources of water, vegetation, forest and animals. In times of personal demands man degrades the natural resources with a subconscious promise to himself that one day he well rebuild and replenish that source by any means but that day rarely sees a sunshine in his lifetime. Continuous encroachment of natural resources by anthropogenic agents may degrade environment that it no longer regenerates naturally. All biotic interference must come to a halt if environment has to replenish, however such situation rarely occurs in reality because it is a matter of co-existence of man to his environment. Man must learn to live in synergy with his environment (Table 2).

Table 2. Environment and human progress threats

Indicators	World	Arab states	East Asia and Pacific	Europe and Central Asia	Latin America & the Caribbean	South Asia	Sub Sahara Africa	India
Energy supply through fossil fuel	81.4	96.8	-	89.4	74.2	76.3	-	72.3
Energy supply through renewable sources	18.6	3.2	-	10.5	25.8	23.7	-	27.6
Electrification rate (% of population)	-	87.8	-	-	-	72.0	-	75.0
CO_2 emission/capita/year (tons)	4.6	4.6	4.9	5.4	2.9	1.7	0.9	1.7
CO_2 emission average annual growth (%) 1970-2010	-	-	-	-	-	-	-	0.5
Natural resources depletion (% of GNI)	5.3	24.7	-	7.2	6.9	6.1	14.8	4.9
Forest area % of total land area	31.0	5.9	29.7	9.1	46.7	14.6	28.3	23.1
Forest area % change	-3.5	-22.5	2.6	7.7	-9.2	3.3	-10.8	7.3
Fresh water withdrawn (% of total removable water resources)	7.6	71.1	-	34.8	1.5	26.8	1.6	33.9
Environmental threats								
I)Death of children under 5								
(A) Outdoor air pollution (2008) (per 100000 children under age 5)	5	6	2	2	1	7	8	5
(B) Indoor air pollution (per 100000 children under age 5)	140	73	28	63	22	153	428	131
(C) Unsafe water, unimproved sanitation or poor hygiene	258	214	90	169	86	328	576	316
II) Population living on degraded lands (%)	10.2	24.3	-	10.7	5.3	10.0	22.3	9.6
III) Impact of natural disasters								
(A) Number of deaths (per year per million people)	12	2	15	1	44	7	4	1
(B) Population affected (per million people)	24203	10933	54689	5389	12252	14621	22382	11130

Women's Multidimensional Role

Gender roles shape men and women's decision making in all areas of household and community life, from agricultural decision such as what crops to grow or when to harvest, how to earn or spend income, what food to eat and how to raise their children. Depending on the context it may be typical for men and women to have different spheres of decision making or they may share decision making. In traditional rural societies, commercial agricultural production is mainly a male responsibility, they plant, irrigate, partially harvest and trade products in market, they also trade and own large animals and land but rural women have primary responsibility to maintain households, rear children, collect fuel wood and water besides grazing animals. Women may also earn a small portion of income for themselves by selling vegetables and collecting forest produce. Women still have to work in agriculture and livestock with all constraints because to feed the family is their primary concern and hidden responsibility.

a. Women and food security

Food security is essentially built on availability, accessibility and its utilization. An individual must have access to sufficient food of the right dietary mix at all times to be food secure. Individuals who normally have enough to eat but become food insecure in the face of disaster triggered by economic, climatic and civil shocks are transitionally food insecure. Agricultural production depends on natural resources; land, soil, water and plant genetic resources. Women often have a unique perception on understanding local biodiversity and can be key partners in higher food production. But women are not target for agricultural technology transfer, credit and input supply such as fertilizer and improved seed. Environmental vagaries created due to climate change have increased the household food insecurity concerns of women, they subconsciously remain engaged with calculations relating to food security, climate and households. Women can play significant role in conserving the agro-biodiversity and also establishment of community seed bank.

b. Women and agricultural land ownership

Land ownership to the women can ensure them to participate in climate change adaptation strategies. In most of the developing countries right of land is with the men and therefore, decision remain with them to operate it, however decision of whole family is also considered. Generally household operates within a large family system. Six categories of legal rights to land affect women. When possible encourage the legal norms that provide women with access and control over land rights. In fact land ownership rights empower women to go green.

- Shared tenure through marriage.
- Marital household change through polygamy, divorce or separation.
- Right through inheritance.
- Right to purchase land.
- Co-ownership of land and property acquired due to marriage.
- Registration: providing rights to use and control land by the women.

c. Women and rural finance

Women are vulnerable in different ways, gender inequality affects the way women relate to institutions of all types, women face specific risks because of gender discrimination.

- Unequal control of property makes women extremely vulnerable in cases of divorce or widowhood.
- Women's lower income makes them less able to invest in risk reducing technology such as hybrid seeds, veterinary care for animal etc.
- Women's physical vulnerability makes their property vulnerable to theft and crime.
- Rural finance is a game challenge between traditional money landers and banking sector.
- Policies of banks are also threatening and sometimes impossible without a middle man who himself makes women vulnerable. If these policies are relaxed women are likely to opt for climate friendly technologies.

d. Women and agricultural markets

In most of the developing countries, women are engaged in supplying of traditional and high value produce in the national and international markets, but compared to men, women farmers face a number of disadvantages such as specifically lower mobility, less access to training, market information and to productive resources. Evidence suggests that women tend to lose income and control as a product moves from the farmer to the market. Women's economic empowerment requires targeted analysis and intervention. Women can be facilitated to develop kisan markets near to their farms thereby reducing transport costs and green production.

e. Women and capacity building opportunities

Climate resilient training and capacity development programme are needed to ensure that woman entrepreneurs participate effectively in markets.

- Entrepreneurial skills programmes should be adapted to local cultural contents.
- Confidence building measures such as skills in literacy and numeracy may be required in conjunction with entrepreneurial skills development.
- Women trainers perform more amicably with women trainees.
- Gaining confidence of men who are relatives of women trainees is a necessary prerequisite.
- Multi-disciplinary teams need to be engaged for capacity development.
- Appropriate infrastructural facilities like building, electricity, computer, child care centre may be necessary for capacity building of women.

f. Women and transport burden

Transport takes up a large amount of time and physical burden in rural areas. Because women have to take up multitask simultaneously and balance between competing claims on limited times, women's labor time and flexibility are much more constrained and less adjustable. Head loading and back loading transport activity have direct costs in human energy and time as well as health and opportunity cost. Head loading adds an estimated 20% to women's travel time. Risk is greatest where women have low status and widespread poverty. Transport situation is partially guided by scarcity of natural resources and climate change science.

g. Women and energy crises

Domestic energy tasks are disproportionately women's responsibility especially when the main source of energy comprises collected fuel wood and animal waste and where women and girls do most of the cooking, heating, cooling, lighting and food processing. Switching to gender sensitive energy interventions is pragmatically simpler than restricting to age old life style. These are some of the ways by which women can be safeguarded from energy crises.

- Reforestation with specific focus on establishing mass fuel wood plantation.
- Helping household to have better energy efficient cook stoves and LPG option.
- Developing renewable energy resources for electricity and biogas.
- Sensitization of women consumers to save and grow trees.
- Use of solar energy in house hold and agricultural activities.

h. Women and economic employment

The degree to which women benefit from economic opportunity is highly dependent on the prevailing gender relations. Economic empowerment achieved through agricultural production entails several indicators dominating the process. They may include autonomy in production, ownership of assets, control over income use and decision related to credit etc. Climatic adversities reduce women's income.

i. Women and natural resource management

Agricultural livelihood of poor rural women and men depends on the condition of natural resources. Unsustainable and uneven consumption of natural resources has resulted in stressed environment, desertification and biodiversity losses endangering human, and animal and plant life. Gender responsive action is an appropriate approach to manage climate change impacts and natural resource base. Rural women play important roles in natural resource management, as they possess local knowledge that serves as critical livelihood asset for securing food, shelter and medicine.

Knowledge of climate change research tools

Tool I: Village resource map

It is a tool to learn about a community and its resource base. This map is not made with cartographic precision, rather it is based on acquiring useful information with local precision. Map may highlight

- Infrastructure (Roads, Housing and Buildings)
- Water sites and sources
- Agriculture land
- Soils, slope, elevation (Agro-eco zones)
- Forest lands
- Grazing lands
- Shops, markets, houses
- Health clinic, schools, anganwadi, temple
- Waste sites
- Special services (Bus stop, post office, hospitals).

Comparative maps for past and present are made for last thirty years to understand climate change patterns.

Tool II: Seasonal calendar

It is a tool to learn about farmers perception of typical conditions such as rain fall, amount, time, dimension of food production in different seasons and related livelihood activities. It reflects:

- Activity schedule of people
- Gender differences in activity
- Past thirty years analysis study of climate and activity differential patterns.

Tool III: Daily activity clocks

Unlike seasonal calendar daily activity tool helps illustrate different types of activities carried out by an average individual. This will reveal the approach of action to be carried out in changing climatic variables which will also give a road map for planning of future activities.

Tools IV: Farming systems diagram

It helps classify how rural household livelihoods are assembled. It works with input-output diagram.

- Daily activity of men and women on farm
- Who is responsible for what

- Non agriculture activity of men and women
- Income generating activity of men and women
- Which activity generates more income
- Key linkages between major activities.

Tool V: Capacity and vulnerability analysis matrix

It supports long term planning as per needs of specific groups based on sex, age, disability, social groups, health status, location, ethnicity in different socio-economic groups.

Tools VI: Venn diagram for institution

It is drawn to document the key local groups and institutions that are utilized by the target population that constitute part of providing a specific service (ball and stick diagram is drawn).

Tool VII: Institutional Profiles

Institutional profiles that can help us to learn about local organization, their purpose and how they are functioning. (A matrix is to be drawn along with organization).

Tool VIII: Changing farming practices

It is a tool to document how a change in farming practices (additional tree plantation, soil management modification, and changes in external input) impacts the activities of men and women. (Matrix is drawn).

Tool IX: Seasonal food security calendar

It is a tool to document connection between seasonal climate condition and food security over the course of year. Normal and abnormal climate year are drawn to document how security shifts (seasonal calendar is drawn).

Tool X: Climate Related Risk Management Practices

It is a tool to capture the farming practices of cause and effects of a major past climate related event as well as its impacts and responses (Combination of time line, Venn, seasonality, and other tools).

Problems arising due to climate change can be combated by adopting cost effective strategies. On resolution to climate lens, equity to all the category of the society should be prime concern. It is assumed that a large fraction of anthropogenic climate change from CO_2 emission is irreversible on a multi century to millennial time scale. Depending on the scenario, about 15 to 40 percent of emitted CO2 will remain in the atmosphere longer than 1000 years. A prerequisite for sustainability in the context of climate change is addressing the underlying causes of vulnerabilities including the structural inequalities that create and sustain poverty and constrain access to resources. Swaminathan the veteran agriculture scientist and reformer opines through Human Development Report (2014) that Indian food security act has several features worthy of emulation. Climatic vagaries therefore need to be addressed to for poor in light of human rights of equality, particularly more important is gender equality. Since principles of climate justice are derived from the subject of gender justice its impact assessment and plan of action require further deeper brainstorming and thought provoking in three areas women, climate and sustainable livelihood generation.

References

Christiana Figueres 2014. *Why women are the secret weapon to tackling climate change.* *UNCC* Secretariat, Germany.

FAO. 2013. *Training guide: Gender and climate change research in agriculture and food security for rural development*. CCAFS & FAO www. fao.org

Gender in Agriculture 2008. Source book (2008) *Agriculture and Rural Development series.*

Human development report 2007-08. *Fighting climate change: Human solidarity in divided world*. www.hdr.undp.org

Human Development Report. 2014. *Sustaining human progress, reducing vulnerabilities and building resilience*. www.hdr.undp.org

Terry Geraldine. 2009. No climate justice without gender justice: An overview of the issues *Gender and development*.17(1):5-18, World bank, FAO, IFAD pg 1-729.

□□□

Chapter 6

Biodiversity and its Conservation in Climate Change Scenario

SURESH KUMAR

ICAR-Central Arid Zone Research Institute, Jodhpur-342003, Rajasthan,
E-mail: sk_ecology@yahoo.co.in

Introduction

Conservation of plants has been continuing since time immemorial in view of its importance in saving the environment and humanity as a whole. However, last century has seen a very rapid degradation of environment resulting in wiping out of many species from the face of the earth and pushing many more to the brink of disappearance and vulnerability. While uses of many species are now getting known, it is imperative that we can't afford to make species vulnerable to such threats until we discover their economic importance. Hence, all countries have started conservation programmes, more so after adoption of Convention on Biological Diversity. This threat has become more and more enormous by variability and changes in global climate. The arid zones all over the world have been experiencing the impact of such extreme events that are believed to be consequence of global climate change. An attempt is made here to discuss these impacts, their scale, enormity and extent in the deserts at global scale so as to review and validate these issues in the Indian arid zone in respect of biodiversity.

Biodiversity

Biodiversity is the variety of life on earth. It comprises the variability within species, among species and of ecosystem. Biodiversity provides essential ecosystem services and is basis of ecosystem functioning and life support. In view of this, all nations came together between 1989 to 1992 to negotiate "The Convention on Biological Diversity" (CBD) under the auspices of the United Nations Environmental Programme. The treaty was opened for signature during the United Nations Conference on Environment and Development in Rio de Janeiro in June 1992 and came into force on 29th December, 1993. May 22 each year is celebrated as international Biodiversity Day. CBD has following three objectives:

- The conservation of biological diversity
- The sustainable use of its components and
- The suitable benefits arising out of the utilization of genetic resources
- The 188 member countries and European Union which are signatories to CBD including India, follow global framework for action on biodiversity (CBD Website online 2011).

Biodiversity in desert and climate change: Global scenario

Change in biodiversity at biome level is defined as "the changes in number and relative abundance of species that occurs naturally in that biome". Within biome variation in climate may cause the greatest biodiversity change near climatically determined boundaries of organism distribution. IPCC (2012) predicts more droughts in deserts due to climate change. Loarie *et al.* (2009) stated that the velocity of climate change will be fastest in the deserts. Hence, an urgent need has arisen to re-focus desert plants for understanding their adaptive mechanisms. In view of impending change in biodiversity due to climate changes in deserts, an intensive and deeper understanding of baseline information on biodiversity trends and threats is urgently required including locally available tools and strategies for their management by local people for their up scaling (Daries, 2012).

Global scenarios of biodiversity change as impacted by climate have been modelled for ten terrestrial biomes in the world for year 2100, so as to assess sensitivity of biodiversity to changes in land use, temperature, atmospheric CO_2 concentration, nitrogen deposition and acid rain, climatic and biotic exchanges (Sala *et al.*, 2000). It was found that habitat modification would be modest in deserts while temperature rise will be moderate in these areas. Of all the five factors, the land use change emerged as the most crucial driver that bring changes in biodiversity. Climate change mediated impacts on biodiversity will be maximum in areas like deserts which experience extreme climates. Besides, habitat loss, degradation and hunting are identified as key drivers of species loss in deserts (Safriel *et al.*, 2005).

This does not contradict the already identified and widely agreed 25 biodiversity hotspots for conservation priority (Myers *et al.*, 2000); in fact deserts are emerging as additional micro hotspots. Increased CO_2 concentration due to climate change will affect all desert plants but response of C_3 and C_4 species will differ. More importantly, it is likely to change competitive balance between species having different rooting depths, woodiness and associated below ground organisms and *inter alia* the grasslands and savannas will be the most affected (Sala *et al.*, 2000). Since these very sites are also hotspots of development such as industrialization, urbanization, oil and gas exploration and irrigated cropping, the loss of biodiversity will also be maximum in these areas. Jarvis *et al.* (2008) predicted that some 16% of total wild relatives could be lost by 2050 due to climatic changes. Thomas *et al.* (2004) modelled the probability of extinction for mid range climate warming scenarios for 2050 and predicted that 15-37 % species are likely to become extinct. Many of these could be wild relatives of crop plants. Jarvis *et al.* (2008) predicted that up to 60% groundnut, 12% potato and 8% cowpea wild relative species may be lost by 2050.

Floristic diversity in Indian arid zone

Despite inhospitable and harsh climate (50° C in summer to – 4° C in winter, low and erratic rainfall (100-420 mm/year) with high coefficient of variation, high wind speed (30-40 km/hr), high evapo-transpiration (1500-2000 mm/year), as well as anthropogenic pressure (108 people/km^2) (CAZRI, 2007), Indian arid zone has 682 species belonging to 352 species genera and 87 families. Some 63 species belonging to 37 genera and 9 families are introduced in this desert (Bhandari, 1990). This floristic wealth is contributed from various regions of the world: 37% species are the African elements, 20.6% species are oriental, 14% are tropical and 10.3% are cosmopolitan. Compared to other deserts of the world, this desert has higher number of endemic species, i.e., 6.4%. Poaceae and Fabaceae are the largest families. Some 331 species of 44 families are ephemerals. Of these 52 species are winter annuals, 17 are winter and monsoon annuals and 282 species (38.7% of total flora) monsoonal, belonging to 184 genera and 44 families. Of these, maximum monsoonal ephemerals are from family Poaceae (50), followed by Fabaceae (25), Cyperaceae (20), Asteraceae (20), Euphorbiaceae (15), Convolvulaceae (13) and Acanthaceae (11) (Kumar, 2005). All these species have typical habitat plant cover relationship resulting in major vegetation types on hills and rocks outcrops, piedmonts and pediments, alluvial plains, saline flats, fresh water lakes/ talavs, river and streams banks, sandy hummocky plains and sand dunes (Table 1) of varying shapes and sizes (Kumar, 1998).Thus arid zone of India is a cradle of diverse communities in a large variety of environmental situations (Kumar, 1994) as well as immense variability at genetic level making it a new-found micro-hot spot of diversity (Kumar, 1997).

Economic importance of diversity for life support

Though vegetation is sparsely distributed, it is an important source of life support for local people. These local plants are used locally in a variety of ways such as vegetables (40 species), seeds (27 species), fruits (27 species), fibres (8 species), ropes (3 species), gums and resins (7 species), medicinal uses (131 species) (Kumar *et al.*, 2004) and many provide fuel wood and forage (Saxena, 1979, 1981). Besides, immense variability has also been found in many semi domesticated species, *Aloe vera* (123 lines), *Cucumis callosus* (558 lines), *Citrullus lanatus* (193 lines), *Cucurbita melo* var. *momordica* (90 lines), Salad Kakkri (*Cucumis* species- one line) and undocumented variability in grasess, trees and shrubs (Kumar, 1997).

Climate change scenario in Indian arid zone

It has been estimated that due to climate change, northern part of the desert regions of India may receive lesser rainfall by up to 30%, while it may increase in its southern and eastern parts up to 15% and gradual rise in temperature by 4-5° C everywhere (CAZRI, 2007). The abnormal monsoon rains will be more common. Consequently, there will be more recurrence of cloud burst and floods in south and south eastern Rajasthan and southern Gujarat and droughts in arid Rajasthan. Such a scenario would result in high wind and water erosion. Dust storms with more dust loads will increase. Soil moisture status will experience a change. Coastal areas of arid Gujarat are likely to have sea level rise and spoilage of underground water with sea water ingress as well as loss of land. The above changes in climate will reduce summer rainfall up to 25%, resulting in more frequent draughts and more severe droughts, increased incidence of wind erosion and more frequent floods (CAZRI, 2007). Quest for bringing more land under plough will accelerate the severity and rate of desertification. Industrialization, mining, intensification of agriculture, irrigation, tourism and urbanization with enhanced pace have posed a major threat to desert vegetation. Climate change coupled with these factors become synergistically adverse to species survival, abundance and distributional extent.

Table 1. Vegetation type of Indian arid zone under degraded and non-degraded conditions

Rainfall (mm)	Habitats	Grass cover type	Associated herbage species Non degraded sites	Associated herbage species degraded sites	Associated woody perennial		Area of occurrence
					Non-degraded sites	Degraded sites	
300- 400	Sand clay Loam to clay Loam alluvium	*Dichanthium annulatum, Desmostachya bipinnata*	*Cynodon dactylon, Digitaria adscendens, Brachiaria ramose, Tetrapogon tenellus*	*Aristida mutabilis, A. funiculata, Goniogyna hirta, Zornia diphylla, Indigofera cordifolia*	*Acacia nilotica, Prosopis cineraria, Balanites aegyptiaca, Capparis decidua, Salvadora oleoides*	*Mimosa hamata, Cassia auriculata, Crotalaria burhia, Calotropis procera*	N. Shri Ganganagar, Nagaur, SE Sikar, pockets in Bali, Jalore, Sirohi
	Sandy loam Soil alluvium	*Ochthochloa compressa, Dichanthium annulatum, Desmostachaya bipinnata*	*Eragrostis poaoides, Cymbopogon jwaracusa, Eremopogon foleolantus, Indigofera cordifolia*	*Aristida sp., Cenchrus biflorus, Tragus biflorus, Dactyloctenium sindicum*	*Prosopis cineraria, Salvadora oleoides, Capparis decidua*	*Calotropis procera, Cassia auriculata, Indigofera oblongifolia, Crotalaria burhia*	Nagaur, Jodhpur, Pali, Jalore, Sirohi
100-450	Saline clay loam	*Sporobolus marginatus, Dichanthium annulatum*	*Chlorisvirgata, Aleuropus lagopoaedes, Cressa creatica, Cynodon dactylon, Ochthochloa compressa, Cyperus sp.*	*Aeluropus lagopoades, Cressa cretica, Oropetium hromaeum t*	*Salvadora oleoides, Acacia nilotica, Tamarix aphylla*	*Prosopis juliflora, Indigofera oblongifolia*	Saline low land in entire Indian arid zone

contd...

300-400	Sandy-sandy loam fossilized and stabilized dunes	*Saccharum spontaneum, Saccharum benghalese*	*Desmostachaya bipinnata, Cenchrus sp.*	*Aristida sp., Dactyloctenium sp., Indigofera cordifolia*	*Prosopis cineraria, Acacia jacquemontii*	*Sericostoma pauciflorum, Aerva pseudotomentosa, Crotalaria burhia*	Sikar, Nagaur, Churu, Jhunjhunun
250-350	Well-drained sandy alluvial soil	*Cenchrus ciliaris, Cenchrus setigerus*	*Ochthochloa compressa, Eragrostis sp.*	*C. biflorus, Aristida sp.*	*Prosopis cineraria, Acacia nilotica*	*Calotropis procera, Crotalaria burhia, Aerva pseudotomensa*	Churu, Jhunjhunun, Sikar, part of Nagaur, Jodhpur, Pali
Upto200	Loose sandy soil	*Lasiurus sindicus*	*Ochthochloa compressa, Indigofera cordifolia, Tribulus alatus*	*Aristida funiculata, Cenchus biflorus, Mollugo cerviana*	*Prosopis cineraria, Tecomella undulata, Calligonum polygonoides, Z. nummularia*	*Crotalaria burhia, Dipterygium glaucum, Lycium barbarum*	Bikaner, Jodhpur, Jaisalmer, Barmer, west Jodhpur and west Nagaur
125-350	Stabilized sand dunes with reactivated surface	*L. sindicus, Panicum turgidum*	*Cymbopogon jwarancusa, Cyperus arenarius.*	*Aristida funiculata, Dactyloctenium sindicum*	*Prosopis cineraria, Calligonum polygonoides*	*Dipterygium glaucum, Lycium barbarum*	Barmer, Jaisalmer, Bikaner, Jodhpur part of Jalore and Nagaur
150-300	Sandy gravelly plains	*Ochthochloa compressa, Dactyloctenium sindicum*	*Tribulus terrestris, Aristida sp.*	*Oropetium thomaeum, Aristida mutabilis, Tephrosia sp.*	*Prosopis cineraria, Salvadora oleoides, Acacia leucoploea, Capparis decidua*	*Barleria prionitis, Rhus mysorensis, Lycium barbarum*	Jaisalmer, Bikaner, Jodhpur
125-350	All types of soils except saline	*Aristida sp., Eragrostis sp., Cenchrus biflorus*	*Oropetium thomaeum*	*Oropetium thomaeum*	*Capparis decidua, Ziziphus sp.*	*Prosopis juliflora*	Jalore, Jaisalmer, Jodhpur, Barmer, Sirohi

Impacts of climate change

(a) Entire vegetation

A poor pulse event of rain results in varying degree of drought. The drought will accentuate moisture stress, delaying germination and poor survival of germinated saplings. The vegetative phase will be shortened or telescoped resulting in early flowering, less/poor seed set culminating in poor seed bank. This will ultimately adversely affect regeneration of species causing a steady decline in populations over time. The biotic pressure will further decline the already poor population of plants. The net result will be increase in vulnerability of entire vegetation. Differential grazing pressure in Jaisalmer caused a shift in phenology of perennials and annuals differently: more grazing pressure resulted in earliness of flowering in perennials while delay in annuals (Kumar *et al.*, 1999). Such shifts will be more frequent, affecting larger area and magnified to become more accentuated under climate change scenarios.

Increased and frequent wind erosion will also enhance the moisture stress. The seeds are likely to be blown or buried and finally lost resulting in poor regeneration and population build up. The seeds may be dispersed to longer distances by the enhanced wind velocities because of higher ambient temperatures and ultimately get lost in unfavourable niches due to burial or washing away. Leaf folding, shedding, changed leaf architecture and altered phylotaxy could be other affected adaptive strategies.

Likewise, extreme events of flood water will submerge the seed bank ultimately killing them because of anaerobic conditions. Seeds will also be washed away and lost. At landscape level the number of favourable habitats will decline and inhospitable niches will increase, which ultimately result in poor or low germination, subnormal growth of saplings, low seed sets and finally enhanced vulnerability. Ultimately, this will cause shifts in their distributional range. For example, *Cullen plicata* occurs in the Indian desert on the embanked bund of fresh water bodies such as Khadins that impounds rain water. This taxa has already become rarer and likely to become threatened because it has specialised niche preference (Kumar and Purohit, 2015). Compositional changes on major habitats will be evident. It is likely that successional pathways will be either deflected or shortened with a change in climax formation. With this many species of economic and medicinal value will be insularised and become less and less available. This will also affect animal biodiversity along the food chain resulting in their compositional shift in many situations. With increase in vicariates, invasives and weedy species having immunity to changes in climate, the existing climax of *Prosopis cineraria- Salvadora – Acacia spp*. amongst woody perennials and *Dicanthium- Lasiurus – Cenchrus spp* among grasses may become sub dominant and give way to these stubborn invasives. Spread of *Prosopis juliflora* in Bhuj (Jadhav *et al*., 1993) and Jamnagar district (Kumar *et al*., 2002) is already evident to this effect. In view of above impacts, the annuals and perennials in the desert will follow different adaptive strategies.

(b) Annual species

The annual species life cycle being mediated by moisture availability will strategise its germination so as to have staggered germination. Such trends are known in respect of many species as their seeds are viable but don't germinate till next rain event (Sen, 1977). They may also have prolonged dormancy of different durations, so as to achieve germination at different intervals. They may also develop seeds with longer viability to match prolonged dormancy. Unfavourable environmental conditions are known to impose enforced dormancy which might prolong during climatically changed scenario so much so that they are permanently lost. Developing apomictic seeds bypassing pollination and fertilization can also be one strategy. Autogamy or agamospermy, a well documented adaptive strategy in many desert weeds could be affected either way. Sen (1981) noted that "... viability of many arid zone seeds is very high, which enables them to tide over several uncongenial environmental conditions. Seeds of members of Gramineae (Cenchrus, Dactyloctenium), Leguminoseae (Indigofera, Lathyrus), Boraginaceae (Heliotropium, Arnebia), Cucurbitaceae (Citrullus, Miukia) can remain viable for long periods varying from 1-15 years....". Continuous prolonged occurrence of extreme events (high temperature, dryness) mediated due to climatic change at seed formation and maturation level could adversely affect this capacity of desert plants and ultimately results in compositional change. Plasticity and polymorphism, common properties of desert plants (Sen, 1982) allow them to face and survive a variety of situations from drought, high temperature to inundation, frosts and high winds. The change in climate might reduce this adaptive width in some of the desert plants while allow others to enhance this capacity, survive these vagaries and spread at the expense of others. In either case, end result will be change in vegetation composition. Interestingly, it is already known that some species when assume gregariousness, smother understory species like *Cyperus rotundus* (Sen, 1981). In order to cope the heat stress and dry conditions, the seeds may also develop hard seed coat. The vegetative growth will essentially be rain driven which is also the case even now in all deserts. Though survival could depend on moisture taken from air or dew, much of the survival will depend on the ecological amplitudes of these species. Thus, these will either find newer niches by way of dispersal or will have narrow specialized occurrences.

(c) Perennial Species

The impact of climatic change on perennials and their adaptive strategies will be entirely different. The climate change will alter their phenology and shorten their vegetative phase and end up in untimely leaf fall. There could be altered photosynthesis resulting in less photosynthate and consequently the storage of food in root zone will decline. Anaerobic condition under floods may be injurious to their life cycle strategies. The most profound impact will be in phenological events resulting in poor seed set and consequently less recruitment in population build up. In order to cope with these stresses, perennials may have more number of flowering event which may be small in duration and quantity. This will also result in multiple seed setting events affecting their survival possibility. In fact, off-season flowering has already been reported in *P. cineraria,* more being in younger trees (15 years old) than older ones (25 years old). Seeds obtained from off-season fruiting showed only 6.67% germination as compared to 45.5% in seeds from normal in-season flowering and fruiting .This indicates that if there is climatic change resulting in off season flowering, the germinating seeds will greatly decline (Ratha Krishnan and Jindal, 2015).

They may also retain seed bank on mother plant itself. Branches may dry/die and resurrects during favorable climate as is seen after frosts in *Calotropis procera* and *Prosopis juliflora.* This is so because rate of adaptive changes in response to climate changes, especially in perennial species due to their intrinsic physiology, longer life and poor dispersal mechanism will all combine to slacken the intergenerational selection for evolutionary progress (Pearson and Dawson, 2003). Such variation in rainfall has evoked concomitant responses in flowering and fruiting of *Ceropegia bulbosa* var *bulbosa* in Jodhpur conditions (Table 2). Though more rainfall resulted in more flowers in the year 2010 and 2011, the number of fruits produced showed only marginal increase, from 26% in 2009 to 30% in 2010 and declined in 2011 to 22%.

Table 2. Flowering and fruiting in *Ceropegia bulbosa* in different years in Jodhpur

Year	Rainfall (mm)	Fl/Fr
2009	187	45/12
2010	560	75/23
2011	360	78/17

Peterson *et al.* (2001) also mentioned that those species which are highly dispersive will fill all the space created by loss of species due to climatic change but poor dispersive species will continue to be locked up in an area affording suitable habitat conditions. This is the reason that *Prosopis juliflora* has become an invasive species and the extent of *Commiphora wightii* is declining (Kumar and Mathur, 2014; Kumar and Kulloli, 2015).

Many of the desertic species are known to have secondary metabolites. The quantity and quality of these metabolites is likely to be adversely affected because many of these are produced in response to heat and drought stress. Oleo-gum-resin yield of *C. wightii* has been adversely affected if there were rain events during March – April when a cut is given for its collection. Similarly, gum yield in *A. senegal* has been reported to be less due to untimely rains during its gum oozing period of April- May. We have also observed that if there is rain event during Sept – October it washes away the pollens of *Lasiurus sindicus* resulting in failure of pollination and consequently less or no seeds are produced in that season. Decline or failure of rains during monsoon is known to adversely affect germination of desert grasses eg. *Lasiurus sindicus* (Lahiri and Kharabanda, 1963) which are known to be nature's 'rain-gauges' as they need certain amount of water to wash down germination inhibitors (located in their spikelets) from their seeds. Dislocation of rain events becoming more frequent due to climate change will therefore predictably affect many such species adversely in their life cycle completion and ultimately result in change in vegetation composition. Such events would also cause a decline in biomass production/yield of grazable species in the desert.

Bartholomeus *et al.* (2011) reported that process based simulation of oxygen and drought stresses, due to climates getting warmer and more variable, predicted 20% enhanced severity and co-occurrence at same rate and can cause a decline of about 16% in endangered species while commonly occurring species will apparently have no risk. Individual stress had no effect on endangered species. *Caralluma edulis* a threatened species when reintroduced in Jaisalmer and Jodhpur conditions did survive in Jodhpur but grew very well in Jaisalmer. In fact, it flowered and set seeds both in captivity and in nature (Singh *et al.*, 2016) at Jaisalmer but failed to do so in Jodhpur. This shows, though circumstantially, that any drastic change in the functional niche of threatened species due to climate change or any other reason would make them more endangered. The endangerment of species will increase with reduction in population size and extent of occurrence, decline in area of occupancy, reduction in mature individuals and finally increased probability of extinction in wild. A typical example of such a sequence of events is now seen in respect of endemic species, *Ziziphus truncata* (Kumar and Purohit, 2015).

Research needs-modeling

The climate change induced impacts on biodiversity are so intertwined and complex that it is often difficult to point any one factor responsible for affecting a species, its population, or vegetation as a whole at landscape level. The development of 'Dynamic Global Vegetation Models' (DGVMs) based on mechanistic representation of physiological, biophysical and biogeochemical processes can effectively model vegetation - climate interaction at global scale but would require a hierarchical modelling approach to scale down to a regional level. The down scale of hierarchy might include different ecosystem processes as key components at different spatial and temporal scale (Sykes, *et al.*, 2001). Ecologists are now faced with this challenging task of translating global predictions of climate change so as to fine-tune these impacts at landscape and species level for their management and safe survival. Maintaining range distribution by a species in the newer stress of climate change is possible only if it keeps pace in terms of evolutionary change and this is yet poorly understood. Shifting distribution range in response to climate change would necessarily require matching dispersal capability of these species and how many species have or can develop this feature is a question yet to be answered. Besides, it will be limited by natural barriers, fragmentation of area by land use changes and anthropogenic factors. These need to be validated specially in respect of endemic and threatened species. Besides, the impact induced changes also need to be isolated from so well known effects of biotic interaction and changed pest and disease scenario. These issues also need to be incorporated while predicting future spatial distribution of a species based on present climate conditions and projected climatic conditions (Reutter *et al.*, 2003). Until then, simultaneous approach of mitigating these risks therefore, is catching the attention of biodiversity managers.

Mitigation – future research

Given that biodiversity will be adversely affected in its various facets by climate change, mitigating its impacts would need to be considered along with its habitat/landscape. Climate change induced impacts on biodiversity will therefore, require new productive system to endure changing environment and these have to be energy and cost efficient. These should fulfil gap in demand, promote human welfare and be ecologically sustainable. For this, both *in-situ* and *ex-situ* conservation approaches would need to be employed. Ecological guidelines for *in-situ* conservation include preparing check list of plants of the area, in relation to landscape maps of the area. Then these landscapes need to be monitored for changes and patterns of biodiversity. Area selected for *in-situ* conservation has to be larger so as to encompass all possible plants communities. It should also have fairly good number of rare species and condition of the area should be such as recoverable and its cost feasible. Ecotones such as periphery of saline water bodies (ranns), stream banks and piedmonts are most appropriate habitats for *in-situ* conservation if other conditions are favourable.

In order to do conservation of these species, participatory mode would be most desirable. Farmers could be encouraged to grow these species as hedges around croplands and/or vegetative barriers. Many of these species can also attract pollinator, host insects that eat aphids and provide additional income besides serving as wind break. Boundary plantation of such species as *Moringa concanensis* will help in its conservation. Besides, sustainable land management to build soil carbon and/or reduce loss of organic matter will need to be followed. Some of the proven practices that build soil carbon in arid area include stubble retention, use of cover crops, growing green manuring crops, application of organic manure, controlled grazing, correcting nutrient deficiencies by applying lime or gypsum and managing acidity or sodicity. Practices that reduce loss of organic matter are - zero tillage, contour terracing and integrated crop-livestock-forestry system which are possibly discussed elsewhere in the book.

References

Bartholomeus, R.P., Witte, Jan-Philip M., van Bodegom, P.M., van Dam, J.C. and Aerts, R. 2011. Climate change threatens endangered plant species by stronger and interacting water-related stresses. *Journal of Geophysical Research* 116. Doi:10.1029/2011JG001693.

Bhandari, M.M. 1990. *Flora of the Indian Desert*. Scientific Publishers, Jodhpur.

CAZRI, 2007. *CAZRI Perspective Plan Vision - 2025*. Central Arid Zone Research Institute, Jodhpur, 94 p.

Davies, I., Poulson, L., Schulte-Herbruggen, B., Mackinnon, K., Crawhall, N., Henwood, W.D., Dudley, N., Smith, J. and Gudka, M. 2012. *Conserving Dryland Biodiversity*, IUCN. www.iucn.org/publication.

Durant, S.M., T. Wacher., S. Bashir., R. Woodroffe., P. De ornellas., C. Ransom., J. Newby., T. Abaigar., H. El Alqamy., J. Baillie., M. Beddiaf., F. Belbachir – Bazi., A.A. Berbash., N.E. Bemadjim., R. Beudels – Jamar., L. Boitani., C. Breitenmoser., M. Cano., P. Chardonnet., B. Collen., W.A. Cornforth., F. Cuzin., P. Gerngross., B. Haddane., M. Hadjeloum., A. Jacobson., A. Jebali., F. Lamarque., D. Mallon., K. Minkowski., S. Monfort., B. Ndoassal., B. Niagate., G. Purchase., S. Samaila., A.K. Samna., C.Sillero – Zubiri., A.E. Soultan., M.R. Stanley Price and N. Pettorelli 2014. Fiddling in biodiversity hotspots while deserts burn? Collapse of Sahara's megafauna. *Diversity and Distributions* 20: 114-122.

IIED, 2013. Global public policy narratives on drylands and pastoralism. http:// 10.1080/01431169308904422.

IPCC, 2012. *Managing the Risks of Extreme Events and Disasters to Advance Climate Change Adaption: A Special Report of Working Groups I and II of the Intergovernmental Panel on Climate Change* (eds. C.B. Field, V .Barros, T.F. Stocker, D. Qin, D.j. Dokken, K.L. Ebi, M.D. Mastrandrea, K.j. Mach, G.K. Plattner, S.K. Allen, M. Tignor and P.M. Midgley). Cambridge University Press, Cambridge.

Jadhav, R.N., Kimothi, M.M. and Kandya, A.K. 1993. Grassland mapping/ monitoring of Banni, Kachchh (Gujarat) using remotely-sensed data. *International Journal of Remote Sensing* 14 (17): 3093-3103.

Jarvis, A., Lane, A., and Hijimans, R. 2008. The Effect of climate change on crop wild relatives. *Agr. Ecosystem and Environment* 126:13-23.

Kumar, S. and Kulloli, R.N. 2015. Drivers of vegetation demography for ensuring success in conservation of *C. wightii*. *Taiwania*. MS No-2016-001-RA (Communicated).

Kumar, S. and Purohit, C.S. 2015. *Conservation of Threatened Desert Plants*. Scientific Publishers, Jodhpur 148 p.

Kumar, S. 1994. Current status of desertification in India and future research priorities from ecological viewpoint. In *Towards Solving the Global Desertification Problem* (eds. T Miyazaki and A. Tsunekawa). pp: 19-46. National Institute for Environmental Studies, Japan.

Kumar, S. 1997. Conservation of biodiversity in arid and semi-arid areas. In *Research and Development priorities for Arid Zone beyond 2000 AD: Souvenir* (eds. S.K. Verma, R.R. Bhansali, S.K. Jindal and N.M. Nahar). pp. 39-41. CAZRI, Jodhpur.

Kumar, S. 1998. Vegetation Resources in the Indian Arid Zone. In *Fifty years of Arid Zone Research in India* (Eds. A.S. Faroda and M. Singh). pp: 89-119 CAZRI, Jodhpur.

Kumar, S. 2005. Monsoon Herbage – Their Ecology and Potential. *National Seminar on Conservation, Processing and Utilization of Monsoon Herbage for Augmenting Animal Production.* pp 1-15. Indian Society for Sheep and Goat Production and Utilization.

Kumar, S. and Farzana Parveen 2004. *Traditional Ethnomedicinal Plants in the Indian Thar: Their Status in Natural and Possibilities of their Cultivation and Trading*. CAZRI, Jodhpur. 216 p.

Kumar, S. and Mathur, M. 2012. Invasion of *Prosopis juliflora* in native arid grazing lands: competition and dominance. *Range. Mgmt. & Agroforestry* 33 (2): 162-165.

Kumar, S. and Mathur, M. 2014. Impact of invasion by *Prosopis juliflora* on plant communities in arid grazing lands. *Tropical Ecology* 55 (1): 33-46.

Kumar, S. Jadhav, R., Thakkar, P.S., Sastry, K.L.N. 2002. *Grassland Mapping in Gujarat using Remote Sensing and GIS Techniques: Jamnagar District*. SAC & CAZRI, 36 p.

Kumar, S. Mertia, R.S. Narita, K., Ichikawa, S. and Furukawa, A. 1999. Seasonal shifts in phenology and palatability under various grazing pressures: implications for rangeland health. In *People and Rangelands – Building the Future* – Proc. of the VI International Rangeland Congress (eds. David Eldridge and David Freudenberger). pp. 550-551. Townsville, Queensland, Australia.

Loarie, S.R., Duffy, P.B., Hamilton, H., Asner, G.P., Field, C.B. and Ackerly, D.D. 2009. The velocity of climate change. *Nature* 462, 1052-1055.

Myres, N., Mittermeier, R.A., Mittermeier, C.G., deFonseca, G.A.B and Kent, J. 2000. Biodiversity hotspots for conservation priorities. *Nature* 403(24.2.2000): 853-858.

Pearson, R.G. 2006. Climate change and migration capacity of specieas. *Trends Ecol. Evol.* 21: 111-113.

Pearson, R.G. and Dawson, T.P .2003. Predicting the impacts of climate change on distribution of species: Are bio climate envelop model useful? *Global Ecology and Biogeography* 12: 361-371.

Prentice, I.C., Cramer, W., Harrison, S.P., Leeman, R., Monserud, R.A. and Solomon, A.M. 1992. A global biome model based on plant physiology and dominance. Soil properties and climate. *Journal of Biogeography* 19:117-134.

Ratha Krishnan, P. and Jindal, S.K., 2015. Khejri, the king of Indian Thar desert is under phenophase change. *Current Science* 108 (11):1987-1990.

Reutter, B.A. Helfer, V. ,Hirzel, A.H. and Vogel, P. 2003. Modelling habitat suitability using museum collections: an example of three sympatric Apodemus species from Alps. *J. Biogeogr.* 30: 581-590.

Safriel, U., Adeel, Z., Niemeijer, D., Puigdefabregas, J., White, R., Lal, R., Winslow, M., Ziedler, J., Prince, S., Archer, E., King, C., Shapiro, B., Wessels, K., Nielsen, T., Portnov, B., Reshef, I., Thonell, J., Lachman, E. and McNab, D. 2005. *Dryland System. Millennium Ecosystem Assessment: Ecosystems and Human Well-being: Current State and Trends: Findings of the Condition and Trends Working Group.* (eds. R.M. Hassan, R. Scholes and N. Ash), pp 623-662. Island Press, Washington, D.C.

Sala, O.E, Chapin III, F.S, Armesto, Juan. J., Berlow. E, Bloomfield, J, Dirzo, R., Sanwald, E.H., Huenneke, L.F., Jackson, R.B., Kinzig, A., Leemans, R., ,Lodge, D.M., Mooney, H.A., Oesterheld, M., Poff, N.L., Sykes, M.T., Walker, B.H., Walker, M., Wall, D.H. 2000. Global biodiversity scenario for the year, 2100. *Science* 287:1170-1174.

Saxena, S.K. 1979. Plant Foods of western Rajasthan. *Man and Environment* 3: 34-43.

Saxena, S.K. 1981. Economic plants of Indian arid zone. *Man and Environment* 5: 33-40.

Sen, D.N. 1977. *Environment and Germination of Indian Plants.* The Chronica Botanica Co. India. 116 p.

Sen, D.N. 1981. *Ecological Approaches to Indian Weeds.* Geobios International. Jodhpur.

Sen, D.N. 1982. *Environment and Plant Life in the Indian Desert.* Geobios International. Jodhpur.

Singh, J.P., Kumar, S., Venkatesan, K. and Kulloli, R.N. 2016. Conservation status and utilization of *Caralluma edulis*: an important threatened medicinal plant species of the Thar desert, India. *Genetic Resources and Crop Evolution.* Doi10.1007/s10722-016-0366-3.

Sykes, M.T., Prentice, I.C., Smith, B., Cramer. and Venevsky, S. 2001. An introduction to the European terrestrial ecosystem modelling activity. *Global Ecology and Biogeography* 10:581-593pubs.iied.org/pdfs/10040 IIED.

Thomas, C.D., Alison Cameron., Rhys E. Green., Michel Bakkenes., Linda J. Beaumont., Yvonne C. Collingham., Barend F.N. Erasmus., Marinez Ferreira de Siquiera., Alan Grainger., Lee Hannah., Lesley Hughes., Brian Huntley., Albert S. Van Jaarsveld., Guy F. Midgley., Lera Miles., Miguel A., Lera Miles, Miguel A. Ortega-Huerta, A. Townsend Peterson, Oliver L.Phillips and Stephen E. Williams. Extinction risk from climate change. *Nature* 427:145-148. Doi:10.1038/nature02121.

□□□

Chapter 7

Impact of Climate Change on Some Fruit Crops and their Managment Strategies

P.R. MEGHWAL

ICAR - Central Arid Zone Research Institute, Jodhpur, Rajasthan - 342 003
E-mail: PR.Meghwal@icar.gov.in

Introduction

Indian horticulture is the core sector of agriculture, representing the broad spectrum of crops and production of wide range of horticultural crops including fruits, vegetables, nuts, ornamental, plantation crops, tuber crops, spices, medicinal and aromatic plants and mushrooms. All horticultural crops put together cover about 23.70 million hectares area with production of 269 million tonnes (Anon, 2012-13). India with a total annual production of more than 81 million tonnes of fruits and 162 million tonnes of vegetables is the second largest producer of fruits and vegetables in the world next only to China. These crops make a sizeable contribution to Indian economy in terms of rural employment generation and farmers' income.This sector accounts for 30% of India's agricultural GDP from 13.08% of cropped area. Over the years it has emerged as an engine of growth in agriculture making a significant contribution. It also contribute to about 37% of the total export of agricultural commodities. Fruit growing is an integral component within the broad sector of horticulture. Fruit crops occupy about 29.46% of total area under horticultural crops and contribute 30.23% of horticultural production in India. The area and production of fruits during 2013-14 was 7.2 m ha and 89.97 m tonnes respectively.

Over exploitation of fossil fuels ever since the beginning of industrial era led to increased concentration of greenhouse gases (GHGs) viz., carbon dioxide, methane, nitrous oxide, sulphur dioxide, etc in the atmosphere. The increase in the concentration of these gases is responsible for global climate change. The natural variation in earth's climate is brought about by many geological and cosmological processes. However, climate change refers to additional and relatively rapid changes induced by human activities. Significant variation in either mean state of climate or its variability persisting for an extended period is referred as climate change, which may be due to natural internal processes or external forcing, or to persistent anthropogenic changes in the composition of atmoshpere or in land use. However, United Nations Framework Convention on Climate Change (UNFCCC), in its article 1, defines climate change "as a change of climate which is attributed directly or indirectly to human activities that alters the composition of the global atmosphere and which is in addition to natuaral climate variability observed over comparable time periods." Climate change projection studies indicate a general increase of 3–6°C with more warming in the northern parts than the southern parts of India (Lonergan, 1998; The Energy Resources Institute 2001). Climate change would have negative impacts on tropical regions where the prevailing temperature is already high. The knowledge of research work about the impact of climate change on horticultural crops is limited. Addressing the problems of climate change is more challenging in horticulture crops in general and fruit crops in particular as most of the fruits crops are perennial in nature and face the impact of changing climate all through the year. More over, most of the fruits crops being cross pollinated depend on insect pollinators for its fruit set. The activities of the pollinators are also affected by changing climate which in tern affect fruit set and ultimately fruit production. The issues of climate change and solution to the problems arising out of it requires thorough analysis, advance planning and improved management. The crop productivity is subjected to number of stresses and potential yields are seldom achieved with stress. Climate change is predicted to cause an increase in average air temperature from 1.4-5.8°C, increases in atmospheric CO_2 concentration, and significant changes in rainfall pattern (Houghton *et al.*, 2001). Impact of climate change on four sectors of the economy, namely agriculture, water, natural ecosystems and biodiversity and health in four climate sensitive regions of India, i.e. the Himalayan region, the Western Ghats, the Coastal area and the North-East Region. The present challenges like global climate change, water and soil pollution, less water availability, urbanization etc adds up to the situation. In combination with elevated temperatures, decreased precipitation could cause reduction in availability of irrigation water and increase in evapotranspiration, leading to severe crop water-stress conditions. Vegetable production is threatened by increasing soil salinity particularly in irrigated crop lands which provide 40% of the world's food. Fruits, vegetables, flowers, medicinal plants and tubers are grown from tropical to temperate, while some horticultural crops like spices and plantation crops are location specific. In order to sustain our horticultural production with present day challenges we have to have packages to manage abiotic stresses.

Impact of climate change on fruit crops

The vulnerability of any sector to climate change depends on both expected regional climate change and the sector's ability to adapt. The projected increases in temperature, variability in precipitation patterns, increase in frequency of extreme weather event such as heat wave, cold waves, frost days, drought, floods etc would severely affect the production of horticultural crops. These stresses at different crop developmental phases in varying intensities would ultimately determine the productivity and quality of the produce. The enrichment of carbon dioxide due to anthropogenic activities causes rise in temperature which in turn might influence carbon fixation and crop productivity. There have been instances of loss of horticultural crops due to climate related extreme events at regional scale such as high temperature during March, 2004 in north India which caused significant losses to horticultural crops such as tomatoes, onion, garlic and other crops (Samra and Singh, 2004). Cold wave during December, 2002, January, 2003, caused considearble damage to horticultural crops such as mango, guava, papaya, brinjal, tomato, potato (Samra and Singh, 2003). However numerous extreme weather events which are localized, but not recorded have been causing immense loss to fruits and vegetable crops and rural houshold income. As growing of horticultural crops is taken as income enhancing activity by household, it becomes imperative to provide a scientifically tested low cost and low carbon adatpted strategies so as to not only improve the livelihood of the farmers but also to make them resilient to climatic risks. The gross impact of climate change/climate variability on different kinds of fruit crops are summarized below.

Temperate fruit crops

The IMD monitoring reveals that temperatures are increasing in both Jammu region and Kashmir valley, with significant increase in maximum temperature of 0.05 °Celsius per year. The average mean temperature in Kashmir has risen by 1.45°C in last 28 years while in

Jammu region; it has increased by 2.32°C. The production of apple has gradually increased but the productivity per unit area has fallen from 10.8 to 5.8 t/ha (Awasthi *et al.*, 2001). The reasons for increased production are attributed to area expansion, crop improvement and intensive input management while reduction in productivity per unit area may be the gross result of climate change/climate variability. Among all the productivity reducing factors, climate is difficult to manage. The changes in climate in the form of erratic precipitation, increased temperature, lesser days serving as the chilling period have started affecting the mountain agricultural production systems and ultimately the food security of the people (Datta, 2013).

The apricot and cherries are fast disappearing from some areas of Kashmir Valley as a result of rise in temperature and decline in rainfall. Due to general rise in temperature and less availability of water, the quality of apples in valley and mid temperate region of Jammu is fast deteriorating. Over the last few years, there has been distinct slow growth in production and productivity in rain-fed Kashmir's Karewas areas. Due to unusual hail storms and windstorms in summer, fruits like cherry, apple, plum, peach and apricot are getting damaged heavily.

Marked change in the pattern of snowfall in Kashmir is affecting all the pome and stone fruits in recent years. It has been observed that the snowfall and flowering in some years is coinciding leading to great loss in yield and quality. Due to shortage of water for agronomic crops like rice shift has been observed from agronomic crops to temperate fruits and nut in J&K as fruit crops are more remunerative and require less water as compared to field crops.

In Himachal Pradesh, the impact of climate change in recent years on apple indicated its cultivation shifting towards higher altitudes (Rana *et al.*, 2004). In general, the temperature in apple growing regions of Himachal Pradesh showed increasing trends whereas precipitation declined over the years. The chill units calculated showed decreasing trends upto 2400 msl from Bajaura in Kullu at 1221 msl to Sarbo in Kinnaur at 2400 msl. The Dhundi station situated at 2700 msl showed increasing trend of chill unit at the rate of 25.0 CUs per year. The increasing trends of chill unit at 2700 msl suggested that area is becoming suitable for apple cultivation in higher altitude. These findings have also been supported by the farmers' perceptions which clearly reflected that apple cultivation is expanding to higher altitude in Lahaul & Spitti. The average land use per farm in Lahual and Spitti showed more than two percent shift towards apple cultivation but it showed reverse trend in other apple growing regions. The income of the farmers increased more than 10 percent in Lahul & Spitti whereas it showed a decline of more than 27 percent in Kullu and Shimla districts from fruits in recent decade compared to 1995. In Uttarakhand the area under apple cultivation has reduced. It might be due to less rainfall and higher winter temperature. It is causing major problem of chilling requirement which is very important to meet quality production of apple.

Effects on sprouting

The impact of temperature change is most in apple and almond where trees sprout 2-3 weeks earlier which normally starts in mid April. As a result of such temperature rise during recent years, about 70 per cent of trees began to open their buds in mid March. At the end of March weather become very cold again. At this time most trees have their buds open are very susceptible to frost damage which cause severe loss to apple growers.

Effect on fruit colour

In Kashmir valley, there is an occasional failure of apples to change into their cultivar specific red shades, or an increased instances of sunburn has been observed due to elevated temperature. The deep red colour is favoured by low temperatures during the night in autumn, just before harvesting. If the temperatures are not low enough, most apples fail to turn into their specific red shades. Red colour is a trademark of quality in many apple varieties. It is due to this reason that Ladakh province is becoming potential area for apple cultivation in recent years.

Effect on chilling requirements

Most deciduous fruit trees need sufficient accumulated chilling, or vernalization to break winter dormancy. Inadequate chilling due to enhanced greenhouse warming may result in prolonged dormancy, leading to reduced fruit yield and quality. The warming scenario (less than 1°C) is unlikely to affect the vernalization of high-chill fruit crops (Apple, walnut, apricot, almond, cherry varieties), and if it exceeds 1.5°C it would significantly increase the risk of prolonged dormancy for both stone and pome-fruits. Japanese plums have a chilling requirement (CR) between 500-800 hours, while European cultivars have a CR in excess of 1,000 hours. Periods of mild weather can upset the accumulated CR requiring further periods of cold weather to achieve sufficient hours. Mild winters may result in delayed or irregular flowering, reduced fruit set and an extended flowering period. The CR is a major concern in the marginal temperate area of North Western Himalayas where fruit trees with a low CR have to be grown. Therefore, in some areas, it is impossible to grow cherries. For example, sweet cherry requires the accumulation of 1000 chill units at 3.8°C in order to complete or 'break' dormancy. If chilling is inadequate, the development and/or the later expansion of leaf and flower buds may be impaired. Problems have already been experienced with poor cropping of black currant after mild winters and the same might happen with raspberry, apple and other fruits if winter temperatures continue to rise.

Effect on pollination

More than 70 per cent of orchards have less than 20 per cent pollinizer proportion, whereas a minimum of 30-33 per cent is required in our agroclimatic conditions for good fruit set. Moreover there is lack of diversity in pollinizing cultivars as mainly Golden Delicious and Red Gold are being predominantly used which have attained biennial bearing tendency and their bloom seldom coincides with the flowering period of Delicious cultivars. The population of natural pollinators has gone down due to indiscriminate use of pesticides and deterioration in ecosystem. Managed bee pollination is very limited and available bee hives during bloom hardly meet 2-3 per cent of the demand. All these factors have led to poor fruit setting of Delicious (Kjohl *et al.*, 2011).

Fruit ripening

High temperatures on fruit surface caused by prolonged exposure to sunlight hasten ripening and other associated events. However in grapes exposure of berries to high temperature (above 35^{0}C) delays the ripening process. For fruits exposed to direct sunlight, pulp temperatures reached 35^{0}C and required 1.5 days longer to ripen than those that grew in the shade (Woolf *et al.*, 1999). Cell wall enzyme activity (cellulose and polygalacturonase) was negatively correlated with fruit warmness, indicating that sun exposure, i.e., higher temperatures during growth and development, can delay ripening. However this delay did not occur via a direct effect on the enzymes associated with cell wall degradation. The high temperatures and irradiance caused berries not only to ripen more slowly but also contributed a severe incidence of sunburn and shrinkage on the berries, especially on the western side of the canopy (Greer and Weedon, 2013). A conceptual model accounts for this effect, that of high temperatures in activating the CO_2 fixing enzyme Rubisco, thereby reducing photosynthesis and limiting the supply of sugar for transport to the berries. Sugar loading into the berries may also be impacted on by high temperatures; further restricting the supply of sugar to the berry, with the outcome that ripening is slowed down as observed. However, the yield was not significantly affected by the high temperatures although the quality of bunches was reduced through damage incurred by exposure of the berries (Greer and Weedon, 2013).

In apples, treatments of 38 and 40^{0}C for 2-6 days did not have marked effects on respiration, although ethylene production was reduced. High temperatures on fruit surface caused by pronounced exposure to sunlight can hasten ripening and other associated events.

Physiological disorders and tolerance to high temperatures

Frequent exposure of apple fruits to high temperatures, such as 40 ^{0}C, can result in sunburn, development of water core and loss of texture. Moreover, exposure to high temperatures on the tree, notably close to or at harvest, may induce tolerance to low-temperatures in post harvest storage (Buescher, 1979).

Effects of higher CO_2 & GHG on fruit yield and quality

Carbon dioxide is important because carbon atoms form the structural skeleton of the plant. A doubling of carbon dioxide levels may increase plant growth by 40-50% though continuous high levels saturate the plant's ability to use carbon dioxide and the benefits decrease with time. If other factors remain favourable, increased carbon dioxide concentrations will lead to greater rates of photosynthesis in plants. Current carbon dioxide concentrations limit plant photosynthesis. Growers of protected horticultural crops are already aware from so many years that artificially raising the concentration of carbon dioxide up to certain stage in greenhouses can substantially increase crop growth and yield.

Effect on timing of bud burst, cessation of growth, altered concentrations of carbohydrates and plant hormones in turn altered the dormancy status of trees thereby changing the timing of bud burst and the length of the active growing period. Flowering and fruiting of trees are likely to be hastened under conditions of elevated carbon dioxide. The evidence for an effect of carbon dioxide concentration on leaf senescence and leaf fall is rather contradictory and may be species dependant. Most predictions of the direct effects of carbon dioxide suggested that average yields will increase by about 40-50% with a doubling of carbon dioxide concentrations. Leaves are able to detect and respond rapidly to carbon dioxide concentration. Stomata opening decreases in response to increased carbon dioxide concentration.

Effect on incidence of insect pest and diseases

Erratic changes in temperature and precipitation might lead to more incidences of pest and diseases. In the last few years, the incidence of apple scab, powdery mildew, flee beetle in almost all the temperate fruit crops has been increased.

Interaction effect of temperature and carbon

The combined effect of increased temperature and carbon dioxide predicted in all climate change scenarios suggests that for some species the growth stimulation may be greater than 40-50%. The doubling of carbon dioxide concentration combined with a 3°C increase in temperature could lead to 56% stimulation in growth (Lal and Ahmed,www.biotecharticles.com)

Subtropical and tropical fruit crops

The extreme weather events of hot and cold waves conditions have been reported to cause considerable damage to many fruit crops. In perennial crops like mango and guava, temperature is reported to have influence on flowering. The percentage of hermaphrodite flowers is more in late emerging panicles, which coincided with higher temperature. Cashew, which is mostly grown as rain fed is vulnerable to climate variability and drought conditions caused due to shift in rainfall pattern and inter seasonal variability. The temperature rise will also influence the survival and distribution of pest populations. The rise in temperature will hasten nutrient mineralization in soils and decrease fertilizer use efficiency. The increase in temperature at fruit maturity leads to fruit cracking in litchi and pomegranate while premature ripening in mango. Low fruit setting is observed due to sudden rise in day time temperature (above 35°C) in mango and ber during peak flowering. Low temperature (4-11.5°C) in January accompanied by high humidity (>80%) and cloudy weather delays panicle emergence. During peak bloom period, high temperature (35°C) accompanied by low relative humidity (49%) and long sun shine hours results in excessive transpiration and dehydration injury to panicles. Untimely winter rains promote vegetative flushes in citrus instead of flowering flushes while dry spell during flower emergence and fruit setting affects flower initiation and aggravate the incidence of pest (psylla). In wine grapes anthocyanin development is influenced by differences between day and night temperatures with high variation (15-20°C), promoting colour development. In coconut, impact of climate change related events like consecutive drought and cyclones adversely affected nut yield in Tamil Nadu, Karnataka and Andhra Pradesh (Singh, 2010). In ber high temperature during flowering and fruit setting (September-October) accompanied by low relative humidity results in flower and young fruit drying leading to very low fruit set and ultimately fruit yield is reduced (Meghwal *et al.*, 2014). Mean maximum temperature during September to November between the decade 1991-2000 and 2001-2010 have shown an increase of about 0.81°C while the minimum relative humidity have decreased during the same period by 0.4-11.50%. These changes are affecting flowering, fruiting and yield of ber. Untimely rain at fruit maturity caused heavy fruit cracking in many varieties of ber in western Rajasthan while some varieties were found resistant to cracking (Meghwal *et al.*, 2014). High fluctuation in day and night temperature and low humidity results in very high degree of fruit cracking in pomegranate in Rajasthan. In aonla, rise in temperature during March-April cause flower drying and flower drop leading to low fruit set. Pomegranate grows both in tropical and subtropical climate in temperature range of 10-35°C. Leaves and young fruits are adversely affected by frost. Normally plants flower and fruit through out the year. However for profitable crop, only one particular bahar should be taken. In arid region, Mrig Bahar is taken with flowering during June–July and fruit maturity from December to March. However, wide fluctuation in day and night temperature and low humidity, heavy fruit cracking occurs during January-February. In Hast Bahar and Ambe Bahar, colour developments in the fruit is very poor.The datepalm is quite resistant to drought and salinity and high temperature. However, the growth of date palm is adversely affected if temperature is above 40°C and below 7⁰C.The plant is also resistant to frost and for optimum flowering a brief exposure to low temperature is essential. The spath emergence in datepalm normally takes place during last week of January to mid of February but if low temperature exposure is longer, the spath emergence as well as flowering get delayed.

Table 1. Fruit cracking in different varieties of ber as a result of rain fall during fruit maturity (Meghwal *et al.*,2014)

S.No.	Variety	% cracking
1.	CAZRI Gola	10
2.	Tikadi	6
3.	Banarasi Pebandi	30
4.	Aliganj	50
5.	Kaithali	15
6.	Chuhara	4
7.	Maharwali	10
8.	Thornless	13.33
9.	Seb	60
10.	Sanaur-5	4
11.	Dandan	10
12.	Gola	10
13.	Varieties resistant to cracking	Rashmi, Katha,Banarasi Karaka, ZG-3, Umran, Kali, Jogia, Illaichi, Bagwadi and Mundia

Mango

Climate and weather play critical roles in the economic success or failure of tropical fruit tree species including commercial mango production. Air temperature and rainfall influence vegetative and phenological phases in mango and are two of the most important factors determining suitability of an area's climate for mango production. Climate-related changes have already brought widespread changes in flowering and fruiting patterns of mango. This is adversely affecting fruit production in some areas. But rising temperature in areas previously too cold for mango production are making them more suitable for mango production. For instance, an increase in temperature during the coldest month has made mango cultivation possible in the valley areas of Himachal Pradesh and Uttarakhand. In several parts of the globe increasing temperature will offer opportunities for mango production in new areas. The increasing temperature would have negative impact in tropical regions where the prevailing temperature is already high. It performs optimally within a temperature range of 24-27°C, though, it can tolerate temperature up to 48°C. Moderately higher temperature during ripening hastens maturity and also improves fruit size and quality. Mango being a perennial fruit crop would respond to increase in temperature differently as compared to annual crops. The temperature, especially in subtropics, is an important environmental stimulus for mango flowering. Several studies have shown that low temperature promote reproductive morphogenesis in mango. There was an increase of 18 to 100 per cent flowering in axillary buds of Hayden when trees were transferred to 31/25°C following 1-3 weeks at 19/13°C (Shu and Sheen, 1987).

Citrus

Citrus fruit plants are considered to be better equipped to deal with a changing climate than other fruit crops. That's largely because they flourish in the heat. According to a 2007 study that examined the effect of rising temperatures on various crops, the cultivation area for orange trees will actually expand by the year 2055. Delay in monsoon, dry spells of rains, and untimely rains during water stress period, supra-optimal temperatures during flowering and fruit growth, hailstorms are some of the most commonly encountered climatic conditions experienced by the citrus growers over the past decade or so. While the lemon cultivation area will shrink by around 10 percent, that is a small setback in comparison to other plants. Strawberry producers, for example, will lose around 32 percent of their arable land, and wheat fields will be eroded by around 18 percent, according to the study. "For the lemon harvest, we're not expecting any major drawbacks due to climate change for the time being," according to Tim Grout from Citrus Research International, a research association that examines the citrus industry. "If however temperatures rise by more than one or two degrees, yields could fall because the plants shed their fruit too soon," he says, adding that seedless lemons are especially at risk. Other analysts believe that the increased risk of pest infestations could also lead to greater harvest losses. Citrus greening, for example, is a bacterial disease that is primarily spread by two types of psyllid insects. It turns a citrus plant's leaves and shoots yellow and makes the fruit bitter, often causing the entire plant to wither away.

But in different case in District Kangra, Himachal Pradesh, citrus fruit production has declined by 1,000 tonnes in the district in the last two years. "In 2009-10, it was 22,184 tonnes while in 2010-11 it decreased to 21,704 tonnes," district horticulture authorities said. "Climate change is one major factor affecting the fruit crop. We are facing problems on this count for the last eight to ten years. It's not raining in time because of which, plants are not getting proper nutrition and fruiting is declining gradually," said Ramesh, an orange grower from Indore.

Banana

Bananas are of tropical origin covering >70% of global land areas (mainly tropical areas), but there could be gains towards subtropics. Potentially huge impact on banana production, both in tropical and subtropical regions. The climate change increases the atmospheric temperature and change of rainfall pattern, as a result, banana cultivation may suffer from high temperature, soil moisture stress or flooding / water logging. Review of the impacts of climate change on Musa (Bergh *et al.*, 2012) suggests that future climates will be less suitable when grown in subtropics, they are subject to strong environmental constraints, mainly low temperature but also drought. Temperature and rainfall are expected to change at unprecedented rate in coming decades (IPCC, 2007). The rate of leaf emergence stops below 10°C hence, the production of banana will be affected more under subtropical condition by low temperature stress.

Litchi

The aberrant weather and extreme events may damage the crop completely. The observed temperature trends in the region of litchi production (Bihar) showed a general increase in temperature in order of 2-3°C over the base period of 50 years. The unusual impact of climate change has been witnessed in litchi production system as noted in flowering pattern (shifted early), fruit growth and harvesting periods. The occurrence and the extent of damage by physiological disorders and resurgence of pest are very much dependent on the temperature and humidity variations in the atmosphere (Kumar and Kumar, 2007). The effects of temperature on litchi growth and productivity in changed season and seasonal changes causing differences in maximum and minimum temperatures for various litchi growing areas have been extensively studied (Menzel *et al.*,1989, Menzel and Simpson,1995). Litchi is adapted to warm subtropics doing best below 20°C, while the optimum temperature for leaf and fruit growth is about 30°C. Temperature below 2°C damage new leaves, while below 0°C kill the trees. With such specific temperature requirement, little variation might affect the production and quality of litchi substantially.

Grape

Grape is one such crop that is highly sensitive to climatic changes. The grape being temperate crop has very well adapted to tropical region, under climate change scenario. With increase in temperature there would be change in growing degree days, which has direct bearing on phenology of the crop. Hence under such situation, varieties have to be identified for production of quality grapes. Excessive high temperature for extended periods of time generally results in delayed fruit maturation and reduction in fruit quality. At 35^0C pigment development is completely inhibited in Tokay and reduces in Cardinal and Pinot Noir, compared to 20-25^0C. The elevated CO_2 levels may increase productivity in arid and semi-arid regions but, the drought stress caused by higher evaporative demand may override beneficial effects of increased CO_2 in the atmosphere unless irrigation can be steeped up to compensate these effects. Higher temperature may advance the ripening of berries and alter the berry composition in both table and wine grapes, thereby affecting the quality of the produce. High night temperature reduces anthocyanin accumulation in berry skin which is due to low expression of anthocyanin biosynthesizing gene and enzyme (Mori *et al.*, 2005). Another dimension to climate change is that the pathogens may become more virulent and/or the plants may become more susceptible, thus increasing disease severity. New pathogens may also emerge or the existing once may mutate/develop resistance. Unseasonable rains may lead to serious downy mildew incidence as evidenced by decrease in productivity during the recent years from more than 25 tons per hectare to 9.2 tons per hectare in the previous season. There is also likelihood of change in the incidence and pattern of insect pests like mealy bug, thrips and mites. Similarly the disease incidence pattern is also likely to be affected with the change in climate as has been observed in case of downy mildew (Anon., 2013).

General impact of climate change on fruit industry

- Changes in the suitability and adaptability of current cultivars as temperatures change, together with changes in the optimum growing periods and locations for fruit crops.
- Changes in the distribution of existing pests, diseases and weeds, and an increased threat of new incursions.

- Increased incidence of physiological disorders such as tip burn and blossom end rot Greater potential for down grading product quality e.g. because of increased incidence of sun burn.
- Increases in pollination failures if heat stress days occur during flowering.
- Increased risk of spread and proliferation of soil borne diseases as a result of more intense rainfall events (coupled with warmer temperatures).
- Increased irrigation demand especially during dry periods.
- Changing reliability of irrigation schemes, through impacts on recharge of surface and groundwater storages.
- Increased atmospheric CO_2 concentrations will benefit productivity of most fruit crops, although the extent of this benefit is unknown.
- Increased risk of soil erosion and off-farm effects of nutrients and pesticides, from extreme rainfall events.
- Increased input costs - especially fuel, fertilisers & pesticides.
- Additional input cost impacts when agriculture is included in an Emissions Trading Scheme (ETS).

Priorities for Indian fruit industry in adapting to climate change

- Identify and build on successful strategies of adaptation by the fruit sector to climate changes already experienced.
- Obtain regional climate change scenarios (downscaling) for all fruit growing regions (to 2030) - update as improved scenarios become available.
- Develop impact assessments for all or major commodities in these regions.
- Assess the vulnerability of all or major regions and/or fruit commodities and identify current "at risk" production sites (regions) and/or industries.
- Identify the long-term (2030 and 2070) opportunities and threats to horticultural regions and cropping systems, as a consequence of climate change - long term adaptation.

- Develop (in consultation with growers and their advisors), adaptation strategies which are appropriate, practical, and economically sound.
- Review and/or develop where necessary, Best Management Practices (BMP), Good Agriculture Practices (GAP) for fruit cultivation, which include adaptation and mitigation components.
- Assess the economic benefits of silvi-horti as well as the benefits it might bring for adaptation and mitigation.
- Document the effects of climate change for major overseas production regions, especially in those countries that are major competitors to India's production.
- Identify additional export opportunities for Indian growers.
- Identify alternative regions that may be suitable for production, to take advantage of these market opportunities.
- Investigate the "food miles" concept and the effects decisions on markets and production opportunities for horticulture.
- Develop horticulture specific forecasting tools that can be used for climate change and climate variability (especially temperature variability) related decision making at a farm and regional scale.

Climate change mitigation strategies

The climate change is now a reality as there is enough evidences to show that the emission of greenhouse gases has caused global warming and associated climate change. In addition to adapting the horticultural production system to adverse impact of climate change, horticulture sector can considerably contribute to mitigations. Mitigation is referred to the process in which emission of greenhouse gases are either reduced to carbon sequestration level or by improved crop management practices.

Carbon sequestration potential of horticultural crops

Mitigation efforts through carbon sequestration help to reduce the adverse impact of climate change. The information about carbon sequestration potential of fruit trees is scanty, though they can contribute greatly in this regard. In a study using PROCOMAP model at IISC, Bangalore, estimated the mitigation potential of farm forestry fruit orchard block planting with 75% of area proposed under *Mangifera indica, Tamarindus indica, Achras sapota, Artocarpus,* neem and guava. The carbon stock change under the base line and mitigation scenario (excluding harvested wood product) and carbon increments/per hectare for various project activities for the 30 years period (2005-2035) worked out to be 47.42 t C/ha. The overall mitigation potential for farm forestry with an area of 5381 ha is 81750 t C. The fruit orchard farm forestry was found to be most profitable at 29.92% IRR.

Strategies to minimize the impact of climate change on fruit crops

I. Breeding strategies

- Pheno-typing of all important fruits genetic wealth to enhancing temperature, moisture stress and genetic enhancement for tolerance to biotic and abiotic stress.
- Varieties and rootstocks will be evaluated under controlled temperate moisture stress etc. gradient to identify suitable cultivars of all major fruit crops. Experiments on varietal evaluation will also be conducted under natural conditions at different altitudes/ conditions (with natural variations in temperature and moisture falling under various agro-climatic zones of the countries.
- Marker assisted selection and development of transgenic having resistance to biotic and abiotic resistance.
- Development of genotypes having resistance to heat and drought, crop diversification.

Table 2. List of some varieties tolerant to abiotic stresses (Bose and Mitra, 1996)

S.No.	Crop	Variety	Tolerance
1.	Pomegranate	Ruby	Drought
2.	Annona	Arka Sahan	Drought
3.	Fig	Deanna, Excel	Drought
4.	Grape(Rootstock)	Dogridge	Salinity
5.	Mango	Bapakai	Salinity
6.	Lime	Rangpur lime Cleopatra mandarin	Salinity

II. Agronomic management strategies

- Assessment of the vulnerability and climate risks associated with temperate fruit production system in temperate, tropical and subtropical region.

- Development of cropping systems under various agro-climatic conditions.
- Improvement in the irrigation and drainage systems.
- Development of appropriate tillage and intercultural operations.
- Integrated nutrient management.
- Integrated pest management.
- Integrated weed management.
- Development of water harvesting techniques.

III. Biotechnological innovative strategies

- Molecular characterization for various traits in relations to biotic and abiotic stress.
- Transformation of plants from C_3 to C_4 plants.
- Gene pyramiding against biotic and abiotic stress.
- In vitro conservation of rare and useful species for future use.
- Mitigation strategies for higher CO_2/GHG
- Assessment the carbon sequestration potential of perennial fruit crops production system.

Table 3. Estimates of standing biomass per area (t/ha or Mg/ha), and calculated CO_2 equivalents per ha for perennial crops.

Crops	Total DW (t/ha)	Total CO_2 (t/ha)	Main source of data for calculation
Apple	36	66	Jemenez and Diaz(2004), Palmer *et al.* (1991)
Peach and Nectarines	40	73	Chalmers and Van Den Ende (1975), Jemenez and Diaz (2003)
Plums, prunes	62	114	Kroodsma & Field (2006)
Almond	100	183	Kroodsma & Field (2006)
Walnut	75	38	Kroodsma & Field (2006)

iv. Sustainable energy development

- Use of biofuel like diesel from Jatropha and Pongamia sp.
- The development of nuclear energy.

- The improvements in the efficiency of electricity generation, transmission and distribution.
- The use of fuels with lower carbon content, e.g., natural gas, CNG, Gobber gas fuel switching, appliance efficiency and use of renewable energy; Tree planting and forest management.
- Waste processing.

Future research strategies for optimizing fruit production under changing climate scenario

i. Crop improvement strategies

- Utilizing the current and future regional climatic scenarios of the tropical, subtropical and temperate region a micro-level survey of agro-climatic zones of country should be conducted to identify sensitive regions with high vulnerability with respect to different fruit crop.
- Utilization of wild species and relatives in all the fruit crops.
- Introduction of low chilling cultivars of pome, stone and nut fruits.
- Diversification with other high value fruit crops like peach, apricot, walnut, kiwi and olive.
- Development of new genotypes having resistance to high temperature and CO_2 concentrations.
- Marker assisted selection and development of transgenic having resistance to biotic and abiotic resistance.
- Development of genotypes having resistance to heat and drought.
- Biotechnological approaches for multiple stress tolerance will be standardized.

ii. Development of agro-techniques

- Assessment of the impact of elevated temperature and CO_2 on growth, development, yield and quality of commercial fruit crop using open top chamber (OTC) facilities and free air CO_2 enrichment system (FACE).
- Sensitive stages of crops to weather aberrations will be identified.

- The phenology of all major fruit crops under changing climate will be monitored.
- In-situ soil moisture conservation practices including indigenous technical know-how will be validated to mitigate the impact of drought.
- Development of suitable agronomic adaptation measures for reducing the adverse climate related production losses.
- Study the impact of climate change and development of technologies on water productivity.
- Identify and develop good practices to enhance the adaptation of crop to increased temperature, moisture and nutritional stress.
- Identification and mapping of climate resilient as well as climatically vulnerable micro-niches in all fruit growing regions.
- Extreme events, such as late spring frost or windstorm, may cause crop failure.
- Future climate may also increase occurrence of extreme impacts on crops e.g.weather conditions resulting in substantial reduction in yield and quality (for example severe drought or prolonged soil wetness).
- To develop a set of high resolution daily based climate change scenarios, suitable for analysis of agricultural extreme events.
- To identify climatic thresholds having severe impacts on yield, quality and environment for representative crops and to assess the risks that these thresholds will be exceeded under climate change.

iii. Plant protection strategies

- Assessment of the pest and disease dynamics, study of disease triangle and development of prediction models.
- Strengthen surveillance of pest and diseases.
- Development of eco-friendly pest- ecologies and management strategies and early warning systems.

iv. Post harvest management strategies

- Development of cost effective storage techniques.
- Development of varieties having longer shelf life.
- Studies on mitigation of post harvest spoilage.

v. HRD & creating awareness

Organize seminars/symposia/trainings and conduct field demonstrations, on effective climate resilient technologies.

References

Anonymous. 2013. *NRC grapes Vision* 2050.

Anonymous. 2013-14. *Indian Horticulture Data Base* (2013-14).

Awasthi, R.P., Verma, H.S., Sharma, R.D., Bhardwaj, S.P. and Bhardwaj, S.V. 2001. Causes of low productivity in apple orchards and suggested remedial measures. In *Productivity of temperate fruits* (Eds. Jindal, K.K. and Gautam, D.R), pp 1-8.

Bergh, Van. Den. I., Ramirez, J., Staver, C., Turner, D.W., Jarvis, A. and Brown, D. 2012. Climate change in the subtropics: the impacts of projected averages and variability on banana productivity. *Acta Horticulturae.* 928:89-99 http://www.actahort.org/books/928/928_9.html.

Buescher, R.W. 1979. Influence of high temperature on physiological and compositional characteristics of tomato fruits. *Lebensmittel-Wissenenshaft,* 26: 237-268.

Datta, S. 2013. Impact of climate change on Indian Horticulture- A review. *International Journal of Agriculture, Environment and Technology.* 2(4): 661-671.

Greer, Dennis. H. and Weedon, Mark. M. 2001. The impact of high temperature on *Vitis Vinifera* cv. Semillon grapevine performance and berry ripening. *Frontiers in plant Sciences* 4(491):1-9.

Houghton, J., Ding, Y., Griggs, D., Noguer, M. and Van. Der, Linden. P. 2001. Climate Change 2001: The Scientific Basis. Published for the *Intergovernmental Panel on Climate Change.* Cambridge University Press, Cambridge, UK and New York. 881p.

IPCC. 2007. Climate change 2007. In The physical science basis. *Contribution of working group I to the fourth assessment report of the intergovernmental panel on climate change*: S. Solomon, D. Quin, M. Manning, Z. Chen, M. Marquis, K.B. Averty, M.Tignor, H.L. Miller (Eds), Cambidge University Press 996p.

Kjohl, M., Nielsen, A. and Christian, S. N. 2011. Potential effects of climate change on crop pollination. *Centre for Ecological and Evolutionary Synthesis (CEES),* Department of Biology, University of Oslo, Norway.

Kumar, R. and Kumar, K.K. 2007. Managing physiological disorders in litchi. *Indian Horticulture* 52 (1): 22-24.

Lal, S. and Ahmed, N. Strategies to minimize the impact of climate change on fruit production. *www.biotecharticle.com.*

Lonergan, S. 1998. Climate warming and India. In *Measuring the Impacts of Climate Change on Indian Agriculture* (Eds. Dinar, A., Mendelsontin, R, Evenson, R.),World Bank Technical Report No. 402. Washington DC, USA.

Meghwal, P.R., Kumar, S., Bhatt, R.K. and Singh, A. 2014. Response of Jujube cultivars to natural climatic variability under rain fed condition in arid zone. *International Symposium on new dimensions in agrometeorology for sustainable agriculture,* organized by GBPUAT and association of agrometeorologist, India, at Pantnagar,16-18 October, 2014.

Mori, K., Sugaya, S. and Gemma, H. 2005. High temperature reduces anthocyanin accumulation in grape berry skin. *Scientia Horticulturae* 105:319-30.

Rana, R. S., Bhagata, R.M., Kalia, V. and Lal, H. 2004. Impact of climate change on shife of apple belt in Himachal Pradesh. ISPRS Archives XXXVIII-8/W3 *Workshop Proceedings*: Impact of Climate Change on Agriculture.

Woolf, A. B., Bowen, J. H., & Ferguson, I. B. 1999. Pre-harvest exposure to the sun influences post harvest responses of 'Hass' avocado fruit. *Post harvest Biology and Technology* 15: 143-153.

Menzel, C.M., Rasmussen, T.S. and Simpson, D.R. 1989. Effect of temperature and leaf water stress on growth and flowering of litchi (*Litchi chinenesis* Sonn.). *Journal of Horticultural Sciences* 64:739-752.

Menzel, C.M and Simpson, D.R. 1995. Temperature above 20°C reduce flowering in lychee (*Litchi chinenesis* Sonn). *Journal of Horticultural Sciences* 70: 981-987.

Samra, J.S. and Singh, G. 2003. Cold wave of 2002-2003: Impact on agriculture, *Indian Council of Agricultural Research* 49 pp.

Samra, J.S. and Singh, G. 2004.Heatwave of March, 2004: Impact on agriculture, *Indian Council of Agricultural Research* 32 pp.

Shu, Z.H and Sheen, T.F. 1987.Floral induction in axillary buds of mango (*Mangifera indica* L.) as affected by temperature. *Scientia Horticulturae* 31:81-87.

Singh, H.P. 2010. Mitigating climate change for sustainable horticultural production. *Indian Horticulture* 55(6):3-9.

❑❑❑

Chapter 8

Tree-Crop Interaction and Productivity of Agroforestry System in Arid and Semi-Arid Ecosystems

N.D. YADAVA AND M.L. SONI

ICAR-Central Arid Zone Research Institute, Regional Research Station, Bikaner-334004, Rajasthan, E-mail: narendra_yadava@yahoo.co.in

Introduction

Globally, arid and semi-arid regions comprise 1641.9 and 1897.7 million hectare (m ha) area, respectively and represent about 74% of total dry lands (FAO, 2002). They are diverse in terms of their climate, soils, flora, fauna, land use, and people. Rainfall variability and drought are significant contributors to low levels of economic development, poverty and food insecurity in these areas (Brown, 2012; de Sherbinin, 2009). Arid lands of the world cover about 10.6% of the land surface. The annual rainfall in these areas is up to 200 mm in winter rainfall areas and <400 mm in summer rainfall areas. Africa and Asia have the largest extension of arid zones. Semi-arid zones are more extensive and occur in all the continents, and cover up to 15.2% of the land surface. They have highly seasonal rainfall regimes and a mean annual rainfall of up to 500 mm in winter-rainfall areas and up to 600 mm in summer-rainfall areas. The farmers of these regions have evolved tree based farming systems as a drought cop up mechanism based on their traditional wisdom, descending from one generation to another. The crops, grasses and legumes are integrated with woody perennials on the same unit of land to improve the productivity per unit area. In broad term, these systems are called agroforestry systems.

Agroforestry is a combination of trees with crops, simultaneously or sequentially in the same area, with a major aim of optimization of beneficial ecological interactions among ecosystem components. These systems comprise of silvi-pastoral/agri-silviculture/agri-silvi-pastoral/livestock husbandry that accrue maximum benefits from their woody components. Agroforestry imparts ameliorating and mitigating effects of harsh and erratic climatic conditions and poor soils, raising soil fertility build-up and enhance symbiotic activities that help crop growth, sustainable production of food and feed, fuel wood, timber, fibres, etc. In addition, it permits higher security and sustainability during droughts and famines that are very frequent in dry regions. It puts much emphasis on species diversity and on interactions between its components, i.e., trees and crops. Jose *et al.*(2004) suggested that deeper roots of trees can act as a "safety net" by capturing nutrients that leach below the rooting zone of the crops and recycle them back into the agroforestry systems. Carvalho *et al.*(2010) suggested that a strong competition occurs between trees and crops for sunlight, nutrients, moisture, and other available resources, which can seriously reduce crop yields. Moreover, high density of tree roots present in the region of crop roots is responsible for belowground interactions. This can also have an effect on soil organisms, which play an essential role in the functioning and productivity of agroforestry systems (Lehmann *et al.*, 1998).

Today in climate change scenario the sustainability of agricultural production may be possible through system approach only and agroforestry may be the only option for stakeholders. Arid and semi-arid zones represent large tracts in north-western India and the interior peninsula. The arid regions of Rajasthan, Gujarat, Punjab and Haryana together constitute the Great Indian Desert, better known as the Thar Desert, which accounts for 89.6% of the total hot arid regions of India, and constitutes the principal hot arid zone of the country. Semi - arid region in India constitutes parts of Gujarat, northern plains and central highlands (Malwa), and the Deccan plateau. Almost 53.4% of India's land area comprises arid and semi-arid regions. These areas suffer from water scarcity due to low rainfall and extreme weather conditions. Climate change is further aggravating the existing problems in both arid and semi-arid regions. It is estimated that the arid and semi-arid regions are set to suffer further from water shortages.

To minimize the adverse effects of frequent droughts and water scarcity, efforts are being put to develop agroforestry systems in these regions. Woody perennials have a very important role, both from productivity as well as resource conservation point of view. The most important challenge for acceptable agroforestry system is designing a sustainable tree crop combination because many species suffer from root competition and thus selection of tree species with either less root competitiveness or with complementary root interaction is of strategic importance in agroforestry systems. Thus, much time and expense need to be invested in finding species or genetic varieties that combine in more diverse agro-ecosystems to improve total yield and system productivity.

Effect of tree crop interaction on yield and biomass production

In agroforestry systems, interaction in coactions and competition in higher plants greatly influences the productivity of field crops, fruit trees and forest trees in both the natural and managed ecosystems. The silvi-pastoral system (trees with grasses) has been a traditional practice in the Thar Desert. By virtue of deep root system of trees (Ahuja, 1984) and extensive root system of grasses (Soni *et al.*, 2006), they bind the soil and reduce erosion. Farmers protect the trees of *Acacia albida, Pterocarpus erinaceus,* and *Khaya senegalensis* in sub-humid zones of Mali and *Prosopis cineraria* in arid and semi arid regions of Rajasthan (FAO, 1997). A combination of *Acacia tortilis* and *Cenchrus setigerus* was found best in a silvi-pasture study carried out with four tree species viz. *Acacia tortilis, Azadirachta indica, Albizia lebbeck,* and *Holoptelelia integrifolia* and four grass species viz. *Cenchrus ciliaris, C. setigerus, Dichanthium annulatum* and *Panicum antidotale,* in the transition zone between arid and semi-arid regions of Rajasthan (Faroda, 1997). *Ailanthus excelsa* is a relatively fast growing tree and was found highly suited for agroforestry systems on sandy loam soils of semi-arid areas (Mann *et al.*, 1989; Mann, 1994). The *C. ciliaris* is a common grass forage which volunteers successfully in marginal lands. The *Ailanthus excelsa* had high total dry biomass (10.6 kg $tree^{-1}$) and 90% allocation to stem.

In arid region of Nagaur (Rajasthan), Arya (2006) observed that *Z. mauritiana* and *C. ciliaris* was the best silvi-pastoral combination for maximum dry fodder yield (3633 kg ha^{-1}). *C. ciliaris* had mean 4-year yield of 1840 kg $ha^{-1}$with *Zizyphus mauritiana,* followed by 1510 kg ha^{-1} with *Acacia nilotica,* and 1240 kg ha^{-1} with *A. excels* (Table 1).The root system of *Z. mauritiana* was better developed than the other species. Mean maximum grass yield (1843.5 kg ha^{-1}) was obtained with *Z. mauritiana* followed by *A.nilotica* (1507 kg ha^{-1}). Species-wise and year-wise rainfall variation resulted in significant differences in grass yield.

Table 1. Mean yield of *Cenchrus ciliaris* dry matter (kg ha $^{-1}$) and total fodder yield (kg ha $^{-1}$) in a silvi-pastoral system in the Nagaur district of Rajasthan, India

Species	Grass yield					Total
	1997–1998	1998–1999	1999–2000	2000–2001	Mean	fodder yield
Zizyphus mauritiana	1017(100)	624(41.5)	2655.5(23.5)	3077(307.7)	1843.5[a]	3633[a]
Ailanthus excelsa	900(100)	633(33.3)	1504(23.5)	1923(76.9)	1240[b]	2231[b]
Acacia nilotica	817(150)	833(20.00)	2361(176.3)	2019(50.00)	1507[c]	2931[c]
Mean	911[a]	697[a]	2173[b]	2340[b]	--	--

Source: Arya, (2006) Values carrying different letter are significantly different at p<0.01%. Standard error of means is given in parenthesis. Total fodder yield=Dry grass+dry leaf yield.

Yadava *et al.* (2013) reported that the intercropping of *Cyamopsis tetragonoloba* (clusterbean) and *Vigna mungo* (green gram) positively affected the growth of trees over no intercropping system (Table 2). The increase in plant height of *Citrus aurontifolia, Colophospermum mopane* and *Dalbergia sissoo* in intercropping system was 15.38, 9.50 and 11.67% higher over their sole planting, respectively. Intercropping of green gram increased the stem girth by 45.24 and 18.18% and canopy by 39.48 and19.54% in *C. aurontifolia* and *C.mopane,* respectively over sole planting. Among the three species tried, highest total biological yield of green gram (1412.5 kg ha^{-1}) and cluster bean (1352 kg ha^{-1}) was recorded in intercropping with *C. aurontifolia* which was 9.70 and 172.5% higher in green gram and 21.31 and 161.3% higher in cluster bean over intercropping with *D. sissoo* and *C. mopane,* respectively (Table 3). Increased yield of cucumber fruit and minimal interference in the growth of *C. aurontifolia* seedlings was also reported by Natarajaa and Nairk (1992). The same trend was observed in seed yield of both the intercrops. Highest seed yield of greengram (471.3 kg ha^{-1}) and cluster bean (419.8 kg ha^{-1}) was recorded with *C. aurontifolia,* which was 50.8, 84.8, 54.2 and 184.0% higher over intercropping with *D. sissoo* and *C. mopane,* respectively. Thus, *C. mopane* and *D. sissoo* showed more competitive effect on intercrops than *C. aurontifolia.* Dwivedi *et al.* (2007) have also reported the maximum wheat yield under citrus plantation followed by aonla and minimum crop yield was recorded under guava plantation.

Table 2. Growth of trees as affected by different intercrops in agri-horti-silvi system

Trees	Greengram	Cluster bean	Sole tree	Mean
		Plant height(cm)		
Citrus aurontifolia	281.67	300.00	260.00	277.08
Colophospermum mopane	380.00	402.33	367.00	383.17
Dalbergia sissoo	590.00	578.33	528.33	573.33
Mean	417.22	426.89	385.11	
CD(0.05)	T=18.52	C=21.39		
		Stem girth(cm)		
Citrus aurontifolia	46.00	45.00	31.67	40.83
Colophospermum mopane	78.00	74.33	66.00	71.25
Dalbergia sissoo	59.00	65.67	45.67	56
Mean	61.00	61.67	47.78	
CD(0.05)	T=4.33	C=4.99		
		Tree canopy(m^2)		
Citrus aurontifolia	13.67	13.46	9.80	12.53
Colophospermum mopane	25.10	24.53	21.00	23.35
Dalbergia sissoo	15.33	18.76	16.20	17.15
Mean	18.06	18.92	15.67	
CD(0.05)	T=2.46	C=2.84		

Source: Yadava *et al.*(2013)

Table 3. Growth and yield of intercrops with trees under agri-horti-silvi system

Trees	Plant height (cm)	Branches plant^{-1}	Pods plant^{-1}	Biological yield (kg ha^{-1})	Seed yield (kg ha^{-1})
			Green gram		
Citrus aurontifolia	57.50	3.00	47.50	1412.50	471.30
Colophospermum mopane	34.75	1.00	10.50	775.00	255.00
Dalbergia sissoo	55.25	2.50	22.75	1287.50	312.50
Mean	49.17	2.16	26.91	1158.33	346.25
CD5%	10.03	1.00	2.29	331.62	116.47
			Cluster bean		
Citrus aurontifolia	42.50	3.25	16.25	1352.00	419.80
Colophospermum mopane	38.50	1.50	14.00	517.30	147.80
Dalbergia sissoo	54.00	3.00	25.50	1114.50	271.50
Mean	45.00	2.58	18.53	994.58	279.75
CD(0.05)	NS	NS	NS	244.30	58.29

Source: Yadava *et al.* (2013)

In agri-horti-pasture system under arid condition of Bikaner, Rajasthan, Yadava *et al.* (2014) reported that the growth, yield and yield attributing characters of *Vigna aconitifolia* (mothbean) and cluster bean were highest in intercropping with *C. aurontifolia* as compared to *Cordia myxa* and *Aegle marmelos.* The *Lasiurus sindicus* grown as rainfed intercrop with *C. aurontifolia* gave the highest plant height (92.3cm) and total dry matter yield (6.5 t/ha) which was 259.6, 185.58 and 437.19, 245.7% higher over intercropping with *A. marmelos* and *C. myxa,* respectively. *Aloe vera* intercropped with citrus gave highest green pad yield of 94.4 t ha^{-1} which was at par with *A. marmelos.*

Agroforestry could contribute to livelihoods improvement in India due to accumulated local knowledge. Puri and Bangarwa (1992) reported that *Azadirachta indica* and *P. cineraria* did not show any significant difference in the wheat yield while the other two species (*D. sissoo and A. nilotica*) showed a reduction in wheat yield. *A. nilotica* had the most significant and prominent effect, and a reduction of nearly 40 to 60% in wheat yield was observed. The effect of this tree species was observed even beyond the spread of the crown. *D. sissoo* reduced yield by 4 to 30% but the reduction was only up to a distance of 3 m. In general, the impact of trees on wheat yield was observed up to 3 m distance and there is little, if any, impact up to 5 m distance and almost no impact at 7 m distance. Crop maturity was observed to be delayed by three weeks under *A. nilotica,* by 9-10 days under *D. sissoo,* and only by 6-7 days under *P. cineraria* and *A. indica.*

India is particularly notable for ethno forestry practices and indigenous knowledge systems on tree growing. In order to maximize the trade-off in yield of crops and wood some new models are now emerging. For instance in regions such as Andhra Pradesh, where annual rainfall is around 1,000 mm and soils are fairly good, eucalyptus at a density of 1,666 plants ha^{-1} can be planted in uniformly spaced wide-rows (6 m) or paired rows at an inter-pair spacing of 7-11 m for improving intercrop performance without sacrificing wood production (Prasad *et al.,* 2010). Likewise, in Rajasthan, yield of the annual crops can be optimized in combination with *P. cineraria* at optimum tree densities of 278 trees ha^{-1} at 6 and 7 years, 208 trees ha^{-1} at 10 year and <208 trees ha^{-1} at 11 years of age. Studies on *Tecomella undulate* (Rohida) intercropped with cluster bean, *Vigna radiata* (mungbean), *Pennisetum glaucum* (pearl millet) suggested that seedling density of 833 stem ha^{-1} and 417 stems ha^{-1} were optimum for total production at the age of four and five years, respectively (Singh *et al.,* 2005).

Beyond that age, 287 stems ha^{-1} was most favourable for crop production at the age of 6-7 years and 208 stem ha^{-1} at 10-11 years (Singh *et al.,* 2004). The experiments on *A. indica* and *Phaseolus mungo* agroforestry system suggest that crop yield under the tree canopy decrease but are compensated by increase in wood volume and fruit yield of *A. indica* and thus giving higher economic returns (Pandey *et al.,* 2010).

Tree-crop- root interactions in agroforestry

Lateral root spread of trees especially in the arid and semi-arid regions is often much faster than the crown (Jones *et al.,*1998). However, tree root systems being highly plastic, proximity of two or more trees/species generally favours diminished lateral spread and or deeper root penetration of the woody perennial components (Lehmann *et al.,* 1998).Tree root systems are involved in some of the major favourable effects on soils and crops expected from the application of agroforestry techniques. On the other hand, trees can compete with associated crops through their root systems, and this may lead to yield depressions and may contribute to the economic failure of land-use systems.

Special consideration should be paid to the source of water used by trees, the rate of water depletion below the crop rooting zone and the prospects for deep recharge as using water at faster than replacement rates demonstrates a serious ecological consequence of not mimicking the resource use patterns of locally adapted species. It suggests that to reduce the negative effects of competition, tree spacing should be increased to minimize competition for water with crops (Singh *et al.,* 1989). In southern India, eucalyptus plantations have been reported to not only extract all the rainfall that enters the soil, but also to utilise an additional 100 mm of water from each meter depth of soil that the roots penetrate. This is a matter of concern as recent studies in Karnataka, India, have shown that eucalyptus roots can reach to a depth of 8 m in three years. Consequently, when planning trees in association with crops in dry lands, it is essential to consider the implications of increased water use on the medium and longer-term water budgets.

In western Kenya, *Sesbania sesban* has been shown to utilize nitrogen below the rooting zone of maize (Mekonnen *et al.,* 1997) and to make this available to crops through its leaf litter. Evidence also indicates that the network of tree roots in agroforestry practices are able to reduce the leaching of nutrients from the surface soil layers, so reducing the leakiness of the nutrient cycle (Buresh and Tian, 1998). The study of roots of agroforestry systems where crops and trees are grown concurrently is much more difficult and tends to be confined to species which are easily distinguishable on the basis of visual features. In semiarid environments, there was still considerable overlap between the roots of trees and crops especially in the top 0.5–1.0 m of the soil, implying that most trees will be competing with crops for both water and nutrients in the top 0.5 m of most soils (Rao *et al.*, 1993).

Mechanisms of root interactions and perspectives of their management

The main theme of agroforestry is that the trees must acquire resources that the crop would not otherwise get (Cannell *et al.*, 1996). In the dry season, the proportion of roots under the tree row compared to the alley was higher than during the wet season; the same was found for the proportion of roots in the subsoil compared to the top soil. Pruning decreased the total root length density of sole cropped trees by 47%. The highest root length density was found when the pruned trees were intercropped with sorghum. If the trees were not pruned, combining trees and crops did not increase root length density. Intercropping resulted in a spatial separation of the root systems of trees and crops between the hedge rows, sorghum having more roots in the topsoil and the trees having more roots in the subsoil under alley cropping than in monoculture.

(i) Root distribution and root interactions at the root systems level

A fundamental hypothesis of agroforestry is that different plant life forms such as trees and herbaceous crops or pastures occupy to some extent different soil strata with their root systems when grown in association, leading to a degree of complementarity in their use of soil resources. Such differences in root distribution between plant species, growing at the same site and under similar environmental conditions, illustrate the genetic factors of root development. These interact with environmental factors in determining the shape and functioning of root systems, including soil strength, soil temperature, soil aeration, water and nutrient availability, and microbial and faunal interactions.

(ii) Interactions at the individual root level

The spatial separation of the roots of associated plants may reduce competition and negative rhizospheric interactions between plants, but also the possibility of positive interactions. These may occur when shallow-rooted crops take up water which has been hydraulically lifted into the topsoil by deeper-reaching tree root systems (Emerman and Dawson,1996), or when one species takes up nutrients which have been made available by rhizospheric processes of another species (Jones and Darrah, 1996).

(iii) Factors of root activity and competitiveness

Fine root length density is an important parameter for the acquisition of water and nutrients when the transport to the root, and not the uptake into the root, is the limiting step in resource acquisition. It is therefore especially relevant in dry soil and for nutrients of low to intermediate mobility such as P and K (Eissenstat, 1992).

Allelopathic effect of agroforestry

The allelo-chemicals produced by the species used in agro-forestry systems may play an important role in crop production, soil quality, pest management strategy, animal and human nutrition. *A. indica* (neem) is an indigenous tree of the Indian subcontinent and is considered as a tree for solving global problems. However, the results of some trials indicate its allelopathic effects on companion crops. The forage yield of *Avena sativa* (oats), *Sorghum bicolour* (sorghum) and *Carthamus tinctorius* (safflower) was reduced by 26%, 7 to 33% and 3 to 16% under neem tree than in open plots (Hazra and Tripathi,1989; Srivastava and Rammohanrao, 1989). Contrary to these, Puri and Bangarwa (1992) have found that neem tree has no adverse effect on the yield of *Triticum aestivum* (wheat) if grown 5 m apart from the main stem. Similarly *Eucalyptus* has been promoted on a large scale due to its fast growth behaviour, adaptability to various climatic conditions and its industrial value. However, due to its antiphytosocial nature (Kohli, 1987) and ill effects on the ecology (Poore and Fries, 1985), it is losing its popularity on agricultural field. It releases a number of volatile and non-volatile allelochemicals that affect growth of the associated vegetation (Kohli, 1990). The germination, seedling vigor, and seedling length of *Phaseolus aureus, Hordeum vulgare,* and *Avena sativa* were significantly reduced when placed in chambers flushed with eucalyptus oil (Kohli and Singh, 1991). Similarly, the presence of a non-protein amino acid along with some phenolic compounds in the leaves and seeds of *Leaucaena* is a cause of concern to allelopaths and ecologists.

The study conducted by Krishna *et al.* (2003) on the allelopathic influence of leaves of *Acacia auriculiformis, Casuarina equisetifolia, Bambusa auradinaceae and Tectona grandis* on Tomato, Brinjal and Chilli revealed that leaf leachates of all trees significantly inhibited germination percentage and growth of vegetable crops (Table 4). Response indices revealed that inhibition of radical and plumule growth was more pronounced. There are reports on the allelopathic effects of various plants on nitrogen fixation. The activity of nitrogenase was reduced to the tune of 28% when plants were grown in *Leucaena*-canopy soil (Rizvi, 1996).

Table 4. Allelopathic effect of agroforestry tree species on different vegetable crops

Tree species	Brinjal			Chilli			Tomato		
	Germination (%)	Shoot length (cm)	Root length (cm)	Germination (%)	Shoot length (cm)	Root length (cm)	Germination (%)	Shoot length (cm)	Root length (cm)
Acacia auriculiformis	79.50	3.25	2.45	76.5	3.30	2.82	68.50	4.82	2.12
Casurina Equisetifolia	77.00	2.40	1.48	75.75	2.95	2.40	71.75	4.52	1.72
Bombusa auradinaceae	80.25	2.77	2.07	71.75	3.32	2.72	69.25	4.10	2.25
Tectona grandis	81.50	3.02	2.30	73.25	2.85	2.22	67.00	5.42	1.70
Control	87.25	4.37	5.0	80.25	4.97	5.05	76.50	8.47	2.97
S.Em.±	2.48	0.25	0.176	2.48	0.32	0.11	2.42	0.50	0.17
C.D. at 5%	NS	0.75	0.51	NS	NS	0.33	NS	1.50	NS

Source: Krishna *et al.* (2003)

Moisture competition and roots manipulation

In arid and semi-arid environments, it may be more worthwhile to focus attention on the selection of trees that provide more direct and immediate benefits to farmers(rather than the selection for soil enrichment), with minimum loss of crop productivity. It is perhaps not surprising that farmers are already beginning to experiment with such systems. For example, in the drylands of eastern Kenya, farmers have recently developed an intensive parkland system, using a fast-growing indigenous species, *Meliavolkensii (Mukau)*, which provides high-value timber in 5–8 years and fodder during the dry season without apparent loss in crop productivity (Ong *et al.*, 2002). The study conducted at Machakos, Kenya showed that when rainfall was low (250 mm) maize yield was linearly and negatively related to the amount of water used by the trees. However, this relationship broke down when rainfall exceeded 650 mm, illustrating that the trees can then use water from the same soil profile as the maize without major negative effects on crop yield (Ong and Leakey, 1999).

Beniwal *et al.* (2008) reported that the total soil moisture depletion from the profile was maximum in *C. mopane* followed by *D. sissoo* and *C. aurontifolia*. The extensive root system of *C. mopane* resulted in higher soil moisture extraction from cropping zone as compared to citrus and *D. sissoo* (Table 5). They further reported that the root biomass density of *C. mopane* at a radial distance of 1.0-2.5 meter was more as compared to *D. sissoo* and citrus. The fibrous and thin roots of *C. mopane* were 2.4 and 16.3 times more than citrus, which compete with intercrops for moisture. Since root biomass and soil moisture extraction by citrus in cropping zone (i.e. 2.5 meter radial distance) was less, it can perform well in agri-horti system with minimum effect on crop growth and yield (Table 6).

Table 5. Profile soil moisture upto 60 cm soil depth under different tree species after a rainfall of 67 mm

Species	Lateral distance(meter)	Dates		
		19.09.05 (3 DAR)	23.09.05 (7 DAR)	3.10.05 (17 DAR)
C. aurontifolia	1.00	40.36	20.86	11.58
C.mopane		28.59	17.68	12.63
D.sissoo		32.85	25.56	16.54
C. aurontifolia	2.50	41.08	29.32	24.95
C. mopane		35.10	22.61	11.87
D.sissoo		39.26	29.68	19.39

Source: Beniwal *et al.* (2008) *DAR = days after rainfall

Table 6. Root biomass density (g m^{-3}) of various root grades upto 90 cm soil depth at two radial distances

Lateral distance(m)	Species	Root grades			
		<0.2 cm	0.2-0.5 cm	0.5-1.5 cm	>1.5 cm.
0.5-1.0	*C. aurontifolia*	57.22	34.23	69.19	0.00
	C. mopane	62.04	25.66	98.09	0.00
	D. sissoo	40.54	29.99	41.27	82.22
1.0-2.5	*C. aurontifolia*	15.68	1.00	0.00	0.00
	C. mopane	36.32	16.37	87.13	0.00
	D. sissoo	32.72	8.25	33.77	0.00

Source: Beniwal *et al.* (2008)

Severe reduction in crop yield often observed in alley cropping in the semi-arid tropics is likely to be mainly a consequence of competition for water. Soil water depletion was higher under the tree row than in the alley and higher in alley cropping than in monoculture systems (Lehmann *et al.*, 1998). Recent studies have shown that it is worthwhile to manage below-ground competition by root pruning. Pruning of lateral roots could redirect root function and be a powerful tool for improving spatial complementarity, provided that there are adequate resources at depth (Ong *et al.*, 2002). Singh *et al.* (1989) demonstrated that root barriers to 50 cm depth are extremely effective in reducing competition between 4-year-old *Leucaena leucocephala* hedgerows and associated cowpea and sorghum crops in semi-arid India. However, for long duration crops like castor, the presence of a root barrier was not sufficient to fully compensate for the yield reduction probably because a large part of the reproductive period occurred in the dry season when water was drawn from below 0.5 m. However, the beneficial effects lasted only one season because tree roots reinvaded the crop rooting zone from beneath the root barriers. In contrast, studies in Bangladesh (Hocking, 1998; Hocking and Islam, 1998) revealed that below-ground competition from a wide range of tree species (mainly fruit trees) was virtually eliminated by pruning the lateral roots off the trees.

Effect of agroforestry systems on soil quality

The land use systems comprising of tree, crops and pastures play an important role in improving soil fertility and its quality by several ways. Agri-silvi, agri-horti and agri-pastoral systems have the potential to reduce erosion and runoff, maintain soil organic matter, improve soil physical properties and augment nitrogen fixation and promote efficient nutrient cycling. The importance of alley cropping, agri-horticultural and agroforestry systems on soil properties have been emphasised by many workers (Kessler and Breman 1991; Mac Dicken, 1990; Das *et al.*, 1993).

Soil Physical properties

Agroforestry systems maintain more favourable soil physical properties than agricultural systems through maintenance of organic matter and the effects of roots. Tree roots may penetrate and possibly break up compact soil layers such as stone lines and nodular laterite and improve both physical properties and nutrient intake from the B/C horizons. Conventional cropping management was unable to maintain soil physical quality on the same levels as agroforestry and natural vegetation (Silva *et al.*, 2011). More research evidence on the improvement of physical properties by agroforestry systems is still needed, while the effects of litter on soil physical properties is substantial through mulching and zero tillage studies.

Soil chemical properties

Out of the several benefits accrued from agroforestry systems in terms of soil quality, nutrient cycling is the most predominant. The quantity and quality of litter produced by trees depend on the species, age and density of trees (Soni *et al.*, 2013; Singh *et al.*, 2007). Plants take up nutrients from the soil and use them for metabolic activities. In-turn, these nutrients are returned back to the soil either naturally as litter fall or as pruning in some agro-forestry systems or through root senescence. These plant parts are decomposed as a result of microbial activities and release the nutrients held in them into the soil. The nutrient then becomes available for plant uptake once again. Agroforestry land use systems like agri-horticulture, agri-pastoral system, agri-silvi-pastoral system, etc. are efficient ways of restoring soil organic matter (Pandey, 2007). The leaf litter from agroforestry practices, forms humus after decomposition and improves various soil properties. Tree species have variable impact on soil organic carbon, nutrient status and other soil properties (Patel and Singh, 2000; Soni *et al.*, 2008). In poplar based agroforestry system, Singh *et al.* (2010) observed that the average contents of soil organic carbon(SOC), N, P and K were higher by 22.2, 10.2, 33.6 and 4.5% respectively, than pure pearl millet-wheat rotation (Table 7).

Table 7. Soil organic carbon (SOC) and available nutrients in agroforestry and pure crops (pearl millet-wheat) after growing them for 6 years

Soil depths (cm)	Organic carbon (gkg^{-1})		Available N (kg/ha)		Available P (kg/ha)		Available K (kg/ha)	
	AF*	Pure crops	AF	Pure crops	AF	Pure crops	AF	Pure crops
0-15	3.73	2.82	165	145	15.0	9.86	204	189
15-30	2.74	2.16	139	124	10.98	7.39	186	177
30-45	2.05	1.78	110	104	9.41	6.72	188	181
45-60	1.61	1.53	106	99	9.18	9.41	147	146
Mean	2.53	2.07	130	118	11.14	8.34	181	173
CD (P=0.05)	System (S) = 0.28 Soil Depth (SD) = 014 SxSD=0.20		S=9.3 SD=9.2 SxSD=NS		S=1.32 SD=1.24 SxSD=1.76		S=4.7 SD=11.9 SxSD=NS	

Source: Singh *et al.* (2010)*AF : Agroforestry (poplar with pearlmillet-wheat rotation)

Soil biological properties

Agroforestry systems have shown beneficial effects on soil flora and fauna with consequent improvements in soil fertility. Soil microbial biomass acts as a source and sinks for the plant nutrients cycling and soil organic matter dynamics. It is the prime agent involved in plant residue decomposition, nutrient conservation and cycling processes in the soil. The microbial biomass has therefore been used as an index of soil fertility, which depends on nutrient fluxes. Plant cover through its effect on quantity and quality of organic matter inputs influences microbial biomass. Yadav *et al.* (2011) observed significantly higher amount of soil microbial biomass C, N and P under MPTs (Table 8) which could mainly be attributed to addition of varying quantity of organic matter inputs over the years through litter fall (Yadav *et al.* 2008). The range of microbial biomass carbon in semi-arid Rajasthan (186–320 lg g^{-1} soil) has been reported by Yadav *et al.* (2011) which was comparable to 85.5–537 lg g^{-1} soil under *P. cineraria* under arid conditions (Nagarajan and Sundaramoorthy, 2000), 203–255 lg g^{-1} soil under Bactris, Hevea and polyculture based agroforestry (Kurzatkowski *et al.*, 2004) and 193–267 lg g^{-1} soil under Indian subtropical soils (Suman *et al.*, 2006).

Table 8. Microbial biomass carbon(C_{mic}, ug g^{-1} soil), microbial biomass nitrogen(N_{mic}, ug g^{-1} soil), microbial biomass phosphorus(P_{mic}, ug g^{-1} soil), C_{mic}/N_{mic} and C_{mic}/P_{mic} ratio under MPTs based agroforestry system land uses.(Values not marked with the same superscript letter are significant(P<0.01) using Tukeys HSD test)

Land uses	C_{mic}	N_{mic}	P_{mic}	C_{mic}/N_{mic}	C_{mic}/P_{mic}
Sole cropping	186^{c}	23.2^{c}	8.4^{d}	8.02^{a}	22.4^{a}
P. cineraria + crop	320^{a}	42.4^{a}	15.6^{a}	8.08^{a}	21.2^{a}
D. sissoo +crop	283^{ab}	35.8^{ab}	13.6^{b}	8.12^{a}	20.9^{a}
A. nilotica+ crop	262^{b}	32.1^{b}	11.6^{c}	8.53^{a}	22.9^{a}
A. leucophloea + crop	272^{b}	32.8^{b}	11.9^{bc}	8.56^{a}	23.6^{a}

Source: Yadav *et al.* (2011)

Arbuscular Mycorrhiza Fungi (AMF) colonization under tree-crop combination

AMF are the important component of agroforestry, forming symbiotic associations with the majority of plants and showing multifunctional roles in these systems. The role of perennial trees in maintaining AMF inoculums and in sustaining mycelia networks for shortlived intercrops may be an unintended benefit of agroforestry systems and provides an alternative approach to the use of cover crops to build up soil inoculums (Ingleby *et al.,* 2007; Kumar *et al.,* 2007). Several reports are available on the influence of agroforestry systems on the AMF which suggested that agroforestry has positive effects on the AMF community (Muleta *et al.,* 2008; Bainard *et al.,* 2010; Pandey and Tarafdar, 2004). Rajshekhara *et al.* (1989) reported higher colonization and sporulation of AMF in *Cajanus cajan* and *Helianthus annus* with decreasing distance from the trees, viz., *D. sissoo, Dendrocalamus strictus, T. grandis, L.leucocephala* and *C. equisetifolia.* The tree roots acted as a source of inoculum for AMF colonization of intercrops as AMF are known to be nonspecific in their selection of hosts (Kumar *et al.,* 2007). Fallowing has reported to reduce AMF population in the fields (Gosling *et al.,* 2006; Oehl *et al.,* 2003; Troeh and Loynachan 2003). In *P. mungo* and *S. orientale* the root colonization was higher under canopy of *A. indica, E. officinalis, A. leucophloea, A. indica, E. officinalis, M. latifolia,* and *Z. mauritiana* (Shukla *et al.,* 2012).

Effect of shoot pruning on AMF colonization of trees

Tree shoot management can reduce the light competition occurring between trees and intercrops (Namirembe, 1999), which gives opportunity to intercrops, particularly those beneath trees, to get more light. Hartley and Amos (1999) reported reduced mycorrhizal colonization in pruned trees of *Nardus stricta* and *Calluna vulgaris* as compared with un-pruned trees. Peter and Lehmann (2000) have reported that this effect might be due to the lower photosynthates in the roots of pruned trees than in the unpruned trees. The colonization index of unpruned trees was significantly higher than the colonization index of pruned trees in *A. pendula, H. binata, A. procera,* and *D. sissoo* (Shukla *et al.,* 2012).

Agroforestry for enhanced carbon sequestration

The carbon storage capacity in agroforestry varies across species and geography (Newaj and Dhyani, 2008). Carbon sequestration in different agroforestry systems occurs both belowground, in the form of enhancement of soil carbon plus root biomass and aboveground as carbon stored in standing biomass. Some of the earliest studies of potential carbon storage in agroforestry systems and alternative land use systems for India had estimated a sequestration potential of 68-228 $MgCha^{-1}$ (Dixon *et al.,* 1994) and 25 $MgCha^{-1}$ over 96 m ha of land (Sathaye and Ravindranath, 1998)]. But this value varies in different regions depending on the biomass production (Pandey, 2007). Studies done by Jha *et al.* (2001) showed that agroforestry could store nearly 83.6 MgC ha^{-1} up to 30 cm soil depth, 26% more carbon compared to cultivation in Haryana plains.

There is also clear evidence to suggest that the type of agroforestry system greatly influences the source or sink role of the trees (Montagnini and Nair, 2004). For example, agri-silvi-cultural systems where trees and crops are grown together are net sinks while agro silvi-pastoral systems are possibly sources of GHGs. Practices like tillage, controlled burning, manuring, application of chemical fertilizers and frequent soil disturbance for raising crops in agroforestry systems can lead to significant emissions of GHGs (Montagnini and Nair, 2004). The Greening India mission under the National Climate Change Action Plan targets 1.5 m ha of degraded agricultural lands and fallows to be brought under agroforestry; about 0.8 m ha under improved agroforestry practices on existing lands and 0.7 m ha of additional lands under agroforestry (Puri and Nair, 2004). Much of the opportunity to store carbon through afforestation in India will occur through agroforestry on agricultural lands due to the fact that majority of arable land in India is being cultivated (Ravindranath, 2007).

Agroforestry provides the best example of promoting mitigation and adaptation synergy in addressing climate change (Garrity *et al.*, 2006). The carbon in the aboveground and belowground biomass in an agroforestry system is generally much higher than the equivalent land use without trees (i.e. crop land without any trees). The estimates of total carbon storage in different agroforestry systems are provided in Table 9. The carbon sequestration potential of agroforestry systems has been established theoretically; however field measurements to validate these concepts are limited. Few studies of specific agroforestry practices have proved potential for carbon sequestration (Roshetko *et al.*, 2002; Takimoto *et al.*, 2008).

Carbon sequestration in agri-silviculture systems

Maikhuri *et al.* (2000) estimated species wise annual carbon sequestration potential of planted tree species on abandoned agricultural (3.9t ha^{-1} yr^{-1}) and degraded forest land (1.79t ha^{-1} yr^{-1}). The carbon sequestration in monocropping of trees and food crops were 40 and 84% less than agri-silvi-culture indicating that agroforestry systems have more potential to sequester carbon (Dhyani *et al.*, 2009). In an agri-silvi-cultural system, *D. sissoo* at age of 11 years was able to accumulate 48-52t ha^{-1} of biomass (Newaj and Dhyani, 2008). Carbon dynamics involving different pruning treatments were studied in an agri-silvi-cultural system where tree biomass was 23.61 to 34.49t Cha^{-1} with black gram-mustard. In a study on poplar based agri-silvicultural system, total biomass in the system was 25.2t ha^{-1}, 113.6% higher than sole wheat cultivation, whereas net carbon storage was 34.61t Cha^{-1} compared to 18.74t Cha^{-1} in sole wheat cultivation (Chauhan *et al.*, 2010). In a system comprising of *Albizia*and mixed tree species like Mandarin, tree and crop biomass accumulated 1.3t biomass ha^{-1} storing 6939 kg ha^{-1} (Sharma *et al.*, 1995).

In a study conducted on intercropping of trees with crops, Singh *et al.* (1989) reported an enhancement in soil organic carbon (SOC) by 33.3 to 83.3% for *Populus deltoids* and *Eucalyptus* hybrid with *Cymbopogon* spp., with a greater increase in SOC under *Populus deltoids* plantation. Soil organic carbon has been reported to have improved for agroforestry plantations ranging in age from 6 years (Maikhuri *et al.*, 2000) to 20 years (Saha *et al.*, 2007). In a Poplar based agroforestry system, trees could sequester higher soil organic carbon up to 30 cm depth during the first year of plantation (6.07t $ha^{-1}yr^{-1}$) than in subsequent years (1.95-2.63t $ha^{-1}yr^{-1}$) with greater soil carbon storage in sandy clays than loamy sand (Gupta *et al.*, 2009). Traditional *P. cineraria* based systems lead to a 50% increase in SOC largely due to leaf litter (Venkateswaralu, 2010). Samra and Singh (2000) observed an increase in soil organic carbon status of surface soil by 0.39 to 0.52% under *Acacia nilotica+ Sacchram munja* and 0.44 to 0.55% under *Acacia nilotica+Eulaliopsis binate* after 5 years.

Carbon sequestration in silvi-pastoral systems

Rai *et al.* (2009) studied the effect of introducing a silvipastoral system in a natural grassland in semi arid Uttar Pradesh, where introduced species of *Albizia procera, Eucalyptus tereticornis, Albizia lebbeck, Embilica officinalis* and *Dalbergia sissoo* accumulated 8.6, 6.92, 6.52, 6.25 and 5.41 t $ha^{-1}yr^{-1}$ of biomass. Here, the carbon storage in the system was 1.89-3.45t Cha^{-1} in silvi-pasture and 3.94t Cha^{-1} in pure pasture. Similar studies have reported higher organic carbon levels in dry sub-humid and arid ecosystems when grass species are intercropped with annual crops in a silvi-pastoral system with no increase in organic carbon with grasses in an arid ecosystem (Venkateswaralu, 2010). In a silvipastoral system, carbon flux in net primary productivity increased due to the integration of *Prosopis juliflora* and *D. sissoo* with grasses. Narain (2008) reported that planting trees and grasses in degraded lands in arid areas can help increase soil carbon stock from 24.3 Pg to 34.9 Pg. Studies conducted at RRS, Kukma have reported a total belowground carbon stock of 1.6 tha^{-1}, 23.4% of the total carbon stock under silvi-pastoral system involving *A. tortilis* and *Cenchrus setigerus*. Similar studies conducted in 28-year-old *A.tortilis, Acacia senegal* and *C. mopane* in silvi-pasture systems at Bikaner, Rajasthan revealed that, tree based silvipasture systems could sequester 6.84, 6.20 and 5.11 kg m^{-2} more atmospheric CO_2 into the soil equivalent to 18.98, 16.91 and 13.95t Cha^{-1} (Soni *et al.,* 2014).

Table 9. Total C storage under agro-forestry systems in different regions of the country

Region	Agroforestry system and components	Total C storage (t Cha^{-1})	References
Arid region (Rajasthan)	Agri- silvicultural system (age 8 years)		
	Emblica officinalis+ Vigna radiate	12.7 -13.0	Singh (2005)
	Hardwickia binata+ vigna radiate	8.6 - 8.8	
	Colophospermum mopane+ Vigna radiata	4.7 - 5.3	

Semi- arid region	Silvi-pastoral system (age 5 years)		Rai *et al.*(2001)
	Acacia nilotica+ natural pasture	9.5-17.0	
	A. nilotica + established pasture	19.7	
	Dalbergia sissoo+ natural pasture	12.4	
	D. sissoo+ established pasture	17.2	
	Hardwickia binata+ natural pasture	16.2	
	H. binata+ established pasture	17.0	
	Agri-silvicultural system (age 11 years)		NRCAF(2005)
	Dalbergia sissoo+ crop	26.0	
North-western India	Silvipastoral system (age 6 years)		Kaur *et al.* (2002)
	Acacia/Dalbergia/Prosopis + *Desmostacya*	6.8-18.5	
	Acacia/Dalbergia/Prosopis + *Sporobolus*	1.5-12.3	

Way Forward

Agroforestry systems have been known and practiced for hundreds of years by the farmers of arid and semi-arid regions. There are conclusive evidences that these systems improve the total system productivity and provide an insurance against complete crop failure especially in adverse climatic situations of drought and famine. Agroforestry have greenhouse gas–mitigation strategy through carbon sequestration because of greater efficiency of integrated systems in resource (nutrients, light, and water) capture and utilization than single-species systems. They could contribute to livelihoods improvement of the people of arid and semi-arid regions and have great potential to improve soil quality and mitigate the adverse effect of climate change. However, the issue of resource competition among crops and trees are most important. Hence, special consideration should be paid to the tree architecture, root characters and allelo-chemicals produced by root exudates and leaf litters to minimize the resource competition.

References

Ahuja, L.D. 1984. Range management in agro-forestry system in arid regions of India. In: *Agro-Forestry in Arid and Semi-Arid Zones*(ed. Shankarnarayan). CAZRI publication No.24, Central Arid Zone Research Institute, Jodhpur, pp. 161–166.

Arya, R. 2006. Studies on growth and biomass production of different fodder tree species in a silvi-pastoral trial under hot arid conditions in India. *Journal of Arid Environment* 65 (1):179-191.

Bainard, L.D., Klironomos, J.N.and Gordon, A.M. 2010. Arbuscular mycorrhizal fungi in tree-based intercropping systems: a review of their abundance and diversity. *Pedobiologia* 54:57–61.

Beniwal, R.K., Soni, M.L. and Yadava, N.D. 2008. Soil moisture extraction pattern of arid zone trees under agri-horti-silvi system in north western Rajasthan. In *Natural Resource Management for Sustainable Development in Western India* (eds. S.N. Prasad, R.K. Singh, Ashok Kumar, A.K. Pandiyal, Shakir Ali, Somsundaran, V.K. Sethy and V.N. Sharda). Allied publishers Pvt. Ltd, New Delhi,pp.197-198.

Brown, C. 2012. Water, ecosystems, and poverty: roadmap for the coming challenge Integrating Ecology and Poverty Reduction. In *Ecological Dimensions* (ed. Ingram, J.C.), Springer, New York, pp 151–61.doi:10.1007/ 978-1-4419-0633-5 9.

Buresh, R.J. and Tian, G. 1998.Soil improvement in sub-Saharan Africa .*Agroforestry Systems* 38: 51–76.

Cannell, M.G.R., van Nordwijk, M., Ong, C.K., 1996. The central agroforestry hypothesis: the trees must acquireresources that the crop would not otherwise get. *Agroforestry Systems* 34: 27-31.

Chauhan, S.K., Sharma, S.C., Beri, V., Yadav, S. and Gupta, N. 2010. Yield and carbon sequestration potential of wheat (*Triticumaestivum*)–poplar (*Populusdeltoides*) based agri-silvicultural system. *Indian Journal of Agricultural Sciences* 80: 129- 35.

Das, S. K., Sharma, S., Sharma, K.L., Saharan, N., Nimbole, N.N., and Reddy, Y.V.R. 1993. Land use options in a semi-arid Alfisols. *American Journal of Alternative Agriculture* 8: 34-39.

deCarvalho, A.M.X., de Castro, T.R., Cardoso, I.M. and Kuyper, T.W. 2010. Mycorrhizal associations in agroforestry systems.In *Soil biology* (ed. P. Dion), vol. 21. *Soil biology and agriculture in the tropics.* Springer, Heidelberg, pp 185–208.

Dhyani, S.K., Newaj, R. and Sharma, A.R. 2009. Agroforestry: its relation with agronomy, Challenges and opportunities. *Indian Journal of Agronomy* 54: 249- 266.

Dixon, R.K., Brown, S., Houghton, R.A., Solomon, A.M. and Trexler, M.C. 1994. Carbon pools and fluxes of global forest ecosystems.*Science* 263: 185-190.

Dwivedi, R.P., Tewari, R.K., Kareemulla, K., Chaturvedi, O.P. and RaiP. 2007. Agri-horticultural system for household livelihood - A case study. *Indian Research Journal of Extension Education* 7(1): 22-26.

Eissenstat, D. M. 1992. Costs and benefits of constructing roots of small diameter.*Journal of Plant Nutrition* 15: 763-782

Emerman, S.H. and Dawson, T.E. 1996. Hydraulic lift and its influence on the water content ofthe rhizosphere: an example from sugar maple, *Acer saccharum*. *Oecologia* 108: 273–278

FAO, 1997. Role of Forestry in Combating Desertification YafongBerthe, vol. 2, topic 10, XI World

Faroda, A.S. 1997. Silvipastoral approach for arid and semi-arid areas: issues and strategies. In *Silvi-pastoral Systems in Arid andSemi-Arid Ecosystems*(eds. Yadav,M.S., Singh, M., Sharma, S.K., Tewari, J.C. andBurman, U.). Central Arid Zone Research Institute, Jodhpur, pp. 1–10.

FAO. 2002. *Report of the World Food Summit: Five Years Later*. FAO, Rome.

Garrity, D., Okono, A., Grayson, M. and Parrott, S. 2006. *World Agroforestry into the Future*. World Agroforesty Centre,Nairobi.

Gosling, P., Hodge, A., Goodlass, G. and Bending, G.D. 2006. Arbuscularmy-corrhizal fungi and organic farming. *Agriculture, Ecosystem and Environment* 113:17–35.

Gupta, N., Kukal, S.S., Bawa, S.S. and Dhaliwal, G.S. 2009. Soil organic carbon and aggregation under poplar based agroforestry system in relation to tree age and soil type. *Agroforestry Systems* 76: 27-35.

Hartley, S.E. and Amos, L. 1999. Competitive interactions between *Nardusstricta L.* and *Calluna vulgaris*(L.) Hull: the effect of fertilizer and defoliation on above and below ground performance. *Journal of Ecology* 87:330–340.

Hazra, C.R. and Tripathi, S.B. 1989.Performance of oat genotypes under different tree covers and open. *Agricultural Science Digest* 9: 146–148.

Hocking, D. 1998. Trees in wetland rice fields: a successful tree management technology developed through participatory action-research in Bangladesh. *Agroforestry Today* 10: 4–6.

Hocking, D. and Islam, K. 1998.Trees on farms in Bangladesh.Growth of top and root-pruned trees in wetland rice fields and yield of understorey crops.*Agroforestry Systems* 39: 101–115.

Ingleby, K., Wilson, E.J., Munro, E.R.C. and Cavers, S. 2007. Mycorrhizas in agroforestry: spread and sharing of arbuscularmycorrhizal fungi between trees and crops: complementary use of molecular and microscopic approaches. *Plantand Soil* 294:125–136.

Jama, B., Kwesiga, F. and Niang, A. 2006. Agroforestry innovations for soil fertility management in sub-Saharan Africa: Prospects and challenges ahead. *World Agroforestry into the Future*: 60.

Jha, M.N., Gupta, M.K. and Raina, A.K. 2001. Carbon Sequestration: Forest soil and land use management. *Annals of Forestry* 9: 249-256.

Jones, D.L. and Darrah, P.R. 1996. Role of root exudates in plant nutrient acquisition in low inputagricultural systems. *Agroforestry Forum* 7: 7–9.

Jones, M., Sinclair, F.L. and Grime, V.L. 1998. Effect of treespecies and crown pruning on root length and soil watercontent in semi-arid agroforestry.*Plant and Soil* 201: 197–207.

Jose, S., Gillespie, A.R. and Pallardy, S.G. 2004.Interspecific interactions in temperate agroforestry.*Agroforestry Systems* 61:237–255.

Kaur, B., Gupta, S.R. and Singh, G. 2002. Carbon storage and nitrogen cycling in silvipastoral system on sodic soil of North-western India.*Agroforestry Systems* 54: 21-29. 64.

Kessler, J.J., and Breman,H. 1991. The potential of agro forestry to increase primary production in the Sahelian and Sudanian zones of West Africa.*Agroforestry Systems* 13:41-62.

Kohli, R.K. 1987. *Eucalyptus* – an antisocial tree for Social Forestry.In*Social Forestry for Rural Development*(eds. Khosla, P. K. and Kohli, R. K.).pp. 235–241. ISTS Publications,India.

Kohli, R.K. 1990. *Allelopathic Potential of Eucalyptus*.Project Report MAB-DOEn Project, India.

Kohli, R.K. and Singh, D. 1991.Allelopathic impact of volatile components from *Eucalyptus* on crop plants.*Biol. Plant.* 33:475–483.

Krishna, A., Manjunath, G.O., Rathod,R. and Kannur, S.2003. Allelopathiceffect of four agroforestry tree species leaf leachates on seed germination of certain vegetable crops. *Karnataka Journal of Agriculture Science 16: 430-433.*

Kumar, A., Shukla, A., Hashmi, S. and Tewari, R.K. 2007. Effect of tree on colonization of intercrops by vesicular arbuscularmycorrhizae in agroforestry system.*Indian Journal of AgriculturalScience* 77:291–298.

Kurzatkowski,D., Martius, C., Hofer, H., Garcia, M., Forster, B., Beck, L. and Vlek, P. 2004. Litter decomposition, microbial biomass and activity of soil organisms in three agroforestry sites in central Amazonia. *Nutrient Cycling in Agro-ecosystem* 69:257–267.

Lehmann, J., Peter, I., Steglich, C., Gebauer, G., Huwe, B., Zech, W. 1998. Below-ground interactions in dryland agroforestry.*Forest Ecology and Management* 111: 157-169.

Maikhuri, R.K., Semwa, R.L., Rao, K.S., Singh, K. and Saxena, K.G. 2000. Growth and ecological impacts of traditional agroforestry tree species in Central Himalaya, India. *Agroforestry Systems* 48: 257-272.

Mann, J.S. 1994. Role of Ardu(*Ailanthus excelsa*) and other trees in agroforestry in semi-arid areas ofRajasthan.In *Agroforestry Systems for Degraded Lands*, vol. 1 (eds. Singh, P., Pathak, P.S., Roy, M.M.). pp. 289–293. Oxford & IBH Publishing Co, New Delhi.

Mann, J.S., Hassan, B. and Beniwal, R.K. 1989. Adopt silvi-pasture system for more fodder in arid and semiarid regions of Rajasthan. *Indian Farming* 39: 33–35.

Mekonnen, K., Buresh, R.J. and Jama, B. 1997. Root and inorganic nitrogen distribution in sesbania fallow, natural fallow and maize fields.*Plant and Soil* 188: 319–327.

Montagnini, F. and Nair, P.K.R. 2004. Carbon sequestration: An underexploited environmental benefit of agroforestry systems. *Agroforestry Systems* 61: 281- 295.

Muleta, D., Assefa, F., Nemomissa, S. and Granhall, U. 2008. Distribution of arbuscularmycorrhizal fungi spores in soils of smallholder agroforestry and monocultural coffee systems in south western Ethiopia. *Biology and Fertility of Soils* 44:653–659.

Murthy, I.K., Bhat, D.M., Hegde, G.T., Nagaraj, B.C. and Murali, K.S. 2006. Tree diversity and carbon sinks: role of village agro-ecosystems in the Western Ghats, Karnataka, India. *International Journal of Agricultural Sustainability* 3: 114-121.

Nagarajan, M., Sundaramoorthy, S. 2000. Effect of *Prosopis cineraria* canopy sizes on soil microbial biomass C, N and P in arid zone.*Current Agriculture* 24:75–80.

Namirembe, S. 1999. Tree shoot pruning to control competition for below ground resources in agroforestry. PhD thesis, University of Wales, Bangor, UK.

Narain, P. 2008. Dryland management in arid ecosystem.*Journal of the Indian Society of Soil Science* 56: 337-347, 65.

Natarajaa, M. and Nairk, D.M. 1992.Competitive effects of short duration, bush type cowpea when intercropped with cotton in Zimbabwe. *Experimental Agriculture* 28: 41-48.

Newaj, R. and Dhyani, S.K. 2008. Agroforestry for carbon sequestration: Scope and present status. *Indian Journal of Agroforestry* 10: 1-9.

NRCAF.2005. *Annual Report*, NRCAF, Jhansi, Uttar Pradesh.

Oehl, F., Sieverding, E., Ineichen, K., Mader, P., Boller, T., Wiemken, A. 20 Impact of land use intensity on the species diversity of arbuscul mycorrhizal fungi in agro-ecosystems of Central Europe. *Appl Environment Microbiology* 69: 2816–2824.

Ong C.K. and Leakey, R.B.1999. Why tree-crop interactions in agroforestry app at odds with tree-grass interactions in tropical savannahs *Agrofores Systems* 45: 109–129.

Ong, C.K., Black, C.R., Marshall, F.M. and Corlett, J.E., 1996. Principles of resou capture and utilization of light and water.In *Tree-Crop Interactions*(e Ong, C.K. and Huxley, P.) pp. 73-158. CAB International, Wallingford.

Ong, C.K., Wilson, J., Deans, J.D., Mulatya, J., Raussen, T. and Wajja-Musuk N. 2002. Tree–crop interactions: manipulation of water use and r function. *Agricultural Water Management* 53: 171–186.

Pande, M. and Tarafdar, J.C. 2004.Arbuscularmycorrhizal fungal diversity neem based agroforestry systems in Rajasthan. *Applied Soil Ecology* 233–241.

Pandey, A.K., Gupta, V.K. and Solanki, K.R. 2010. Productivity of neem-bas agroforestry system in semi-arid region of India, *Range Management Agroforestry* 31: 144-149.

Pandey, D.N. 2007. Multifunctional agroforestry systems in India.*Current Scie* 92: 455-463.

Patel, N.L. and Singh, S.P. 2000. Effect of different tree species on amelioration.*Indian Journal of Forestry* 23: 192-96.

Peter, I. and Lehmann, J. 2000. Pruning effects on root distribution and nutri dynamics in an *Acacia* hedgerow planting in northern Kenya. *Agrofore Systems* 50:59–75.

Poore, M.E.D. and Fries, C. 1985. The Ecological Effects of Eucalyptus. *FAO Fore paper no.* 59, Rome, Italy.

Prasad, J.V.N.S., Korwar, G.R., .Rao, K.V., Mandal, U.K., Rao, C.A.R., Rao, G Ramakrishna, Y.S., Venkateswarlu, B., Rao, S.N., Kulkarni, H.D. and R M.R. 2010. Tree row spacing affected agronomic and econo performance of *Eucalyptus*-based agroforestry in Andhra Prade Southern India. *Agroforestry Systems* 78: 253-267.

Puri, S. and Bangarwa, K.S. 1992. Effect of trees on the yield of irrigated wh crop in semi-arid regions. *Agroforestry Systems* 20:229–241.

Puri, S.and Nair, P.K.R. 2004. Agroforestry research for development in India years of experiences of a national program. *Agroforestry Systems* 61: 4 452.

Rai P, Ajit, Chaturvedi, P.O., Singh, R. and Singh, U.P.2009. Biomass production in multipurpose tree species in natural grasslands under semi-arid conditions. *Journal of Tropical Forestry* 25: 11-16.

Rajshekhara, E., Sreenivasa, M.N. and Bhat, R.S. 1989. Effect of cropping system on vesicular–arbuscularmycorrhizal development in red gram and sunflower. *Karnataka Journal of Agricultural Science* 2:231–233.

Rao, M.R., Muraya, P. and Huxley, P.A. 1993. Observations of some tree root systems in agroforestryintercrop situations and their graphical representation. *Experimental Agriculture* 29: 183–194.

Ravindranath, N.H. 2007. Mitigations and adaptation synergy in forest sector. *Mitigation and Adaptation Strategies for Global Change* 12: 843-853.

Rizvi, S.J.H. 1996. Role of allelopathy in improving crop productivity. In *Abstr. 1st World Congress of Allelopathy*, Spain, p. 76.

Roshetko, J.M., Delaney, M., Hairiah, K. and Purnomosidhi, P. 2002. Carbon stocks in Indonesian homegarden systems: Can smallholder systems be targeted for increased carbon storage? *American Journal of Alternative Agriculture* 17: 138-148.

Saha, R., Tomar, J.M.S. and Ghosh, P.K. 2007. Evaluation and selection of multipurpose tree for improving soil hydro-physical behaviour under hilly eco-system of north east India. *Agroforestry Systems* 69: 239-247.

Samra, J.S. and Singh, S.C. 2000. Silvopasture systems for soil, water and nutrient conservation on degraded land of Shiwalik foothills (sub-tropical northern India). *Indian Journal of Soil Conservation* 28: 35-42, 60.

Sathaye, J.A., Makundi, W.R., Andrasko, K., Boer, R. and Ravindranath, N.H. 2001. Carbon mitigation potential and costs of forestry options in Brazil, China, India, Indonesia, Mexico, The Philippines and Tanzania. *Mitigation and Adaptation Strategies for Global Change* 6: 185-211.

Sharma, R., Sharma, E. and Purohit, A.N. 1995. Dry matter production and nutrient cycling in agroforestry systems of mandarin grown in association with *Albizia* and mixed tree species. *Agroforestry systems* 29: 165-179.

Shukla,,A., Kumar, A., Jha,A., Dhyani, S.K. and Vyas, D. 2012. Cumulative effects of tree-based intercropping on arbuscularmycorrhizalfungi. *Biology and Fertility of Soils* 48:899–909.

Silva, G.L., Lima, H.V., Campanha, M.M., Gilkes, R.J. and Oliveira, T.S. 2011. Soil physical quality of Luvisols under agroforestry, natural vegetation and conventional crop management systems in the Brazilian semi-arid region. *Geoderma* 61–70.

Singh, B., Gill, R. and Kaur, N. 2007. Litterfall and nutrient return in poplar plantation varying in row directions and spacing. *IndianJournal of Agroforestry* 9: 33-37.

Singh, B., Gill, R., and Gill, P.S. 2010. Soil fertility under various tree species and poplar-based Agroforestry system. *Journal of Research Punjab Agricultural University* 47: 160-164.

Singh, G. 2005. Carbon sequestration under an agrisilvicultural system in the arid region.*Indian Forester* 131: 543-552.

Singh, G., Bala,N.,Mutha, S., Rathod, T.R. and Limba, N.K. 2004. Biomass production of *Tecomella undulata* agroforestry systems in arid India.*Biological Agriculture and Horticulture* 22: 205-216.

Singh, G., Mutha, S., Bala, N., Rathod, T.R., Bohra N.K. and Kuchhawaha, G.R. 2005. Growth and productivity of *Tecomellaundulata* based on an agroforestry system in the Indian desert. *Forests Trees and Livelihoods* 15: 89-101.

Singh, K., Chauhan, H.S., Rajput, D.K. and Singh, D.V. 1989. Report of a 60 month study on litter production, changes in soil chemical properties and productivity under Poplar (*P.deltoides*) and Eucalyptus (*E. hybrid*) interplanted with aromatic grasses. *Agroforestry Systems* 9: 37-45, 56.

Singh, R.P., Ong, C.K. and Saharan, N. 1989. Above- and below-ground interactions in alley cropping in semiarid India. *Agroforestry Systems* 9: 259–274.

Soni, M.L., Beniwal, R.K., Talwar, H.S., Yadava, N.D., Singh, J. P. and Sunil Kumar. 2006. Root distribution pattern of sewan(*LasiurussindicusHenr.)* and buffel grass (*CenchrusciliarisL.*) of arid ecosystem of western Rajasthan in relation to their soil binding capacity. *Indian Journal of Agriculture Sciences* 76: 716-720.

Soni, M.L.,Beniwal, R.K., Yadava, N.D. and Talwar, H.S. 2008. Spatial distribution of soil organic carbon under agro forestry and traditional cropping system in hyper arid zone of Rajasthan. *Annals of Arid Zone* 103-106.

Soni, M.L. Yadava, N.D. and Bhardwaj, S.2013. Decomposition and nitrogen release dynamics of fruit tree leaf litters in arid western Rajasthan. *Annals of Arid Zone* 52:31-37.

Soni, M.L. Yadava, N.D. and Bhardwaj, S. 2014. Growth, biomass production and C-sequestration by three predominant tree species in North-Western Rajasthan. In *Compendium of Abstracts, World Congress on Agroforestry-2014,* 10-13 Feb, 2014, New Delhi.pp.357.

Srivastava, A.K. and Rammohanrao, M A.S. 1989. Agroforestry for sustainable agricultural production. *Indian Farming* 39: 33–35.

Suman, A., Lal, M., Singh, A.K. and Gaur, A. 2006. Microbial biomass turnover in Indian Subtropical soils under different sugarcane intercropping systems. *Agronomy Journal* 98:398–404.

Takimoto, A., Nair, P.K.R. and Nair, V.D. 2008. Carbon stock and sequestration potential of traditional and improved agroforestry systems in the West African Sahel. *Agriculture, Ecosystems and Environment* 125: 159-166.

Troeh, Z.I., Loynachan, T.E. 2003. Endomycorrhizal fungal survival in continuous corn, soybean and fallow. *AgronomyJournal* 95:224–230.

Venkateswaralu, J. 2010. *Rainfed agriculture in India:Research and Development Scenario*. Indian council of Agricultural Research, New Delhi: 508p.

Walker, B.H. and Noy-Meir, I. 1982. Aspects of stability and resilience of savannah ecosystems. In *Ecology of Tropical Savannas* (eds. Huntley BJ and Walker BH), pp. 556–590.

Yadav, R.S., Yadav, B.L., Chhipa, B.R., Dyani, S.K. and Munna Ram. 2011. Soil biological properties under different tree based traditional agroforestry systems in a semi-arid region of Rajasthan, India. *Agroforestry Systems* 81:195–202.

Yadav, R.S., Yadav, B.L., Chhipa, B.R. 2008. Litter dynamics and soil properties under different tree species in a semi-arid region of Rajasthan, India. *Agroforestry Systems* 73:1–12.

Yadava, N.D., Soni, M.L., Nathawat N.S. and Birbal 2013. Productivity and growth indices of intercrops in agri-horti-silvisystem in arid western rajasthan. *Annals of Arid Zone* 52: 61-65.

Yadava, N.D., Soni, M.L., Birbal andNathawat, N.S.2014. Crop productivity and soil moisture variation pattern of agri-horti system in the arid zone of western Rajasthan.In *Compendium of Abstracts, World Congress on Agroforestry-2014*, 10-13 Feb, 2014, New Delhi.pp.372.

□□□

Chapter 9

Integrated Farming System for Sustainable Agriculture in Climate Change Perspective

S.P.S. TANWAR

ICAR - Central Arid Zone Research Institute, Jodhpur, Rajasthan - 342 003

E- mail: spstanwar @gmail.com

Introduction

Agriculture is the most important economic activity in India. Although contributing only to about 14% in national GDP, its performance strongly influence other sectors of the economy like livestock, industry, trade and commerce due to strong linkages. Besides, it is the income generating activity of more than two third population of the nation. The progress of Indian agriculture is impressive from a net importer to food surplus nation producing 260 million tonnes of food grains. To keep pace with the increasing population and their expectations we need to maintain at least 4-5% annual growth rate in agriculture sector.However, challenges exist at mega scale, mainly the shrinking land and water resources, decreasing factor productivity and threats of climate change. Global climate change induced by increase in concentrations of greenhouse gases is likely to increase temperatures, alter precipitation trends and cause more of extreme events affecting almost every aspect of agriculture (Table 1). In Indian context, its impact would be high as more than 60% of agricultural land is rainfed and by now about 50% of land holdings are below 1 hectare and about 80% are less than 2 hectares. In the fragile ecosystems like arid zone where average land holding size is though large (>4 ha), extreme weather uncertainties and low production levels have made all categories of farmers vulnerable. The objective of any climate change adaptation strategy and hence to imparting resilience in agriculture would be to spread the risk and better utilization of available resources. A judicious mix of cropping system with associated enterprises like dairy, poultry, piggery, fishery, sericulture etc. suited to the given agro-climatic conditions and socio economic status of farmers has immense potential to address this instability of income, food and nutrition, unemployment and vulnerability. Similarly, integration of trees with crops would help in moderating/mitigating climate change through carbon sequestration, buffering effect on moderation of organic matter, halting soil and water erosion, augmenting availability of timber, fuel wood, fodder and other minor forest products, proofing for drought and cyclones and increase in farmers' income.

Table 1. Impact of climate change on various sectors of Indian agriculture

Sector	Impact
Crop	Increase in ambient CO_2 may increase photosynthesis mainly in C_3 plants but the yield of major cereal crops especially wheat is likely to be reduced due to reduction in grain filing duration, increased respiration, and/or reduction in rainfall/irrigation.
	Increase in extreme weather events such as floods, droughts, cyclones and heat waves will adversely affect agricultural productivity.
	Reduction in yield in the rainfed areas due to changes in rainfall patterns during monsoon season and increased crop water demand.
	Cold waves and frost events could decrease in future due to global warming.
	Quality of fruits, vegetables, tea, coffee, aromatic and medicinal plants to be affected.
	Pest and diseases of the crops to be altered because of more enhanced pathogen and vector development, rapid pathogen transmission and increased host susceptibility.
	Threat to agricultural biodiversity.
Water	Demand for irrigation to increase with increased temperature and evapo-transpiration. This may also result in lowering of ground water table at some places.
	The melting of glaciers will increase water availability in some rivers in short run but in the long run availability of water will reduce considerably.
	A significant increase in runoff during wet season, but that may not be very useful unless storage infrastructure could be vastly expanded. This extra water in wet season, on the other hand, may lead to increase in frequency and duration of floods.

	The water balance in different parts of India will be disturbed and the quality of ground water along the coastal track will be more affected due to intrusion of ocean waters.
Soil	Organic matter content which is already quite low would become still lower. Quality of soil organic matter may be affected.
	The residues of crops under elevated CO_2 concentration will have higher C:N ratio, and this may reduce their rate of decomposition and nutrient supply.
	Increase of soil temperature will increase N mineralization but its availability may decrease due to increased gaseous losses through processes such as volatilization and denitrification.
	Change in rainfall volume and frequency, and wind may alter the severity, frequency and extent of soil erosion.
	Rise in sea level may lead to salt –water ingression in the coastal lands turning them less suitable for conventional agriculture.
Livestock	Climate change will affect feed production and nutrition of livestock. Increased temperature would increase lignification of plant tissues reducing the digestibility. Increased water scarcity would also decrease food and fodder production.
	Major impacts on vector borne diseases through expansion of vector population in cooler areas. Changes in rainfall pattern may also influence expansion of vectors during wetter years, leading to large outbreaks of disease.
	Global warming would increase water, shelter and energy requirement of livestock for meeting projected milk demands.
	Climate change is likely to aggravate the heat stress in dairy animals, adversely affecting their reproductive performance.
Fishery	Increasing sea and river water temperature is likely to affect fish breeding, migration and harvest.
	Impact of increased temperature and tropical cyclonic activity would affect the capture, production and marketing costs of marine fishes.
	Coral bleaching is likely to increase due to higher sea surface temperature.

Source : Aggarwal *et al.* (2009)

Integrated farming system:concept, scope and methodology

Traditionally agriculture in India at farmers level is not mono-commodity based rather it is a bundle of interrelated enterprises with diverse outputs mainly referred as farming system. This is a multi-disciplinary whole farm approach aimed at achieving income, food and nutritional security at household level. It is a combination of enterprises in which the product and the byproduct of one enterprise serve as the input for production of other enterprises. The most common enterprises are, cropping systems, horticulture, livestock, fishery, forestry,poultry, etc. The integration focuses on synergistic recycling of plant nutrients and *in situ* management of byproducts. Hence it is a potential adaptation strategy to deal with impact of climate change. Integrated farming system meets demands for food, income and diverse requirements of food grains, vegetables, milk, egg, meat, etc. The choice of enterprises for integration in a farming system depends on many factors viz.

- Soil and climatic features
- Availability of the resources, land, labour and capital
- Present level of utilization of resources
- Return/income from the existing farming system
- Economics of proposed integrated farming system
- Farmer's managerial skill
- Social customs prevailing in the region.

With a view to mitigate risks and uncertainties of income from crop enterprises and to reduce the time lag between investment and returns it is essential that farmers include multi-enterprises in their production programme that yield regular and evenly distributed income throughout the year and are not subjected to vagaries of nature. The choice of components in a farming system must ensure that interaction among the components is complementary with least competitiveness.

Farming system research has a great potential for experimentation, learning and exchange of experiences. This is an operational research which first identifies technical, biological and socio economic constraints for improved and sustainable production. The whole farm serves as the framework for analysis, but specific components, sub-systems or interactions are targeted for interventions. The FSR draws on the 'body of knowledge' of technologies and management strategies, generated by discipline and commodity research and adapts them to the specific environmentand socio economic circumstances of a target group of relatively homogeneous farmers (Sands, 1986).

Farmers participation is ensured at different stages of technology generation and transfer processes such as system description, problem diagnosis, design and implementation of on–farm trials, and providing feedback through monitoring and evaluation (Mahapatra and Behera, 2011). There exists a chain of interactions among the components within the farming systems, and it becomes difficult to deal with such interlinking complex systems manually. This problem could be overcome by construction and application of suitable whole farm models. On the other hand optimization techniques such as linear programming, goal programming and compromise programming proved useful for efficient resource allocation under various constraints (Taha, 2005).

Arid zone experiences

The hot Indian arid zone,popularly known as the Thar Desert, occupies over 31.7 m ha area of which 61.0% lies in Rajasthan state. The region is characterized by frequent droughts and extremes of aridity due to low and erratic rainfall (100-420 mm/year), extreme temperatures (-5.7 to 50.0°C), long sunshine duration (6.6-10 hours), low relative humidity (30%-80%), high wind velocity (9-13 kmph) and high evapo-transpiration (1500-2000 mm). Simulation studies by Hadley Centre (UK) and IITM (Pune) suggest that during the 21st century annual temperature across arid Rajasthan is likely to increase by 2-5°C. Both summer and the winter temperatures will witness a gradual increase, which will affect the soil moisture regime and crop growth pattern. In the next 50 years monsoon rainfall is likely to decline gradually by 20-30% in the northwestern part of Rajasthan and adjoining Punjab, while the eastern fringe of arid Rajasthan and adjoining Haryana state may experience an increase of up to 25%. At the same time winter rains may gradually increase by 20-40%. Arid Gujarat and adjoining south Rajasthan, on the other hand are likely to experience ~25% higher monsoon rain, as well as higher winter rain. There is also the probability of high magnitude droughts and floods. Such a scenario would most likely lead to higher wind and water erosions, as well as changes in vegetation composition and its degradation (Rupa Kumar *et al.*, 2006; IPCC, 2007).

In spite of the environmental sensitivity and perceived fragility, arid zone of India is one of the densely populated regions. Dryland farming is the main occupation of the people. The most common crops grown are pearl millet, clusterbean, green gram and dew gram. The crop yields are meager and unstable and consequently the income from existing cropping alone is hardly sufficient to sustain the farm family. To minimize the risk and uncertainty, farmers in this region practice farming systems in one form or other. In fact traditional farming systems represent a crop-livestock-tree/shrubs-human continuum. The areas falling in <250 mm rainfall zone have predominance of grasses and shrubs; hence range/pasture development with livestock rearing is the major proposition (Bhati and Joshi, 2007). In areas receiving rainfall between 250-350 mm besides grasses and shrubs, multipurpose tree species dominate and mixed farming encompassing agroforestry system, mixed cropping, livestock and pasture management are main livelihood options. In areas receiving rainfall >300 mm, crops and cropping system diversification, agroforestry and livestock rearing are major systems of sustenance of arid zone farmers.These local survival systems some 80 years back were sustainable but now become inadequate to fulfill the needs of ever increasing population and their aspirations. Over the years, economic considerations have overtaken the sustainability issues. This has resulted in over-exploitation of the resources causing rapid and widespread land degradation and decline in productivity. Climate change adds another layer of risk to this precarious situation.

Crop diversification

Diversification of crops and cropping systems seems to be a viable option to impart resilience in agricultural production system. To cushion the adverse effect of drought and uncertainty, mixed sowing of seeds of various dryland crops viz., pearl millet (*Pennisetum typhoides*), clusterbean (*Cyamopsis tetragonoloba*), moth bean (*Vigna aconitifolia)* and sesame (*Sesamum indicum*) in an approximate ratio of 8:2:2:1 (by weight) is a common practice. If drought commences, they cut down pearlmillet plants and feed to the animals. However, this practice of mixed cropping hampers efficient management of crops from weeding to harvesting. During timely onset of monsoon farmers usually take pearlmillet on more than 60% area and if monsoon delays (beyond July 15) *kharif* pulses (*Vigna radiata, Vigna aconitifolia*) get preference. Based on climate analysis, vulnerability to drought and changing food habits, we advocate reduction of the area under pearlmillet to 40% and allocate rest of the area to kharif legumes (30%), oilseeds (15%) and forage crops (15%).

Agroforestry

Keeping trees with agricultural crops is an age old practice in arid and semi-arid regions of India. Rainfall appears to be the governing factor for evolution of traditional agroforestry systems. *Prosopis cineraria* (Khejri) based agroforestry system is most popular and widespread covering 60% area of arid zone. In 300-400 mm rainfall zone it dominates the scene. In general *P. cineraria - Acacia nilotica* system had maximum tree density (Narain and Tiwari, 2005). In *Zizyphus* spp. - *P. cineraria* based traditional agroforestry system *Z. nummularia* was the main system forming species with few trees of *P. cineraria* and *Z. rotundifolia*. However, newer establishment of natural occurring *P. cineraria, Tecomella* and other trees has almost ceased mainly due to uprooting of seedlings and offshoots during mechanized field operations, excessive lopping and insect attack mainly in 200-300 mm rainfall zone. *Tecomella undulata, Hardwickia binnata, Ailanthus excelsa* and *Holoptelea integrifolia* are the other important tree components that may find place in IFS.

Trees, by virtue of their perennial nature, impart stability in production along with improvement in microclimate and soil fertility. Tree crop interaction studies revealed that Cowpea and pearlmillet perform better with *Prosopis* tree species than mung bean and moth bean. Besides good yield of dry land crops, bonus yield of dry leaves and twigs (650-1050 kg ha^{-1}) and fuel wood (1.8-2.6 t ha^{-1}) could be obtained from tree through annual lopping (Bhati *et al.*, 2008). A tree density of 100-200 plants ha^{-1} was found optimum for minimum interference with yield of dry land crops under *P. cineraria* canopy shade.

Agri–horticulture

Horticulture based production system is considered effective strategy for improving productivity, employment opportunities, economic condition and nutritional security. Several drought hardy fruit crops like *Capparis decidua, Salvadora oleoides, Cordia dichotoma, Zizyphus nummularia* var. *rotundifolia, Z. mauritiana* are suitable for the area receiving rainfall <300 mm. Several other fruits such as *Emblica officinialis, Punica granatum, Aegle marmelos, Phoenix dactylifera, and Tamarindus indica* can be grown in the area having irrigation facilities. Agri-horti system comprising *Zizyphus* + mung bean was found very promising in 250-300 mm rainfall zone.

Silvi-pastoral system

It has been found to be best suited for areas receiving <200 mm rainfall, or in degraded rocky-gravelly areas. The highly compatible trees with grasses are: *Acacia senegal, A. tortilis, Albizia lebbek, Tecomella undulata, Colophospermum mopane, Dychrostasis nutans, Hardwickia binnata, Z. nummularia and Z. rotundifolia.* Among the pasture legumes *Clitoria ternatea* and *Lablab purpureus* showed good compatibility with *L. sindicus* and *C. ciliaris* (Narain & Tiwari, 2005). Under goat grazing, *Z. nummularia* with grass strips in 1:2 ratio led to higher economic returns due to weight gain of the animals and higher wool production (Bhati, 1997a). Under cattle grazing the silvi pasture system with *Z. rotundifolia* and *C. ciliaris* could sustain 554 animal days/ha (Tharparker breed) with 60% pasture utilization (Bhati, 1997b).

Climate resilient integrated farming system model for arid zone- A case study

Success of integrate farming system lies in proper enterprise mix and optimum utilization of resources. Various possible enterprises under arid situation have been discussed above. On these considerations a 7 ha integrated farming system model has been developed at CAZRI.This is based on twin strategy of system as well as crop diversification (rainfall zone 300-400 mm). This includes arable cropping (15%), agroforestry (35%), agri-horticulture (20%), agripasture (10%) and silvipasture (20%). Crop diversification of pearl millet (40%) with pulses (30%), oilseed (10%) and clusterbean (20%) was found more profitable and sustainable. The perennial components are *Prosopis cineraria, Hardwickia binnata* and *Acacia tortilis* in agroforestry, *Zizyphus mauritiana* in agri-horticulture, *Colophospermum mopane* and *Zizyphus rotundifolia* in silvipasture. Ten years study of this model indicate that amongst farming system components, B:C ratio was highest in silvipasture (3.11) and was the lowest in arable farming (1.62). The system as per family and herd size was surplus in pulses, fruits, fuel-wood and milk whereas; it was marginally deficient in cereals, fodder and concentrate but substantially deficient in oilseeds. Hence the diversification of FSC should be in tandem with growing family needs and natural resource sustainability. Over the decade (2001-10) the IFS model had showed reasonable IRR (21.6), and B:C ratio (2.05). Besides, a huge amount of carbon was sequestered in tree components and into the soil. This system has shown resilience in low rainfall years (2004, 2009), long dry spells of midterm drought (year 2008, 2011), delayed onset of monsoon (2006, 2012) besides recurring terminal droughts. In all such situations, the perennial component like, fruit, fuel wood and fodder trees as well as perennial grasses provided the much needed cash as well as fodder for sustenance of livestock.

The system was validated for resilience during the year 2012 which was an abnormal year with respect to much delayed onset of monsoon. The first effective rainfall (19.2 mm) received on 6th August caused much delayed sowing of crops. However, no dry spell was experienced afterwards with a seasonal rainfall of 453.6 mm. This delay in sowing severely affected the productivity of pearl millet compared to green gram, dew gram and clusterbean. A positive association of *P. cineraria* on grain yield of pearl millet and green gram was clearly evident compared to sole cropping. Under such situations, this system of 7 ha has generated more than 800 man days employment and met the fodder requirements of six adult cattle units (4 cows, 8 bucks and 4 rams) producing 7712 liter milk and 291.8 kg meat. Overall, the system has generated gross returns of Rs. 4,59,826 with a B: C ratio of 1.76 (Table 2). When compared to sole cropping of annuals (pearl millet, green gram, dew gram, clusterbean) giving a net profit of Rs. 57,435 with a B:C ratio of 1.68, this system had generated a net profit of Rs. 1,99,136. The system had also generated 6500 kg FYM that was used for improving the fertility of soil. Hence, it may be inferred that under arid farming situations with no scope for even supplemental irrigation, diversification of farming with perennial components viz., trees, grasses and animals provide higher income, employment and resilience.

Table 2. Economics of integrated farming system during extra delayed monsoon situation

System	Cost		GrossReturns		
	Unit area (ha)	**Cost (Rs.)**	**Commo-dity**	**Quantity (kg)**	**Amount (Rs.)**
Crops alone	1	12021	Grain	1058	44293
Crops +*P. cineraria*	0.75	14515	Fruits (Ber)	2437	48740
Crops + *Z. mauritiana*	0.75	32963	Grass seed	506	101200
Acacia tortilis(Farm forestry)	0.75	3400	Fuel wood	9991	22607
Crops + *Hardwickiabinnata*	0.75	10710	Milk	7712	200512
Grass (*C. ciliaris*) + *C. mopane*	0.75	11228	Meat - Ram	227	34050
Agri-pasture	0.75	12052	Meat-Bucks	64.8	8424
C. ciliaris + *Z. mauritiana*	1.5 ha	36690*	Total		459826/-
Animals (4 cows, 8 rams, 4 bucks)		127110			
Total		260690/-			
Net Returns		199136			
B:C ratio		1.76			

*Cost of rejuvenation of ber also included

References

Aggarwal, P.K., Singh, A.K., Samra, J.S., Singh, G., Gogoi, A.K., Rao, G.G.S.N. and Ramakrishna, Y.S. 2009. Introduction.In *Global Climate Change and Indian Agriculture* (Ed. P.K. Agarwal). ICAR, New Delhi, p.1-56.

Bhati, T.K. 1997a. Integrated farming systems for sustainable agriculture on drylands. In *Sustainable Dryland Agriculture*, Central Arid Zone Research Institute, Jodhpur, pp.102-105.

Bhati T.K. 1997b. Management of dryland crops in Indian arid ecosystem. In *Desertification Control in the Arid Ecosystem of India for Sustainable Development* (Eds. Surendra Singh and Amal Kar). Agro Botanical Publishers (India), Bikaner, pp. 298-307.

Bhati, T.K. and Joshi, N.L. 2007. Farming systems for the sustainable agriculture in Indian Arid zone.In *Dryland Ecosystem: Indian Perspective* (Eds. K.P.R. Vittal, R.L. Srivastava, N.L. Joshi, A. Kar, V.P. Tewari and S. Kathju,). Central Arid Zone Research Institute, Jodhpur and Arid Forest Research Institute, Jodhpur, pp. 35-52.

Bhati, T.K., Tewari, J.C. and Rathore S.S. 2008.Productivity and dynamics of integrated farming systems in western Rajasthan.In *Diversification of Arid Farming System* (Eds. PratapNarain, M.P. Singh, A. Kar, S. Kathju and Praveen-Kumar). Arid Zone Research Association of India, Jodhpur and Scientific Publisher (India) Jodhpur, India, pp. 23-30.

IPCC 2007.*Climate Change 2007: The Physical Science Basis. Contribution of the Working Group I to the Fourth Assessment Report of the Intergovernmental Panel on Climate Change* (Eds. S. Solomon, D. Qin, M. Manning, Z. Chen, M. Marquis, K.B. Averyt, M. Tignor and H.L. Miller). Cambridge University Press, Cambridge, UK, pp.996.

Mahapatra,I.C. and Behera, U.K. 2011. Rice (*Oryza sativa*) – based farming systems for livelihood improvement of Indian farmers. *Indian Journal of Agronomy* 56: 1-19.

Narain, P. and Tiwari, J.C. 2005. Trees on agricultural fields: a unique basis of life support in TharDesert. In *Proceedings of International Conference on Multipurpose Trees in the Tropics: Assessment, Growth and Management*,organized at Arid Forest Research Institute, Jodhpur during 22-25 November,2004,pp.523 516.

Rupakumar, K., Sahai, A.K., Krishna Kumar, K., Patwardhan, S.K., Mishra, P.K., Revadekar, J.V., Kamala, K. and Pant, G.B. 2006. High-resolution climate change scenarios for India for the 21st century.*Current Science* 90: 334-345.

Sands, M.D. 1986.Farming system research: clarification of terms and concepts.*Experimental Agriculture* 22:84-104

Taha, H.A. 2005.*Operational Research: An Introduction*. Prentice Hall of India Pvt, Ltd., New Delhi.

□□□

Chapter 10

Prospects of Underutilized Crops Under Changing Climate Scenario

H.R. MAHLA, J.P. SINGH AND B.R. CHOUDHARY[1]

ICAR-Central Arid Zone Research Institute, Regional Research Station, Jaisalmer-345 001 [1] Director Research, Agriculture University, Mandor, Jodhpur-342 304, E-mail: hrmahla2010@gmail.com

Introduction

Despite recent advances in agriculture, out of the estimated global wealth of 80,000 edible plant species only150 plant species have been commercially cultivated in one part or other part of world (Paroda, 1988). Further, only 25-30 plant species contributes 90 per cent in total food security. The narrowing of the number of crops upon which global food security and economic growth depend has placed the food and livelihoods at risk (Mlakar*et al.*, 2010). The mentioned facts with profound environmental consequences under climate change scenario stimulate organizations and scientists worldwide in retrieving, researching and disseminating the knowledge in production and utilization of neglected, disregarded, underexploited and new plant species, or so called alternative crops or future crops or potential crops. Alternative crops are those that are used traditionally for food, fiber, fodder, oil and medicinal properties. They have an under-exploited potential to contribute to food security, nutrition, health, income generation and environmental services.

Indigenous traditional crops are perceived as potentially useful for income generation and food security insurance.With the development of cash crops, traditional crops are usually confined to marginal areas with poor soil fertility or are reduced in terms of cultivated area. This situation happens whilst farmers face difficulties in transmitting knowledge to the young generations who are rather oriented towards off-farm activities (Meagher, 1997).The erosion of traditional subsidiary crops may put local agro-biodiversity at risk, which is otherwise very important for sustainability in fragile ecosystem. The value of such crops in the changing agricultural system of many developing countries is therefore suggested by their persistence in the farming system where alternative cash crops dominate (Mertz *et al.*, 2005). Thus, in a changing climate scenario, awareness about the importance of these indigenous crops is vital for sustainable production and their contribution to people's livelihood in arid region.

Under the umbrella of All India Coordinated Research Network on Underutilized Crops (Table 1) efforts are being made in different agro-climatic zones for standardization of agro-techniques and to develop high yielding varieties for the specific niche areas.This paper presents information on three economically important underutilized crops of arid zone, viz., seed purpose watermelon (Kalingada), tumba (Colocynth) and grain amaranth.

Table 1. Nutritional composition and yield potential of underutilized crops.

No.	Crop	Protein (%)	Oil (%)	Fe (mg/100g)	Grain yield potential (kg/ha)
1	Amaranths Grain	9.6 - 15.3	4.3 - 9.2	8.1 - 24.3	1500 - 2000
2	Rice bean	17.4 - 21.8	-	4.4 - 8.8	1000 - 1200
3	Faba bean	20.9 - 27.7	-	-	1500 - 2500
4	Buck wheat	10.5 - 12.5	-	3.0 - 7.6	350 - 700
5	Chenopodium	14.0 - 17.8	-	7.3 - 13.0	-
6	Adjuki bean	20.0 - 23.6	-	-	-
7	Kalingada	18.2 - 21.0	27.6 - 40.3	14.1 - 16.6	600 - 1000
8	Tumba	13.2 - 14.9	15.0 - 21.0	9.3 - 19.7	150 - 300

Seed purpose watermelon (Kalingada)

Watermelon [*Citrullus lanatus* (Thunb.) Matsum and Nakai] belongs to family Cucurbitaceae. Cucurbits are among the economically most important vegetable crops worldwide and are grown in both temperate and tropical regions. In sub-Saharan Africa, the native species of cucurbits are prized for their oleaginous seeds consumed as thickeners of a traditional soup called *egusi* soup in Nigeria. Cucurbits cultivated for seed consumption are reported to be rich in nutrients (de Mello *et al.*, 2001), well adapted to extremely divergent agro-ecosystems and various cropping systems characterized by minimal inputs (El Tahir and TahaYousif, 2004). In Indian arid zone, cucurbits are also regarded as an important source of vegetables and desserts. Their seeds, which contain oil and protein as reserve foods are eaten and seed oil of a few cucurbits is also used in Ayurvedic system of medicines (Singh, 1964).

Cultivation of watermelon is now widespread in all tropical and subtropical regions of the world and is mostly grown for fresh consumption of the juicy and sweet flesh of mature fruit. In Indian sub-continent, it is being grown during summer months as a cash crop under assured irrigation as well as on conserved moisture in river beds of Indo-Gangetic plains for its fresh dessert fruits commonly known as *'Tarbuj'* (Fig. 1 a & b). While, *'Mateera'* (Fig. 1 c & d) represents the landrace of watermelon widely grown in north western parts of Rajasthan under mixed cropping during *kharif* season. It is an integral part of mixed cropping with dominant *kharif* crops valued for highly priced nutritious seeds in sub-Sahara region of Africa (Munshee *et al.*, 2011; Matanyaire, 1998) and Indian Thar desert (Mahla and Singh, 2013). The concept of seed purpose watermelon in Indian Thar desert has been highlighted by Mahla and Singh (2013) and Mahla and Choudhary (2013) following the rigorous evaluation of watermelon germplasm under rainfed condition of Jaisalmer. This seed purpose watermelon locally known as *'Kalingada'* (Fig. 1 e & f) in Gujarat and Rajasthan has the potential to provide seed yield upto 10 quintals per hectare (Table 2) with more than 30 per cent crude protein and oil (Raiger *et al.*, 2014).

Table 2. Promising genotypes identified for seed yield and related traits

No.	Entry	Seed yield (q/ha)	Fruit yield (q/ha)	Days to fruit setting	100-seed weight (g)	Fruit size (cm)	Number of fruits per plant
1	SKGPK - 21	10.0	197.4	40	7.67	43.8	7.5
2	SKGPK - 22	11.3	250.3	42	6.79	39.9	10.9
3	SKGPK - 23	7.1	211.4	38	7.93	43.7	7.0
4	SKGPK - 24	6.9	293.7	41	7.95	42.2	10.7
5	SKGPK - 26	6.5	159.1	34	6.49	37.8	6.2
6	SKGPK - 27	9.8	229.8	46	7.43	40.2	10.4
7	SKGPK - 28	7.6	202.0	42	7.68	43.4	7.3
8	SKGPK - 29	6.1	244.0	42	6.93	42.6	11.0
9	SKGPK - 30	7.1	220.4	44	6.43	43.4	9.0
10	SKGPK - 31	8.0	157.3	42	7.59	42.2	6.8
11	SKGPK - 33	6.6	152.9	35	7.89	41.4	6.7
12	SKGPK - 34	8.8	218.2	37	7.41	42.5	9.4
13	SKGPK - 35	6.3	202.8	39	7.37	44	7.3
14	MGPK - 10-2	7.7	230.9	36	6.94	43.6	9.8
15	MGPK - 10-4	8.0	208.8	40	7.62	41.6	10.7
16	MGPK - 10-5	8.0	225.4	42	7.84	41.4	9.4
17	GK - 1 (C)	6.8	179.6	40	7.70	41.6	7.6
	CD (5 %)	1.23	44.93	-	-	-	-

Fig. 1. Watermelon fruits: *Tarbuj* (a & b), *Mateera* (c & d) and *Kalingada* (e & f)

Uses of Mateera/Kalingada

Mateera can be used in different ways and forms at various stages of development and post harvest.

1. *Green vegetable:*Young fruits of *Mateera* locally known as *'loia'* are relished by the inhabitants as green vegetable (Fig.2a). In local market, its tender fruits fetch very premium prices (Rs. 30-40 per kg) and also have the enormous demand. Hence, farmer can get good income after 40-50 days of crop sowing particularly under aberrant weather situations where crop failure is a very common phenomenon.

2. *Dessert fruit:* Since ancient times,its ripe fruits have been an integral part of peasants' diet. The fruits are supposed to have cooling effect in the body. Besides, providing instant energy, it provides nutrients and quenches thirst under field conditions (Fig. 2b) and hence increasing efficiency of farmers.

(a) (b)

(c) (d)

Fig. 2. Utilization of Mateera as (a): Green vegetable; (b) Fresh fruit; (c) Seed & (d) Magaj

3. *Seed as source of protein and fat:* The nutritionally rich seeds contains crude protein and oil besides other minerals in appreciable quantity (Table 3). The seed oil contains > 80.0 % unsaturated fatty acids with linoleic acid (18:2) being the dominant fatty acid (Table 4). Its roasted seeds are generally taken as a common snack (Fig. 2c) in arid region. Dehulled kernel

Table 3. Proximate composition of dehulled, undehulled and flour of watermelon seed

Nutrients	[A]Dehulled Watermelon seeds	[A]Undehulled seeds	[B]Seed flours
Crude Protein (%)	38.59±0.04	28.50±0.01	30.11±0.32
Fat / Crude lipid (%)	17.78±0.01	10.65±0.01	45.05±0.12
Carbohydrate (%)	33.08±0.03	48.35±0.01	17.62±0.02
Total Ash (%)	3.17±0,02	2.23±0.02	3.75±0.11
Crude fibre(%)	7.32±0.01	10.30±0.01	3.47±0.009
Calcium (mg/100 g)	-	-	86.75
Phosphorus (mg/100 g)	-	-	1073.3
Potassium (mg/100 g)	705.37+5.28	421.86+1.70	598.95
Magnesium (mg/100 g)	526.58+2.26	594.11+3.61	1118.0
Iron (mg/100 g)	1.94+0.03	4.27+0.05	10.70
Zinc (mg/100 g)	19.55+0.03	18.62+0.03	9.65

(*Source:* [A]Ibeanu, *et al.*, 2012 and [B]Samia *et al.*, 2012)

Table 4. Fatty acid composition of watermelon seed.

No.	Linoleic	Oleic	Palmitic	Stearic	Reference
1	68.3	13.0	8.8	5.6	Nolte and Loesecke, 1939
2	48.7	35.3	7.6	6.1	Dhingra and Biswas, 1945
3	62.0	19.0	10.5	6.1	T. Sao and Potts, 1952
4	48.7-73.0	7.9-25.7	5.1-11.5	9.7-14.4	Jarret and Levi, 2012
5	67.3	11.1	10.7	10.6	Das *et al.*, 2002
6	45.1-51.2	20.2-23.0	11.2-13.8	14.3-16.2	Raziq*et al.*, 2012
7	59.6	18.1	11.3	10.2	El Adawy and Taha, 2001
8	68.0	11.0	13.0	8.0	Sabahelkhier*et al.*, 2011
9	68.3	13.2	11.4	7.0	Baboli and Kordi, 2010
10	62.1	13.7	10.6	8.3	Edidiong*et al.*, 2013

locally known as '*Magaj*' (Fig. 2d) is mainly used in sweet preparation and also in cooling drinks. Presently the kernels of the seeds are used in restaurants/hotels as source of flavor and thickener in various dishes. It not only increases the viscosity and luster of the preparation but also the taste of the dish.

4. *Livestock feed:*Animal husbandry is the integral part of the arid zone agriculture. Traditionally, green as well as dry rind portion of the fruits locally called as '*Khuparia*' is used as live stock feed (Fig. 3). The dried rind of fruits is stored for long periods. The dried rind is crushed and boiled along with other feed concentrates as guar *churi*, and oil seed cakes. The boiled material is fed to milch animals particularly cattle and buffalo in morning and evening before milking, which is supposed to increase the milk yield (Mahla *et al.*, 2014). On dry weight basis *Khuparia* contains 12-15 per cent crude protein, 22–30 per cent ether extract and 9-14 per cent ash. It may substantially replace the costly concentrate and thereby reduce the feeding cost.

Fig. 3. Khuparia (rind), fresh (left) and dried (right)

5. *Medicine:*Traditionally, fruit juice is used against sun strokes by the desert dwellers whereas regular use of fruit cures the stone formation (Singh *et al.*, 2004). In case of snakebite, the juice of fruit kept for two years in glass bottle is given to affected person, which helps in vomiting and neutralize the effect of poison (Kumar *et al.*, 2003). Regular use of *Mateera* fruits relieve constipation, and is also useful in cardiac and kidney troubles. The seeds regulates blood sugar levels, increase energy, maintain the nervous system and promote healthy skin. The seeds are also used in pharmaceuticals.

Economic importance

After harvest of the main crop, the left over fruits of '*Mateera*'are harvested for extraction of seeds (Fig. 4). Seed price varies from Rs. 5000-8000 per quintal (sometimes > Rs. 10000) in local market, which is remunerative among rainfed kharif crops (Table 5).

Table 5. Comparative price of Kharif (Rainfed) crops in Rajasthan (as per Daily News Paper)

No.	Crops/Commodities	Price (Rs.) per quintal
1	Til	8000 - 10000
2	MateeraBeej	5500 - 6000
3	Mung	5300 – 6000
4	Guar	4200 - 4700
5	Moth	4000 – 4500
6	Jowar	1600 – 2000
7	Bajra	1250 - 1400
8	Maize	1200 - 1350

Fig. 4. A farmer family extracting the seed from Mateera fruits

It has been very well recognized that *Mateera* (watermelon) is an integral component of rainfed mixed cropping that supports the livelihood of desert dwellers in a big way. But little attention has been paid for its commercial cultivation. Therefore, the targeted breeding program and in-depth diagnostics of quality traits of seed will help to identify factors that would boost its production. In the changing climate scenario, concerted scientific efforts on following aspects are very much required to support the arid zone economy (Mahla *et al.*, 2014).

- Development of high seed yield varieties and availability of seeds to farmers

- Biochemical analysis for detailed composition of watermelon seeds for multifarious uses viz., food, pharmaceuticals and industrial applications
- Standardization of package of practices for higher seed production
- Designing and fabrication of seed extractor for easy and quick seed extraction
- Evaluation of dry rind as livestock feed supplement to replace costly concentrate, and
- Value addition to Mateera derived products for its wider acceptability.

The cultivation of improved genotypes of seed purpose watermelon in arid region not only increases the income of the farmers but also diversifies the cropping system which leads to sustainable rainfed production. Further, higher seed production and assured supply of raw material to processing units will provide congenial environment for industries and opportunities for employment generation at the door step in the arid region.

Tumba (Colocynth)

Tumba [*Citrullus colocynthis* (L.) Schrad.] a close relative of watermelon, is a drought hardy perennial creeper originated from tropical Asia and Africa. In Indian arid zone it is found growing wild with luxuriant growth and heavy bearing in western Rajasthan and north Gujarat on sandy wastelands (Fig. 5). This xerophytic species has rich history of medicinal value and valuable oil content. Besides, stabilizing sand dunes through soil binding habit it provides oil (20 %) and protein (8 %) rich seeds (Choudhary and Solanki, 2008). In ancient times it was among the oils permitted to be used for candle light (Palevitch and Yaniv, 1991). Its oil resembles in composition with sunflower and is being used as cooking oil in Africa and Middle Eastern countries (Schafferman *et al.*, 1998) and for human consumption (Dane *et al.*, 2006). The palmitic and stearic acid constituted the two principal saturated fatty acids while linoleic and oleic acids are main (77 %) unsaturated fatty acids in tumba oil. Since, a large area (51,434 km^2) is under wasteland category (Culturable or otherwise) in six arid zone districts (Barmer, Jaisalmer, Bikaner, Churu, Jodhpur and Jalore),

Fig. 5. Tumba fruits on vine on wasteland (left) and collected tumba fruits (right)

it can be utilized and managed through systematic *Tumba* cultivation.This leads to greening of desert as well as stabilization of rolling and shifting sand dunes besides providing livelihood support and employment to local inhabitants.It has a potential to produce around one million ton of oil rich seeds from the arid districts of Rajasthan (Joshi and Solanki, 1995).

Uses

Tumba seeds are mixed with bajra, floured and eaten by rural people for saving life in famine periods (Yadav and Singh, 1992).The colocynth fruit after cutting into small pieces is given to cattle, sheep and goats. In Western Rajasthan (Barmer, Bikaner and Jaisalmer) fruits harvested for seed collection are major source of income from wasteland. Oil extracted from Tumba seed is used for manufacturing soap, candles etc. During biblical times, fruits of tumba were gathered and considered as deadly poison. The pulp contains glucosides, such as colocynthin that is considered as a drastic hydragugue and laxative (Dafini *et al.*, 1984). The fruits have been commonly used as catharsis and antidiabetic in traditional medicines. Its seed has been used in treating malaria (Ali *et al.*, 2004). In Iran, it has been used to treat constipation, oedema, bacterial infections, cancer and diabetes and as an abortifacient (Madari and Jacobs, 2004). However, its toxicity has been reported on liver cells which may induce hepatocyte necrosis and liver fibrosis (Dehghani and Panjehshahin, 2006).

Keeping in view its immense potential from sandy wastelands where any vegetation hardly grows efforts are being made for increasing seed yield of Tumba at Agricultural Research Station, Agricultural University Mandor, Jodhpur and one variety 'RMT-59' has been developed for general cultivation in Rajasthan and Gujarat. Further, efforts are being made to obtain perennial elite material with thin testa so as to increase Tumba germination from 10-15 per cent to 85-90 per cent and seed oil content from 18-22 per cent to 30-35 per cent. Interspecific hybridization between *C. colocynthis* and *C. lanatus* (Matira as well as Kalingada) and intra specific crosses between Indian (Tumba) and Nigerian (Egusi) colocynth did provide elite material presently under advanced generations.

Amaranth grain

Amaranth belongs to family Amaranthaceae, sub-family Amaranthoideae and genus *Amaranthus* (Sauer, 1967). The name – amaranth signifies in Greek "immortal", "everlasting" or "non wilting". The genus *Amaranthus* includes approximately 60 species, most of which are cosmopolitan weeds (*A. retroflexus* L., *A. hybridus* L., *A. powellii* S. Watt., *A. spinosus* L.) and some are being cultivated and used as food grain, leafy vegetables, forage and ornamentals. Flour, starch, bran and oil can be produced from its seed. The people of India, China and Central Africa use separate forms of amaranth as vegetable crops, primarily for preparing salads rich in carotene, vitamin C, iron, calcium and microelements (Zheleznov *et al.*, 1997).

The grain amaranths are presently receiving a lot of research attention because of high lysine content, wide adaptability to various ecological zones and due to some anatomical characteristics of C4-photosynthesis pathway (Gupta and Gudu, 1990). It is now grown as a minor grain crop in parts of Central and South America and India and considered as a pseudo-cereal which contains 2-3 times more lysine than wheat, rice, ragi and jowar (Table 6). The grain is nutritious food especially for growing children, older people and for pregnant and lactating mothers.

The three principal species considered for grain production include:

1. *Amaranthus hypochondriacus* L. (sin. *A. leucocarpus* S. Watts, *A. frumentaceous*) - prince's feather;
2. *A. cruentus* L. sin. *A. paniculatus*L. - bush greens, red amaranth and
3. *A. caudatus* L. of two subspecies: subsp. *caudatus*; and subsp. *Mantegazzianus* Passerini syn.: *A. edulis* Spagazzini, named love-lie bleeding and Inca wheat, respectively.

Fig. 6. Variability for morphological traits in grain amaranthus.

These species can be distinguished by inflorescence type and some characters of pistillate flower structures: bract length, sharpness and position, as well as tepal and utricle shape, leaf morphology and gross plant morphology (Fig. 6). The colour of the seed in amaranth varies from white, gold, brown and pink to black.

Table 6. Nutritive value of Amaranth grain in comparison to other cereals (per 100 g basis).

No.	Constituents	Amaranth	Rice	Wheat	Maize
1	Energy (kcal)	365-370	330 - 350	370-390	390-410
2	Protein (g)	11.0-18.0	6.8	14.0	10.0
3	Fat (g)	7.0-8.5	0.5	1.7-2.5	4.0-5.0
4	Carbohydrate (g)	65.0-70.0	78.2	75.0-77.0	80.0-82.0
5	Crude fibre (g)	6.5-9.0	0.6	2.0-3.0	3.0-4.0
6	Ash (g)	2.8-3.5	0.6	1.9-2.2	1.5-1.8
7	Calcium (mg)	160-212	10	33-35	5.0-7.0
8	Iron (mg)	7.0-15.0	3.2	2.5-3.3	2.0-3.0
9	Lysine (mg/g)	301	230	170	

(*Sources:* Bressani, 1993; Caselato-Sousa and Amaya-Farfan, 2012; Barba *et al.*, 2009; Valcarcel-Yamaniet *al.* 2012)

Keeping in view the nutritional status and wider adaptability a number of varieties of amaranth have been developed under all India

Table 7. Details of grain amaranthus varieties released under AICRN on Underutilized crops

Variety	Year of release	Average yield (q/ha)	Characteristics	Recommended areas
1. Anna purna	1984	22.5	High yield potential, high protein (15 %), drought tolerant and wider adaptability	Mid and high Himalayan region
2. GA – 1	1991	19.5	High seed yield and drought tolerant	Gujarat, Maharashtra
3. Suvarna	1992	16.0	High seed yield and drought tolerant	Peninsular region (Karnataka, Orissa, Gujarat)
4. PRA – 1	1997	14.5	High grain yield potential	Uttranchal hills
5. PRA – 2	2001	14.5	High grain yield potential	North-west Himalayan region (except J&K)
6. GA – 2	2002	15.5	High grain yield	Gujarat
7. PRA – 3	2003	16.5	High grain yield potential	North-west Himalayan region (except J&K)
8. Durga (IC – 35407)	2006	21.0	High grain yield and early maturity	North west hill zone comprising states of Himachal Pradesh, Uttranchal and J&K
9. Kapilasa (BGA – 2)	2006	13.3	High grain yield and early maturity	Karnataka, Orissa and Tamil Nadu
10. VL Chua 44	2006	13.2	Early maturity, escapes from leaf webber and has no spiny bract for easy threshability	Mid and higher hills of Uttranchal
11. GA – 3 (SKNA 21)	2008	12.6	High grain yield	Gujarat and Jharkhand
12. RMA – 4	2008	13.9	High grain yield	Rajasthan, Jharkhand and Orissa
13. RMA - 7	2010	14.7	High grain yield	Rajasthan, Gujarat,

Research Network on Underutilized Crops for different climatic conditions (Table 7) and recommended for different agroecological conditions. The production technology for higher productivity is also standardized. Therefore, grain amaranth could be fascinating crop not only for researcher but also for nutritional and livelihood security for the growing population under resource constraint situation of arid zone.

References

Ali, Awadh A.N., Al-rahwi, K. and Lindequist, U. 2004. Some medicinal plants used in Yemeni herbal medicine to treat malaria. *African Journal of Traditional Medicine* 1: 72-76.

Baboli, Z.M. and Kordi, A.A.S. 2010. Characteristics and composition of watermelon seed oil and solvent extraction parameters effects. *Journal of American Oil Chemist Society* 87: 667-671.

Barba de la Rosa, A.P.; Fomsgaard, I.S. Larsen, B.; Mortensen, A.G.; Olvera-Martýnez, L., Silva-Sanchez, C., Mendoza-Herrera, A., Gonzalez-Castaneda J. and De Leon-Rodrýguez A. 2009. Amaranth (*Amaranthus hypochondriacus*) as an alternative crop for sustainable food production: Phenolic acids and flavonoids with potential impact on its nutraceutical quality.*Journal of Cereal Science* 49: 117–121.

Bressani, R. 1993. Amaranth. In: Encyclopaedia of Food Science, Food Technology and Nutrition.(Eds. Macrae, R., Robinson, R.K., Sadler M.J.), Vol. 1, Academic Press, London, 135-140pp.

Caselato-Sousa, V.M. and Amaya-Farfan J. 2012. State of knowledge on amaranth grain: a comprehensive review.*Journal of Food Science* 4: 93-104.

Choudhary, B.R. and Solanki, Z.S. 2008. Tumba (*Citrullus colocynthis*) – A Potential Plant of Arid Land. Agricultural Research Station, Rajasthan Agricultural University, Mandor, Jodhpur, India, 17p.

Dafini, A., Yaniv, Z. and Palevitch, D. 1984. Ethnobotanical survey of medicinal plants in northern Israel. *Journal of Pharmacology* 10: 295-310.

Dane, F., Liu, J. and Zhang, C. 2006. Phylogeography of the bitter aple, *Citrullus colocynthis*. *Genetic Resources and Crop Evolution* 54: 327-336.

Das, M., Das, S. K. and Suthar, S. H. 2002. Composition of seed and characteristics of oil from karingda (*Citrullus lanatus* (Thumb) Mansf.). *International Journal of Food Science and Technology* 37: 893"896.

Dehghani, F. and Panjehshahin, M.R. 2006. The toxic effect of alcoholic extract of *Citrulluscolocynthis* on rat liver. *Iranian Journal of Pharmacology and Therapticals* 5: 117-119.

De Mello, M.L.S., Bora, P.S. and Narain, N. 2001. Fatty and amino acids composition of melon (*Cucumismelo* var. *saccharinus*) seeds. *Journal of Food Composition and Analysis* 14: 69-74.

Dhingra, D.R. and Biswas, A.K. 1945. Component fatty acids of oil of *Citrullus vulgaris*Schrad. (water melon) seed. *Journal of Indian Chemical Society* 22: 119-122.

Edidiong, A. Essien and Ubong, M. Eduok. 2013. Chemical analysis of *Citrullus lanatus* seed oil obtained from Southern Nigeria. *Elixir Organic Chemistry* 54: 12700-12703.

El-Adawy, Tarek A. and Taha Khaled, M. 2001. Characteristic and composition of watermelon, pumpkin and paprika seed oil and flours. *Journal of Agriculture and Food Chemistry* 49: 1253-1259.

El Tahir, I.M. and TahaYousif, M. 2004. Indigenous melons (*Cucumismelo*L.) in Sudan: a review of their genetic resources and prospects for use as sources of disease and insect resistance. *Plant Genetic Resources Newsletter* 138: 36-42.

Gupta, V.K. and Gudu, S. 1990. Inheritance of some morphological traits in grain amaranthus. *Euphytica* 46: 79-84.

Ibeanu, V.N, Onyechi, U.A. and Ugwuanyi, G.U. 2012. Nutrient and dietary fibreprofile of dehulled and undehulledseeds of sweet princess watermelon (*Citrullus lanatus*) consumed in Nigeria. *International Journal of Basic and Applied Sciences* 12: 249-252.

Jarret, R.L. and Levy, I.J. 2012. Oil and fatty acid contents in seed of *Citrullu-slanatus* Schrad. *Journal of Agriculture and Food Chemistry* 60: 5199–5204.

Joshi, P. and Solanki, Z.S. 1995. Genetic improvement through interspecific hybridization in genus Cirullus. Proceeding of Golden Jubilee Symposium on Genetic Reseach and Education: Current Trends and the Next Fifty Years. 116-121 pp.

Kumar,S., Goyal, S. and Parveen, F. 2003. Ethno-medico-botany of household remedies of Kolayat tehsil in Bikaner district of Rajasthan. *Indian Journal of Traditional Knowledge* 2: 35-36.

Madari, H. and Jacobs, R.S. 2004. An analysis of cytotoxic botanical formulations used in the traditional medicine of ancient Persia as abortifacient. *Journal of Natural Products* 67: 1204-1210.

Mahla, H.R. and Choudhary, B.R. 2013. Genetic diversity in seed purpose watermelon (*Citrullus lanatus*)genotypes under rainfed situations of TharDesert. *Indian Journal of Agricultural Sciences* 83: 300-303.

Mahla, H.R. and Singh J.P. 2013. Seed purpose watermelon – a livelihood security option for rainfed mixed cropping in western Rajasthan. *Den News* 15 (1): 1-3.

Mahla H R, Singh J P and Roy M M. 2014. Seed Purpose Watermelon in Arid Zone. Central Arid Zone Research Institute, Jodhpur, 44p.

Matanyaire, C.M. 1998. Sustainability of pearl millet (*Pennisetumglaucum*) productivity in northern Namibia: current situation and challenges. *South African Journal of Science* 94: 157-166.

Meagher, K. 1997. Shifting the imbalance: the impact of structural adjustment on rural-urban population movements in northern Nigeria. *Journal of Asian African Studies* 32: 81-92.

Mertz, O., Wadley, L.R. and Christensen, E.A.2005. Local land use strategies in a globalizing world: subsistence farming, cash crops and income diversification. *Agricultural Systematics* 85: 209-215.

Mlakar, S.G., Turinek, M., Jakop, M., Bavec, M. and Bavec, F. 2010. Grain amaranth as an alternative and perspective crop in temperate climate. *Journal for Geography* 5: 135-145.

Munisse, P., Andersen, S.B., Jensen, B.D. and Christiansen, J.L. 2011. Diversity of landraces, agricultural practices and traditional uses of watermelon (*Citrullus lanatus*) in Mozambique. *African Journal of Plant Science* 5: 75-86.

Nolte,A.J. and Loesecke,H.W. 1939. Characteristics and composition of watermelon seed oil (Cuban Queen variety). *Journal of American Chemists Society* 61: 889-891.

Palevitch, D. and Z. Yaniv. 1991. Medicinal plants of the Holyland. (in Hebrew) Tamus Modan Press, Tel Aviv, 56–58p.

Paroda, R.S. 1988. The need for life support species: An Indian Perspective. In: Life Support Plant Species – Diversity and Conservation. (Eds. R.S. Paroda, P. Kappor, R.K. Arora and Bhagmal), NBPR, New Delhi, 190 p.

Raiger, H.L., Phogat, B.S., Singh, M.C. and Bansal, K.C. 2014. Annual Report 2013. All India Coordinated Research Network on Underutilized Crops, NBPGR, New Delhi, 401 p.

Raziq, S., F. Anwar, Z. Mahmood, S.A. Shahid and R. Nadeem. 2012. Characterization of seed oils from different varieties of watermelon [*Citrullus lanatus* (Thunb.)] from Pakistan. *Grasasy Aceites* 63: 365-372.

Sabahelkhier, M.K., Ishag, K.E.A. and Sabir Ali,A.K. 2011. Fatty acid profile, ash composition and oil characteristics of seeds of watermelon grown in Sudan. *British Journal of Science* 1: 76-80.

Samia, El-Safy F., Salem Rabab, H. and Abd El-Ghany, M.E. 2012. Chemical and nutritional evaluation of different seed flours as novel sources of protein. *World Journal of Dairy and Food Sciences* 7: 59-65.

Sao, T.C. M. and Potts, W. M. 1952. The analysis and characterization of the oil from the seed of *Citrullus vulgaris*. *Journal of American Oil Chemist Society* 29: 444-445.

Sauer, J. D. 1967. The grain amaranths and their relatives: a revised taxonomic and geographic survey. *Annals of Missouri Botanical Garden* 54: 103-137.

Schafferman D., Beharav, A., Shabelsky, E. and Yaniv, Z. 1998. Evaluation of *Citrullus colocynthis*, a desert plant native in Israel, as a potential source of edible oil. *Journal of Arid Environment* 40: 431-439.

Singh, Dalbir. 1964. Some useful cucurbits for the arid zone of Rajasthan. In: *Symposium on Problems of Indian Arid Zone*, Jodhpur. Ministry of Education, Govt. of India, New Delhi, 48-50p.

Singh, J.P., Beniwal, R.K. and Yadava, N.D. 2004. Ethnobotanical importance of cucurbits in India.In: *Advances in Arid Horticulture,* Vol. I *Present Status* (Eds. Saroj, P.L, Vashishtha, B.B., and Dhandhar, D.G.). International Book Distributing Co., Lucknow. 203-219p.

Valcarcel-Yamani, B. and Caetano da Silva Lannes S. 2012. Applications of Quinoa (*Chendopoium quinoa* Willd.) and Amaranth (Amaranthus spp.) and their Influence in the Nutritional Value of Cereal Based Foods, In: Food and Public Health 2, 265-275 p.

Yadav, N.D. and Singh, P.M. 1992. Tumba and Mateera Cultivation in Indian Arid Zone. Central Arid Zone Research Institute, Jodhpur, 21p.

Zheleznov, A.V., Solonenko, L.P. and Zheleznova, N.B. 1997. Seed proteins of the wild and the cultivated *Amaranthus* species. *Euphytica* 97: 177-182.

❑❑❑

Chapter 11

Sustainable Management of Rangeland of Hot Arid Region with Special Reference to Climate Change

J.P. SINGH[1] AND V.S. RATHORE*

[1] ICAR-Central Arid Zone Research Institute, Regional Research Station, Jaisalmer

* ICAR-Central Arid Zone Research Institute, Regional Research Station, Bikaner

E-mail: singh_jpbkn@yahoo.com

Introduction

It is well accepted that atmospheric changes have occurred worldwide, including increased CO_2 (Diaz *et al.*, 1998; IPCC, 2001), increased temperatures and changed precipitation patterns (Smit and Yunlong, 1996; IPCC, 2001). Increasing atmospheric concentration of carbon-di-oxide (CO_2) and other greenhouse gases (GHGs) are increasing air temperatures and changing precipitation patterns globally (Hatfield *et al.*, 2008,2011). There is largely consensus, that over the next 30 – 50 years, CO_2 will increase beyond 450 µ molmol^{-1}, global mean temperature will increase by at least 0.8 to 1.0 °C, and precipitation will become more variable (IPCC, 2007a). The climate change is increasingly recognized as one of the greatest challenges to humankind and all other life on the earth (Fitzpatric *et al.*, 2008). It is projected that, by the end of the century, climate change and its impacts may be the most important causes of biodiversity loss and changes in the ecosystem services at global level (Phillip *et al.*, 2006).

Rangelands occupy ~ 51 % of the terrestrial surface of the earth (Lean *et al.*, 1990, Prentice *et al.*, 1992), supports approximately 50 % of the world's livestock, provide forage for both domestic and wildlife population (Briske and Heitshmidt, 1991), contains 36 % of the total biomass carbon (Solomon *et al.*, 1993), and sustain millions of people. Thus, rangelands provide a multitude of biological and social benefits that are a key component of global sustainability (Reynolds *et al.*, 2007; Munasinghe, 2009). The future of products and services provided by rangelands is uncertain as are impacts expected from climate change. Low and variable precipitation, extreme temperatures and high evaporative demands render most rangelands inherently vulnerable to degradation under climate change (Reynolds *et al.*, 2007; Korner, 2000; Le Houérou, 1996; Sala *et al.*, 2000). Thus, understanding climate change impacts on rangelands are fundamental for sustainability of goods and services provided by these ecosystems.

There have been relatively few studies of climate change effects on pasture and rangelands systems compared to those on crop production systems, despite the economic and ecological importance of former. Furthermore, the complex characters of pasture and rangelands comprised of multiple interacting plants (annual and perennial) and animal species challenge our understanding of response of these systems to climate change (Izaurralde *et al.*, 2013).

Response of Rangelands to Elevated CO_2 and Climate Change

Primary production

Majority of research on plant response to elevated CO_2 has confirmed positive effects in terms of photosynthesis, biomass and yield (Kimball *et al.*, 2002; Ainsworth and Long, 2005; Hatfield *et al.*, 2011). On average, a doubling of CO_2 increases reproductive yield by 30 % in C_3 species and by 10 % in C_4 species (Hatfield *et al.*, 2011). Results of majority of experiments showed that elevated CO_2 increase photosynthesis, water use efficiency, productivity and nitrogen-use efficiency and decrease stomatal conductance, transpiration, and whole-plant respiration (Kimball *et al.*, 1990, Bazaz and McConnaughay 1992, Zak *et al.*, 1993, Rogers *et al.*, 1994, Morgan *et al.*, 2004). Various environmental factors *i.e.* temperature, precipitation and nutrient availability interact with CO_2 thereby leading to lower than expected plant response under some environmental combinations such as increased temperature and decreased soil moisture (Amthor, 2001; Long *et al.*, 2006). Net primary productivity response to elevated CO_2 in most rangeland depends on precipitation and soil water availability (Izaurralde *et al.*, 2011). Beneficial effects of elevated CO_2 on productivity do not consistently extended to arid regions. For instance, Smith *et al.* (2000) reported that elevated CO_2 increased biomass production in Mojave Desert, USA during a wet year, having precipitation 2.4 times higher compared to long-term average. These findings suggest that minimum level of precipitation or soil water availability exists below which CO_2 has little influence on productivity (Morgan *et al.*, 2004; Nowak *et al.*, 2004). The major part of precipitation is lost through evaporation rather than in transpiration in arid environments due to high evaporative demand of atmosphere, sparse vegetation and vegetation often must recovered from extended period of drought in order to appropriate water for growth and most precipitations are small (Sala and Lauenroth, 1982; Wythers *et al.*,1999). Therefore, the elevated CO_2 have negligible effects in arid ecosystems. Furthermore, low N availability may reduce or even eliminate benefit of CO_2 enrichment on plant growth and production (Owensby *et al.*, 1994; Reich *et al.*, 2006). Soil water content has indirect effect in stimulating decomposition, mineralization, nutrient cycling and mass flow of elements to roots (Dijkstra *et al.*, 2008). Therefore, the positive effects of CO_2 enrichment are negligible in arid ecosystems due to water and nutrient (particularly N) deficit.

Increasing temperature may increase (growth rate and duration of plants in high latitude and altitude region is restricted due to low temperature, and the warming climate may stimulate yields in these regions) or decrease (in hot and dry region) plant productivity, depending on the current climatic conditions and soil resources availability (Hatfield *et al.*, 2011). The AR4 summarizes the impacts on grasslands for different temperature changes. Warming of 2° C suggest positive impacts on pasture and livestock productivity in humid temperate regions. By contrast, negative impacts are predicted in arid and semiarid regions (IPCC, 2007b). The higher temperature will have profound effects on phenology and length of growing season. Temperature is the primary climatic drivers that will determine phenology and length of growing season, but precipitation variability and CO_2 may cause deviation from the overall patterns expected from temperature alone (Izaurralde *et al.*, 2011).

The amount and timing of precipitation are the major determinants of community structure and function in rangelands. In rangelands ecosystems, there is strong linear relationship between above ground net primary productivity and annual precipitation (Rutherford, 1980, Le Houerou, 1984, Sala *et al.*, 1988). The rangeland ecosystem will respond to changes in precipitation; however, the nature and magnitude of response is dependent on many geographic and physiologic variables. The average global precipitation is expected to increase; its positive effects on plant production sometimes nullify or reverse by warmer temperature through heat stress and desiccation (Wang, 2005). Additionally, increased precipitation and warmer temperature may partially negate positive effects of rising CO_2 on plant WUE (Morgan *et al.*, 2004). Thus the ultimate effects of changing CO_2, temperature and precipitation on rangeland production will be determined by the interaction among these factors. Timing and seasonality of precipitation plays an important role in determining the productivity of rangelands.

Plant community species composition

Sustained increase in mean temperature results in significant change in rangeland species distribution, composition, patterns and biome distribution (Hanson *et al.*, 1993).As temperature and CO_2 levels change due to climate change, the optimal growth ranges for different species also changes, species alter their composition dynamics, and the composition of mixed grasslands changes. While future CO_2 levels may favor C_3 plants over C_4 plants, the opposite is expected under associated temperature increases. The net effects are uncertain (IPCC, 2007b). The proportion of browse in rangelands may increase due to elevated CO_2 (Morgan *et al.*, 2007). Plants with C_3 photosynthesis, forbs and legumes show a positive response to rising CO_2 and temperature; however, changes in soil water availability due to precipitation change may temper the response of these functional groups (Polley, 1997; Morgan, 2005). Deep- rooted forbs and shrubs may be at an advantage under warming and rising CO_2 because of the ability of their roots to access deep soil water (Polley *et al.*, 2000; Bond and Midgley, 2000; Morgan *et al.*, 2007).

Quality of forage

Hanson *et al.*, (1993) indicated that mean forage quality decreased under various climate change scenarios studied by modeling. Plant grown under elevated CO_2 decrease mineral concentration (except P) compared to that grown at ambient CO_2 (Owensby *et al.*, 1994) and decreased N content under elevated CO_2 has been demonstrated in number of species (Conory, 1992; Diaz *et al.*, 1993). CO_2 enrichment decrease forage quality due to increasing C: N ratio and higher concentration of unpalatable or/toxic compounds in plants (Fajer *et al.*, 1991; Johnson and Lincoln, 1991). The CO_2 enrichment and climate change will affect forage quality. Increasing CO_2 will alter forage quality, in productive grassland with ample N; forage quality may increase due to more TNC. In N-limited rangeland systems, CO_2 –induced reduction in N and increased fiber may lower quality (Izaurralde *et al.*, 2011). Increased temperatures increases lignification of plant tissues and thus reduce degradation rate and digestibility of plant (Minson, 1990).

Rangelands of Hot Arid Region in India

Rangelands provide a wide range of ecosystem goods and services. In India, vast areas in arid and semi-arid tropics including the Great Indian Desert, temperate pastures and alpine meadows of the Himalaya with seasonal pattern of animal migration and vast areas of degraded forest lands depict the true nature of rangelands. Over 40 % of the land surface (121 million ha) in India serves as seasonal grazing lands besides their multiple uses. They occupy large land areas in the hot arid regions and cold montane tract of the Himalaya. In India rangelands are the backbone of the traditional pasture-based livestock rearing activity which is the mainstay of the economy of the rural often nomadic or semi-nomadic population. It is estimated that grasslands supply about 40% of the total available forage (Shankar and Gupta, 1992).

The economy of the hot arid region of Rajasthan is largely livestock-based, and animal husbandry is the major occupation of the people (Chaudhry *et al.*, 2011). Rangelands are being an important component of livestock-based animal husbandry and landscapes in Thar Desert, India. They cover~ 5.5 million ha in western Rajasthan (Kar *et al.*, 2009), and a major source of forage supply and habitat of an array of plant, animal and microbes. The majority of rangelands of the region is in a highly degraded state, especially due to unrestricted grazing by a large number of livestock, fuel wood collection (Kar *et al.*, 2009), wind erosion and land degradation (Singh and Singh, 2010).

Bhimaya and Ahuja (1969) reported that during normal years, the excellent, good, fair, poor, and very poor condition rangelands (with productivity of 2.0, 1.5, 1.0, 0.75 and 0.5Mg ha^{-1} productivity, respectively) can safely provide year round grazing to 0.25–0.30, 0.20, 0.17, 0.13, and 0.01–0.06 ACU ha^{-1} blocks, respectively. Tewari and Arya (2004) gave detailed account of degradation of rangelands in arid region of Thar Desert. The grazing pressure was 0.87 ACU ha^{-1} in 1981 against the optimum desirable density of 0.2 ACU ha^{-1} for permanent pasture and other grazing lands of the arid zone of Rajasthan (Gupta, 2000). Estimates indicated that the pressure increased to 0.96 ACU ha^{-1} in 1995 and 1.02 ACU ha^{-1} in 2001. A rough estimate shows excessive grazing pressure – five times of the normal carrying capacity in the arid regions (Shankar *et al.,* 1996). Further, rangelands in hot arid region face natural constraints like fragility, extremes of temperatures, irregular and erratic rainfall, poor soil fertility, erosion hazards and depleted vegetation cover. Rangelands are thus rendered fragile ecosystems and are in degradation stage due to combination of factors viz., large livestock population, free range grazing, lack of management practices, clearing of browse shrubs. Conversion of rangelands in western Rajasthan to arable lands and developmental activities, also affected the livestock productivity in rangeland based animal husbandry system. Moreover, climate change will also show own impact on arid rangeland productivity.

Climate variables such as high temperature, high wind velocity, low and erratic rains, and low relative humidity, often limit the productivity of arid rangelands and aggravate severity of degradation in these lands (Tewari and Arya, 2004). A study by Ragab and Prudhomme (2002) indicated that by 2050 in the Thar Desert (India–Pakistan–Afghanistan), the annual average increase in temperature ranges from 1.75 to 2.5°C (1.5 to 2.25°C in winter, 2 to 2.5 °C in summer), and annual average precipitation decrease by 5–25% in the region. The winter will have values closer to the annual average, but the summer will have a higher decrease and most of the region will see a decrease closer to 25%. Poonia and Rao (2013) analyzed long-term trends in annual rainfall and temperatures for Thar region to understand the climate changes in Thar region. They indicated that by the end of 21st century the air temperatures are likely to increase (+2.5, 2.9, 3.3, and 3.4° C at Pali, Bikaner, Jodhpur and Jaisalmer, respectively). The annual rainfall is likely to be decreased by – 40 mm at Jodhpur and increased by +21, 100, and 124 mm at Pali, Bikaner, and Jaisalmer, respectively. They reported that projected air temperature upto a rise of 4 °C by 21st century increases in evapotranspiration requirements by 9 - 23% during monsoon period and 13-47% during winter period and such increased demand in water will reduce the water and feed resources of Thar region. These studies indicate that the climate in western arid region of India will be warmer with higher evapotranspiration demand. These changes in temperature and soil water availability will influence the structure and functions of rangelands in these regions.

Rangeland Management under Climate Change Scenario

There is largely agreement regarding grazing system and climate change that rangeland response to climate change is very complex; and changing variance in the system is as important as change in changing in means of climatic variables. Climate change i.e. elevated CO_2, warming, and altered precipitation patterns will have profound influence on soil water content and plant water relations (Morgan *et al.*, 2004; IPCC, 2007b), so understanding their combined effects on the functioning of rangeland ecosystems is essential. Selection of suitable species and grazing management (controlling stocking rate as per carrying capacity of rangelands, appropriate grazing system) are the promising strategy to lessen the negative impact of future climate scenarios for sustainable rangeland management. One management option for the future is to actively change species composition of rangelands. Legume-based grassland systems may become more important in the future because legumes reduce the reliance on fertilizer inputs and improve the nutritive value of forage. Thus, selection of suitable legume species and their inclusion in grazing land in planned manner is an imperative to improve sustainability of rangelands of hot arid region.

Managing Perennial components

The perennial woody plant species having forage values are seem to be of immense importance in changing climate in hot arid region. Perennial plant species make the best use of soil and climate in areas with a long dry season (Ben Salem, 1980). Annuals are more vulnerable to climatic fluctuation and year-to-year variation in rainfall and temperatures. These plants depend on specific climatic requirement e.g. on set of rain during their establishment and development (Ben Salem and Eren, 1982). In contrast to annuals, growth of perennial does not depend on rain starting within a few days or weeks of a certain date. Furthermore, most of the perennials growing in arid region possess physiological and morphological characteristics, which assist them to withstand extended period of moisture deficit. Secondly, if fluctuations in climatic conditions adversely affect growth and development of these perennials, this does not result into complete failure as once condition become favorable; these plants are likely to resume their full potential.

Among the woody perennials, the shrubs are dominant in the hot arid region as they cover > 70 % landscape from Aravalli to northwestern limit of Indian part of Thar Desert. Singh *et al.*, (2005, 2008a, 2008b,

2009), Singh and Rathore (2010) and Rathore *et al.* (2011) gave detailed account of economic and ecological significances of shrubs in hot arid region of India. The shrubs of the region possess excellent adaptation to minimize transpiration viz., reduced transpiration surface, modification of leaves into spines, shedding of leaves and leaflessness (*Ziziphus nummularia* (Burm. f.) Wt. &Arn., *Calligonum polygonoides* L., *Capparis decidua* (Forsk.) Edgew.), canopy orientation to minimize radiation load (*C. decidua*) and have extensive and deep root system (*Acacia jacquemontii* Benth., *Z. nummularia, C. polygonoides, Haloxylon recurvum*) that helps to extract the moisture and nutrients from greater volume of rhizosphere. Some shrubs viz., *Salsola baryosma* (Roem. et Schult.) Dandy, *Suaeda fruticosa* (L.) Forsk., *Haloxylon recurvum* (Moq.) Bunge ex Boiss. etc. have ability to survive under saline conditions. Along with these morphological adaptations, these species possess considerable plasticity in phenology, which add to their survival under warmer and drier climatic conditions. Browse shrubs such as *Acacia jacquemontii, Calligonum polygonoides, Cordia gharaf* (Forsk.) Ehrenb. & Aschers., *Grewia tenax* (Forsk.) Fiori, *Haloxylon salicornicum, Heliotropium parviflorum* Stocks,*Indigofera oblongifolia* Forsk., *Ziziphus nummularia* are promising species in this area. In degraded lands of hot arid region, planting of *H. salcornicum* and *C. polygonoides* shrubs alter physical soil traits, enhance soil nutrients, and facilitates the development of herbaceous plant communities (Rathore *et al.* 2015). More emphasis is therefore required to introduce and popularize indigenous rangeland shrubs, which constitute a vital component in sustainability of hot arid rangelands. Optimization of species – specific density of perennial is key for sustainable rangeland production.

Acacia nilotica (L.) Del., *A. senegal* (L.) Willd.,*Prosopis cineraria* (L.) Druce, *Salvadora oleiodes* Decne, *Tecomella undulata* (Sm.) Seemetc.are important multipurpose trees for silvi-pasture and rangeland development in the Thar Desert.They increase rangeland productivity and provide nutritious top feed (as leaves and pods which are rich in proteins and minerals) to livestock during the lean periods. Furthermore, trees planted on boundary of rangeland, and along approach road of grazing blocks acts as shelterbelts which improve microclimatic conditions and hence increase productivity. Ahuja (1984) reported that 50 trees ha^{1} is suitable in pastureland.

Grazing Management

Adequate protection and grazing based on carrying capacity is an important strategy to sustainable rangeland production of hot arid region in changing climate. Ahuja (1977) reported that forage yield of grasslands of Thar Desert increased by 92-148 % with adequate protection and grazing as per carrying capacity of these lands. A study conducted by CAZRI indicated that deferred rotational grazing system increased forage production by 47% over a continuous grazing system (CAZRI, 1992).

Moisture conservation

Soil and water conservation measures are imperative to increase rangeland productivity. Suitable moisture conservation techniques increased forage yield by 639 % in arid rangelands (Ahuja, 1984). Contour furrows (60.96 cm wide, 22.86 cm deep, cross section 929 sq. cm with spacing 8-10 m across slope) are suitable soil conservation option on the rangelands having shallow soil and 2-5 % slope. The contour furrow increased the forage production 4-5 times in a period of 10-15 years. Creation of staggered micro-pits with *Pitting Discer* improves the moisture retention, and creation of these micro-pits is reported to increase the forage yield by 80 -115 % in natural pasture.

Reseeding and nutrient management

Natural succession of high yielding and palatable perennial grasses in the arid regions requires longer periods. Reseeding of site-specific suitable perennial grasses (viz., *Cenchrus ciliaris* L., *C. setigerus* Vahl, *Cymbopogon jwaruncusa* (Jones) Schult., *Dichanthium annulatum* (Forsk.) Stapf, *Lasiurus sindicus* Henr., *Panicum turgidum* Forsk., *Sporobolus helvolus*) is a suitable option to increasing rangeland productivity. Low perennial grass species like *Dactyloctenium sindicum* Boiss and *Ochthochloacompressa* (Forsk.) Hilu are also important for arid rangelands. Studies conducted by CAZRI indicated that reseeding increased forage yield by 30 to 122% in "poor" and 29 to 107% in "fair" class rangelands after 3–5 years of reseeding (CAZRI, 1992). Pre-monsoon sowing of grass seeds give 36% higher forage yield over monsoon sowing. Fertilization may also be considered as a mean to improve rangeland productivity. Application of 22.5 kgNha^{-1} during normal year resulted in additional increase in forage yield by 20 to 70% (CAZRI, 1983). Fertilizer application increased crude protein yield of grasses (26, 58, 82 and 108 % in *P. antidotale, C. ciliaris, L. sindicus* and *C. setigerus*, respectively). Moreover, rangeland forage legumes viz., *Indigofera argentea* Burm. f., *I. cordifolia* Heyne ex Roth., *I. sessiliflora* DC., *Tephrosia uniflora* Pers. are important in hot arid region for small ruminants.

System diversification

Diversified crop–livestock or integrated crop-livestock farming system is an attractive option both to reduce environmental impacts and to improve the profitability and sustainability of livestock production system, enhancing their ecological resilience to climate stressors (Russelle *et al.*, 2007). These diversified production systems are likely to remain more productive and retain greater biological diversity in an increasingly variable climate; and thus better adapt to climate change (Izaurralde *et al.*, 2011).

Rangelands are large carbon sinks and have up to 30% terrestrial carbon stocks (Schuman *et al.*, 2002; Neely *et al.*, 2009). Rangelands have the potential to remove 198 million tons of CO_2 from the atmosphere per year for 30 years. The C-sequestration by rangelands is reported to environmentally friendly, inexpensive and natural process of mitigating climate change (Lal, 2011). Changing climate especially drought, may impact greatly on rangelands SOC pools, changing them from sinks to emission sources because of the xeric nature of soils which directly affect the photosynthetic rates than respiration rates (Balogh *et al.*, 2005). Improved understanding of the processes and mechanisms that affect SOC dynamics on arid and mesic environments, coupled with land management and technology must be the focus for rangeland soils to sequester carbon. The proposed options for C-sequestering emissions includes application of appropriate farming systems, suitable land use through implementation of restorative perennial systems, introduction of species with wide adoption capability, application multiple ecosystem services, conservation of soil, water and nutrients; agroforestry and soil amendments technologies and is prescribed to increase the soil, ecosystem C pools and their residence time (Lal, 2011).

The proposed synthesis for sustainable rangeland management in hot arid region involves selecting species having wider adaptability to warmer and drier environment, grazing management, protection, reseeding of palatable grasses, soil and water conservation techniques, nutrient management and diversified crop-livestock production systems would results in lessen the negative impacts of anticipated climate change and variability.

Rangelands are important component of hot desert landscapes. They improve soil fertility, combat soil erosion, protect water tables, enhance water infiltration in soil and store carbon in soils mitigating climate change effects. Compared to arable farming, rangeland based animal husbandry has positive impacts on the desert environment and provide various ecosystem services and goods. Rangeland based livestock production is very important in Thar Deseret particularly for sheep, goats and camel rearing. Grazing based animal husbandry is not only an economic activity in the desert but also an important component of cultural heritage. Existence of the *Gochars* (grasslands)/ *Orans* (sacred woodlands) in the Thar Desert linked to the villages are the testimony of rich cultural heritage. They are the important forage resource base for livestock in the region. Such resources should also be viewed from the point of biodiversity reserves in the region, with a proper regulatory framework from their utilization aspects.

Although rangelands in hot arid region largely found in degraded state due to combination of environmental and socio-economic factors, they still remain the major forage resource base significantly contributing to the animal husbandry sector. Hot arid rangeland shrubs due to their morpho-physiological characteristics have very important role in climate change scenario. Species like *Haloxylon salicornicum* should be considered as a potential shrub adapted to a future climate warmer and drier than to-day's due to its remarkable abiotic stresses (drought, high temperature) tolerance (Singh *et al.*, 2015). If, somehow, with available technologies and sound management these rangelands are brought back to their potential production level, the gap between forage supply and demand can be at least leveled up in view of climate change scenario. Finally, scientific research and technological innovations are further needed for sustainable utilization and conservation of rangeland biodiversity.

References

Ahuja, L. D. 1977. Improving rangeland productivity. In *Desertification and its Control*. Indian Council of Agricultural Research (ICAR), New Delhi, India. pp. 203–214.

Ahuja, L. D. 1984. Range management in agro-forestry system in arid regions of India. In *Agro-forestry in Arid and Semi-arid Zones* (Ed. K.A. Shankarnarayan). Central Arid Zone Research Institute (CAZRI) publication No. 24.CAZRI, Jodhpur, India. pp. 161–166.

Ainsworth, E.A. and Long, S.P. 2005. What have we learned from 15 years of free-air CO_2 enrichment (FACE)? A meta-analytic review of the responses of photosynthesis, canopy properties and plant production to rising CO_2. *New Phytologist* 165: 351–371.

Amthor, J. 2001. Effects of atmospheric CO_2 concentration on wheat yield: Review of results from experiments using various approaches to control CO2 concentration. *Field Crops Research* 73: 1–34.

Balogh, J.S., Czobel, S., Foti, Z., Nagy, O. and Szirmai, E. 2005. The influence of drought on carbon balance in loess grassland.*Cereal Research Communication* 33: 149-152.

Ben Salem, B. and T. M. Eren. 1982. Forestry in sandy world. *Unasylva* 34: 8-12.

Ben Salem, B. 1980. Arid zone forestry: where there are no forests and everything depends on trees. *Unasylva* 32 : 16-18.

Bhimaya, C. P. and L. D. Ahuja. 1969. Criteria for determining condition class rangelands in western Rajasthan. *Annals of Arid Zone* 8:73–79.

Bond, W.J. and Midgley, G.F. 2000. A proposed CO_2–controlled mechanism of woody plant invasion in grasslands and savannas. *Global Change Biology* 6: 865–869.

Briske, D.D. and Heitschmidt, R.K. 1991. An ecological perspective.In *Grazing Management: An Ecological Perspective* (Eds. R.K. Heitschmidt and J.W. Stuth). Timber Press Inc., Portland, OR, pp. 11-27.

CAZRI. 1983. *Progress in Arid Zone Research.* Central Arid Zone Research Institute (CAZRI), Jodhpur, India.

CAZRI. 1992. *Forage 2000AD: The scenario for arid Rajasthan.* Central Arid Zone Research Institute (CAZRI), Jodhpur, India.

Chaudhry, P., Bohra, N.K. and Choudhary, K.R. 2011.Conserving biodiversity of community forests and rangelands of a hot arid region of India. *Land Use Policy* 28: 506 -513.

Conory, J.P. 1992. Influence of elevated atmospheric CO_2 concentrations on plant nutrition. *Australian Journal of Botany* 40: 445-456.

Diaz, S., Fraser, L.H., Grime, J.P. and Falczuk, V. 1998. The Impact of elevated CO_2 on plant- herbivore interactions: Experimental Evidence of Moderating Effects at the Community Level. *Oecologia* 117: 177–186.

Díaz, S., Grime, J.P., Harris J. and McPherson, E.1993. Evidence of a feedback mechanism limiting plant response to elevated carbon dioxide. *Nature* 364: 616-617.

Dijkstra, F.A., Pendall, E., Mosier, A.R., King, J.Y., Milchunas, D.G. and Morgan, J.A. 2008. Long-term enhancement of N availability and plant growth under elevated CO_2 in a semi-arid grassland. *Functional Ecology* 22: 975–982.

Fajer, E.D., Bowers, M.D. and Bazzaz, F.A. 1991. The effects of enriched CO_2 atmospheres on the buckeye butterfly Junoniacoenia. *Ecology* 72: 751-754.

Fitzpatric, M.C., Gove, A.D., Sanders, N.J. and Dunn, R.R. 2008. Climate change, plant migration, and range collapse in a global biodiversity hotspot: the Banksia (Proteaceae) of Western Australia. *Global Change Biology* 14: 1337–1352.

Gupta, J. P. 2000. Technology approach for greening degraded arid lands. Central Arid Zone Research Institute (CAZRI), Jodhpur, India.

Hanson, J.G., Baker, B.B. and Bourdon, R.M. 1993. Comparison of the effects of different climate change scenarios on rangeland livestock production.*Agricultural Systems* 41: 487–502.

Hatfield, J.L., Boote, K.J., Fay, P., Hahn, L., Izaurralde, C., Kimball, B.A., Mader, T., Morgan, J., Ort, D., Polley, W., Thomson, A. and Wolfe, D. 2008. Agriculture.In *The effects of climate change on agriculture, land resources, water resources, and biodiversity in the United States*. U.S. Climate Change Science Program and the Subcommittee on Global Change Res., Washington, DC.

Hatfield, J.L., Boote, K.J., Kimball, B.A., Ziska, L., Izaurralde, R.C., Ort, D., Thomson, A.M. and Wolfe, D.A. 2011. Climate impacts on agriculture: Implications for crop production. *Agronomy Journal* 103: 351-370.

Intergovernmental Panel on Climate Change (IPCC): 2001. *Climate Change 2001*: The Scientific Basis. Contribution of Working Group 1 to Third Assessment Report of the IPCC, Houghton, J. and Ding, Y. (chairs), Cambridge University Press, Cambridge, 881 pp.

IPCC (Intergovernmental Panel on Climate Change), 2007b.*Climate Change 2007: Impacts, Adaptation and Vulnerability*. Summary for policy makers.<http:// www.ipcc.cg/SPM13apr07.pdf>.

IPCC.2007a. Summary for policymakers. In S. Solomon et al. (ed.) Climate change 2007: The physical science basis. Contribution of Working Group I to the Fourth Assessment Report of the Intergovernmental Panel on Climate Change. Cambridge Univ. Press, New York. p 996

Izaurralde R.C., Thompson A.M., Morgan J.A., Fay P.A., Polley H.W. and Hatfield J.L. 2011. Climate impacts on agriculture: Implications for forage and rangeland production. *Agronomy Journal* 103 (2): 371 – 381.

Johnson, R.H. and Lincoln, D.E. 1991. Carbon allocation patterns and grasshopper nutrition – the influence of CO_2 enrichment and soil mineral limitation.*Oecologia* 87: 127-134.

Kar A., Moharana P.C., Raina P., Kumar Mahesh, Soni M.L., Santra P., Ajai, Arya A.S. and Dhinwa P.S. 2009. Desertification and its control measures.In ; Trends in Arid Zone Research in India (Eds. A. Kar, B.K. Garg, M.P. Singh, and S. Kathju) Central Arid Zone Research Institute, Jodhpur, pp.1-48.

Kimball, B.A., Kobayashi, K. and Bindi, M. 2002. Responses of agricultural crops to free-air CO_2 enrichment. *Advances in Agronomy* 77:293–368.

Kimball, B.A., Rosenberg, N.J. and L.H. Allen, Jr. (Eds.). 1990. *Impact of Carbon Dioxide, Trace Gases, and Climate Change on Global Agriculture.* ASA Spec. Publ. No. 53, American Society of Agronomy, Crop Science Society of America, and Soil Science Society of America, Madison, WI, 133 pp.

Korner, C. 2000. Biosphere responses to CO_2 enrichment.*Ecological Applications* 10:1590–1619.

Lal, R. 2011. Sequestering carbon in soils of agro-ecosystems.*Food Policy* 36: S33-S39.

Le Houerou, H.N. 1984. Rain use efficiency: a unifying concept in arid-land ecology. *Journal of Arid Environments* 7: 213-247.

Lean, G., Hinrichsen, D. and Markham, A. 1990. *Atlas of the Environment.*Prentice Hall Press, New York, NY, 192 pp.

Long, S.P., Ainsworth, E.A., Leakey, A.D.B., Nosberger, J. and Ort, D.R. 2006. Food for thought: Lower-than-expected crop yield stimulation with rising CO_2 concentrations. *Science* (Washington, DC) 312: 1918–1921.

Minson, D.J. 1990. *Forage in Ruminant Nutrition.* Academic Press, San Diego.

Morgan, J. A., Pataki, D. E., Korner, C., Clark, H., Del Grosso, S. J., Grunzweig, J. M., Knapp, A. K., Mosier, A. R., Newton, P. C. D., Niklaus, P. A., Nippert, J. B., Nowak, R. S., Parton, W. J., Polley, H.W. and Shaw, M. R. 2004. Water relations in grassland and desert ecosystems exposed to elevated atmospheric CO_2 *Oecologia* 140: 11–25.

Morgan, J.A. 2005. Rising atmospheric CO_2 and global climate change: Management implications for grazing lands. In *Grasslands: Developments opportunities perspectives* (Eds.S.G. Reynolds and J. Frame) Science Publ., Enfield, NH. p. 245–272.

Morgan, J.A., Milchunas, D.G., LeCain, E.D.R., West, M.S. and Mosier, A.R. 2007. Carbon dioxide enrichment alters plant community structure and accelerates shrub growth in the shortgrass steppe. *Proceedings of the National Academy of Sciences, USA* 104: 14724–14729.

Munasinghe, M. 2009. *Sustainable development in practice: sustainomics methodology and applications.* Cambridge University Press, Cambridge.

Neely, C., Bunning, S., and Wilkes, A. (edn) 2009. Review of evidence on drylands pastoral systems and climate change: Implications and opportunities for mitigation and adaptation. FAO, Rome, Italy.

Nowak, R. S., Ellsworth, D.S. and Smith, S.D. 2004. Functional responses of plants to elevated atmospheric CO_2 – do photosynthetic and productivity data from FACE experiments support early predictions? *New Phytologist* 162: 253–280.

Owensby, C. E., Auen, L.M. and Coyne, P. I. 1994. Biomass production in a nitrogen-fertilized, tallgrass prairie ecosystem exposed to ambient and elevated levels of CO_2. *Plant and Soil* 165: 105–113.

Phillip, S.J., Anderson, R.P. and Schapire, R.E. 2006. Maximum entropy modeling of species geographic distributions. *Ecological Modelling* 190:231–259.

Polley, H.W. 1997. Implications of rising atmospheric carbon di oxide for rangelands.*Journal of Range Management* 50: 561-577.

Polley, H.W., Morgan, J.A., Campbell, B.D. and Stafford Smith, M. 2000. Crop ecosystem responses to climatic change: Rangelands. In *Climate change and global crop productivity* (Eds. K.R. Reddy and H.F. Hodges).CABI, Wallingford, Oxon, UK. . p. 293–314.

Poonia, S. and Rao A.S. 2013.Climate change and its impact on Thar Desert ecosystem.*Journal of Agricultural Physics* 13(1): 71 – 79.

Prentice, K.C., Cramer, W., Harrison, S.P., Leemans, R., Monserud, R.A. and Solomon, A.M. 1992. A global biome model based on plant physiology and dominance, soil properties, and climate.*Journal of Biogeography* 19: 117-134.

Ragab, R. and C. Prudhomme. 2002. SW-Soil and Water: Climate change and water resources management in arid and semi-arid regions: Prospective and challenges for the 21st Century. *Biosystems Engineering* 81:3–34.

Rathore, V.S., Singh J.P. andRoy M.M. 2011. Shrubs of hot arid Rajasthan: economic and ecological imperatives – a review. *Range Management and Agroforestry*, 32(2): 71-78.

Rathore, V.S., Singh, J.P., Bhardwaj, S., Nathawat, N.S., Mahesh Kumar and Roy,M.M. 2015. Potential of native shrubs *Haloxylon salicornicum* and *Calligonum polygonopides* for restoration of degraded lands in arid western Rajasthan, India.*Environmental Management* 55: 205-219.

Reich, P.B., Hobbie, S.E., Lee, T., Ellsworth, D. S., West, J. B., Tilman, D., Knops, J.M. H., Naeem, S. and Trost, J. 2006. Nitrogen limitation constrains sustainability of ecosystem response to CO_2. *Nature* 440: 922–925.

Reynolds, J.F., Smith, D.M.S., Lambin, E.F., Turner, B.L., Mortimore, M., Batterbury, S.P.J., Downing, T.E., Dowlatabadi, H., Fernandez, R.J., Herrick, J.E., Huber-Sannwald, E., Jiang, H., Leemans, R, Lynam T, Maestre FT, Ayarza M and Walker B.2007. Global desertification: Building a science for dryland development. *Science* 316: 847–851.

Rogers, H.H., Runion, G.B. and Drupa, S.V. 1994. Plant responses to atmospheric CO2 enrichment with emphasis on roots and the rhizosphere. *Environmental Pollution* 83: 155-189.

Russelle, M.P., Entz M.H. and Franzluebbers A.J. 2007. Reconsidering integrated crop-livestock systems in North America. *Agronomy Journal* 99: 325-334.

Rutherford, M.C. 1980. Annual plant production-precipitation relations in arid and semiarid regions.*South African Journal of Science* 76: 53-56.

Sala, O.E., Chapin, F.S., Armesto, J.J., Berlow, E., Bloomfield, J., Dirzo, R., Huber-Sanwald, E., Huenneke, L.F., Jackson, R.B., Kinzig, A., Leemans, R., Lodge, D.M., Mooney, H. A., Oesterheld, M., Poff, N.L., Sykes, M.T., Walker, B.H., Walker, M. and Wall,D.H. 2000. Biodiversity - Global biodiversity scenarios for the year 2100. *Science* 287: 1770–1774.

Sala, O.E. and Lauenroth, W.K. 1982. Small rain events: an ecological role in semi-arid regions. *Oecologia* 53: 301-304.

Sala, O.E., Parton, W.J., Joyce, L.A. and Lauenroth, W.K. 1988. Primary production of the central grasslands region of the United States. *Ecology* 69: 40-45.

Schuman, G.E., Janzen H.H. and Herrick J.E. 2002. Soil carbon dynamics and potential carbon sequestration by rangelands.*Environmental Pollution* 116: 391-396.

Shankar, V. and Gupta, J.N. 1992. Restoration of degraded rangelands. pp. 115-155. In *Restoration of Degraded Lands: Concepts and Strategies* (Ed. J.S. Singh).Rastogi Publications. Meerut, India.

Shankar, V., Gupta,J.N. and Singh, J.P. 1996. Forage Production technology on rangelands. Pp. 41-51. In: *Transfer of forage Production Technologies* (Eds. Bhag Mal and R.J. Hagger). Indo-UK Forage Production Project, IGFRI, Jhansi.

Singh, G., and Singh, B. 2010. Assessment of growth and biomass production of *Cenchrussetigerus* based silvipastoral system in community pasture land in Bhilwara district of Rajasthan. *Indian Forester* 136, 898-909.

Singh, J.P., Mathur B.K. and Rathore V.S. 2009. Fodder potential of Lana (*Haloxylon salicornicum*) in hot arid region.*Range Management and Agroforestry* 30: 34-37.

Singh, J.P. and Rathore, V.S. 2010. Role of shrubs in augmenting fodder availability and environmental restoration in hot arid region.*Range Management and Agroforestry*. Symposium issue (B): 262-263.

Singh, J.P., Rathore, V.S. and Beniwal, R.K. 2008a. Diversity and conservation of shrubs of arid region of Rajasthan, India. In: Biodiversity for Sustainable Development (Eds N.P. Singh and P.K. Chhonkar) Samskriti and ASEED, Vasant Kunj, New Delhi. pp. 51-74.

Singh, J.P., Rathore, V.S., Soni, M.L. and Mathur, B.K. 2008b. Shrubs for diversification of farming system in western Rajasthan. In *Diversification of Arid Farming Systems* (Eds. PratapNarain, M.P. Singh, Amal Kar, S. Kathju and Praveen Kumar), Arid Zone Research Association of India, Jodhpur and Scientific Publishers (India), Jodhpur. pp. 31-38.

Singh, J.P., Soni, M.L. and Rathore, V.S. 2005. Halophytic chenopod shrubs of arid zone. In *Shrubs of Indian Arid Zone* (Eds. P. Narain, M. Singh, M.S. Khan and S. Kumar) Arid Agro-ecosystem Directorate, CAZRI, Jodhpur. pp. 27-32.

Singh, J.P., Mathur,B.K. and Rathore,V.S. 2009. Fodder potential of Lana (*Haloxylon salicornicum*) in hot arid region.*Range Management & Agroforestry* 30: 34-37.

Singh, J.P., Rathore V.S. and Roy, M.M. 2015. Notes about *Haloxylon salicornicum* (Moq.) Bunge ex Boiss., a promising shrub of arid regions. *Genetic Resources and Crop Evolution* 62:451-463.

Smit, B. and Yunlong, C. 1996. Climate Change and Agriculture in China. *Global Environmental Change* 6: 205–214.

Smith, S.D., Huxman, T.E., Zitzer, S.F., Charlet, T.N., Housman, D.C., Coleman, J.S., Fenstermaker, L.K., Seemann, J.R. and Nowak, R.S. 2000. Elevated CO_2 increases productivity and invasive species success in an arid ecosystem. *Nature* (London) 408:79–82.

Solomon, A.M., Prentice, I.C., Leemans, R. and Cramer, W.P. 1993. The interaction of climate and land use in future terrestrial carbon storage and release. *Water, Air, and Soil Pollution* 70: 595-614.

Tewari , V.P. and Arya,R. 2004. Degradation of arid rangelands in Thar Desert, India: A Review. *Arid Land Research and Management* 19: 1-12.

Wang, G. 2005. Agricultural drought in a future climate: Results from 15 global climate models participating in the IPCC 4th assessment. *Climate Dynamics* 25: 739–753.

Wythers, K. R., Lauenroth, W.K. and Paruelo, J.M. 1999. Bare-soil evaporation under semi-arid field conditions.*Soil Science Society of America Journal* 63: 1341- 1349.

Zak, D.R., Pregitzer, K.S., Curtis, P.S., Teeri, J.A. ,Fogel, R. and Randlett, D.L. 1993. Elevated atmospheric CO_2 and feedback between carbon and nitrogen cycles.*Plant and Soil* 151: 105-117.

□□□

Chapter 12

Grasses for High Productivity in Hot Arid Region

M.P. RAJORA AND C.S. SHANTHARAJA

ICAR - Central Arid Zone Research Institute, Jodhpur, Rajasthan-342 003
E-mail:mahavirrajora@yahoo.co.in

Introduction

About 62% of the Thar Desert of India is situated in the arid zone of Rajasthan. Droughts occur frequently in this zone resulting in low production and productivity of grasses, ultimately resulting in shortage of fodder affecting the productivity of the livestock, which are the valuable component of farming system of arid zone. Over exploitation of natural vegetation cover has resulted in significant degradation of grazing lands resulting in reduced productivity. Development of pastures/renovation of grazing lands in the arid areas of Rajasthan may help in amelioration of the economy of arid zone farmers.

Dichanthium-Cenchrus-Lasiurus is the main grass cover located in arid region. In this grass cover, genetic diversity is depleting at faster rate due to various biotic and abiotic factors and human activities. Therefore, conservation of the germplasm which are adapted to arid climate is the need of the hour.Perennial grasses, viz. sewan (*Lasiurus sindicus* Henr.), buffel grass (*Cenchrus ciliaris* L.), birdwood grass (*Cenchrussetigerus* Vahl), blue panic (*Panicum antidotale* Retz), *Panicumturgidum, Cymbopogon* species, *etc.* are high fodder yielding species of hot arid areas of the Thar Desert. These grasses provide fodder to livestock in general and during drought years in particular, protect soil erosion and increase soil fertility. These grass species are the cheaper source of animal feed and helps in reducing the dominance of unpalatable grasses due to heavy grazing. In addition, these grasses offer many more tangible and intangible benefits to farming community and soil environment also. By keeping above points in mind, we discussed about the important perennial grasses of arid zone and their establishment techniques to achieve higher production and productivity under varying climatic conditions in this chapter.

1. Sewan (*Lasiurus sindicus* Henr.)

Sewan is an important perennial grass which is one of the dominant species of *Dichanthium-Cenchrus–Lasiurus* type grass cover of India. It is a sand loving species which survives under extreme hot arid conditions of Rajasthan and gives high forage yield. The grass is drought resistant and grow vigorously even in low rainfall regions receiving 100-300 mm rainfall annually and also has shown moderate tolerance to salinity. The species naturally grows extensively in Jaisalmer, Barmer and Bikaner districts of Rajasthan. Three hectares of the grasscan support one adult cattle throughout the year. From good condition grasslands, 150 to 250 q green fodder and 50 to 75 q dry matter can be harvested from one hectare land. Seed production depends on various factors and ranges from 5 to 40 kg ha^{-1}.

Germplasm

In arid zone, continuous droughts, over grazing and human activities led to rapid deterioration of natural genetic diversity in the species.Therefore, collection and conservation of germplasm is essential. At CAZRI, Jodhpur around 111 accessions of *L. sindicus* collected from western parts of Rajasthan are being maintained and evaluated for fodder yield and its related traits.

Nutritive value

It is highly nutritive, palatable and well relished by the livestock. The grass is main fodder for Tharparkar breed of cattle of western Rajasthan.Under natural conditions crude protein (CP) ranges from 5.9 to 6.7% but when cultivated and fertilised with NPK, CP can reach up to 15%. Sawal *et al.* (2009) reported that *L.sindicus* and *C.ciliaris* are similar in crude protein (9.94 and 9.35%), when the grasses are harvested at the time of inflorescence appearance.

Table 1. Chemical composition of feed ingredients (% DM basis)

Attribute	OM	CP	EE	CF	NFE	ADF	Cellulose	Lignin
Sewan	91.14	9.94	1.84	39.23	40.13	48.31	40.39	7.92
Cenchrus	90.35	9.35	1.86	35.20	43.94	42.26	35.12	7.14

Source: Sawal, *et al.* 2009. Indian Journal of Small Ruminants, 15 (2): 277-280.

Drought resistance

It has well-developed root-rhizomatous system and gives productive yield for years. Its roots penetrate deeper than 1.5 m, thus has capacity to draw moisture from lower strata of the soil which provides the species drought resistance.

Main Issues

The major bottlenecks in its cultivation are non-availability of seeds as the seed collection is very costly and erosion of genetic diversity.

2. Buffel grass (*Cenchrus ciliaris* L.)

This grass is popularly known as 'dhaman' in Rajasthan and 'anjan' in other parts of India. It has wider adaptability to different types of soils -it can be grown on dry sandy to stony soils in arid, sandy-to-sandy loam in semi-arid parts, deep basaltic soils and on red lateritic soils- under varied climatic conditions. It can be cultivated in areas receiving rainfall from 150-1250 mm annually. Its palatability is high when fed at young stage and decreases gradually as it mature. Crude protein varies from 6-10%. The varieties like Marwar Anjan, Bundel Anjan, Biloela and Molopo give 40 to 45 q ha^{-1} dry matter production. Depending on the soil moisture, 3 to 4 harvests of fodder can be taken from August to April. In arid areas, the pasture remains productive for 4-5 years when managed properly.

In a trial at Jodhpur, genotype CAZRI 358 yielded about 100 q green and 25 q dry matter during 2011-2013 followed by CAZRI 2221 (Table 2). Seed production was about 50-70 kg/ha, and high seed yields could be harvested even in off-season during the establishment year.Rajora and Singh (2005) reported seed yield up to 381.8 kg/ha from the winter generated growth of *C. ciliaris*.

Table 2: Performance of *C. ciliaris* genotypes under arid environment

Genotypes/varieties	GFY (q/ha)				DMY (q/ha)			
	2011	2012	2013	Mean	2011	2012	2013	Mean
CAZRI 75	63.6	63.1	97.0	74.6	14.1	17.5	31.1	20.9
CAZRI 358	101.3	101.1	99.1	100.5	22.4	25.4	26.9	24.9
CAZRI 585	38.1	101.4	58.3	65.9	9.9	25.5	15.3	16.9
CAZRI 2178	64.1	94.1	62.1	73.4	14.5	21.9	17.9	18.1
CAZRI 2221	112.9	103.6	77.2	97.9	25.7	24.1	21.9	23.9

Varieties

Variety CAZRI 75 (MarwarAnjan), a selection from exotic collection no. 14369 (Australia), notified in 1985. Erect growth habit, medium tall, long, broad, droopy and green leaves, actively growing and late flowering (Fig. 1). The yield potential is 88 q green fodder and 28 q dry matter per hectare. Some other genotypes, like CAZRI 2178, CAZRI 2221 (Fig.1), CAZRI 585, CAZRI 2170 have also been found promising in the arid zone for fodder production.

Fig. 1. *Cenchrus ciliaris* var. CAZRI 75 and genotype CAZRI 2221

Phenotypic variability of different genotypes of *C.ciliaris* was carried out at CAZRI Research farm. Among quantitative traits, more variability was recorded for flowering characters, like days to spike initiation and days to 50% flowering with the range of 33.0-46.3 and 42.7-69.3, respectively. Similarly genotypes showed significant variation for seeds per spike and the range recorded 60.3-120.2. A high correlation ($r > 0.7$) was found between plant height and leaf width; plant height and seed weight/plant, days to spike initiation and days to 50% flowering.

The initial vigor of plants varies from low to very high and CAZRI-2221 showed very high vigor which is the prime character for influencing fodder yield. The other qualitative character like, waxiness of leaf and leaf sheath was found in genotypes CAZRI-75 and IGFRI-3108, which is important for drought tolerance in arid environment.

Drought resistance

Roots penetrate deeper than 1.0 meter helping the plant to draw moisture from deeper soil levels. This trait makes the species a very drought resistant grass.

Main issues

The main issues are nonexistence of formal seed supply chain; preferential grazing; lack of awareness of production technologies.

3. Bird Wood Grass (*Cenchrus setigerus* Vahl)

It is commonly known as 'modadhaman' in Rajasthan. It tolerates heat and drought and grows in areas where rainfall as low as 200 mm annually making the species good for improvement of the grazing lands of low rainfall regions. It is relished well by all kinds of livestock as well as horses and can be propagated from seeds and may be used for renovation of over grazed pastures or establishment of new pasture. Crude protein content of dry matter varies from 5 to 12%. High yielding varieties of *C. setigerus* produce about 40 to 50 t ha^{-1} of green forage from 2 to 3 cutting during August to April in arid areas, which becomes double in semi-arid areas. The dry matter varies from 4 to 20 q ha^{-1}.

Varieties

Variety CAZRI 76 (Marwar Dhaman) has been released by CAZRI, Jodhpur (Fig. 2). It is a selection from exotic material no. 17655 received from Australia in 1957, notified by the Central Variety Release Committee vide notification number 832 (E) in 1985. Semi-erect growth habit, 50-60 cm tall plants, leaves are medium narrow, long, yellowish green and semi-erect, brown nodal pigmentation, days to 50% flowering around 45-55 days and slow senescence and gives 58 q green fodder yield form one hectare. Very much relished by the small ruminants.

4. Blue Panic (*Panicum antidotale* Retz.)

Blue panic is popularly known as 'gramna' (Fig. 2) in Rajasthan, often grows on sand dunes, an excellent sand binder and drought resistant perennial grass adapted to different kind of soils and climatic conditions. It is a widely adapted, persistent, productive and good seed yielding grass therefore, can be a good fodder source throughout the year on the livestock farms of tropical and sub-tropical countries. It can be either grazed or cut and fed to the animals. The grass is highly nutritive when cut at pre-flowering stage. The grass is palatable to all kinds of livestock. However, it is not liked at maturity stage as the stem becomes hard and woody.

Fig. 2. *Cenchrus setigerus* var. CAZRI 76 (upper) and *Panicum antidotale* (lower)

5. Murat (*Panicum turgidum* Forsk.)

Grown on sand dunes and sandy plains in open rangelands and cultivated fields of Thar Desert of Rajasthan and Gujarat. This species, often play an important role in checking the soil erosion under low rainfall (100-250 mm) condition. Green fodder productivity ranges from 50 to 75-q ha^{-1} with 20 to 30 q ha^{-1} dry matter. Although hard and woody, young stems serve as an excellent nutritive fodder for camel. At flowering, the fodder has 59 per cent dry matter and 4.9%CP.

Pasture Establishment and Management: adaptation strategy for hot arid climate

Climatic conditions in arid areas are too erratic and individual events of rain are too less for regular germination and continued early growth of introduced species. Introduction of new species is risky in the areas below 300 mm average annual rainfall. Therefore, while selecting the species preference should be given to locally adapted species. In arid region, heavy grazing pressure results in deterioration of grazing lands in open areas, hence protection of pasture is essential to minimize the lossesdue to overgrazing. Angle iron post with barbed wire is effective fencing but the installation cost is high. Hence, ditch and core wall fencing may be constructed on 'sharmdan' or 'participatory approach' basis. Top feed species like *Acacia senegal, A. tortilis, A.nilotica* var. *cuprisiformis, Calligonum polygonoides, Capparis deciduas,* etc. can be planted on the wall.

Choice of species

For the areas receiving < 300 mm rainfall annually and having deep sandy soils,*L. sindicus* is the best choice. On sandy loam soils *Cenchrus* species are recommended. In the areas where>300 mm rainfall occurs, *C. ciliaris, C. setigerus, P. antidotale* and *P. turgidum*are recommended. For heavier soils of the higher rainfall zone, *Dichanthium annulatum* and *Cenchrus* species are better species.

Sowing

Pasture seeds are light and chaffy in nature, making the sowing operations difficult. Therefore, a mixture of seeds and wet field soil in 1:5 ratios (by volume) can be prepared. This mixture is drilled in open furrows in the field. Line sowing which makes the inter-culture operations easy is advised. Pastures can be established by planting rooted slips/newly raised seedlings after the first effective rains of summer monsoon. Two to three slips are propagated in ahill. Water is requiredat the time of plantation and also 2-3 watering after planting. Some practical problems are faced with direct seeding for undulated areas and transplanting is costlier,therefore, pelleted seed sowing may be used for undulated areas. Pellets are prepared with 100-125 g seeds, 3-3.5 kg clay soil, 250g sand and 250g FYM. For one hectare 60-80 kg pellets are required.

The caryopsis of grasses is small and should not be sown deep otherwise seedlings will fail to emerge out of the soil. Seeds of the desert grasses have shown optimum germination under field conditions when sown at shallow depth of 1-2 cm with a light soil cover. *C.ciliaris, C. setigerus* and *D. annulatum* with 50-75 cm row-to-row spacing proved to be the best for higher forage production in arid zone. *L. sindicus, P. antidotale* and *P. turgidum* exhibit more radial expansion. The row-to-row spacing for these grasses varies from 75-100 cm depending upon soil characteristics and rainfall. The time of sowing is limited to period in which soil moisture and temperature are sufficiently high to permit fast germination and better establishment. These grasses may be sown after first effective rains of summer monsoon. The seed rate for grasses vary considerably, however the seed required for one hectare is 6-7 kg for *L. sindicus,* 5-6 kg for *C. ciliaris* and *C. setigerus,* 2-3 kg for *D. annulatum, P. antidotale* and *P. turgidum.*

Pasture Utilization

Proper utilization of grasses enhances the productivity and longevity of the pastureland. Normally grasses are harvested at 50% flowering stage and it gives very good palatability to the animals. In a normal rainfall year, two to three harvests of fodder can be taken. In the establishment year, grazingshould not be allowed and grasses may be harvested and fed to animals whereasexcess fodder can be conserved as hay for utilization for future use. From second year and onward, pasture starts giving higher fodder yield; hence utilization of pasture in judicious way is advised for proper maintenance and regeneration in coming years. To harness the potential productivity of pasturelands, controlled grazing as per the carrying capacity should be practiced. Appropriate grazing system also increases the longevity of the grazing land. Excess fodder of good rainfall year can be converted into 'hay' for scarcity period. Fodder banks may be created to meet the requirement of the local farmers.

References

Sawal. R.K., Ratan, R. and Chadra, S.2009. Nutritive evaluation of *Lasiurus-sindicus* and *Cenchrus ciliaris* hays in sheep. *Indian Journal of Small Ruminants* 15 (2): 277-280.

Rajora, M.P. and Singh M. 2005. Seed yield and its components in winter regenerated flush of buffel grass (*Cenchrus ciliaris* Linn.) under rainfed conditions of Rajasthan. *Seed Research* 33 (1): 65-69.

❑❑❑

Chapter 13

Effect of Elevated CO_2 on Growth, Biomass Production and Photosynthesis of Crops

R. K. BHATT

ICAR-Central Arid Zone Research Institute, Jodhpur-342 003, Rajasthan
E-mail:researcher.rkb@gmail.com

Introduction

Human activities such as fossil fuel combustion, industrialization, land clearing and agricultural operations increased atmospheric concentrations of greenhouse gases (GHG) such as carbon dioxide, methane, chlorofluorocarbons, sulphur dioxide and nitrous oxides. This has caused global warming. Climate change have major impacts on ecosystems and human systems on all the continents and across ocean (IPCC 2014). Inter-Governmental Panel on Climate Change (IPCC, 2007) reported that due to increase of greenhouse gases the global mean annual temperature was recorded higher by 0.40-0.76 °C at the end of the 20th century than was at the end of the 19th century. It has also been projected that by the end of the 21st century the temperature may increase 1.1 to 6.4 °C. The increased atmospheric temperature have impacted on plant species distribution, plant community composition and functioning of ecosystem. Out of the total anthropogenic CO_2 emission almost half of it accumulates in the atmosphere and rest is absorbed by sinks in the ocean and in the terrestrial ecosystems. Natural CO_2 sink strengths are varying with time and space accordingly change in weather and climate. The assessment of impacts, adaptation, and vulnerability in the Working Group II contribution to the IPCC's Fifth Assessment Report 2014 (WGII AR5) evaluates how patterns of risks and potential benefits are shifting due to climate change. It considers how impacts and risks related to climate change can be reduced and managed through adaptation and mitigation. The report assesses needs, options, opportunities, constraints, resilience, limits, and other aspects associated with adaptation.

Increase in CO_2 concentration and temperature may alter the plant's ability to compete with individuals with in a given environment. Each one degree Celsius rise in mean temperature, there will be heavy losses in crop productivity. Enhanced levels of CO_2 in the atmosphere affect the carbon balance in the rhizosphere and photosynthetic carbon assimilation in plants, thereby affecting the agricultural productivity. Predicting climate change and evaluating the effects of changing levels of CO_2 in the atmosphere on agro-ecosystems, requires the use of models. However, these models must be developed and evaluated with data obtained under realistic field condition. In addition to predicted productivity increases under elevated CO_2, species composition of plant communities are likely to change, although the exact nature of these changes are difficult to predict (Polley *et. al.,* 2000). But it can be assumed that it will bring serious implications for land use and land cover in different eco-zones.

Increased atmospheric CO_2 concentration and associated global warming are expected to alter growth rates and competitive relationships in pasture crops. Many studies have been conducted in fully controlled environment but there is need to take such type of experiments in the natural field conditions where realistic diurnal and seasonal temperature and radiation fluctuation occur. There have been a few studies on the effects of elevated CO_2 on fodder crops (Gorisson and Cotrufo 2000, Morgan *et al.,* 2001). Forage species particularly perennial grasses cover large land area and serve an important role as source and sinks for atmospheric carbon. Therefore, it is important to understand the impacts of increased CO_2 on productivity and nutritive quality of these species.

Atmospheric CO_2 enhancement

Inter-governmental Panel on Climate change (IPCC 2007) estimates that the current atmospheric CO_2 concentration has increased by about 40% from preindustrial level of about 280 ppm to 398 ppm today, and it is projected to rise at the rate of 2 ppm per year. The annual growth rate of atmospheric CO_2 was 2.5 ppm in 2007-2009, one of the largest growth rates in the past decade. The average increase in atmospheric CO_2 decade wise was 1.9±0.1, 1.5±0.1 and 1.6±0.1ppm per year for 2000-2009, 1990-1999 and 19980-1989, respectively. Current emissions are growing much faster than rates required for stabilization at either 450 or 650 ppm. As per Mauna Loa observation, CO_2 in October 2011 was 388.92 ppm, whereas it was 387.18 ppm in October 2010 indicating the growth rate of 1.74 ppm in one year

(Dr. Pieter Tans, NOAA/ESRL (www.esrl.noaa.gov/gmd/ccgg/trends) and Dr. Ralph Keeling, Scripps Institution of Oceanography (scrippsco@.ucsd.edu/). As reported from various reports that the global CO_2 emissions from fossil fuel burning decreased by 1.3% in 2009 because of the global financial and economic crisis. The emissions increased to 2.5% in 2013-14.

Forest degradation and deforestation are considered to be an important source of GHG contributing to 17.4% of the global emissions (IPCC 2007). However, it is estimated that recent CO_2 emissions from deforestation and other land-use changes (LUCs) have declined compared with the 1990s, primarily because of reduced rates of deforestation in the tropics and may be a smaller contribution of forest regrowth. As world population has crossed 7 billion and further projected to more than 8.3 billion by 2030, it has been expected that the global carbon emissions will increased by 55%. The accumulation of atmospheric CO_2 in 2010 was 5.0±0.2 Pg C, with a total cumulative of 157.5 Pg C since the beginning of atmospheric high precision measurements in 1959 and 237 Pg C since 1750. The countries with highest absolute values of emissions are China (2.2 Pg C), US (1.5 Pg C), India (0.5 Pg C), Russia (0.5 Pg C), and Japan (0.3 Pg C) although the emissions per capita in China (1.7 t C/person/y) and India (0.5) are still a fraction of the emissions contributed by US (4.8), Russia (3.3) and Japan (2.5) (Peter *et al.,* 2011). As the global temperature is inceasing, it has been assumed that the vulnerable carbon pools may release enough CO_2 to cause an increase of around 200 ppm in atmospheric CO_2 over the CO_2 increases from fossil fuel combustion in the next 100 years.

Projection scenario of climate change and its impact on agriculture

The projected increase in atmospheric CO_2 and temperature and their effect on agriculture is given in Table 1.

Table 1. Effect of climate change on agriculture (with no climate policy interventions)

	2025	2050	2100
CO_2 Concentration	405 – 460 ppm	445-640 ppm	540-970 ppm
Global mean temperature changes from 1990	0.1-1.1 °C	0.8-2.6 °C	1.4-5.8 °C
Global mean sea level rise from 1990	3-14 cm	5-32 cm	9-88 cm

contd...

Agricultural effects			
Average crop yields	Cereals crop yield increase in many mid and high altitude regions. while it may decrease in most tropical and sub-tropical regions	A mixed effect on cereal yield in mid latitude regions. More pronounced decreases in tropical and subtropical regions.	General reduction in cereal yields in most mid latitude regions for warming of more than a few ^{0}C.
Extreme low and high temperatures	Reduced frost damage to some crops. Increased heat stress damage to some crops. Increased heat stress in livestock.	Effects of changes in extreme temperatures amplified.	Effects of changes in extreme temperatures amplified.
Income and prices	-	Income of poor farmers in developing countries decreases.	Food prices increase relative to projected that exclude climate change

(*Sources*-Anonymous)

Climate change scenario for India

In India agriculture contributes to the global warming through emission of carbon dioxide (CO_2), methane (CH_4) and nitrous oxide (N_2O). The major sources of emission of these GHGs in the agriculture sector are enteric fermentation, rice cultivation, agricultural soils, manure management and burning of waste materials crop residue. Soil management practices such as tillage and other faulty agricultural practices also increased carbon dioxide emission.

As per report of The Indian Network for Climate Change Assessment (INCCA, 2010), that the total net Greenhouse Gas (GHG) emissions from India in 2007 were 1727.71 million tons of CO_2 equivalent that include 1227.76, 20.56 and 0.24 million tonnes CO_2 equivalent through CO_2, CH_4 and N_2O emissions, respectively. On a per-capita basis, India is one of the lowest GHG emitters in the world. Its emission of 1.18 tonnes of CO_2 equivalent per capita in 2008 was nearly one-fourth of the corresponding global average of 4.38 tonnes. The major contributors of GHG emissions in India were energy sector (57.8%), industry (21.7%), agriculture (17.6%) and waste sectors (3%). In the agricultural sector major sources are enteric fermentation (63.4%), rice cultivation (20.9%), agricultural soils (13.0%), manure management (2.4%) and on-field burning of crop residues (2.0%) (Pathak and Aggarwal, 2012). Thus, rice cultivation, soils, and field burning of crop residues contributes 35.9% of the total emissions from agriculture (INCCA, 2010).

The most reliable climate change projections for India under A2 and B2 scenarios have so far been made by the Hadley Centre, UK, and the IITM, Pune, using a regional climate model, PRECIS. It suggests that by the end of 21st Century the mean annual surface temperature of the country may increase by 3.0-5.0 °C under A2 scenario and by 2.5-4.0 °C under B2 scenario, with warming more pronounced in the northern part (Rupa Kumar *et al.*, 2006). A 20% rise in all India summer monsoon rainfall is also projected, but Punjab and Rajasthan may experience decline by 10-30%. It has been assumed that rainy days are expected to be fewer in some northern part of the country. By the last quarter of this century annual temperature would most likely increase in Gujarat, Rajasthan, Haryana and Punjab. Both summer and winter temperatures are expected to increase. Across western Rajasthan wind speed may increase by 10-20%. Since temperature is expected to rise by 4 to 5°C across Rajasthan, Punjab and Haryana, the slight rainfall gain in the eastern half of Rajasthan and adjoining Haryana could be partially offset by high evaporation losses, with factors like sunshine, net radiation, wind strength, relative humidity in play, and therefore increase dryness. The most hard hit will be western part of Rajasthan and adjoining Punjab plains where a combination of higher summer and winter temperatures, reduced rainfall and fewer rainy days will make the environment very dry, plant cover more sparse and potential for wind erosion greater than at present.

Temperature rise over India is expected to be uniform over most of the country while slightly more warming over NW region is expected. The expected warming by 2070 is likely to be 1.5 to 2.0 °C over most parts of the country. Extreme rainfall events are likely to increase along the west coast, west central India and NE region.

Effect of elevated CO_2 on growth and crop production

The increasing CO_2 level has initiated to re-evaluate our knowledge of plant responses to increasing CO_2 research particularly the potential impact of elevated CO_2 on photosynthesis and crop production. As CO_2 is the sole source of carbon for growth in plants, the response of photosynthesis and growth to increasing CO_2 has been the subject of a number of studies. In recent years researchers have been predicting that the greenhouse effect will dramatically change global weather pattern in the next century, i.e. air temperature and CO_2 level will rise in the coming years. Therefore, it is very essential at this stage to assess the effect of this CO_2 and take necessary measures to avoid any adverse impact that this change may have.

Amongst the two major photosynthetic pathways, it is assumed that the growth of C_3 plants should be stimulated more by CO_2 enrichment than C_4 plants (Imai and Murata, 1979). Kimball (1983) reported biomass increase of 10 to 43% in C_3 crop plants and 24% in C_4 plants in response to doubling of ambient CO_2. However, there may be interactions with CO_2 that modify or even cancel temperature effects on vegetation. For instance, while warming may favor warm season C_4 species, some of the effects of CO_2 enrichment on C_3 photosynthesis will tend to counter that response, rendering C_3/C_4 distribution changes relatively insensitive to increases in temperature (Polly *et al.*, 2000).

The responses of grasses to CO_2 were earlier reviewed by Wand *et. al.*, (1999). They report strong growth enhancements of both C_3 and C_4 grasses from elevated CO_2, although greater responses were with C_3 grasses. While photosynthesis in C_4 plants can respond directly to increases in CO_2 above present atmospheric concentrations (Le Cain and Morgan, 1998), the response is considerably more limited compared to that of C_3 species. Average photosynthetic rates of the youngest fully expanded leaves of *L. perenne* increased by 30 to 80% under elevated CO_2 as compared to ambient CO_2 (Rogers *et. al.*, 1998, Isopp, 2000). Total symbiotic N_2 fixation increased by 66% in the grass legume mixture under elevated CO_2 (Zanetti *et al.* 1997, Lüscher *et al.*, 2000). *Paspalum dilatatum* was stimulated by elevated CO_2. The mechanism for this has not been identified, but Ghannoum *et. al.* (2000) reported that increased intercellular CO_2 partial pressure, changes in diurnal CO_2 fixation patterns, improvements in shoot water relations and increased leaf temperatures as possible ways in which C_4 species might respond to elevated CO_2. Chem *et al.* (1994) observed that the maximum carboxylation rate of Rubisco, potential electron transport rate and quantum efficiency were increased by CO_2 enrichment, while the maximum phosphoenol pyruvate carboxylase activity was decreased.

They also predict that the light saturated leaf photosynthesis will increase in atmospheric CO_2 and increasing atmospheric CO_2 significantly decreased the stomatal conductance. Production increases resulting from growth at elevated CO_2 are often accompanied by decreases in plant N concentration, and there is some evidence that N may limit plant response to rising CO_2 (Ghannoum and Conroy, 1998; Poorter *et al.*, 1996). Similarly, there is some indication that plants capable of fixing their own N may have a greater capability to respond to rising levels of atmospheric CO_2 (Poorter, 1993).

The effect of CO_2 enrichment and irradiance on the growth and gas exchange of two tropical grasses (*Panicum lexum* (C_3) and *P. antidotale* (C_4) were investigated under either 350 (low) or 700 (high) u11-1 CO_2 conc. under 40% (low) or 100% (high) of direct sunlight and elevated CO_2 enhanced plant dry weight at both irradiance in the C_3 species (1.41fold and 1.71fold increase at low and high light, respectively) but only at high light in the C4 species (1.28-fold increase). CO_2 enrichment had no effect on the dry weight of *P. antidotale*, when stem developed was suppressed by growth under artificial light. The assimilation rates were similar in the low and high CO_2 grown plants, for both species. High CO_2 strongly reduced the stomatal conductance of both species, while it affected the RuBisco content (30% decrease) of the high light in C_3 species only (Oula Ghannoum *et.al.*, 1997).

Uprety and Rabha (1999) studied the response of *Brassica juncea* CV Pusa Bold to elevated CO_2 under water stress. They found that carbon accumulation in leaves, stems and roots was significantly higher at elevated CO_2 concentration. The water stress decreased the carbon content in these plants and this adverse effect was reduced by CO_2 enrichment. On the contrary, nitrogen content of leaves, stem and roots was significantly reduced at elevated CO_2. Water stress caused reduction in nitrogen content in these plant parts, similar at ambient as well as elevated CO_2 concentration. Elevated CO_2 markedly increased net photosynthetic rate, internal CO_2 concentration and reduced variable and maximal chlorophyll fluorescence under moisture stress. Elevated CO_2 increased water potential, relative water content and reduced transpiration rate. The greater allocation of biomass to the roots, which serve as strong sink for assimilated carbon under elevated CO_2, helped in better root growth as also reported by Rabha and Uprety (1998). An attempt has been made by Uprety and Mahalaxmi (2001) to study the response of *Brassica juncea* var. Bio-183-92 to elevated CO_2 under increased nitrogen treatment. The interactive effect of CO_2 and nitrogen nutrition demonstrated that by adding more nitrogen, plants have tried to sustain the positive effect of CO_2 enrichment by utilizing the additional carbohydrates for the development of new sinks. Excess carbohydrate makes plants flexible and responsive to additional nitrogen application for sustaining this CO_2 enrichment effect.

Chakraborty *et al.* (2015) also reported that CO_2 enrichment showed significant increase in growth, leaf area, dry matter production and rate of photosynthesis in Brassica cultivars. The increased rate of net photosynthesis resulted higher accumulation of sugar, non-structural carbohydrates and starch in leaves. Madanpal *et al.* (1997) experimented on the effect of CO_2 enrichment on growth and protein content of Berseem and they concluded that the elevated level of CO_2 caused increased leaf growth both interms of area and dry mass. It also increased stem length and total biomass, this increase in growth was apparently due to increase in photosynthesis. The increase in photosynthesis although increased the total carbon gain, declined the protein content in the leaf, which may alter the nutritional quality of the fodder.

Long term effect of elevated CO_2 on biomass production and photosynthesis

1. *Cenchrus ciliaris L.:* Long term effect of elevated CO_2 (600 ±50 ppm), in open top chambers was assessed in *Cenchrus ciliaris* (C_4) under semi-arid environment. Rate of photosynthesis increased by 25% under elevated CO_2 than that of the ambient grown plants indicating the fact that *C. ciliaris* having the potential to fix more CO_2 even at higher concentration of CO_2 (Bhatt *et al.*, 2007, Table 2). The canopy photosynthesis (PN x LAI) increased significantly under elevated CO_2. The intercellular CO_2 concentration was also observed higher under elevated CO_2. The stomatal conductance decreased at elevated CO_2 which may be due to partial closer of stomata at higher level of CO_2. Rogers *et al.* (2004) have also reported decrease in stomatal conductance at elevated CO_2 in soybean. The rate of transpiration decreased slightly at elevated CO_2 than the plants grown in open field condition. The carboxylation efficiency (P_N/Ci ratio) also decreased at elevated CO_2 which may be due to higher intercellular CO_2 concentration (Farkuhar and Sharkey 1982). The P_N/E ratio (photosynthetic water use efficiency) increased significantly under elevated CO_2 over open grown crops indicating that there is improvement in the photosynthetic water use efficiency under elevated CO_2. Similar to these findings, Baker *et al.*

(2005) also indicated that CO_2 enhancement significantly increased canopy, net photosynthetic rate, Chl a,b, Chl a: Chl b ratio and water use efficiency while reducing evapo-transpiration (Table 2). Moreover Wang *et al.* (2004) also reported that elevated CO_2 increased chloroplast number /unit cell area.

Table 2. Effect of elevated CO_2 on photosynthesis and transpiration in *C. ciliaris*

S.No.	Parameters	Open	OTC with elevated CO_2 (600 ± 50 ppm)	CD (P=0.05)
1	Photosynthesis (P_N, μ moles $m^{-2}s^{-1}$)	34.02	42.46	3.065
2	Intercellular CO_2 concentration (Ci, ppm)	250.26	457.06	7.655
3	Stomatal conductance (g_s, cm / sec)	1.23	1.79	0.214
4	Transpiration (TR) (m moles m^2s^{-1})	10.21	9. 52	NS
5	P_N x LAI	71.95	299.51	13.473
6	P_N /Ci	0.1	0.09	0.026
7	P_N/TR	3.33	4.46	0.254

Long term exposure of *Cenchrus ciliaris* to elevated CO_2 (600 ±50 ppm), in open top chambers increased the growth and biomass production (Bhatt *et al.,* 2007). Plant height increased by 43.72% in elevated CO_2 grown plants (Table 3). The tiller production per square meter increased significantly under elevated CO_2. The leaf area index (LAI) also increased 3.34 fold in the crops grown in open top chambers with elevated CO_2 over the open grown field crops. The specific leaf weight per unit leaf area decreased under elevated CO_2 whereas the specific leaf area on dry weight basis increased. The increase in specific leaf area at elevated CO_2 indicates that there is more expansion of the leaf area per unit dry weight accumulation. The biomass production in term of fresh and dry biomass accumulation significantly increased (Table 3). The fresh biomass increased by 134.35% in the crop grown under elevated CO_2 over the open (1754.70 g/m²). Similarly the dry matter accumulation was recorded to three folds in the crop grown under elevated CO_2 over the open grown crops (223.9 g/m²). Booker *et al.* (2004) reported that biomass, leaf area, water use efficiency and yield were increased significantly by elevated CO_2.

Table 3. Effect of elevated CO_2 on growth and biomass production of *C. ciliaris*.

Parameters	Open	OTC with Elevated CO_2 (600 ± 50 ppm)	CD (P=0.05)
Plant height (cm)	107.50	154.50	18.649
Tiller / m^2	116.25	154.50	9.345
Leaf length (cm)	36.47	44.78	6.288
Leaf width (cm)	0.63	0.95	0.118
Leaf Area Index	2. 12	7. 05	0.828
Specific Leaf Weight (mg/cm^2)	5.30	4.02	NS
Fresh biomass (g/ m^2)	1754.70	4112.20	219.02
Dry biomass (g/ m^2)	223.90	656.80	53.53

(*Source:* Bhatt *et al.*, 2007)

2. *Panicum maximum* Jacq.: Rate of photosynthesis and stomatal conductance increased under elevated CO_2 (600 ±50 ppm) over open grown crops in *P. maximum* (C_4) (Table 4) indicating the positive influence of elevated CO_2 on assimilatory functions of this pasture grass (Bhatt *et.al,* 2005). The rate of transpiration increased in the elevated CO_2. The photosynthetic water use efficiency (P_N/TR) has not shown any remarkable variation in the crops grown under these environmental conditions. The decrease in P_N/CINT may be due to increase in intercellular CO_2 concentration. The canopy photosynthesis (P_N x LAI) increased by 40.77% under elevated CO_2 and 28.13% in open top chambers without elevated CO_2 over the open field grown crops Bhatt *et al.,* 2005).

Table 4. Effect of elevated CO_2 on physiological characters.

Treatment	P_N (µ moles $m^{-2}s^{-1}$)	CS (mol $m^{-2}s^{-1}$)	TR (µmoles m^2s^{-1})	P_N/TR	P_N/ CINT	P_N X LAI
P. maximum						
OPEN	19.21	0.164	8.24	2.33	0.086	97.97
OTC	22.42	0.202	9.42	2.38	0.105	181.62
OTC+CO_2	24.26	0.212	10.63	2.28	0.057	219.06
CD (P=0.05)	4.635	0.024	1.26	-	-	-
S. hamata						
OPEN	13.02	2.266	10.214	1.275	0.052	40.39
OTC	15.13	2.463	8.979	1.685	0.045	75.74
OTC+CO_2	20.49	1.794	9.519	2.153	0.047	114.93
CD (P=0.05)	4.490	0.245	1.439	-	-	-

(where: OTC-Open Top Chambers, PN-net photosysnthesis, TR-Transpiration, CS-Stomatal conductance, CINT-Intercellural CO_2 Concentartion, LAI-Leaf Area Index)

The *P. maximum* (C_4) responded significantly to its growth performance and biomass production. Main tiller height increased to 234.5 cm under elevated CO_2 as compared to open grown crops. The leaf area index (LAI) increased to 1.7 times under elevated CO_2 over the control (5.10). SLW were also influenced under the elevated CO_2. The fresh and dry biomass production was recorded higher in the first cut as compared to the second cut of the crop. The cumulative fresh biomass production of two cuttings increased by 61.25% under elevated CO_2 over the open field grown crops. The dry matter accumulation was also increased to the tune of 59.24% under elevated CO_2 over the open field grown corps.

3. *Stylosanthes hamata* Taub.: In *S. hamata* (C_3) the increase in photosynthetic rate was to the tune of 36.45% under elevated CO_2 (600 ±50 ppm) over open grown crops (Bhatt *et. al.,* 2005). The canopy photosynthesis increased up to 64.86% with elevated CO_2 over the control (Table 4). The rate of transpiration decreased slightly in elevated CO_2. The stomatal conductance decreased significantly in elevated CO_2. The P_N/TR (photosynthetic water use efficiency) increased significantly with the higher value (2.153) in elevated CO_2 indicating the interacting effect of high level CO_2 in the efficient utilization of water for higher productivity. The data revealed that the intrinsic water use efficiency of this C_3 crop improved significantly under elevated CO_2 and the plants can sustain the productivity under higher level of CO_2 and warmer environmental condition. The LAI was recorded to 5.61 in the plants grown under elevated CO_2 (600 ±50 ppm) over open grown crops (3.10). The increase in fresh biomass was 39.33% in C_{600} and 23.06% in C_{OTC} and the dry matter increase was to the tune of 39.79% under elevated CO_2 over the open field conditions. The increase in biomass under elevated CO_2 indicating that being the C_3 legume it has responded significantly and therefore, this species can be promoted for higher biomass production and carbon sequestration in the semiarid tropics.

Grass-legume intercropping systems: Perennial grasses are the potential to assimilate increased atmospheric carbon dioxide and stabilize it in the soil. C_4 grasses (*Cenchrus ciliaris* and *Panicum maximum*) and C_3 legumes (*Stylosanthes hamata*) were evaluated under elevated CO_2 (600±50 ppm) in side open top chambers (OTC) for their growth, photosynthesis and dry matter accumulation. *Cenchrus ciliaris, Panicum maximum* and *Stylosanthes hamata* responded significantly with respect to their growth, biomass production and rate of photosynthesis as grown under elevated CO_2 (600±50ppm) for long period in open top chambers (OTC, Bhatt 2007a, 2008, 2010). The canopy photosynthesis (Photosynthetic rate (PN) x Leaf Area Index (LAI)) increased significantly in all the plant species under elevated CO_2 (600±50 ppm) over the open field grown crops (Fig.1).

Elevated CO_2 stimulated total dry biomass accumulation which is a common physiological response to high CO_2 (Atkinson *et al.*, 1997). Pal *et al.*, (2004) also observed increased leaf size, plant height and dry mass of shoot in *Trifolium alexendrinum* under elevated CO_2. In *Lolium. perenne* average photosynthetic rates of the youngest fully expanded leaves increased by 30 to 80% under elevated CO_2 as compared to ambient CO_2 (Rogers *et. al.*, 1998, Isopp, 2000).

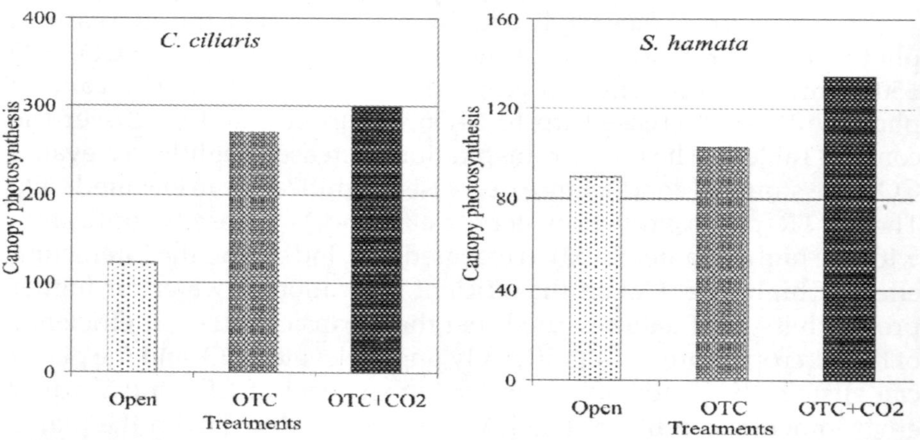

Fig. 1. Canopy photosynthesis (PN x LAI) of *C. ciliaris + S. hamata* in intercropping system under elevated CO_2 (600±50 ppm).

In *Panicum maximum* and *Stylosanthes hamata* intercropping system, the canopy photosynthesis of *P. maximum* and *S. hamata* increased by 152% and 128% respectively in elevated CO_2 over the open grown crops (Fig. 2). The increase in canopy photosynthesis in these crop species may be due to increase in the rate of photosynthesis and leaf area index indicating more assimilation of CO_2 and there by higher biomass production. Plant height, biomass production and chlorophyll accumulation in both the species were significantly influenced under elevated CO_2 (600 ±50 ppm) in open top chambers (Fig. 3). The plant height increased by 32.0 and 49.0% over the control in *P. maximum* and *S. hamata*, respectively, in intercropping system under elevated CO_2

over open field grown crops. The accumulation of chlorophyll a and b in these crop species increased under elevated CO_2 as compared to the open field grown crops indicating the adaptation potential of these species.. Total chlorophyll (chlorophyll a+b) was estimated higher by 1.67 fold in *P. maximum* and 1.23 fold in *S. hamata* under elevated CO_2. The increase in total chlorophyll accumulation predicts that there could be higher PN under elevated CO_2 leading to higher CO_2 assimilation and hence increased growth and biomass production. Thomas and Sean (2005) also reported increase in leaf chlorophyll content under elevated CO_2.

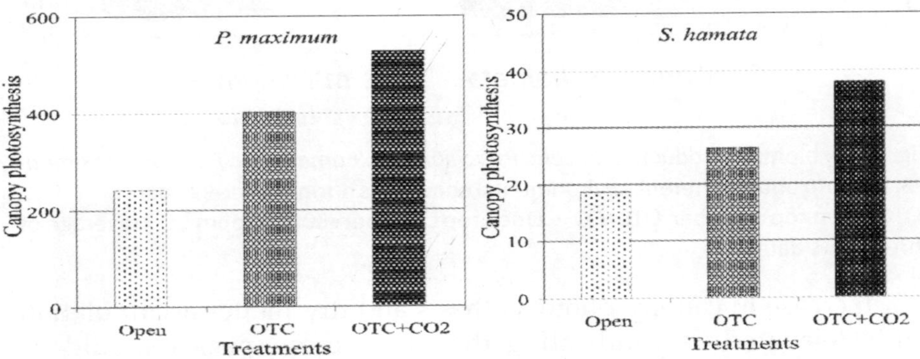

Fig. 2. Canopy photosynthesis (PN x LAI) of *P. maximum+S. hamata* in intercropping system under elevated CO_2 (600±50 ppm).

In *C. ciliaris* + *S. hamata* intercropping system, as per our experimentations for three years dry biomass accumulation increased by 2.41 folds under elevated CO_2 in open top chambers over the open field grown crops, where as it increases by 1.85 fold in *P. maximum* + *S. hamata* (Fig.). The biomass production (fresh and dry) in both the crop species under intercropping systems increased significantly under elevated CO_2. *P. maximum* + *S. hamata* produced 4159.07 g/m^2 fresh biomass under elevated CO_2 which was 67% higher over the control (2491.46 g/m^2) (Fig.4). Similarly, dry biomass production increased by 85% under elevated CO_2 (600±50 ppm) over the crops grown in open field (684.70 g/m^2, 4). The increase in biomass production may be due to enhanced rate of photosynthesis and higher fixation of carbon and its allocation to the plant components.

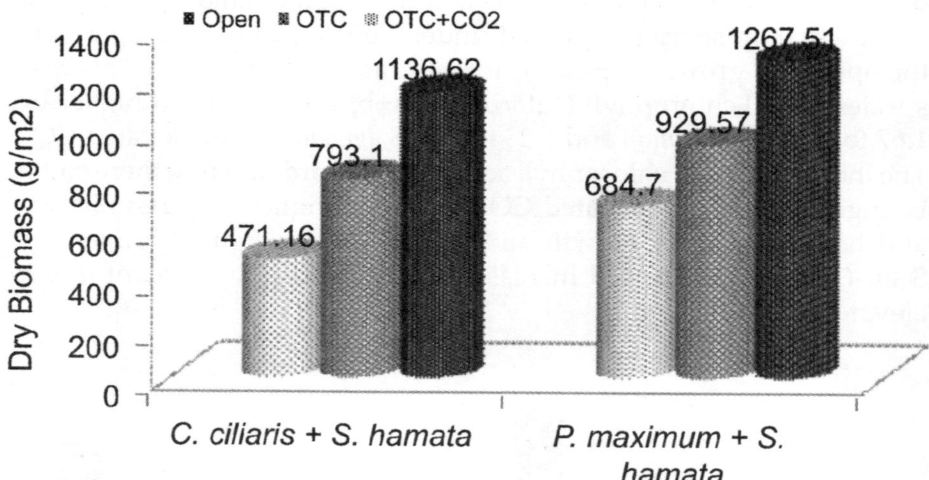

Fig. 3. Dry biomass production in *Cenchrus ciliaris + S. hamata* and *Panicum maximum + S. hamata* under different environmental conditions during different years of growth (OTC-Open top chamber, OTC+CO_2 – Open top chamber+600±50 ppm CO_2) (average of three years data).

Increase in canopy photosynthesis and dry matter accumulation under elevated CO_2 indicating that these crop species should be promoted for carbon sequestration in the tropical environment. These results revealed that these fodder crop species having the potential to assimilate atmospheric CO_2 even under increasing level of CO_2 (600±50ppm) and therefore, will help in understanding the relationships between climate resilience of crops and CO_2 sequestration in different agro-ecosystems in monoculture and intercropping systems.

CO_2 assimilation in grasses

The CO_2 sequestration poptential of pasture grasses with stylo (*C. ciliaris+S. hamata* and *P. maximum+S. hamata)* were also calculated under elevated CO_2 (600±50ppm) in open top chambers (Bhatt *et al.*, 2008). *C. Ciliaris + S. hamata* sequestered carbon 41.67 tonnes CO_2 e/ ha/yr and *P. maximum+S. hamata* sequestered carbon 46.47 t CO_2e/ ha/yr under elevated CO_2 which were 141.3% and 85% over the crops grown in open field indicating their resilience to climate change (Figure 5).

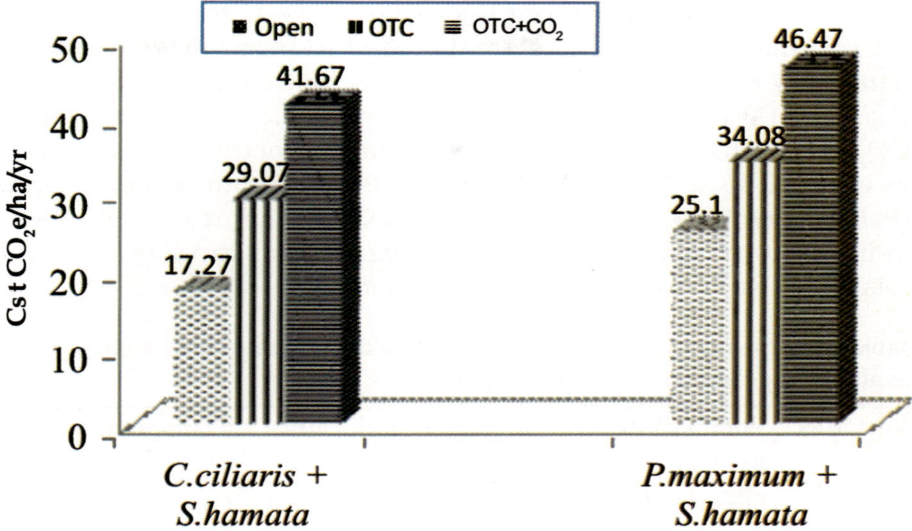

Fig. 4. Carbon sequestration in *C. ciliaris* + *S. hamata* and *P. maximum* + *S. hamata* under different environmental conditions (average value of three years) (OTC- Open Top Chamber (without elevated CO_2), OTC+CO_2 (with elevated-600±50 ppm)

Fig. 5. Effect of elevated carbon dioxide (600±50) ppm on growth, biomass production and physiological adaptation of *P. maximum* + *S. hamata* under open top chambers (OTC).

Effect of elevated CO_2 on photosynthesis, growth and biomass production of Oat (*Avena sativa* L.)

Three Oat (*Avena sativa* L) cultivars viz., JHO 822, JHO 851 and Kent were grown inside the open top chambers under elevated CO_2 (600 ±50 ppm). Rate of photosynthesis increased by 37% in Kent followed by JHO 822 under elevated CO_2 as compared to open grown crops, however, no significant change was observed in JHO 851(Bhatt *et al.*, 2010, Table 5). Increased photosynthesis during the growth period of the crop could be interpreted in terms of high CO_2 induced transient activation of photosynthesis as a stress response (Lichtenthaler, 1996). The stomatal conductance followed similar pattern as photosynthesis. Lesson and Rozema (1990) and Hertog *et al.* (1993) also reported increased photosynthetic rate due to elevated CO_2. According to Harley *et al.* (1992) stomatal conductance decreases in elevated CO_2. Of course these effects depend on water supply (Palanisamy, 1999). Plants at elevated CO_2 did not reach saturation, indicating that net photosynthetic rate regeneration capacity increased relative to RuBP carboxylase regeneration.

Table 5. Photosynthesis, stomatal conductance and transpiration rate in oat varieties as affected by elevated CO2 (600±50 ppm).

Varieties	Photosynthesis (μ moles $m^{-2}s^{-1}$)		Stomatal conductance (moles $m^{-2}s^{-1}$)		Transpiration (m moles $m^{-2}s^{-1}$)	
	Open	Elevated CO_2	Open	Elevated CO_2	Open	Elevated CO_2
Kent	22.39	35.96	0.811	0.958	10.52	12.64
JHO 822	23.73	28.79	0.899	1.327	14.99	15.54
JHO 851	18.98	18.24	0.775	0.570	13.48	15.22
CD (P=0.05)	T=10.23, V=7.12 VxT=7.96		T=0.308, V=0.377 VxT=NS		T=3.88, V=2.63 VxT=2.94	

Plant growth in term of plant height, leaf area, specific leaf mass and LAI increased as gown inside open top chambers under elevated CO_2 (Bhatt *et al.*, 2010). This increase in growth may be due to higher rate of CO_2 assimilation. The biomass production in terms of fresh and dry matter accumulation of two successive cuttings showed an increase of 196.22% (fresh) and 140.97% (dry) under elevated CO_2 (600 ±50 ppm) over open grown crops (Table 6).

Table 6. Fresh and dry matter yield in Oat cv. JHO 822 under elevated CO_2 Treatments

	1st cut		2nd cut		Cumulative	
	Fresh wt. (gm)	Dry wt. (gm)	Fresh wt. (gm)	Dry wt. (gm)	Fresh wt. (gm)	Dry wt. (gm)
OPEN	3197.3	746.8	3177.8	1071.7	6375.1	1818.5
OTC	4621.9	1061.5	4811.1	947.9	9433.0	2009.4
OTC +CO2	6387.3	1130.5	6122.2	1258.5	12509.5	2389.0
CD (P=0.05)	547.18	247.31	1061.7	268.2	1000.4	349.44

Effect of elevated CO_2 on photosynthesis, growth and biomass production of berseem (*Trifolium alexandrinum* L)

Long term effect of elevated CO_2 (600 ±50 ppm) on morpho-physiology and assimilatory functions was observed in berseem (*Trifolium alexandrinum*) cv wardan. Rate of photosynthesis, stomatal conductance and canopy photosynthesis (rate of photosynthesis x leaf area (P_N X LAI) increased significantly in all the cuttings with elevated CO_2 (Bhatt *et al.*, 2004). The increase in canopy photosynthesis was 37.06%, 29.20% and 59.01% in 1st, 2nd and 3rd cut respectively over the open condition. The increase in canopy photosynthesis has indicated the fact that there was higher assimilation of CO_2, which has intern the maximum growth and biomass production (Table 7).

Table 7. P_N and P_N x LAI in berseem cv. Wardan in successive three cuttings

Treatments	1st cut		2nd cut		3rd cut	
	P_N	P_N x LAI	P_N	P_N x LAI	P_N	P_N x LAI
OPEN	25.15	55.63	28.31	45.08	24.21	74.23
OTC	26.58 (5.37)	64.48 (13.73)	31.42 (9.89)	94.06 (52.08)	26.34 (8.09)	86.06 (13.75)
OTC+CO2	29.82 (15.66)	88.39 (37.06)	34.21 (17.25)	104.85 (29.20)	30.12 (19.62)	109.97 (59.01)
CD (P=0.05)	8.423	-	6.24	-	5.84	-

(Data in parentheses indicates percent increase over open)

Leaf area index (LAI) also increased significantly under the elevated CO_2 (600± 50 ppm) over open grown crops under all cutting stage of crop growth. The expansion of the leaf was more under elevated CO_2 through which the assimilatory surface was increased and intern maximization of biomass production. The biomass production (fresh and dry) increased significantly in OTCs with or without elevated CO_2 but the maximum biomass production was recorded under elevated CO_2. The fresh and dry matter accumulation increased to double fold in OTC with elevated CO_2 over open grown crops (Table 8) (Bhatt *et al.*, 2004). Elevated CO_2 increased leaf size, plant height and dry mass of shoot in *Trifolium alexendrinum* as also reported by (Pal *et al.*, 2004).

Table 8. Fresh and dry matter yield in Berseem cv. wardan under elevated CO_2.

Treatments	1st cut		2nd cut		3rd Cut		Cumulative	
	Fresh wt. (gm)	Dry wt. (gm)	Fresh wt. (gm)	Dry wt. (gm)	Fresh wt. (gm)	Dry wt. (gm)	Fresh wt. (gm)	Dry wt. (gm)
OPEN	418.33	38.90	1305.54	182.36	672.16	141.42	2396.03	362.68
OTC	644.44	47.11	1405.54	177.32	1349.53	244.50	3399.51	468.93
OTC +CO2	911.10	82.98	1727.76	196.17	1366.53	242.42	4005.39	521.57
CD (P=0.05)	221.23	16.09	388.16	43.16	334.23	61.23	-	-

The increasing GHGs in the atmosphere is causing global warming at an alarming rate and now it is evident in terms of increasing extreme weather events and large climate variability. Its impacts over next 100 years will be more vulnerable than past 100 years. Increasing level of all GHGs caused increase in atmospheric temperature. Agricultural production is more sensitive because of their dependence to climate. There is urgent need to develop adaptation and mitigation technologies and policies to combat climate change. The increasing concentration of CO_2 and temperature have direct effect on growth, morphology and assimilatory functions of the plants. These growth physiological and assimilatory traits are adapted to the changing environment and therefore, plants resilient to a particular climate. The crop species viz., Berseem (*Trifolium alexandrinum*) and oat (*Avena sativa*) are the dominant cultivated multicut fodder and *Cenchrus ciliaris, Panicum maximum* and *Stylosanthes hamata* are perennial crops adapted under elevated CO_2 (600±50 ppm). The higher canopy photosynthesis has maximized the growth and biomass production in these crop species under elevated CO_2 and therefore, recommended as major sink of the increasing level of CO_2.

Way Forward

There is urgent need to develop the technologies for sustainable production under changing scenario of climate. The following way forward are suggested for mitigation and adaptations of climate change:

1. Assessment and monitoring of drought in different agro eco-zones for planning and modelling.

2. Sensitivity analysis of climate variability and change on productivity of different crops in the most climatically vulnerable regions
3. Screening and development of genotypes resilient to climate change and assessment of production potential under natural and simulated controlled conditions (water, temp. and CO_2).
4. Development of short duration, drought and terminal heat tolerant varieties.
5. Development of crop diversification and integrated farming systems as adaptation strategies.
6. Enhancing water and nutrient use efficiency adopting climate resilient technologies.

References

Atkinson, C.J., Taylor, J.M., Wilkins, D. and Besford, R.T. 1997. Effect of elevated CO_2 on chloroplast components. Gas exchange and growth of oak and cherry. *Tree Physiology*. 17: 319-325.

Baker, J.T. and Allen, L.H. Jr. 2005. Rice growth, yield and photosynthesis responses to elevated CO2 and drought. *Journal of Crop Improvements* 13: 7-30

Bhatt, R.K., Baig, M.J., Tiwari, H.S. and Sharmila Roy. 2005. Effect of elevated CO_2 on growth, biomass production and photosynthesis of *Panicum maximum* and *Stylosanthes hamata* under semi arid tropics. Research paper presented in 7th International Carbon Dioxide Conference organized by "National Oceanic and Atmospheric Administration (NOAA) held at Boulder, Colorado, USA, during Sept. 24 to October 4, 2005.

Bhatt, R. K., Baig M.J., and Tiwari, H.S. 2007. Growth, biomass production, and assimilatory characters in *Cenchrus ciliaris* L. under elevated CO_2 condition. *Photosynthetica* 45 (2): 296-298.

Bhatt, R.K., Baig, M.J., Tiwari, H.S., Jyoti Dubey and Yadava, R. B. 2008. Effect of Elevated CO_2 on Biomass production, photosynthesis and Carbon Sequestration of *Cenchrus ciliaris* and *Stylosanthes hamata* under Intercropping System. *Indian Journal of Agroforestry* 10(2): 40-43.

Bhatt, R.K., Baig, M.J. and Tiwari, H.S. 2004. *Effect of enhanced level of CO_2 on physiology and productivity of tropical grasses and legumes*. ICAR A P Cess Fund Scheme, Final Report sumitted to ICAR 117 p.

Bhatt, R.K., Baig, M.J. and Tiwari, H.S. 2010. Elevated CO_2 influences photosynthetic characteristics of Avena sativa L cultivars. *Journal of Environmental Biology* 31 (5): 813-818.

Bhatt, R.K., Baig M.J, Tiwari H.S. and Sharmila Roy, 2010. Growth, yield and photosynthesis of *Panicum maximum* and *Stylosanthes hamata* under elevated CO_2. *Journal of Environmental Biology* 31(4): 549-552.

Bhatt, R.K. Yadava, R. B. Baig, M. J. Sharmila Roy Tiwari H.S. and Jyoti Dubey 2006. Sequestration of CO_2 in rangelands: towards realistic mitigation. Research paper presented in International Conference on Regional Carbon Budgets held at Beijing, China during 16-18 August.

Chakraborty, K., Uprety, D.C. and Bhaduri, D. 2015. Growth physiology and biochemical of two different Brassica species to elevated CO_2. Proc. Natl. Acad. Sci., India (Sec. B Biol. Sci.): DOI 10.1007/s40011-015-0615-9 (Published online 25 July 2015).

Chen, D.X., Coughenour, M.B., Knapp, A.K. and Owensby, C.E. 1994. Mathemathetical simulation of C_4 grass photosynthesis in ambient and elevated CO_2. *Ecological modeling* 731-2:,63-80.

Contribution of Working Group II to the Fifth Assessment Report of the Intergovernmental Panel on Climate Change 2014. [Field, C.B., V.R. Barros, D.J. Dokken, K.J. Mach, M.D. Mastrandrea, T.E. Bilir, M. Chatterjee, K.L. Ebi, Y.O. Estrada, R.C. Genova, B. Girma, E.S. Kissel, A.N. Levy, S. MacCracken, P.R. Mastrandrea, and L.L. White (eds.)]. Cambridge University Press, Cambridge, United Kingdom and New York, NY, USA, pp. 1-32.

Farquhar, G.D. and Sharkey, T. D. 1982. Stomatal conductance and photosynthesis. *Annual Review of Plant Physiology*. 33: 317-345.

Ghannoum, O., Von Caemmerer S, Ziska, L.H. and Conroy, J.P. 2000. The growth response of C_4 plants to rising atmospheric CO_2 partial pressure: a reassessment. *Plant, Cell and Environment* 23: 931-942.

Ghannoum, O. and Conroy J.P. 1998. Nitrogen deficiency precludes a growth response to CO_2 enrichment in C_3 and C_4 *Panicum* grasses. *Australian Journal of Plant Physiology* 25: 627-636.

Ghannoum, O., Von Caemmerer S., Ziska L.H. and Conroy J.P. 2000. The growth response of C_4 plants to rising atmospheric CO_2 partial pressure: a reassessment. *Plant, Cell and Environment*, 23(9): 931-942.

Global Carbon Project 2011. *Carbon budget and trends of 2010*. [www.globalcarbonproject.org/carbonbudget] released on 4 December 2011.

Ham, J.M., Owensby, C.E., Coyne, P.I. and Bremer, D.J. 1995. Fluxes of CO_2 and water vapor from a prairie ecosystem exposed to ambient and elevated atmospheric CO_2. *Agriculture & Forest Meteorology* 77: 73-93.

Harley, P.C., Thomas, R.B., Reynold, J. F, and Strain, B.R. 1992. Modeling photosynthesis of cotton grown in elevated CO_2. *Plant Cell Environment* 15: 271-282.

Hertog, J.D., Stulen, I. and Lambers, H. 1993. Assimilation, respiration and allocation of carbon in Plantago major as affected by atmospheric CO_2 Levels-A case study. In: *CO_2 and Biosphere* ,.Rozema, J., Lambers, H., Van de geijn, S.C. and Cambridge, M.L. (eds.) Kluwer Academic Publishers, Belgium. 369-378.

Imai, K. and Murata, Y. 1979. Effect of carbondioxide concentration on growth and dry matter production on crop plants. VII. Influence of light intensity and temperature on the effect of carbondioxide environment in some C_3 and C_4 species. *Japanese Journal. of Crop Science* 48:409-417.

INCCA (Indian Network for Climate Change Assessment) 2010. *Assessment of the Greenhouse Gas Emission: 2007*, The Ministry of Environment & Forests, Govt. of India, New Delhi.

Interim Report of the Expert Group on Low Carbon Strategies for Inclusive Growth , Planing Commisssion, Govt. of India (2011).

IPCC, 2007. *Climate change: the scientific basis*. Oxford University Press, Oxford.

IPCC, 2000. *Land use, land-use change, and forestry. A special report of the IPCC. Cambridge University Press*, Cambridge, p 375

IPCC, 2014. Summary for policymakers. In: *Climate Change 2014: Impacts, Adaptation, and Vulnerability*.

Isopp, H., Frehner, M., Almeida, J.P.F., Blum, H., Daepp, M., Hartwig, U.A., Lüscher, A., Suter, D. and Nösberger, J. 2000. Nitrogen plays a major role in leaves when source-sink relations change: C and N metabolism in *Lolium perenne* growing under free air CO_2 enrichment. *Australian Journal of Plant Physiology* 27: 851-858.

Kimball, B.A. 1983. Carbondioxide and agricultural yield: an assemblage and analysis of 430 prior observations. *Agronomy Journal* 75:779-788.

Lapitan, R.L., Mosier, A.R. and Morgan, J.A. 2000. Fluxes of CO_2, N_2O and water in a short grass steppe under ambient and CO_2-enriched atmospheres. *American Society of Agronomy Abstracts*, Minneapolis, MN.

Le Cain, D.R. and Morgan, J.A. 1998. Growth, photosynthesis, leaf nitrogen and carbohydrate concentrations in NAD-ME and NAD-ME C4 grasses grown in elevated CO_2. Physiol. Plant., 102, 297-306.

Lenssen, G.M. and Rozema, J. 1990. The effect of atmospheric CO_2 - enrichment and salinity on growth, photosynthesis and water relations of salt marsh species. In: *The greenhouse effects and primary productivity in European agro-ecosystems*, J., Gouriaan, H., Van Keulen and H.H. Val Laar, (eds.). Prodoc, Wageningen, 64-67.

Lichtenthaler, H. K. 1996. Vegetation stress, an introduction to the stress concept in plants. *Journal of Plant Physiology 148*: 4-14.

Lüscher, A., Hartwig, U.A., Suter, D. and Nösberger, J. 2000. Direct evidence that symbiotic N_2 fixation in fertile grassland is an important trait for a strong response of plants to elevated atmospheric CO_2. *Global Change Biology* 6: 655-662.

Morgan, J.A., Lecain, D.R., Mosier, A.R. and Milchunas, D.G. 2001. Elevated CO_2 enhances water relations and productivity and affects gas exchange in C_3 and C_4 grasses of the Colorado Shortgrass Steppe. *Global Change Biology* (in press).

Morgan, J.A., Lecain, D.R., Read, J.J., Hunt, H.W. and Knight, WGg. 1998. Photosynthetic pathway and ontogeny affect water relations and the impact of CO_2 on *Bouteloua gracilis* (C_4) and *Pascopyrum smithii* (C_3). *Oecologia* 114: 483-493.

Oula Ghannoum, Susanne von Caemmerer, Edward, W.R. Barlow and Jann, P. Conroy 1997. The effect of CO_2 enrichment and irradiance on the growth, morphology and gas exchange of C_3 (*Panicum laxum*) and a C_4 (*Panicum antidotale*) grass. *Australian Journal of Plant. Physiology* 24:227-237.

Pal, Madan, Karthikeyapandiam, V., Jain, Vinita, Srivastava, A.C., Anupam Raj, Sengupta, U. K. 2004. Biomass production and nutritional levels of berseem (*Trifolium alexandrinum)* grown under elevated CO_2. *Agriculture, Ecosystem and Environment*. 101:31-38.

Palanisamy, K. 1994. Interactions of elevated CO_2 concentration and drought stress on photosynthesis in *Eucalyptus cladocalyx* F. Muell. *Photosynthetica 36*: 635-638.

Pathak, Hand Aggarwal, P.K. 2012. Low Carbon Technologies for Agriculture: A Study on Rice and Wheat Systems in the Indo-Gangetic Plains. Indian Agricultural Research Institute, p. xvii + 78, ISBN : 978-81-88708-83-3.

Peters, G. P., Marland, G., Le Quéré C., Boden, T., Canadell, J.G. and Raupach, M. R. 2011. Rapid growth in CO2 emissions after the 2008-2009 global financial crisis. *Nature Climate Change*, 2: 2-4.

Polley, H.W., Morgan, J.A., Campbell, B.D. and Stafford Smith, M. 2000. Crop ecosystem responses to climatic change: Rangelands. In: Reddy, K.R. and Hodges, H.F. (eds) *Climate Change and Global Crop Productivity*, CABI Publishing Co., Wallingford, Oxon, UK, pp 293-314.

Poorter, H. 1993. Interspecific variation in the growth response of plants to an elevated ambient CO_2 concentration. *Vegetation*, 104/105: 77-97.

Poorter, H., Roumet, C. and Campbell, B.D. 1996. Interspecific variation in the growth response of plants to elevated CO_2: A search for functional types. In: Korner, C. and Bazzaz, F.A. (eds), *carbon dioxide, populations and Communities*, Academic Press, New York, pp. 375-412.

Rabha, B.K. and Upreti, D.C. 1998. Effects of elevated CO_2 and moisture stress on *Brassica juncea*. *Photosynthetica 35 (4)*: 597-602.

Rogers, A., Fischer, B.U., Bryant, J., Frehner M., Blum, H., Raines, C.A. and Long, S.P. 1998. Acclimation of photosynthesis to elevated CO_2 under low-nitrogen nutrition is affected by the capacity for assimilate utilization. Perennial ryegrass under Free-Air CO2 enrichment. *Plant Physiology* 118: 683-689.

Uprety, D.C. and Rabha, B.K. 1999. Effect of elevated CO_2 and moisture stress on the carbon and nitrogen contents in *Brassica juncea*. *Biologia Plantarum* 42 (1): 133-136

Uprety, D.C. and Mahalaxmi, V. 2001. Effect of elevated CO_2 and nitrogen nutrition on photosynthesis, growth and carbon-nitrogen balance in *Brassica juncea*. *Journal of Crop Science and Agronomy (Germany)* 184: 271-276

Wand, S.J., E., Midgley, G.F. Jones, M.H. and Curtis, P.S. 1999. Responses of wild C_4 and C_3 grass (Poaceae) species to elevated atmospheric CO_2 concentration: A meta-analytic test of current theories and perceptions. *Global Change Biology* 5:723-741

Wang Xianzhong, Roger Anderson, Kelvin O. and Griffin, L. 2004. Chloroplast numbers, mitochondrion number and carbon assimilation physiology of *Nicotiana sylvestris* as affected by CO_2 concentration. *Environmental and Experimental Bot*any 51: 21-31.

Zanetti, S., Hartwig, U.A., Van Kessel, C., Lüscher, A., Hebeisen, T., Frehner, M., Fischer, B.U. Hendrey, G. Blum, H. and Nosberger, J. 1997. Does nitrogen nutrition restrict the CO_2 response of fertile grassland lacking legumes. *Oecologi*a 112: 17-25.

❏❏❏

Chapter 14

Physiological Approaches for Climate Change Mitigation and Adaptation

UDAY BURMAN

ICAR-Central Arid Zone Research Institute, Jodhpur-342003, Rajasthan,
E-mail: burmanuday821@gmail.com

Introduction

Human beings have been engaged in an intense and sustained assault on the environment in their pursuit for material comforts and high standard of living. There is need to understand response of the terrestrial ecosystems to global change so as to improve our ability to protect our future climate and identify opportunities for mitigation. In this context, it is appropriate to mention here that though normally plants adjust to changing environment but recent rapid global climate change appears to have outpaced that ability in some cases consequently impacting both productivity and diversity worldwide.

Indian arid zone comprises of 12% of the total geographical area (329 M ha) of the country and is characterized by very low and erratic rainfall (<400mm). About 62% of arid zone in India covering an area of about 20 Mha lies in western Rajasthan, thus forming the principal arid zone of the country. Analysis of rainfall data (1901-1999) for this zone indicated that out of 99 years, it experienced agricultural drought once in three years to alternate years (Narain *et al.*, 2000). Apart from water deficits, crop productivity in arid regions is also affected by low soil fertility. Therefore, arid region faces the twin problems of 'water thirst' and 'plant nutrient hunger'. Thus, marginal soils cultivated with marginal inputs produce marginal yields and consequently support marginal living.

The challenge of feeding ever growing hungry world is far too important to rely on any one approach or technology as a solution. In practice no single intervention may be able to provide maximum achievable benefit and thus there is an imperative need to integrate and adopt a holistic approach. In other words there is need for synergy of management practices which encompasses all aspects of Genotype × Environment × Management interactions (G × E × M) as a package. This has led to emergence of concept of resilience in response to the need to manage interactions between humans and ecosystems sustainably. Sustainable agriculture is a form of natural resource management for the production of food, fuel, and fibre. As such, it depends on the resilience of both social and ecological systems. Agricultural resilience can be build by reducing vulnerability and at the same time generating income and reducing poverty. The Intergovernmental Panel on Climate Change (IPCC) defines vulnerability as a function of the character, magnitude, and rate of climate variation to which a system is exposed; its sensitivity and its adaptive capacity. Mathematically, this can be denoted as

Vulnerability = f (Exposure, Sensitivity, Adaptive capacity)

Interplay of different aspects associated with ecosystem influencing both productivity and resource use efficiency (Fig.1) is of prime importance in today's agriculture.

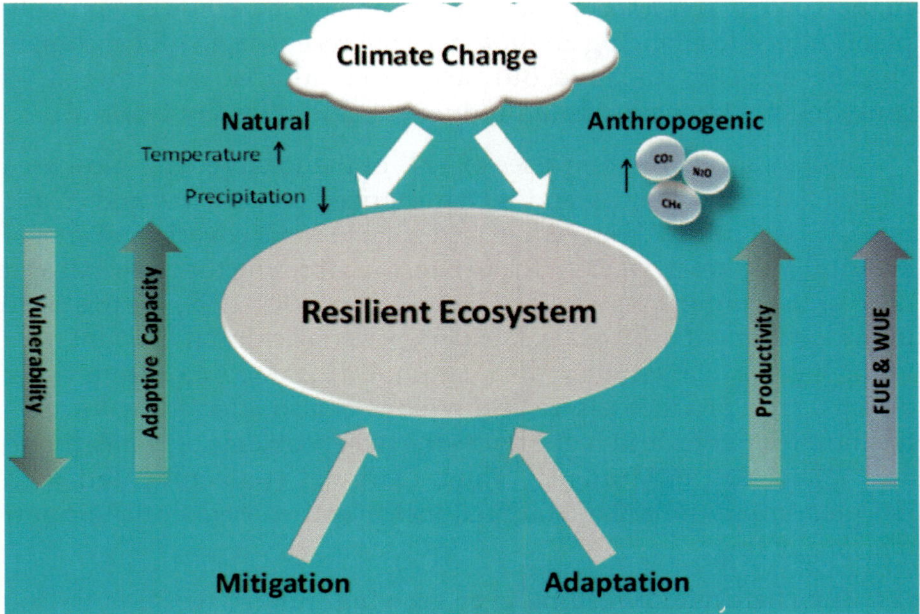

Fig. 1. Depiction of interplay of different aspects associated with resilient ecosystem.

Many changes in agricultural and water management practices, as well as crop productivity improvements, contribute to adaptation while also fulfilling mitigation objectives. Examples include low-till land management practices and soil and water conservation techniques. These practices can help build ecosystem resilience and generate income, and subsequently ensure food security in the region.

Influence of environmental stresses depends on the frequency, intensity, type and duration besides the vulnerability and resilience of the species exposed to the stress factor. There exists specific stress tolerance or sensitivity to different stresses in the crops which requires specific ameliorative actions. Generalization at these steps may not work. Therefore, there is need to clearly understand that in years to come sustenance and growth in a fragile arid ecosystem shall depend on scientific basis of the stress impacts, alleviation mechanisms, and subsequent development of do-able technologies for the region.

The distinction between adaptation and mitigation still confuses many. Mitigation refers to actions taken to lower the concentration of greenhouse gases in the atmosphere.This amounts to comparison of extent of their reduction in global atmosphere *vis a vis* the trend in recent past. Mitigation involves reducing or eliminating emissions at the source (e.g. building a wind farm instead of a coal-fired power plant, avoiding deforestation, creating more energy efficient farm machinery etc.), and/ or sequestering greenhouse gases out of the atmosphere (e.g. through planting more trees that absorb CO_2, carbon capture and stimulating additional sequestration into the ocean via plankton growth, etc.).

Adaptation, in the field of climate change, refers to actions taken to reduce the negative consequences of changes in the climate (e.g. switching to drought and temperature resistant crops, developing an effective early warning system, building flood barriers, extending insurance, etc.). One may adapt agriculture to the changing climate by adopting any one or more ways listed in Fig. 2.

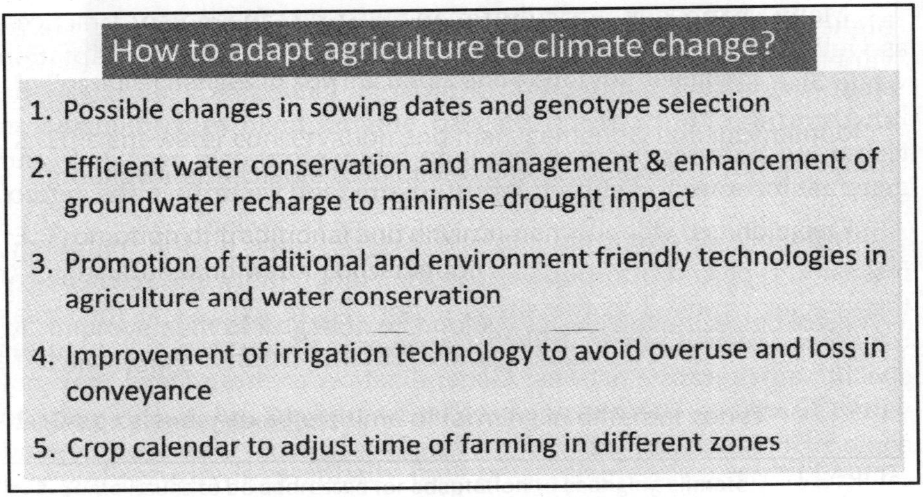

How to adapt agriculture to climate change?

1. Possible changes in sowing dates and genotype selection
2. Efficient water conservation and management & enhancement of groundwater recharge to minimise drought impact
3. Promotion of traditional and environment friendly technologies in agriculture and water conservation
4. Improvement of irrigation technology to avoid overuse and loss in conveyance
5. Crop calendar to adjust time of farming in different zones

Fig. 2. Some issues to be addressed for adaptation to changing climate.

Physiological basis of response to relevant interventions

In most rainfed environments, both productivity and sustainability of agricultural lands depend mainly on temperature and precipitation and these are likely to be challenged by predicted warmer climates (Singh *et al.*, 2011). Having understood the trend of changes in weather variables and likely future climatic scenario in this arid part of our country we may focus primarily towards alleviation of stress due to water scarcity and/or rise in ambient temperature. In other words plants are likely to be exposed to different intensities and duration of water and temperature stresses during different growth phases. Thus, all interventions that help in ameliorating the adverse effects of these stresses shall play crucial role in maintaining productivity. However, to sustain productivity and to cope with the increasing food demand, in terms of both quantity and quality, newer approaches need to be adopted.

Crop management strategies

Genotype

Crops differ in their maturity period, root density, spacing, height, leaf orientation, reflection coefficients, photosynthetic efficiencies etc., and thus naturally in their water use efficiency (WUE). The C-3 plant type, particularly legumes have a low photosynthetic rate and thus have low WUE. The C-4 (millet, sorghum) plants, on the contrary, have a higher photosynthesis rate and their WUE is twice as high. Different varieties of the same crop can be markedly different in WUE. Kathju *et al.* (2001) have reported a twofold difference in the WUE of different pearl millet genotypes more so under low rainfall condition. A good understanding of crop management effects on WUE may provide researchers with opportunities to identify and select appropriate crop management practices for improved WUE. Single genotype may not be suitable in all the situation and old genotypes are being continuously replaced by more water efficient genotypes and fast growing genotypes, as a result average pearl millet productivity has increased from 188 kg ha^{-1} in 1950s to over 350 kg ha^{-1} in 2005. Similarly in arid legumes e.g. moth bean early flowering (30-31 days) genotypes such as RMO-40 and RMO-257 consumed soil moisture at a slower rate during vegetative stage and conserved it for the critical flowering stage. In contrast late flowering (45-50 days) genotypes consumed water at a faster rate during the vegetative stage. Consequently early flowering genotypes produced higher seed yield and exhibited greater WUE than late types. Maintenance of higher photosynthetic rates and better metabolic efficiency in early genotypes under water stress at either growth stages have led to significantly less reduction in their seed yield and dry matter production as compared to late flowering genotypes (Garg *et al.*, 2004). Early genotypes also displayed higher nitrogen harvest index as well as higher nitrogen use efficiency (NUE) than late genotypes under low rainfall, while these were comparable under high rainfall condition indicating a better partitioning and utilization of nitrogen by early genotypes under arid conditions as in clusterbean (Garg *et al.*, 2003). Many a times, as an insurance to aberrant weather geno-cropping is often advocated wherein higher yields are recorded under both low and high rainfall situations (Table 1).

Plant population and spacing

Wider row spacing (50 to 75 cm) and optimum plant density (150,000 plants ha^{-1}) also enhanced grain yield, WUE and NUE under varying rainfall conditions in pearl millet (Garg *et al.*, 1993; Kathju *et al.*, 2001). If drought occurs during the middle of a growing season, corrective measures can be adopted which vary from reducing plant population to fertilization or weed management. However, improved performance of individual pearl millet plants under wider spacing could not compensate for the losses accrued due to a decrease in plant population per unit area beyond a point. A moderate population of

88888 plants ha^{-1} (equal proportion of pearl millet and clusterbean) was optimum for row intercropping (1:1) and strip cropping (4:4) in the moisture sufficient season, and low (40,000 plants ha^{-1}) to moderate populations produced satisfactory yields in the moisture stressed season. Intercropping systems used 8 to 14 mm more water than sole pearl millet, but the improvement in WUE was conspicuous, more so in the moisture stressed season.

Table 1. Relative performance of geno-cropping in mothbean and clusterbean compared to early and late maturing genotypes (Adapted from Kathju *et al.*, 2003 and Vyas *et al.*, 2003)

Planting pattern	Seed yield (kg ha^{-1})	
	266mm	424mm
MOTH BEAN		
Early (RMO-40)	744	825
Late (Maru moth)	528	1011
Early + Late	755	1188
CLUSTERBEAN		
Early (Suvidha)	833	986
Late (Maru guar)	717	1122
Early + Late	928	1172

Compared to 45x10 cm, the spacing of 30x15 cm was more conducive for better utilization of available soil moisture which augments more favourable plant water status leading to higher photosynthesis in moth bean and clusterbean. The improved performance of geno-crop, particularly under low rainfall situation could be attributed to more efficient use of water, low initial canopy development and higher photosynthetic rates. This also ensures that species diversity would buffer environmental fluctuations and enhance resistance and resilience of arid croplands. Additionally, with respect to arid zone, the approaches at physiological plane can also be narrowed down to :

1. Adoption of early maturing genotypes to evade terminal drought
2. Adoption of high temperature tolerant genotypes
3. Emphasis on 'stay green' character

Nutrient management strategies

Crops in the arid and semi-arid regions suffer from both moisture and nutrient stresses and these limit the yields. Therefore, addressing the soil fertility constraints shall also alleviate drought stress effects on crop productivity in arid environments. This results in improved plant vigor as compared to plants those are raised under low soil fertility because application of appropriate fertilizers results despite desiccation in an efficient enzyme activity and higher chlorophyll content. Possibility of alleviation of water stress effects by P nutrition in clusterbean, at least up to moderate stress level also exits (Burman *et al.*, 2009). N-fertilization also improved plant vigour and seed yield in different categories of pearl millet genotypes (Garg and Burman, 2011). Emphasis should also be on micro dosing of fertilizers for higher nutrient use efficiency.

Introduction of perennial

Plants remove CO_2 – a major greenhouse gas (GHG) that is driving today's global warming – from the atmosphere during photosynthesis. This leaf-level process when scaled up to the global level removes ~130 billion or giga tonnes of carbon (Gt-C) each year. This is estimated to be 13 times more than the ~10 Gt-C emitted each year from human activities such as fossil-fuel burning (~9 Gt-C/yr) and deforestation (~1 Gt-C/yr). Thus, introduction of perennial is a viable option which is also likely to check soil erosion and reflect net gain in soil organic carbon from 0.14 to 0.20% over the years. According to an estimate done at arid zone (Anonymous, 2007) maximum soil loss of 33 tons/ ha occurred in arable cropping system. This eroded the soil surface of its nutrients which in turn, as reflected in preceding paragraph, lowers the production potential of the field. Thus, perennial component also provides insurance against harsh weather condition. When tree and crops are grown together, some above ground and below ground interactions may occur between the species. Above ground interspecies interactions are mainly reflected by the improved stereo-structure that influences light capture between the composite plants, thus influencing photosynthesis. Below ground interspecies interactions are mainly reflected by the effective nutrient transformation between the rooting zones of the two species besides increased water use and nutrient uptake.

Recent concepts and their potential in arid zone

Phenomics

Plants are known to be affected by complex genome x environment x management interactions and phenomics is an area of biology concerned with the measurement of physical and biochemical traits of organisms as they change in response to genetic mutation and environment. Phenomics is typically performed by running multiple phenotypic assays on a large set of genetic variants. This essentially is an approach to study how genetic instructions from a single gene or the whole genome translate into the full set of phenotypic traits of an organism. Through this technique complex traits can be broken down to individual components of physiological traits.

The major benefits of high-throughput phenotyping are that one can accurately measure a large number of phenotypes and these measurements can be taken repeatedly over time. Routine image processing coupled with data analysis can help with basic analysis of characteristics such as growth rates, water usage and stress response. This allows generation of robust physiological predictors to bridge the knowledge gap that could be used for application in future breeding and related basic research programs. In the process of crop domestication natural selection for higher yield is done. This leads to pyramiding of related traits (Fig. 3).

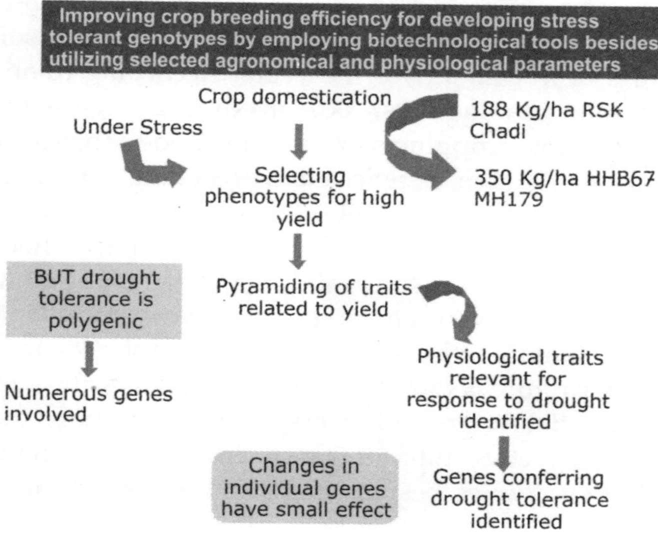

Fig. 3. Flow chart depicting approach for developing drought tolerant genotypes

However as drought tolerance is polygenic there is need to critically evaluate genetic differences. But changes in individual genes have small effects. This involves contrasting requirement for field evaluation which often acts as speed breakers (Figure 4).

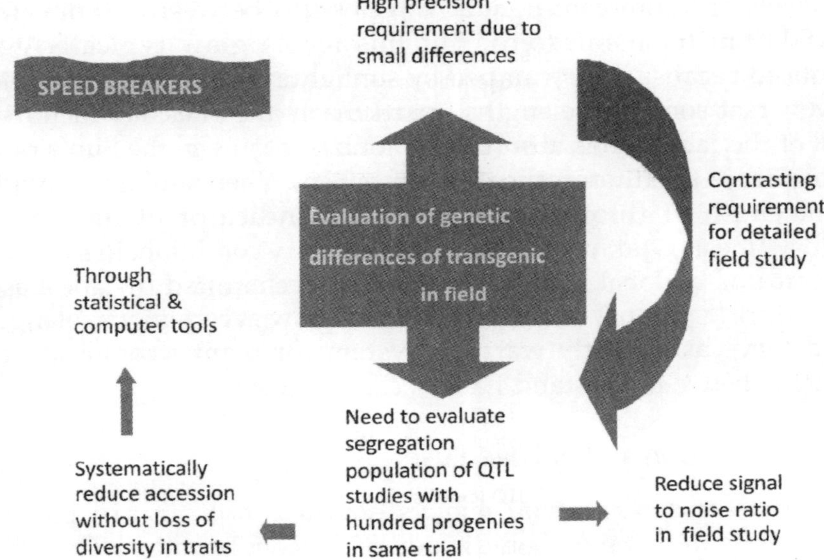

Fig. 4. Depiction of 'speed breakers' in the evaluation of genetic material for drought tolerance

Many researches follow another approach of correlation of different biological traits with yields besides determining their heritability (Table 2; Burman *et al.*, 2011) primarily to narrow down the parameters for field screening of drought tolerant genotypes.

Table 2: Physiological traits associated with grain yield in pearl millet (From Burman *et al.*, 2011)

Physiological traits	Correlation**	Heritability
Plant water potential	-0.778	49.3
Relative water content	0.810	61.8
Net photosynthetic rate	0.763	48.7
Chlorophyll	0.631	75.8
Stomatal conductance	0.838	47.4
Fv/Fm ratio	0.687	59.3
Nitrate reductase activity	0.729	49.9

Mapping photosynthesis

During photosynthesis, plants use chlorophyll and solar energy to convert carbon dioxide, water and other nutrients into sugars and other organic compounds. Together, these chemical reactions emit low levels of fluorescent light at wavelengths between 650 nm (red) and 800 nm (near-infrared). But this feeble glow typically goes unnoticed because it is swamped by sunlight reflected from the plant. However, at some wavelengths - particularly the ones at which most or all of the radiation is absorbed by ionized atoms in the Sun's outer atmosphere - the fluorescence is not masked. When sunlight is bright, the strength of fluorescence is a good indicator of how much photosynthesis is taking place under laboratory condition. Researchers have created a global map of the fluorescence emitted by land-based plants during photosynthesis. This subtle glow at certain wavelengths could serve as an early warning system for plant stress and help scientists better understand Earth's carbon cycle.

Balancing carbon and nitrogen use

Plants and terrestrial ecosystems are not simply passive respondents to rising carbon dioxide (CO_2) concentration. They play an important role in determining the rate of global change and are subsidizing our use of fossil fuels by absorbing approximately 30% of anthropogenic CO_2 emissions. Therefore, rising CO_2 concentrations offer a potential boon for crop production, which if fully exploited, could offset many of the projected negative impacts of global climate change on crop productivity. Sustained stimulation in photosynthesis and growth is dependent on the capacity to utilize the extra photosynthate produced at elevated CO_2 and on an adequate nitrogen supply. There is need to focus on understanding and elucidating the interactions between carbon and nitrogen metabolism that underlie many of the commonly observed responses of plants to climate change.

Flexible conservation agriculture

Rain-fed cropping systems usually practised on erosion-prone, structurally-unstable soils influences physiology of crops with respect to soil water plant relationship and nutrition of plants derived from the fractions available in soil. Thus there is need to conserve soil for sustainable productivity. In other words, innovations to generate and maintain some minimum level of soil protection must be crafted, as holding soil in place is a necessary pre-requisite in any sustainable system. Therefore, there is need to adopt system that makes sense so as to optimize both economic and environmental outcomes.

In undisturbed no-till systems problem of nutrient stratification is associated, whereby immobile nutrients (especially P) remain in surface mulch layers and may be unavailable, poorly accessed and inefficiently used by plants during extremes of temperature or dry periods. This also has implications for C-sequestration. Though no (or reduced) till systems coupled with the maintenance of surface cover make sense in drylands however in export-focussed, unsubsidized commodity futuristic markets profit needs to generated considering labour input and resource use efficiency.Thus, strategic tillage designed to generate some surface roughness to reduce wind and water erosion and increase water infiltration, is often practiced when soil cover is sparse after long periods of drought. Loosening of the soil using tillage can temporarily increase porosity and improve mineralization of nutrients where chemical fertilisers are rarely used as in arid region.

Synchronization of fertilizer application and plant growth

In preceding paragraph we have highlighted the importance of nutrient management strategies for optimizing production potential of crops in arid soils. Thus, sustainability of such a system lies in the fact that low nutrient use efficiency, a major problem associated with fertilizers, can be addressed by site-specific nutrient management that ensures spot targeted application of nutrient elements at specified quantity. Among the resources, fertilizer being one of the costly inputs, its judicious use would trigger the process of accelerated growth in production. The current gap between nutrient removal and application is 10 Mt, and this is likely to grow further. To meet the demand of food grain needs of the projected population of 1.4 billion by 2025, it is necessary to use 30–35 Mt of NPK from fertilizers and10 Mt from organic and biofertilizer sources. Thus, it is necessary to raise fertilizer production by 2025. Estimates based on the sufficiency approach show that the requirement for zinc will be 324, iron 130, copper 11, boron 3.9 and manganese 22 thousand tonnes by 2025. Moreover, most conventional N fertilizers are water soluble and mineralize rapidly in soil. Consequently about 50% of applied N escapes either as gaseous emission or nitrate leaching. This could be controlled to some extent by (1) reducing solubility (2) through urease or nitrification inhibitors and (3) use of urea derivatives. Therefore, more effective synchronization between crop growth and the amount and timing of nitrogen application is required which involves calibration of indigenous soil nitrogen, crop variety, weather, and other factors that may affect growth rates. However, this is knowledge-intensive and requires substantial investments in terms of up-gradation of soil testing infrastructure, manpower and extension services.

In the context of Indian arid zone though the use of fertilizers in drylands is limited, its applications in recommended quantum and integration with nutrient management practices have shown to increase the productivity of several dryland crops up to 15–32%. Integrated nutrient management wherein plant's requirement is met through both organic and inorganic sources has been reported to be a viable option for pearl millet cultivation in arid zone which resulted in high nitrogen use efficiency (Praveen *et al.*, 1998). In arid farms, soil health can vary dramatically from one end of field to the other. Thus the development of N fertilization strategies and tools for precision N fertilization to maximize N use efficiency could be interesting from both economic and environmental point of view. Precision placement of nutrients to depth for more efficient use and slow-release formulations is one possibility. This is likely to influence the surface accumulation of nutrients also. Utilization of alkylated ureas to obtain N mineralization at the desired time due to varying delay period in N release also hold promise for controlled release of fertilizers. Additionally efficiency of N fertilizers could also be increased by sulphur mixed urea that resulted in reducing the loss due to ammonia volatilization by 50% (Aggarwal *et al.*, 1987).

Our recent experience

Heat stress due to increased temperature could cause major shift in productivity in the coming years. Transitory or constantly high temperatures cause an array of morpho-anatomical, physiological and biochemical changes in plants. Therefore, a thorough understanding of physiological responses of plants to high temperature, mechanisms of heat tolerance and possible strategies for improving crop thermotolerance is imperative. At whole plant level high temperature results in flower drop and pod abortion in legumes (Siddique *et al.*, 1999). At the physiological plane high temperature also adversely affects photosynthesis resulting in decline in relative growth rate (RGR) and net assimilation rate in pearl millet (Asharaf and Hafeez, 2004). This temperature induced adverse effect was reported to be due to altered photo function of thylakoid membrane in wheat (Murthy and Prasuna, 2006). Adverse effects on respiration, water relations, membrane stability along with modulated levels of hormones and primary and secondary metabolites were also reported to be responsible for the adverse response to high temperature. In order to cope with heat stress, plants implement various mechanisms, including maintenance of membrane stability, scavenging of reactive oxygen species and production of antioxidants. However, crops and genotypes differ in their response to high temperature.

High temperature coupled with micronutrient deficiency especially of iron and zinc in sandy soils along with low organic matter further complicates the issue. Iron is involved in the production of chlorophyll, and iron chlorosis is easily recognized on iron sensitive crops growing on calcareous soils. Iron is also a component of many enzymes associated with energy transfer, nitrogen reduction, and fixation. Zinc, however, is an essential component of various enzyme systems for energy production, protein synthesis, and growth regulation. At CAZRI, we attempted to address two important abiotic stresses, drought and high temperature, through different ameliorative measures (Garg and Burman, 2011) and also through identifying genotypes tolerant to theses stress (Burman *et al.*, 2011). Experiments revealed that adverse effects of simulated high temperature could be partially alleviated by application of iron and zinc in pearl millet, clusterbean, and wheat through their effect on membrane stability index, nitrate reductase activity and total chlorophyll content in leaves. The response was however species and dose dependant. In another experiment conducted under National Initiative on Climate Resilient Agriculture (NICRA) project, genotypes of pearl millet, clusterbean and mung bean were screened under four different environments in rain-out facility and promising genotypes were identified (on the basis of variation in relative water content of leaves) to be used as such or as parents' in the future breeding program.

Different physiological approaches discussed in the chapter provide insight in to the kind of experiments being conducted to understand response of crops plants to changing climate and means to alleviate adverse effects of the consequent environmental stresses. Understandably "one size fit all" concept shall not apply for all agro-climatic zones and for all crops. Thus, there is need to fine tune the experiments so as to address to location specific objectives. Interventions identified through similar experiments done throughout the country are likely to provide alternatives for mitigation and adaptations to changing climate. However, they shall be required to be 'farmer friendly' so as to enhance their feasibility considering all the socio-economic aspects of the target region. Such interventions are likely to influence the "pace of change" and also the "face of threat" due to climate change. "Business as usual" concept shall not hold the regional agriculture related resources for long and there is strong need for more concentrated research efforts in this direction. This shall require huge investments on both research and infrastructure development which is possible if benefits from global carbon markets are made available for mitigation in agriculture. It will therefore be important to incorporate agriculture, forestry, and other land uses into carbon markets through global commitments for generation of monetary resources. Furthermore, policy makers shall also need more region specific information as all interventions, whether for mitigation or adaptation, need to go along with societal development.

References

Aggarwal, R.K., Raina, P. and Praveen-Kumar. 1987. Ammonia volatilization losses from urea and their possible management for increasing nitrogen use efficiency in an arid region. *Journal of arid Environment* 13: 163-168.

Anonymous 2007. Annual Report. Central Arid Zone Research Institute (ICAR), 172 p.

Ashraf, M. and Hafeez, M. 2004. Thermotolerance of pearl millet and maize at early growth stages: growth and nutrient relations. *Biologia Plantarum* 48: 81–86.

Burman, U., Garg, B.K. and Kathju, S. 2009. Effect of phosphorus application on clusterbean under different intensities of water stress. *Journal of Plant Nutrition* 32: 668-680.

Burman, U., Garg, B.K., Yadav, O.P. and Kathju, S. 2011. Effect of Terminal Water Stress on Certain Physiological Parameters in Pearl Millet Genotypes. *Indian Journal of Plant Physiology* 16 (3): 275-283.

Garg, B.K. and Burman, U. 2011. Physiological basis of yield improvement of arid zone crops through nutrient management. *Plant Stress* (Special Volume on Abiotic Stress) 73-81.

Garg, B.K., Burman, U. and Kathju, S. 2003. Genotypic differences in growth, mineral composition and leaf metabolites of cumin under salt stress. *Journal of Spices and Aromatic Crops* 12:113-119.

Garg, B.K., Burman, U. and Kathju, S. 2004. Influence of spacing on seed yield, water relations and photosynthetic parameters of clusterbean genotypes under arid conditions. *Journal of Arid Legumes* 1: 128-131.

Garg, B.K., Kathju, S., Vyas, S.P. and Lahiri, A.N. 1993. Effect of plant density and soil fertility on pearl millet under drought and good rainfall situations. *Annals of Arid Zone* 32: 13-20.

Kathju, S., Burman, U. and Garg, B.K. 2001. Influence of nitrogen fertilization on water relations, photosynthesis, carbohydrate and nitrogen metabolism of diverse pearl millet genotypes under arid conditions. *Journal of Agricultural Science* 137: 307-318.

Kathju, S., Garg, B.K., Vyas, S.P. and Lahiri, A.N. 2003. Sustainable production of moth bean through genotype management under arid environments. *Journal of Arid Environments* 53: 137-143.

Murthy, S.D.S. and Prasuna, B.A. 2006. Altered thylakoid membrane photofunctions under high temperature stressing wheat primary leaves. *Journal of Phytological Research* 19: 11-13.

Narain, P., Sharma, K.D., Rao, A.S., Singh, D.V., Mathur, B.K. and Ahuja, U.R. 2000. *Strategy to Combat Drought and Famine in the Indian Arid Zone.*CAZRI, Jodhpur, 1-65 p.

Praveen-Kumar, Singh, Y.V., Lodha, S. and Agarwal, R.K. 1998. *Efficient Management of Resources for Sustainable Crop Production in Arid Region,* Central Arid Zone Research Institute, Jodhpur, India. 72 p.

Siddique, K.H.M., Loss, S.P., Regan, K.L. and Jettner, R.L. 1999. Adaptation and seed yield of cool season grain legumes in Mediterranean environments of south-western Australia. *Australian Journal of Agricultural Research* 50: 375–387.

Singh, D.V., Purohit, R.S., Mertia, R.S., Laxminarayan, Poonia, S. and Kar, A. 2011. *Agrometeorological Data Handbook of Jodhpur (1971-2010).* Central Arid Zone Research Institute, Jodhpur, India, 110 p.

Vyas, S.P., Garg, B.K., Kathju, S. and Lahiri, A.N. 2003. Genotypic management of clusterbean for sustainable production under arid environment. *Indian Journal of Pulse Research* 16: 104-107.

□□□

Chapter 15

Silvopastoral Systems and Carbon Sequestration

M. M. ROY* AND R. K. BHATT

ICAR-Central Arid Zone Research Institute, Jodhpur-342003, Rajasthan

* ICAR-Indian Institute of Sugarcane Research, Lucknow-226002

E-mail: mmroyster@gmail.com

Introduction

Increasing demands of fuel, fodder and timber at house hold level in many developing countries, including India have led to integration of trees into agricultural and pastoral areas. Such tree-crop intercropping systems have a great potential to conserve the natural resources and biodiversity in semi-arid and arid regions of tropics and subtropics (Castro *et al.*, 2003; Schroth *et al.*, 2004; Shibu Jose, 2009). The term silvopasture essentially refers to an integrated system of growing and managing the tress/ shrubs in association with grass or grasses and/ or grass - legume mixtures simultaneously on the same piece of land, where animals may or may not become a part of the system (Singh and Roy, 1993; Roy, 1996). These multi-tiered systems are close to natural systems - comparatively more stable and ensure better solar energy flow through the food chain. Thus, silvopasture structured on the functional pattern of natural ecosystem is a viable proposition (Paroda and Jatasra, 1985). Their role is soil fertility enhancement (Schroth and Sinclair, 2003) and carbon sequestration is also well documented through case studies around the world (Montagnini, 2006).

These systems are used for the livestock either through grazing or in a cut and carry mode. Since silvopastures are tree-crop/grass inter-cropping systems of perennial trees, grasses and legumes; there are ecological and economic interactions between the biophysical components. Being a multi-strata system, there are opportunities to optimize yields by managing the crop geometry and functional attributes in its temporal and spatial dimensions. The physical parameters such as canopy temperature, relative humidity, light transmission, soil temperature, soil moisture and energy balance influence the physiology of species under the canopy.

In arid and semi-arid region of India silvopastoral systems raised on degraded lands are extremely important in providing fodder, fuel and timber in one hand and soil and water conservation on the other. Tree-crop in Silvopastoral systems is a mixed community composed of tree and crops, which form entirely different pattern for the utilization of light, water and nutrients(Conner,1983). If water and mineral nutrient are plentiful than light is the main environmental resource limiting plant productivity. The unfiltered light affect the productivity and other biological phenomenon of the crop and the physical parameters under the canopy and to the surroundings.

The semi-arid and arid lands with mostly marginal soil cover the largest area of land in the tropics and subtropics. The use of trees and shrubs to enhance organic matter production to maintain soil fertility, and to reduce erosion and to create more even micro climate has been found very effective. The crop cultivation in the rainfed areas especially inarid and semi-arid regions is a big gamble due to low and erratic rainfall and associated crop failures. Thus, for sustainable production, the best alternative is to promote appropriate integrated agro forestry systems in the arid and semi-arid areas. The development of waste lands through the establishment and management of silvopasture not only checks land degradation but also provide diverse products for human and animals use. Trees and shrubs are used as for age either through browsing or lopping when other for age like grasses, legumes and cultivated fodder fail to grow. The tree species also provide timber and other minor products to the in habitants as well as shade and shelter to the grazing animals. The tree and shrubs grow luxuriantly even during the drought years as their deep tap root system enables them to grow independently of surface moisture. Trees improves oil through leaves and other litter in addition to decomposition of old roots. Nitrogen fixing trees have added advantage of fixing the atmospheric nitrogen. They act as a nutrient and moisture-pumping machine by drawing nutrients and moisture from deeper layers of soil. The presence of trees acts as a protective barrier against soil erosion or as wind breaks. Tree reduces the soil moisture evaporation by lowering the soil surface temperature. Grasses while conserving soil and water also utilize off-season precipitation. The legumes enrich soil by nitrogen fixation and in the mixture they help growth of the grasses and trees besides improving the for age quality. Compared with other land uses, in marginal and sub marginal and the silvopastoral system provides for age/grazing availability and its quality for alonger period of the year.

In India, silvopasture research was initiated during the early sixties at the Central Arid Zone Research Institute (CAZRI), Jodhpur. This institute worked out the technologies for establishment of shelter belts, stabilization of sand dunes and development of agri-silvoculture systems. Later on comprehensive studies on such system were taken up around 1970s at Indian Grass land and Fodder Research Institute(IGFRI), Jhansi. The studies included on designing and development of silvopastoral system on degraded lands with special emphasis on fodder production on one hand and the conservation on the other hand. Gradually the other institutes viz., Central Soil and Water Conservation Research and Training Institute (CSWCRTI), Dehradun and Central Institute for Dry Land Agriculture (CRIDA), Hyderabad, initiated work on such systems to develop appropriate land use system in their respective region. Besides these, an All India Coordinated Research Projecton Agro forestry (AICRPAF) with 20 centers was initiated by ICAR during the year 1983. Silvopasture research in India was further strengthened during the seventh plan (1985-90) by adding 11 more centers to the AICRPAF and by creating a National Research Center for Agro forestry (NRCAF) at Jhansi during the year 1988. Now studies on various aspects of silvopasture are also under taken by most of the state Agricultural Universities (SAUs) and some traditional universities as well.

Microclimatic Changes

The availability of photosynthetically active radiation(PAR) under tree canopies determines the yield patterns. However, variation in light transmission under the tree canopy is governed by the tree geometry, canopy structure and canopy density contributed by branch architecture. The temporal and spatial distribution and quantity of PAR determines the synthesis of the farming technique to develop the productive tree-crop inter cropping systems. Therefore, there is need to concentrate on tree-crop interface, the micro- environmental changes and resultant physiological processes to understand the adaptation behavior, sustainability, productivity and economic feasibility of the system.

The law of temporal and spatial distribution of the tree shading and variation in light intensity may help in improving the farming technique in order to bring into full play the productive potential of tree-crop inter cropping system (Liu 1991).

The solar radiation received by crop varied with the shading effect of trees. To see the effect of the light it is best to account the daily sum of shady area and shade light intensity. The vertical and horizontal distribution of radiation is also important for crop productivity. Radiation intensity of a point varies greatly from place to place, depending on the sun elevation angle, which leads to corresponding changes in the horizontal distribution in the daily course.

Studies carried in a 7 year old *Acacia tortilis* plantation during the monsoon season at Jodhpur showed that during the morning hours air temperature beneath the tree canopy was lower by 0.1°C to 0.7°C than those recorded in the open, whiled during the afternoon period the temperatures underneath the tree canopy remained lower by as much as 0.6°C to 2.0°C than those recorded in the open (Rama Krishna and Sastri, 1977).

Micro-metrological parameters reported by Hazra and Tripathi (1994) in different silvopastoral systems revealed that under the *A.nilotica* canopy, the light infiltration was minimum (59.2% of PAR) while it was maximum in association with *H. binata* (83.1%). The relative humidity and the difference of air temperature and leaf temperature were higher in all the silvopastoral system as compared to open pasture. Roy and Pathak (1994) reported that in general, soil and leaf temperature, relative humidity and light intensity showed decreasing value with increasing intensity of tree canopies. However, ambient air temperature showed increasing trend with increasing intensity of tree canopy. Under *A. lebbek*; soil, leaf and air temperature were found higher in summer whereas values of light intensity and relative humidity were found higher during monsoon.

Mishra and Bhatt (1993) studied interaction of infiltered light and canopy temperature on the productivity of grasses viz., *Sehima nervosum* and *Heteropogon contortus* under the tree canopies of *Hardwickia binata, Leucaena leucocephala, Acacia torttils, Albizia lebbek* and *Albizia amara. Acacia tortilis* and *Hardwickia binata* allowed 60 percent and 55 percent PAR infiltration, whereas *Leucaena leucocephala* infiltered the least amount (33%). Higher canopy air temperature difference (-6.8°C) was recorded under *Leucaena leucocephala.* About

98 per cent dry matter yield of grasses was obtained under *Hardwickia binata* and *Acacia tortilis* canopy when compared with the dry matter yield in open field (Bhatt *et al.*, 1994).

Biomass Production

Forage production under silvopastoral system is directly influenced by its establishment, subsequent management, edaphic and climatic conditions, tree density, canopy development, etc. The information is available on forage production under different silvopastoral systems in different agro-climatic conditions. It is generally based on 4 to 6 years studies (Singh and Roy, 1991; Pathak and Roy, 1994). Lopping/ pruning of trees also contribute towards forage production. Long term studies conducted on silvopastoral systems at Jhansi and Datia in red gravelly soils with five trees pecie sat each location (*Albizia amara, Albizia lebbek, Acacia tortilis, Hardwickia binata* and *Leucaena leucocephala*) at Jhansi and (*A.amara, A.lebbek, A.tortilis, Dalbergia sissoo* and *L.leucocephala*) at Datiaunder 3 spacing (4x3, 4x4 and 4x6m) revealed that trees did not have any adverse effect on forage production upto 11-12 years at any spacing (IDRC, 1983; IDRC, 1994).

In arid condition, it has been observed that in addition to indigeno us species like *Prosopis cineraria* and *Zizyphus numularia*, the introduced species such as *Acacia tortilis, Dichrostachy scinerea, Colophospermum mopane* showed very good performance in silvopastoral systems (Ahuja, 1980). Forage production from lower-story crops under 14-18 years old plantation of 4 desert trees (*Prosopis cineraria, Tecomella undulata, Albizia lebbek* and *Acacia senegal*) ranged from 0.6 to 1.5 t/ ha/year (Ahuja, 1980).

In another study on silvopastoral system conducted at CSWRI, Bikaner exhibiting arid climate involving *P.cineraria, Zizyphus mauritiana, A.tortilis and C.mopane* as a tree component and *Lasirus sindicus* (Sewan grass) as the pasture component; the total dry matter and grass yield obtained ranged from 3.50 to 3.97t/ha and 2.64 to 3.10t/ha, respectively (CSWRI,1998). However, silvopasture had to be reestablished after some years due to poor survivability of tree species.

Mishra and Bhatt (1994) reported that the crop growth rate and dry matter yield of grasses were higher (98%) under canopies of *A.tortilis* and *H.binata*. The dry matter yield was closely associated with the available PAR under the tree canopies than the canopy temperature indicating that for production of these grasses about 55 to 60 percent PAR is required.

The relative forage yield under *Albizia procera* was minimum (70.7%) while relative dry forage product ion of 94.6% was recorded higher in association with *A. lebbeck*. It was also observed that reduction in forage yield of under story crop, light transmission through canopy is not the only responsible fact or but moisture content sseems to be another factor as relative humidity under*A. nilotica* was maximum (70%). Lowest relative yield of 70.7% was recorded in association with *A. procera* (Hazraand Tripathi, 1994).

On the basis of more than 20 years of study in shallow red gravelly soils under semi arid condition at Jhansi, Pathak *et.al.* (1995) reported that the degraded lands producing hardly upto 1t/ha/year biomass can be improved upto 10t/ha/year (at a 10 years rotation) through silvopastroal systems. Besides yield improvement by 8 to 10 times, the quality of mixed forage has also improved by 6 to 7 times.

Another 15 years of study in Yamuna ravine near Agra, revealed that a silvopastoral system consisting of *A.nilotica* + *C. ciliaris* and *A. tortilis* + *C. setigerus* planted at 3x3m spacing produced on average the total biomass yield of 5.0t/ha/year (2.5t fuel+2.5t forage) and 6.9t/ha/year (4.2t fuel + 2.7t forage), respectively (Samara *et.al.*, 1999).

Similarly, studies at CIRG, Mathura also suggested that a 3 tier silvopastoral system consisting of *A. nilotica* + *L. leucocephala* + *C. ciliaris* produced 8.0t/ha biomass, 33,866 Mcal/ha gross energy and 1270 kg/ha crude protein in ravine grazing lands, followed by the two tier system. The minimum dry forage of 2.40 t/ha, gross energy of 12,624 Mcal/ha and crude protein yield of 160 kg/ha were recorded in the natural silvopasture (Bhattacharya and Sharma, 1993).

The production potential of pasture alone (*C. ciliaris*), fodder trees alone (*H. binata* + *C. mopane*) and silvopastoral systems *(C. ciliaris + H. binata + C. mopane)* was evaluated at CAZRI, Jodhpur for nine years. In the end, it was observed that silvopastoral system was superior in achieving higher average forage production (1.73 t/ha/year) and livestock maintenance (4.1 acu/ha). It was followed by pure pasture and pure trees block (Harsh *et.al.*, 1992).

Dry matter product ion by grasses under *Albizia lebbek, Albizia procera, Leucaena leucocephala* and *Acacia tortilis* was 665, 621, 565 and 608g/m2, respectively, and ino pen sites the production was 660g/

m2 (Hazra and Patil, 1985). A significant positive relationship between dry matter yield (DMY), leaf area index (LAI), intercepted photosynthetically active radiation (IPAR) and light utilization efficiency (LUE) was developed by Bhatt and Mishra (1990). Genotypes IGFRI-3108 and 678 of *Cenchrus ciliaris* and 76,77 of *C. setigerous* were identified for high LAI, LUE and DMY and were recommended as promising genotype for higher forage productivity in tropical region.

Growth, biomass production, photosynthesis and quality attributes of *Panicum maximum* were studied under the canopies of seventeen year old trees of *Leucaena leucocephala* in rainfed semi-arid tropical environment (Bhatt *et al.*, 2010). The full grown canopy of *L. leucocephala* at the spacing of 4x4 mallowed only 30-35 percent of total photosynthetic photon flux (PPF) which resulted in reduction in under canopy temperature to 4.10 C over the open air temperature. The total biomass production in terms of fresh and dry weight decreased under the tree canopies on account of reduction in leaf area index. On average basis of two years, *P. maximum* had produced 25.56t/ha green and 7.37t/ha dry biomass under the tree canopies of *L. leucocephala*. The dry matter yield reduced to 13.4 percent under the tree canopies over the open grown grasses. The higher soil moisture, organic carbon content and available nutrients under *L. leucocephala* + *P. maximum*; led to sustainable growth and biomass production of the system for longer period. On the basis of overall growth performance, biomass production, chlorophyll accumulation and quality attributes,*P. maximum* is recommended for high shading environment of *L. leucocephala* in the rainfed silvopasture system of semi-arid tropics.

Carbon Sequestration

It has been assumed that the rates of C sequestration will be sufficiently high through development of perennial pasture and silvopasture in the carbon depleted soils. In the terrestrial ecosystem carbon sequestration depends on different land use changes. Afforestation, reforestation and rehabilitation of degraded forests, range lands and wastelands are the effective ways of promoting carbon sequestration. The largest active potential is in tropics and subtropics, but realization of the potential will depend on land water availability (IPCC, 2001). Agroforestry is the ideal land use practice to increase productivity per unit area along with maximization of carbon sequestration indifferent stories of vegetation and at different depth sof soils. The IPCC report (IPCC, 2000) indicates that conversion of unproductive crop lands and grasslands to agro forestry is one of the best options to sink atmospheric carbon. Rapid changes in CO_2 emission has caused to assess the different land use systems for their carbon source and sink potential. There is sufficient evidence to suggest that land use systems like agroforestry, intercropping, multi-cropping, lay farming, mixed cropping, perennial pasture crop sand mulching promote carbon sequestration in the vegetative biomass and at different depths of the soil. The net flux of carbon from land use and land cover changes (LULCC) is significant in the global carbon budget. The estimates of the net emission from LULCC yield 1.10 + 0.25 Pg C/ year for the 1980 sand 1990s (Houghton, 2011). Out of the total emission the ocean is absorbing roughly 2Pg C of anthropogenic carbon per year (Tanhua *et al.,* 2011). Yude Pan *et al.,* (2011) estimated that the total C link in the world's forests outside areas of tropicaland use changes was 2.4Pg C/year or one third of annual fossil fuel emissions. The study also reveals that the global forest play an important role in regulating atmospheric carbon levels. The carbon storage of different tree-crop systems in different eco-regions of the world is given in (Table1).

Table 1. Potential C storage for Agroforestry systems in different eco-regions of the world

Country	Eco-region	System	MgC/ha	Reference
Africa	Humid tropic High	Agro-silvicultural	29-53	Dixon *et.al.*,1994
South America	Humid tropic Low	Agro-silvicultural	39-102a	
	Dry low land		39-195	
South east Asia	Humid tropical	Agro-silvicultural	12-228	Krankina and Dixon, 1994
	dry low lands		68-81	
Australia	Humid tropics low	Silvipastoral	28-51	Krankina and Dixon, 1994
North America	Humid tropics	Silvipastoral	133-154	Schroeder, 1994
	Humid tropics low		104-198	
	Dry low lands		90-175	
Northern Asia	Humid tropics low	Silvipastoral	15-18	Winjum *et.al.*, 1992

[a]Carbon storage values were standardised to 50-year rotation
(*Source:* after Albrecht and Kandjii, 2003)

The silvopsture systems have perennial trees, shrubs and perennial pasture grasses and range legumes and these are the potential components to sequencer atmospheric carbon in above and below ground biomass and also in the soils. Soil have the sink capacity of 0.4– 0.6 Pg C/yr over 50-1000 years (Paustian *et.al.*, 1998, Ingram and Fernandes 2001). Dixon (1995)worked out that 1.1–2.2 Pg C can be sequestered annually over 50 year in above ground and soil. Carbon sequestration through tree-crop inter cropping systems involves the removal and storage of carbon from the atmosphere invegetation and soil through biological processes. The multitier of trees + shrubs + grasses in the systems can increase the amount of carbon sequestered compared to a monoculture field of crop plants or pasture as also reported by Sharrow and Ismail (2004) and Kirby and Potvin (2007). Integration of trees, agricultural crops, with or without incorporating animals into an agro forestry system has the potential to enhance soil fertility, biodiversity and sequester carbon (Garrity 2004, Nair *et al.*, 2009).

Nair *et al.* (2009) showed that the carbon sequestration potential of the vegetation component (above and below ground) varied from 0.29Mg/ha/year in a fodder bank agro forestry system of West African Sahel to 15.21Mg/ha/year in mixed species stands at Puerto Rico. Dixon (1995) estimated a total sof 585–1,215 million ha of land in Africa, Asia and the Americas under agro forestry and a global potential to sequester 1.1–2.2Pg of carbon (vegetation and soil) over 50 years. The existing agro forestry systems if managed effectively could sequester an additional 12,000Mg C/year by 2010 and 17,000Mg C/year by 2040 as estimated by IPCC (2000). Trees play an important role in soil C sequestration (MontagniniandNair 2004; Takimoto *et al.*, 2008) and high tree density in a system, the biomass production per unit area will be higher with more C storage in vegetation and soil. Infact, recent research has shown higher soil C stock under deeper soil profiles in agro forestry systems compared to tree less agricultural or pasture systems under similar ecological settings (Haile *et al.*, 2008, Nair *et al.*, 2009). As reported by other workers the average carbon storage by agro forestry practices has been estimated as 9, 21, 50, and 63 Mg C/ha in semiarid, sub-humid, humid, and temperature regions. In the small holder agro forestry systems in the tropics, the carbon sequestration ranges from 1.5 to 3.5 Mg C/ha (Montagnini and Nair 2004). Lal and Kimble (2000) worked out that restoration of 1x109 ha of land in tropics has the potential to sequestration carbon at the rate of 1.5Pg/year at 1m soil depth with bulk density of 1.5 mg/m^3 with increasing soil organic content (SOC) at 0.01 percent every year.

In India large area is under one or other category of waste lands. Most of the soils of these waste lands are characterized by low soil organic carbon and, therefore, the carbon carrying capacity of such soils will be higher (Gill and Roy, 2001; Roy, 2010). Restoration of such lands will be useful for carbon sequestration and stabilization. Strategies for soil carbon sequestration can be conservation agriculture (reducing tillage, mulching etc.), soil and water conservation etc. The carbon storage under various agricultural systems in different regions of the India is reported in (Table 2).

Table 2: Total C storage under silvopasture systems in semi-arid regions of India

Pasture/Agroforestry systems	Total C storage (t C/ha)	Reference
Silvipastoral system (5years) *Acacia nilotica* + natural pasture	9.5 -17.0	Rai *et.al.*, 2001
A. nilotica+ established pasture	19.7	
Dalbergia sissoo + natural pasture	12.4	
D. sissoo + established pasture	17.2	
Hardwickia binata + natural pasture	16.2	
H. binata + established pasture	17.0	
Agri silvicultural system (11years) *Dalbergia sissoo* + crop	26.02	NRCAF, 2005
Silvopasture systems under highly degraded soils (10 years) (Trees + grasses + legumes)		Bhatt *et.al.*, 2007b
Acacia tortilis + *C. ciliaris* (20 years)	57.37	
Leucaenaleucocephala+ *P. maximum* (20 years)	59.74	

Bhattacharya *et.al.*, (2006) report about 10Pg SOC stock in 0-30cm soil depth in arid tropics of India. In silvopastoral systems of different trees under degraded soils in semi-arid tropics of India, the total carbon stock was estimated to be in the range of 33.77 MgC/ha to 45.27 MgC/ha at the age of ten years rotation (Bhatt *et.al.*, 2007b) (Fig1).

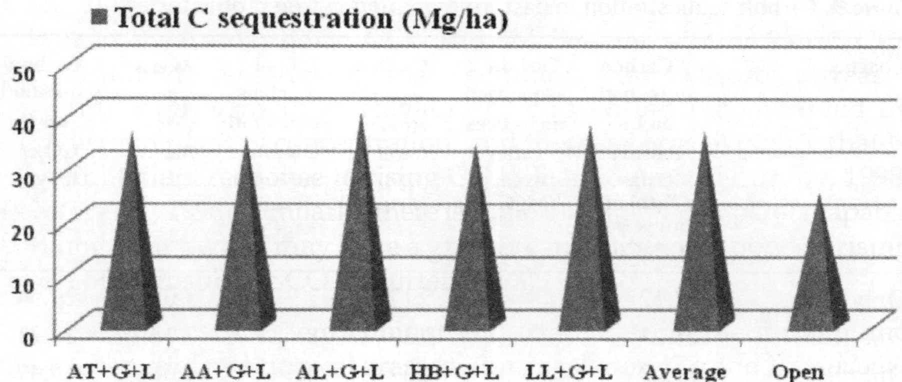

Fig. 1. Carbon sequestration in different silvopastoral systems in degraded range lands in semi-arid tropics (10 year rotation) (AT = *Acacia tortilis*; AL = *Albizia lebbek*; HB = *Hardwickia binat*; LL = *Leucaena leucocephala*; G = grasses; L = legumes)

The pasture species (*Panicum maximum, Cenchrus ciliaris* and *Stylosanthes hamata)* as grown alone and also as intercrops with trees (*Leucaena leucocephala, Grewia optiva, Dalbergia sissoo* and *Acacia tortilis)* in the silvopastoral systems were evaluated for carbon sequestration in sandy clayloam soil in semi-arid tropics of India (IGFRI, Jhansi; 25^0N78^0E, 275 m above msl) for four years (Bhatt *et al.*, 2006; Bhatt *et al.*, 2007a; Bhatt *et al.*, 2007b). The organic carbon stock of soil increased in the range of 3.39Mg C/ha to 5.40Mg C/ha and the carbon pool in vegetation has been recorded as 10.23Mg C/ha to 41.00Mg C/ha under different pasture crops and tree-crop combinations in the four year sof system growth (Table 3). The total carbon sequestration in vegetation + soil under different trees + pasture grasses + legume ranged from 13. 47 to CO_2e/ha/year to 42.24 to CO_2e/ha/year after four years. These results revealed that the C stock in vegetation and soil increased over the years during the active growth stage of the systems and should be considered as the potential systems for carbon sequestration in degraded soils of semi-arid region of the India.

Table 3. Carbon sequestration in pasture crops and in tree-crop interface

Systems	Carbon sequestration in Pasture crops (Total of four years) (MgC/ha)	Carbon sequestration in trees (after 4year) (MgC/ ha)	Carbon sequestration in soil (Mg C/ha)	Total Carbon sequestration in (vegetation +soil) (MgC/ha)	Average (MgC /ha/yr)	Carbon sequestration (t CO_2e/ ha/yr)
Only PM	27.12	-	5.17	32.29	8.07	29.54
Only CC	25.23	-	3.39	28.62	7.16	26.21
Only SH	10.23	-	4.51	14.74	3.68	13.47
PM + SH + LL	36.75	4.25	5.17	46.17	11.54	42.24
PM + SH + GO	35.61	0.39	5.40	41.40	10.35	37.88
CC + SH +DS	29.62	2.78	4.28	36.68	9.17	33.56
CC + SH + AT	29.40	3.32	4.04	36.76	9.19	33.64
CD at 5%	0.86	0.12	0.21	-	-	-

(Notes: PM– *Panicum maximum*, CC – *Cenchrus ciliaris*, SH– *Stylosanththes hamata*, LL – *Leucaena leucocephala*, GO – *Grewia optiva*, DS– *Dalbergia sissoo*, AT–*Acacia tortilis*)

In view of the above information it could be concluded that significant amounts of carbon may be sequestered through promotion of silvopastoral systems on degraded and wastelands that are available in the country. The effect of such interventions is likely to have an important role in regulating atmospheric carbon levels in regional/ global contexts. However, there is still a good scope of additions to our current knowledge, especially on C sequestration and GHG mitigation. The following thrusts are considered relevant:

i) More precise estimations of C stocks using standard methods in order to improve reliability and uniformity of data sets.

ii) Development of predictive models to depict future climate and land use changes and role of silvopasture/agroforestry systems in CO_2 mitigation.

iii) Linking widespread adoption of silvopasture/agroforestry systems on account of its impressive cost benefit analysis and GHG mitigation.

References

Albrecht, A. and Kandjii, S. T. 2003. Carbon sequestration in tropical agroforestry systems.*Agriculture, Ecosystems and Environment* 99: 15-27.

Ahuja, L. D. 1980. Grass production under Khejri. In: *Khejri* (*Prosopis cinererea*) *in theIndian Desert – its Role in Agroforestry* (Eds Mann, H. S, and Saxena, S. K.). Monograph Number 11. CAZRI, Jodhpur 28-30 p.

Bhattacharyya,N. K. and Sharma, K. 1993. Silvipastoral system in relation to goat farming. In: *Agroforestry in 2000 AD for the Semiarid and Arid Tropics* (Eds Gill, A. S., Deb Roy, R. and Bisaria, A. K.). NRCAF, Jhansi 1-12 p.

Bhattacharyya, T., Chandran, P.,Ray, S.K., Mandal, C., Pal, D.K., Venugoplan, M.V., Durge, S.L.Srivastava, P.,Dubey, P.N., Kamble, G.K.Sharma, R.P.,Wani, S.P., Rego, T.J., Ramesh, V.and Manna, M.C. 2006. Estimation of carbon stocks in red and black soils of selected bench mark spots in semi-arid tropics of India. Global Theme on Agro-ecosystems Report No.28. ICRISAT, Patancheru 86 p.

Bhatt, R. K. and Misra, L. P. 1990. Transpiration rate and energy exchange in trees under silvopastoral farming systems. *In*: *Proceedings National Seminar of Indian Society of Plant Physiology*. Bombay 230-234 p.

Bhatt, R. K., Misra, L. P. and Pathak, P. S. 1994.Transpiration and stomatal conductance of grass species under silvipastoral system.*Range Management & Agroforestry* 15: 87-92.

Bhatt, R.K., Baig, M. J., Tiwari, H. S. and Roy, S. 2007a. Effect of high level CO_2 on biomass production and carbon sequestration in silvipastoral systems under semi agri tropic. ICARAP Cess Fund Scheme, FinalReport (April 2004 – March 2007) 60 p.

Bhatt, R.K.2007b. Carbon Sequestration in silvipastoral systems. In: *Compendium of Training Programme* (Jan. 18-24) *on Design Development and Management of Silvopasture on Degraded Dry Lands*. IGFRI, Jhansi.

Bhatt, R.K., Yadava, R. B., Baig, M. J., Roy, S., Tiwari, H. S. and Dubey, J. 2006. Sequestration of CO_2 in range lands: towards realistic mitigation. Paper presented in International Conference on Regional Carbon Budgets (16-18 August) at Beijing, China.

Bhatt, R.K.,Baig, M. J., Tiwari, H. S.and Roy, S. 2010. Growth, yield and photosynthesis of *Panicum maximum* and *Stylosanthes hamata* under elevated CO2. *Journal of Environmetal Biology* 31:549-552.

Castro, K., Sanchez, G. and Rivard, B.2003. Monitoring secondary tropical forests usings pace borne data: implications for Central America. *Int.J. Remote Sensing* 24: 1853–1894.

Conner,D. J.1983. Plant stress factors and their influence on production of agro forestry plant association. *In: Plant Research and Agroforestry* (ed. Huxley, P. A.). ICRAF, Nairobi 401-426 p.

CSWRI. 1988. *35 Years of Research.* Central Sheep and Wool Research Institute, Avikanagar 128 p.

Dixon, R.K.1995. Agroforestry systems: sources or sinks of green house gases? *Agroforestry System* 31:99-116.

Dixon, R. K., Brown, S., Houghton, R. A., Solomon, A. M., Trexler, M. C. and Winiewski. 1994. Carbon pool and flux of global forest ecosystems. *Science* 263: 185-190.

Garrity, D.P. 2004. Agroforestry and the achievement of the millennium development goals. *Agroforestry Systems* 61:5–17.

Gill, A. S. and Roy, M. M. 2001. Improved silvopastures for joint forest management in India. *Indian Forester* 127: 303-314

Haile, S.G., Nair, P.K.R. and Nair, V.D. 2008. Carbon storage of different soil-size fractions in Florida silvopastoral systems. *Journal of Environmental Quality* 37:1789–1797.

Harsh, L. N., Tewari, J. C., Burman, U. and Sharma, S. K. 1992. Agroforestry for arid regions. *Indian Farming* 42: 32-37.

Hazra, C. R. and Patil, B. D. 1985. Forage and seed yield of crops under agroforestry production system – radiation and temperature relation. *Journal of Agronomy and Crop Science* 155: 186-192.

Hazra, C. R. and Tripathi, S. B. 1994. Soil properties, micrometeorological parameters, forage yield and phosphorus uptake of berseem as influenced by phosphate application under agroforestry system of production. *Journal of Agronomy and Crop Science*156: 145-152.

Houghton, R.A.2011. Land use changes, carbon sequestration and emission. Paper presented in Geo- Carbon Conference: carbon in Changing World (24-26October), FAO, Rome.

IDRC. 1983. *Report on Silvipasture Research Project (Phase I).* Indian Grassland & Fodder Research Institute, Jhansi 42p.

IDRC. 1994. *Report on Silvipasture Research Project (Phase II).* Indian Grassland & Fodder Research Institute, Jhansi 63p.

Ingram J. S. I. and Fernandes, E. C. M. 2001. Managing carbon sequestration in soils: concepts and terminology. *Agriculture, Ecosystems and Environment* 87: 111-117.

IPCC. 2000. Landuse, land-use change, and forestry. Special Report of the Inter-governmental Panel on climate Change. Cambridge University Press, Cambridge 375p.

IPCC. 2001. Climate change 2001. Mitigation. Special Report of the Intergovernmental Panel on climate Change. Cambridge UniversityPress, NY (http:www.ipcc.ch/pub/tar/wg2/index.htm).

Jose, S. 2009. Agroforestry for ecosystem services and environmental benefits: an overview. *Agroforestry System* 76:1-10.

Kirby,K.R. and Potvin, C.2007. Variation in carbon storage among tree species: implications for the management of a small scale carbon sink project. *Forest Ecology & Management* 246:208–22.

Krankina, O. N. and Dixon, R. K. 1994. Forest management options to conserve and sequester terrestrial carbon in Russian Federation. *World Resource Review* 6: 88-101.

Lal, R. and Kimble, J. M. 2000. Soil C pool and dynamics in cold eco-regions. In: (eds) *Global Climate Change and Cold Regions Ecosystems* (Eds R. Lal, J. M. Kimble and B. A. Stewart). CRC/Lewis Publisher, Boca Raton, FL 3-28 p.

Liu, N. 1991. Light distribution in tree intercropping area and its agricultural value. In: *Agroforestry Systems in China* (Eds. Zahaohua Z., Mantang C., Shlji W. and Youxu J.). Chinese Academy of Forestry, Bejing& IDRC Regional Office (South East and East Asia), Singapore 14-20 p.

Mishra, L. P. and Bhatt, R. K. 1993. Light transmission and canopy temperature variation of MPTS. *Agroforestry Newsletter* 5: 3-4.

Mishra, L. P. and Bhatt, R. K. 1994. Productivity and adaptation of grasses under silvipastoral system in semiarid region. In: *Agroforestry Systems for Degraded Lands* (Eds Singh, P. S. Pathak and M. M. Roy). Oxford & IBH Publishing Co., New Delhi 727-731 p.

Montagnini, F.2006. *Environmental Services of Agroforestry Systems.* Food Products Press, USA126 p.

Montagnini, F.and Nair, P. K. R.2004. Carbon Sequestration: An under-exploited environmental benefit of agroforestry systems, *Agroforestry Systems* 61:281-295.

Nair, P. K. R., Kumar,B. M. and Nair,V. D.2009. Agroforestry as a strategy for carbon sequestration. *Journal of Plant Nutrition and SoilS cience*. 172:10–23.

NRCAF.2005. AnnualReport, National Research Centre on Agroforestry, Jhansi.

Paroda, R. S., Jatasra,D. S.1985. Forages for arid lands. In: *Current Practices in Dry land Resources and Technology* (Eds. Alani Singh and G.R. Chowdhary). Scientific and Geo environ Academia, Jodhpur 256p.

Pathak, P. S. and Roy, M. M. 1994. *Silvipastoral System of Production*. Indian Grass land & Fodder Research Institute, Jhansi 54p.

Pathak, P. S., Gupta, S. K. and Singh, P. 1995. *IGFRI Approaches: Rehabilitation of Degraded Lands*. Indian Grassland & Fodder Research Institute, Jhansi 23p.

Paustian, K., Cole, C. V., Sauerbeck, D. and Sampson, N. 1998. CO_2 mitigation by agriculture: An overview. *Climate Change* 40: 135-162.

Rai, L., Yadav, R.S. Solanki, K.R., Rao, G.R. and Singh, R. 2001. Growth and pruned biomass production of multi purpose trees pecies in silvopastoral system on degraded lands in semi arid region of Uttar Pradesh, India, *Forest Tree and Livelihood* 11:347-364.

Ramakrishna, Y. S. and Sastri, A. S. R. A. S. 1977. Microclimate under *Acacia tortilis* plantation. CAZRI Annual Report, CAZRI, Jodhpur 69-70 p.

Roy, M. M. 1996. *Hardwickia binata* for silvipaastoral systems in India. *Agroforestry Today* 8: 12-13.

Roy, M. M. 2010. Ameliorating forage availability from non-arable lands: Issues and strategies in arid regions of India. *Range Management & Agroforestry* 31: 39-40.

Roy, M. M. and Pathak, P. S. 1994. Agroforestry interventions for livestock producers. *Range Management & Agroforestry* 15: 217-227.

Samra, J. S., Vishwanatham, M. K. and Sharma, A. R. 1999. Biomass production of trees and grasses in a silvopasture system on marginal lands of Doon valley of north west India. *Agroforestry Systems* 46: 197-212.

Schroeder, P. 1994. Carbon storage benefits of agroforestry systems. *Agroforestry Systems* 27: 89-97.

Schroth, G., and Sinclair, F. 2003. *Trees Crops and Soil Fertility: Concepts and Research Methods*.CABI,Walling ford, UK464 p.

Schroth, G., daFonseca, G. A. B., Harvey, C. A., Gascon, C.,Vasconcelos, H. and Izac, A.N. 2004. *Agroforestry and Biodiversity Conservation in Tropical Landscapes*. Island Press, Washington 537p.

Sharrow, S.H. and Ismail, S. 2004. Carbon and nitrogen storage in agroforests, tree plantations and pastures in western Oregon, USA. *Agroforestry Systems* 60:123–130.

Singh, P. and Roy, M. M. 1991. Forage production through agroforestry: constraints and priorities, *Range Management and Agroforestry* 12: 169-178.

Singh, P. and Roy, M. M. 1993. Silvipastoral systems for ameliorating productivity of degraded lands in India. *Annals of Forestry* 1: 61-73.

Takimoto,A.,Nair P.K.R., Nair,V. D.2008. Contribution of trees to soil carbon sequestration under agroforestry systems in the West African Sahel. *Agroforestry Systems* 125: 159-166.

Tanhua, T., Kortzinger, A.Wallace, D.W.R.2011. Observations of the ocean's role in the carbon cycle. Paper presented in Geo-Carbon Conference: Carbon in Changing World (24-26 October). FAO, Rome.

Winjum, J. K., Dixon, R. K. and Schroeder, P. E. 1992. Enhancing global potential of forest and agroforest management practices to sequester carbon. *Water Air Soil Pollution* 64: 213-228.

Yude P., Birdsey, R. A., Fang, J., Houghton, R., Kauppi, P. E., Kurz, W. A., Phillips, O. L., Shvidenko, A, Lewis, S. L., Canadell, J. G., Ciais, P., Jackson, R. B., Pacala, S. W., McGuire, A. D., Piao, S., Rautiainen, A., Stich, S. and Hayes, D. 2011. A large and persistent carbon sink in the world's forests. *Science* 333: 988-993.

□□□

Chapter 16

Screening and Selection of Pearl Millet Genotypes

V. K. MANGA

ICAR - Central Arid Zone Research Institute, Jodhpur, Rajasthan-342 003
E-mail: Vk_manga@yahoo.co.in

Introduction

Arid regions occupy nearly 3,17,090 sq km area in India, of which 2,46,790 sq km is hot arid region and 70300 sq km is classified as cold arid (Krishanan, 1968). The hot arid zone is spread over the states of Rajasthan, Gujarat, Punjab, Haryana, Maharashtra, Karnataka and Andhra Pradesh. About 80% of the hot arid area falls in the eleven districts of Rajasthan. These areas experience an annual rainfall between 100 mm in the north-western sector of the Jaisalmer district to 450 mm in the eastern boundary of the arid zone in Rajasthan. The arid region also experiences extreme temperatures. The mean highest day time temperatures vary between 36°C and 42.9°C in the east to 38.8°C and 45.5°C in the western most part of the arid region (Rao and Singh 1998). Maximum surface soil temperatures reach beyond 62 °C during May and June. Low erratic rainfall combined with extreme temperatures results in frequent crop failures. Soils of arid western Rajasthan are mostly sandy with a low organic matter content (0.1 to 0.45 per cent) and with poor moisture holding capacity and high infiltration rate of 9 cm/hour. Soil salinity and alkalinity to the extent of 45 per cent of the un-irrigated area and surface crust formation after kharif sowing, following light showers, complicates the situation (Singh *et al.*, 1974).

Pearl millet is a major warm season cereal valued for its food, feed and fodder uses. It is cultivated in the arid and semi-arid tropical regions, where soil surface temperatures can rise above 60°C, which can adversely affect germination and seedling survival, leading to poor crop stand. Seedlings of some of the pearl millet genotypes has been found to be even more heat tolerant than sorghum with several genotypes surviving at as high as 62°C of soil surface temperature (Peacock *et al.*, 1993). Sorghum is a drought tolerant species, and pearl millet is even more drought tolerant with higher water use efficiency than sorghum. When frequently irrigated, water use efficiency of sorghum and pearl millet was comparable to maize, but as number of irrigations reduced and severe stress situations emerged, pearl millet becomes most water use efficient crop (Singh and Singh 1995). Sorghum has been characterized as moderately tolerant to soil salinity and even more salt tolerant than maize (Mass 1985). Pearl millet is even more salt tolerant than sorghum and is the second most salinity tolerant major cereal after barley. Also much larger genetic variability for whole plant response to soil salinity has been reported in pearl millet (Ashraf and McNeilly 1987; Krishnamurthy *et al.*, 2007). Large variability for salinity tolerance has been detected and salinity tolerant germplasm and improved populations and breeding lines have been identified (Ramesh *et al.*, 2005; Kulkarni *et al.*, 2006). Pearl millet is better adapted to dry nutrient deficient soils and is thus cultivated under extreme harsh conditions of high temperature, low erratic rainfall and on soils with poor water holding capacity. It also has strong deep root system and short life cycle and can grow rapidly when moisture is available. As a result it can survive and reliably produce small quantity of grain where mean annual precipitation is as low as 250 mm as compared to minimum water requirement of 400 mm for sorghum and 500-600 mm for maize.

Pearl millet is the only kharif cereal farmers can grow successfully under highly unfavourable and harsh conditions of arid region. This crop has short growing period (70-85 days), can grow under extreme moisture stress and low fertility conditions and has ability to recover quickly after intermittent dry spell. Because of these qualities, this crop has been extensively adopted as a subsistence crop by poor farmers of dry areas, and forms staple diet of the people residing the driest and poorest regions. India is the largest producer of pearl millet, both in terms of area (9.3 m ha) and production (7.97 m t), with an average productivity of 856 kg ha^{-1}. This crop contributes 7.8% to the total food grain area of the country and 3.9% to the total food grain production. Rajasthan constitutes about 50% area and 42% of production of pearl millet in the country. Other principal pearl millet growing states are Maharashtra, Uttar Pradesh, Gujarat and Haryana. The average productivity of A zone (north-western states) is 685 kg ha^{-1}. Within this zone, parts of Rajasthan, Haryana and Gujarat receiving less than

400 mm of rainfall are grouped into a sub zone i.e. A1 zone. This sub-zone is highly drought prone with average annual rainfall below 400 mm, light sandy soils, and high temperatures. Average productivity of the A1 zone (hyper arid areas of Rajasthan, Haryana and Gujarat state) is about 410 kg ha^{-1}.

Pearl millet is cultivated mainly in the kharif (rainy season). However cultivation of pearl millet as a summer crop is also catching up in northwestern India. In the summer season it is cultivated under irrigated conditions during April–June when other crops are not commonly grown. Within the northwestern states of India, the area under summer pearl millet has expanded rapidly in Gujarat and now accounts for the highest area and production in the region. While *Kharif* pearl millet is cultivated under rainfed conditions with low-input-low-output management, summer pearl millet is cultivated under high-input-high-output management. With the higher prices of wheat, rice and maize, cattle and poultry feed and alcohol industries are looking for cheaper alternative sources like pearl millet. Further demand for pearl millet from health conscious food products industry is increasing as it contains more fibre and is good for diabetic and heart patients. Farmers are shifting to summer cultivation under irrigated conditions which gives higher yields (3000-4000 kg/ha) and returns. Thus grain yield of summer pearl millet are much higher since it is grown under irrigation and its grain quality is also superior. It is grown as a commercial crop with the bulk of the crop sold domestically, and also exported to neighboring countries. During summer season crop experiences high air temperatures exceeding 42°C in parts of Gujarat and Rajasthan. Pearl millet hybrids developed for cultivation during the rainy season when planted in the summer season, fail to set seed or have unsatisfactory seed set at temperatures exceeding 42°C during flowering.

Climate change and agriculture

Climate is one of the major physical factors that govern the choice of crops, cropping systems or farming systems of a region. Climate change is occurring due to global warming. Using a regional climate model, PRECIS (Providing Regional Climates for Impact Studies), it has been predicted that by the end of 21st Century mean annual surface temperature of the country may increase by 3.0-5.0°C with warming more pronounced in the northern part. A 20% rise in all India summer monsoon rainfall is also projected, but Punjab and Rajasthan may experience decline by 10-30% (Rupa Kumar *et al.*, 2006). Along with these changes the rainy days are expected to be fewer by about 10 days in a year in the northern half of Rajasthan, Punjab and Haryana states, while the southwestern part of Rajasthan and adjoining arid Gujarat may have rainy days fewer by ~5 days. By the last quarter of this century annual temperature would most likely increase in Gujarat, Rajasthan, Haryana and Punjab by 3-5°C (Rupa Kumar *et al.*, 2006). The IPCC 2007 and PRECIS model have projected hotter days and warm nights and a reduction in rainfall in Thar region by 21st century. A study by Poonia and Rao (2013) showed that air temperatures by the end of 21st century are likely to increase by +3.3 °C at Bikaner, +3.4°C at Jaisalmer, +2.9°C at Jodhpur and +2.5°C at Pali. Jodhpur experienced the highest temperature of 48.3°C on 8th June 2011 and warmest winter in 2008-09. Similarly annual rainfall is likely to be increased by +100 mm at Bikaner, +124 mm at Jaisalmer, -40 mm at Jodhpur and +21 mm at Pali. This climatic variability will affect climate sensitive sectors like agriculture. Since agriculture in India is a source of livelihood for 65 per cent of the population and contributes to 27 per cent of the GDP, adverse impact on agriculture could have serious effect on the economy of the country besides an impact on food security and health.

A study of weather parameters over a long period revealed that both maximum and minimum temperatures showed increasing trend over time. The increasing temperature will most likely to have adverse impact on crop productivity in this region. Crop development will be faster, reducing the total length of crop growing season (Singh et. al., 2011). This would further exacerbate the water stress through higher evapo-transpiration. The projected 4°C rise in temperature due to global warming by the end of the 21st century will increase water requirement of pearl millet by 12.9% from the current levels (308-411 mm). This increased crop water requirements will result in reduction in crop growing period by about 5 days. As a result the crop acreage where rainfall satisfies crop water requirements will be reduced by 23.5% (Rao and Poonia 2011). Climate change would also result in shortage of moisture availability period, increase in dry areas, increase in salinity, change in soil pests and weeds, and increase in insect pests and diseases. As a result the crops/genotypes suitable for cultivation at present may not fit well in the changed scenario of future (Singh *et al.*, 2011). In such a scenario efforts would then be required to enhance crop resilience and adaptation to climate change for sustainable production, food and nutrition security. This may be achieved by developing varieties that are tolerant to high temperature coinciding flowering period, extra early varieties for shortened moisture availability period and varieties with terminal drought tolerance. Developing varieties with in-built developmental plasticity, i.e., that are early-maturing and capable of yielding grain under short dry periods but should have the ability to make best use of high moisture availability during good rainfall years by prolonging their vegetative phase, resulting in higher grain and fodder yield. Concerted efforts are also required to understand the underlying mechanism involved in heat and terminal drought tolerance to identify genetic stocks for the traits correlated with these stresses.

Screening of pearl millet genotypes resilient to climate change

For identifying pearl millet genotypes resilient to climate change experiments were conducted during the rainy seasons (Kharif) of the years 2011, 2012 and 2013. Each year genotypes were evaluated under two or more different environments (different dates of planting) and data were recorded on grain yield and other traits to study resilience of genotypes to climate change. The genotypes were also evaluated during the summers of 2012 and 2013 to study the effect of high temperatures at flowering on seed setting and to identify high temperature tolerant genotypes (Table 1 and 2).

Table 1. Pearl millet genotypes tested during kharif season in different yearsm

Season	No. of genotypes	Constitution	Environments
Kharif 2011	5	5 Populations	Two dates of sowing 7/7/ 2011 and 6/8/2011
Kharif 2012	36	6 pop+ 8 hybrids+ 14 inbreds + 8 ms lines	Two dates of sowing 9/8/ 2012 and 24/8/2012
Kharif 2013	38	6 pop+ 10 hybrids + 14 inbreds + 8 ms lines	Four set of environments in July 2013

Table 2. Number of genotypes of pearl millet tested during summer season in different years

Season	No. of genotypes	Constitution
Summer 2012	46	8 pop+ 6 hybrids + 25 inbreds +7 ms lines
Summer 2013	34	6 pop+ 6 hybrids + 14 inbreds +8 ms lines

Kharif 2011

Five pearl millet open pollinated varieties, CZP 9802, CZP 9603, CZIC 923, CZP2K-3 and CZP 2K-9 were planted at two different dates. First date of sowing (I DOS) was 7th July 2011 and second date of sowing (II DOS) was 6th August 2011. This allowed varieties to experience different climatic conditions at different growth stages. Crop planted on I DOS received higher rainfall and greater number of rainy days as compared to crop planted on II DOS. As a result crop planted on II DOS experienced terminal drought, higher maximum temperature and higher evaporation at and after flowering till maturity. This resulted in reduced plant height, flowering time, grain yield and dry fodder yield.

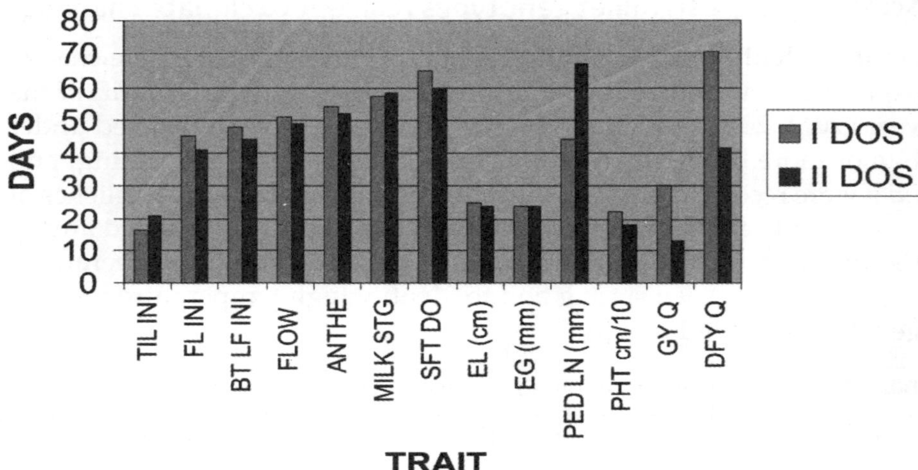

Fig. 1. Effect of Date of Sowing on Expression of Different Traits in Pearl Millet

TILINI = Tiller initiation; FLINI = Flowering initiation; BTLFINI = Boot leaf initiation; FLOW = 50% Flowering; ANTHE = Days to Anthesis; MILKSTG = Days to milk stage; SFTDO = Days to soft dough stage; EL = Ear length (cm); EG (Earhead girth in mm); PEDLN = Peduncle length in mm; PHT = Plant height in cm; GYQ = Grain yield in Quitals/ha; DFYQ = Dry fodder yield in Quintals/ha.

Kharif 2012

Thirty six genotypes of pearl millet consisting of six open pollinated varieties, eight single cross hybrids, fourteen inbred restorers and eight maintainers of ms lines were planted at two different dates. Sowing was delayed due to delayed onset of rains. First planting was done on 9th Aug 2012 and second planting was done on 24th Aug 2012. First planting received 423mm, while second planting received 202 mm of precipitation. As a result first planting received rains up to 40 days of its growth period, while second planting received rains up to 24 days of its growth, as a result experienced stress at flowering and post flowering stage.

Grain yield in first date of planting (213 kg/ha) over all entries was more than double, the average grain yield recorded in II date of planting (80 kg/ha). Among groups of genotypes, hybrids on an average recorded a higher grain yield (465 kg/ha) followed by populations (318 kg/ha) in I DOS. Average grain yield of populations (181 kg/ha) was slightly higher than hybrids (171 kg/ha) in II DOS. Inbreds were poor yielder as such, but among them inbred restorers recorded higher grain yield over ms lines.

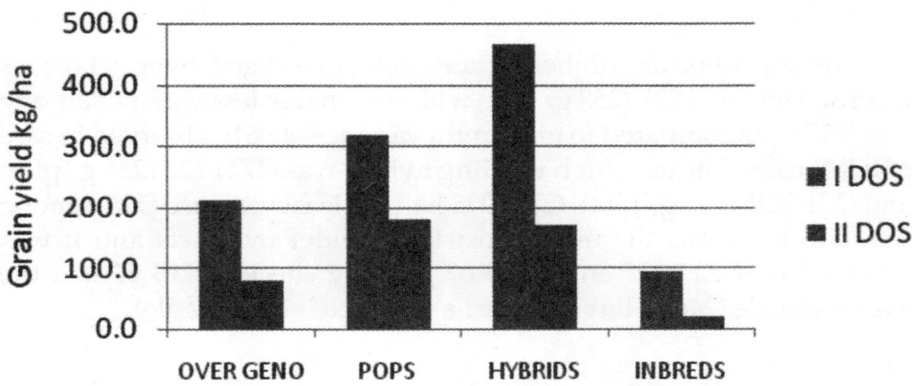

Fig. 2. Effect of Date of Sowing on Grain Yield in Pearl Millet Kharif (Kh) 2012

Hybrid CZH 227 recorded highest grain yield (562 kg/ha) in I DOS, while GHB 538 (217 kg/ha) was highest yielding in II DOS. Among populations CZP-2K 9 recorded the highest grain yield (441 kg/ha) in I DOS followed by CZP 9802 (374 kg/ha). In II DOS CZ-IC 923 recorded the highest grain yield (224 kg/ha) followed by CZP-2K 3 (219 kg/ha).

Considering ranking of genotypes for grain yield in two dates of sowing and mean over dates of sowing promising genotypes during 2012 were CZP 9802 and CZP 923 (populations), CZH 221, CZH 225, CZH 226 and GHB 538 (hybrids), CZI 2011/4, CZI 2011/7 and CZI 2011/5 (inbreds) and 97111A and CZMS 2A (ms lines).

Kharif 2013

Thirty eight genotypes consisting of six populations, ten hybrids, fourteen inbred restorers and eight maintainers of male sterile lines were planted in four sets. In each set, genotypes were subjected to a different environment. Set-I = Shade at vegetative and open at reproductive; Set-II = Shade at vegetative and reproductive stage; Set-III = Open at vegetative and shade at reproductive stage and Set-IV = open at vegetative and reproductive stage. Set-IV that remained open though out received 389.9 mm of rainfall.

Significant differences were observed among sets and genotypes. Set-II recorded the highest mean grain yield (200 g/plot) over genotypes as compared to 151 g/plot by set-III, 133 g/plot by set-I and 134 g/plot by set-IV.

Among hybrids, highest grain yield (averaged over sets) was recorded by CZH 227 (258 g/plot) with reasonably less variance among sets (2274), as compared to maximum variance of 9616 observed in case of ICMH 356. The next high yielding hybrid was CZH 226 (231 g/plot) and CZH 225 (222 g/plot). CZH 228 had the least variance (525) among sets and thus was the most consistent yielder over sets and it was followed by CZH 227 and HHB 67. Among checks HHB 67 had the least variance (2464), but recorded a low yield of 184 g/plot.

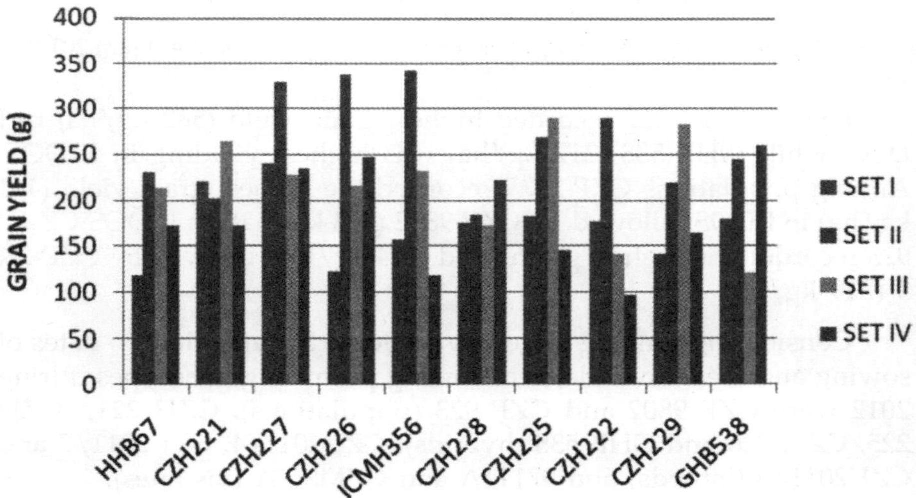

Fig. 3. Performance of Hybrids in Kharif 2013

Among populations, CZP 2K-9 recorded maximum mean grain yield (254 g/plot) followed by CZP 9802 (232 g/plot) and CZP 9603 (228 g/plot). Population CZP 2K-11 (178 g/plot) showed least variance (70) among sets and thus was most consistent in performance. It was closely followed by CZP 2K-3 (204 g/plot) and CZP 9802 (232 g/plot).

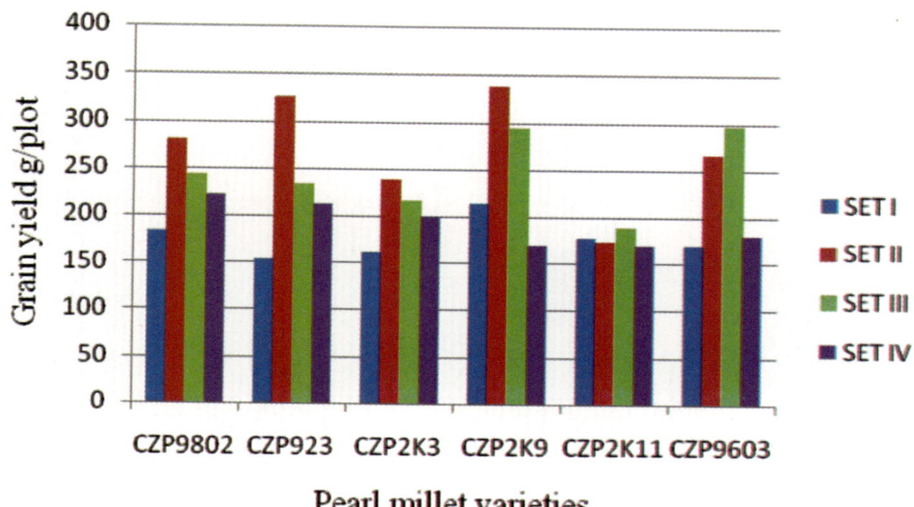

Fig. 4. Performance of Populations in Kharif 2013

Inbreds showing high mean yield over sets were CZI 9621, CZI 2011/7, CZI 2011/4 and CZI 2011/5. However inbreds with less variance among sets were CBI 832, CZI 2000/22, CZI 2004/8, CZI 2000/13, CZI 2011/6 and CZI 2004/7. Among maintainers of ms lines, ICMB 92777, ICMB 97111 and ICMB 93333 recorded higher grain yield. Of these ICMB 97111 had the least variance and hence performed consistent over sets. Among maintainers of CAZRI bred ms lines, CZMS 2B was most promising with least variance followed by CZMS 11A.

Evaluation of genotypes under high temperatures of summer

Summer 2012

Fifty three genotypes comprising of eight open pollinated varieties, seven hybrids, ten B (maintainer) lines and 28 inbred restorers were planted in high temperatures of summer 2012 (March to May). Crop experienced more than 44 °C temperatures during flowering. Variation in seed setting was observed.

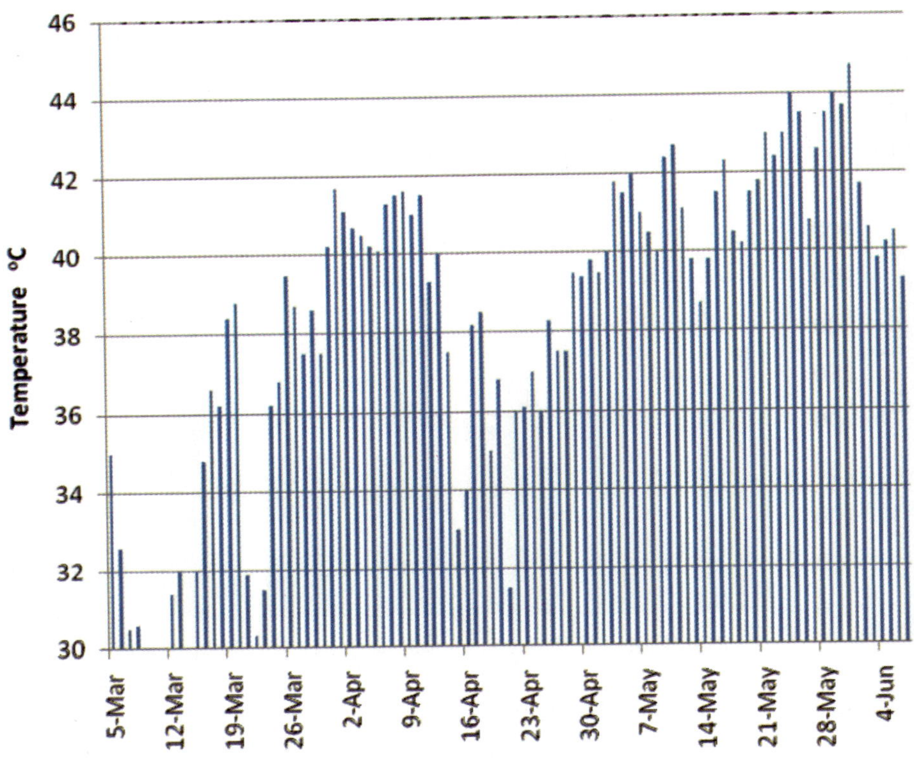

Fig. 5. Maximum Temperature during Summer 2012

Grain yield of 50 random panicles and 50 filled panicles was compared to study effect of high temperatures of summer on seed setting among different genotypes. In case of random ear heads, CZP-IC 923 recorded the highest grain yield (199 g/50 ear heads), followed by CZP 2K-9 (185 g) and CZP 2K-3 (182 g). In selected ear heads highest grain yield per 50 ear heads was recorded by CZP 904 (370 g), CZP 2K-9 (363 g) and CZP 2K-3 (351 g). However when difference in grain yield of selected and random ear heads was considered, pop CZP 9802 showed minimum (18%) yield reduction followed by CZP 9603 (37%) and CZP-IC 923 (38%). Thus CZP 9802 was the most tolerant to heat at flowering, but has low yield potential compared to CZP 2K-9 and CZP 2K-3.

Promising hybrids showing less reduction in yield and high absolute yield were 2A x CZI 2004/7, HHB 67, 5A x CZI 2007/7 and 88004A x CZI 2004/8. Among inbreds CZI 2002/3 recorded high mean yield and less reduction. Other inbreds showing high mean yield and up to 50% reduction in grain yield were CZI 2007/13, CZI 99/8 and CZI 2007/6. In case of ms lines 97111B and 93333B recorded high mean yield and less than 50% reduction in yield along with high yield potential. Among inbreds, promising were CZI 2002/3, CZI 2007/13, CZI 99/8, CZI 2007/6, CZI 2000/24 and CZI 2010/14.

Summer 2013

Thirty four genotypes of pearl millet including six hybrids, six populations, fourteen inbred restorers and eight maintainers of ms lines were evaluated for performance under high temperatures of summer 2013. The crop was planted on 13th March 2013 and harvested during the first week of June 13. Maximum temperature crossed 40 °C from the last week of April, and was between 42°C to 45°C from 15th May to 22nd May 13. This period coincided with the anthesis of most of the genotypes.

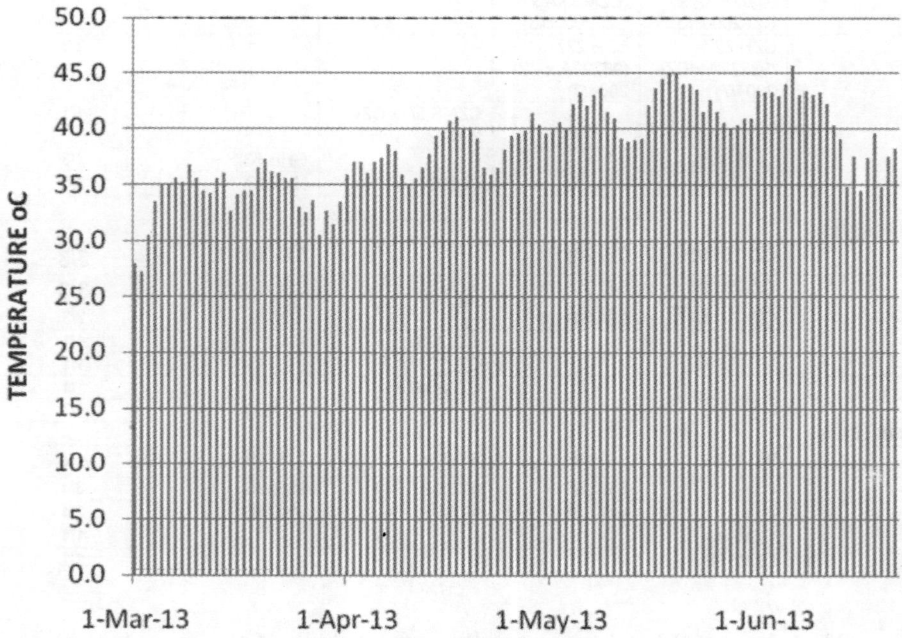

Fig. 6. Maximum Temperature during Summer 2013

Population CZP 9603 recorded maximum mean grain yield followed by CZP 2K-9 and CZP-IC 923. Out of six hybrids evaluated, CZH 225 (88004A x CZI 2004/8) recorded maximum mean grain yield and minimum reduction (29%) in grain yield/panicle. Among inbreds

CBI 834 recorded maximum grain yield and was followed by CZI 2011/5, CBI 832 and CZI 2000/24. Among maintainers of ms lines ICMB 93333 recorded maximum grain yield followed by ICMB 97111 and ICMB 92777.

Table 3. Performance of the promising genotypes over seasons

KH 2011	KH 2012	KH 2013	SUM 2012	SUM 2013	Freqency genotype among top three
POP					
CZP 9802	CZP 9802	CZP 9802	CZP 9802	-	**4/5**
CZP 2K-9	-	CZP 2K-9	CZP 2K-9	CZP 2K-9	**4/5**
CZP-IC 923	CZP-IC 923	-	-	CZP-IC 923	**3/5**
-	-	CZP 9603	-	CZP 9603	2/5
-	CZP 2K-3	-	CZP 2K-3	-	2/5
HYBRID					
-	-	**CZH 225** (88004A x CZI 2004/8)	CZH 225 (88004A x CZI 2004/8)	CZH 225 (88004A x CZI 2004/8)	**3/3**
	CZH 221 (CZMS 4A x CZI 2000/13)	CZH 221 (CZMS 4A x CZI 2000/13)	-	-	2/2
-	CZH 227 (98222A x CZI 2004/7)	CZH 227 (98222A x CZI 2004/7)	-	-	2/3
-	-	-	CZMS 5A x CZI 2007/7	-	1/1
-	GHB 538	-	-	GHB 538	2/3
-	-	-	HHB 67	HHB 67	2/4
-					
INBREDS					
-	**CZI 2011/7**	CZI 2011/7	-	CZI 2011/7	**3/3**
-	**CZI 2011/5**	CZI 2011/5	-	CZI 2011/5	**3/3**
-	CZI 2011/4	CZI 2011/4	-	-	2/3
-	-	-	CZI 2002/3	-	1/1
-	-	-	CZI 2007/13	-	1/1
-	-	-	CZI 2007/6	-	1/1
-	-		-	CBI 834	1/3
MS LINES					
-	97111B	97111B	97111B	97111B	**4/4**
-	-	93333B	93333B	93333B	3/4
-	-	92777B	92777B	92777B	3/4
-	CZMS4A	-	-	-	1/1
-	CZMS2B	-	-	-	1/1

The above table gives overall performance of the genotypes over five seasons. Perusal of the table reveals promising/resilient genotypes on the basis of frequency it performed better. It was observed that promising ones were CZP 9802, CZP 2K-9 and CZP-IC 923 (populations), CZH 225 (hybrid), CZI 2011/7 and 2011/5 (inbreds) and 97111B (maintainer of ms line).

The problems resulting from climate change, like elevated temperatures, drought, reduced growth period, salinity and insect pests etc. have hampered food production in areas where extra food production is needed, such as arid and semi-arid areas. To mitigate the effects of climate change, there is need to develop early maturing, heat tolerant varieties capable of setting and filling seeds at high temperature; drought tolerant varieties adapted to water deficits; salt tolerant varieties and disease and insect pest resistant varieties. Availability of genetic variation and sources of resistance to specific production constraints is prerequisite to any breeding programme.

The genetic structure of a variety plays a decisive role in its potential for adaptation to climate variability. The development of heterozygous and/or heterogeneous cultivars involving carefully selected genotypes forms one option to enhance the cultivar's capacity for buffering against climate variability. The degree of heterozygosity can be influenced through (i) choice of variety type-hybrids, and open-pollinated cultivars of outcrossing species like pearl millet posses high levels of heterozygosity; (ii) exploitation of heterotic groups in both population and hybrid breeding to maximize heterozygosity in these types of cultivars. Pearl millet population hybrids with significant superiority over the parental populations can be obtained *via* crossing of genetically distinct landraces. Genetically heterogeneous cultivar types that can profit from population buffering include varietal mixtures, multi-line varieties in case of autogamous species, open-pollinated population and synthetic varieties, as well as 3-way, 4-way and top cross hybrids. Single-cross hybrid cultivars being genetically uniform would not express population buffering. Such highly uniform variety types should be avoided aiming to improve adaptation to extreme climate variability unless farmers use multiple varieties differing for adaptation traits, like flowering time.

A wide range of crop improvement options for enhancing adaptation to climate variability are available. These include the choice of the type of cultivar, direct selection in multiple environments, farmer-participatory testing, indirect selection for individual adaptation traits using conventional or genomic selection methods, a dynamic genepool management approach, and selection for responsiveness to improved crop and soil management techniques. A better knowledge of the target environments and seasonal predictions may further reduce the potential costs of adaptation. Selection for climate variability adaptation will require testing and selection in many environments which enhances the chances of sampling key stress environments.

Advances in biotechnology and genetic engineering provide great opportunity for using such novel genetic material for development of different type of varieties suitable for different agro-climatic and management conditions and to tackle likely problems posed by climate change. QTLs have been identified in arid region crop like pearl millet for important traits including yield, disease tolerance, and drought tolerance. The identified QTLs could be utilized in backcross breeding program to transfer the desired traits into an adapted cultivar through marker assisted selection (MAS) and breeding (MABC).

References

Ashraf, M. and McNeilly, T. 1987. Salinity effects on five cultivars/lines of pearl millet [*Pennisetum americanum* (L)Leeke]. *Plant and Soil* 103:13-19.

Krishanan, A. 1968. Distribution of arid areas in India. *Proceedings of International Geographical Congress Symposium on Arid Zone*, 1968, pp. 11-19. Central Arid Zone Research Institute, Jodhpur.

Krishnamurthy, L., Serraj, R., Rai, K.N., Hash, C.T. and Dakheel A.J. 2007. Identification of pearl millet [*Pennisetum glaucum* (L.) R.Br.] lines tolerant to soil salinity. *Euphytica* 158:179-88.

Kulkarni, V.N., Rai, K.N., Dakheel, A.J., Ibrahim, M., Hebbara, M. and Vadez, V. 2006. Pearl millet germplasm adapted to saline conditions. *International Sorghum and Millet Newsletter* 47: 103-6.

Mass, E.V. 1985. Crop tolerance to saline sprinkling water. *Plant and Soi*l 89: 273-84.

Peacock, J.M., Soman, P., Jayachandran, R, Rain, A.U., Howarth, C.J. and Thomas, A. 1993. Effects of high soil surface temperature on seedling survival in pearl millet. *Experimental Agriculture* 29: 215-225.

Poonia Surendra and Rao, S. S. 2013. Climate change and its impact on Thar Desert Ecosystem. *Journal of Agricultural Physics* 13: 71-79.

Ramesh, S., Reddy, B.V.S., Sanjana Reddy P., Hebber, N. and Ibrahim, M. 2005. Response of selected sorghum lines to salinity-stress under field conditions. *International Sorghum and Millet Newsletter* 46: 14-17.

Rao, A.S. and Singh, R.S. 1998. Climatic features and crop production. In *Fifty Years of Arid Zone Research in India* (Eds. A.S. Faroda, Manjit Singh), pp 17-38. Central Arid Zone Research Institute, Jodhpur.

Rao, A.S. and Poonia Surendra. 2011. Climate change impact on crop water requirements in arid Rajasthan. *Journal of Agrometeorology* 13(1): 17-24.

Rupa, K. K., Sahai, A.K., Krishna, Kumar. K., Patwardhan, S.K. Mishra, P.K., Revadekar, J.V, Kamala, K. and Pant, G.P. 2006. High resolution climate change scenarios for India for 21st century. *Current Science* 90: 334-345.

Singh, R.P., Daulay, H.S., Prasad, M.V.R. and Singh, H.P. 1974. Hungary drylands of western Rajasthan need a new technology. *Indian Farming* 23(12): 5-6.

Singh, B.R. and Singh, D.P. 1995. Agronomic and physiological response of sorghum, maize and pearl millet to irrigations. *Field Crops Research* 42: 57-67.

Singh, D.V., Purohit, R.S., Mertia, R.S., Laxminarayan, Poonia, S and Kar, A. 2011. *Agrometerorological Data Handbook of Jodhpur* (1971-2010). Central Arid Zone Research Institute, Jodhpur, India, 110p.

❑❑❑

Chapter 17

Genomic Approaches for Development of Climate Resilient Cultivars

A.K. JUKANTI AND R.K. BHATT

ICAR-Central Arid Zone Research Institute, Jodhpur-342003, Rajasthan
E-mail:aravindjukanti@gmail.com

Introduction

The significant increase in the global population coupled with change in the climatic conditions demands the development of climate resilient or stress-tolerant crop cultivars. It is predicted that by 2030, climate change will have a massive impact on the agriculture production of country like India which largely depends upon agriculture as a primary occupation (Reynolds and Ortiz 2010). Development of climate resilient cultivars largely depends upon the availability of natural variability. But, unfortunately natural variability is not too large and needs to be widened for utilization/development of climate resilient cultivars. In addition to widening the genetic base proper care is to be adopted to conserve the available variability. Efforts in the direction of collecting, conserving and evaluating of plant genetic resources have been going on since long time. These efforts directly help the plant breeders in selecting and developing crop cultivars that can tolerate adverse climatic changes. The traditional plant breeding approaches has no doubt contributed significantly towards development of better performing cultivars. But, rapidly changing climatic conditions, water shortages, and global warming poses a serious challenge to agriculture production worldwide (IPCC 2007). Low and high temperature, unpredictable rainfall patterns, salinity, high radiation, non-uniform availability of nutrients in soil are all the manifestations of rapidly changing climate which can have devastating effects on global agriculture production.

Despite significant developments in the field of crop improvement during the past few decades, an approach involving both the traditional and modern technologies is the need of the hour to counteract the impending impacts of climate change. Though several advanced technologies are available they have not been utilized to their fullest potential. Therefore, an integrated approach involving (any or some) molecular breeding (marker assisted plant breeding), genomics (molecular markers, QTL mapping, genome sequencing, genome-wide association studies [GWAS],transcriptome profiling) coupled with other 'omics' (ex. phenomics, proteomics, epigenomics) and transgenics would definitely aid in faster and efficient development of tolerant crop cultivars. Here we present a critical but practical look at different technologies and their application in crop improvement under climate change conditions.

Marker-Assisted Breeding

Molecular markers

Molecular markers are widely being used in several crop improvement programmes, as they are linked to important agronomic and biotic/abiotic resistance traits. Molecular markers include: (i) based on hybridization such as RFLP (restriction fragment length polymorphism), (ii) PCR-based such as RAPD (random amplification of polymorphic DNA), AFLP (amplified fragment length polymorphism) and SSR (simple sequence repeat), and (iii) sequence-based markers such as SNP (single nucleotide polymorphism). Markers can be used for different purposes such as: (i) can be used for preparation or saturating molecular maps (ii) marker association with traits of interest can be utilized in crop improvement through marker-assisted selection (Koebner, 2004;Korzun, 2002) (iii) other important uses of markers include phylogeny and genome organization studies, diagnostics, gene introgression using backcrossing, and germplasm characterization (Varshney *et al.*, 2007).

Though different marker systems are available (as explained above), for crop improvement applications 'microsatellites or SSRs are markers of choice (Gupta and Varshney, 2000). SSR and AFLP markers are efficient at identifying polymorphisms but SSRs are more suited for automation (Shariflou *et al.*, 2001). While RAPDs are not easily reproducible, RFLPs are not suitable for automation. Another advantage of SSRs developed for a particular species can also be used in related species (cross species utility). 'Single nucleotide polymorphisms [SNPs]'are another class of markers that indicates single nucleotide (A, T, C, or G) difference in gene/genome between member species. SNPs are the most common type of nucleotide variations either in plants or humans. SNPs can be found either within a gene or outside. Although theymay not be associated with any character/trait, but if linked they certainly act as marker for that trait. SNPs provide wide marker density because the single nucleotide exchanges are common and numerous among the genomes. Additional advantage with SNPs is that they are not gel-based. In the near future, SNPs are expected to become the markers of choice especially with the availability of plant genome sequences being readily available (Ganal*et al.,* 2009).

'Diversity arrays technology [DArT]' is microarray-based hybridization procedure that used to detect presence versus absence of a particular sequence or fragment in genomic representations. Both SNPs and DArT systems enables the researchers to analyse genomes without prior knowledge of DNA sequence.'Genic or Functional' markers (FMs) from transcribed regions of the genome are possible due to the availability of sequence data through sequencing projects (Varshney *et al.,* 2007). The major advantage with FMs is that they exhibit a true and complete linkages with the potential trait therefore are better suited for crop improvement. Several SSRs, SNPs and FMs have been identified from EST or gene sequences.

Quantitative Trait Loci (QTL)

QTL is part/region of a genome that is linked or contains a gene (s) influencing a particular trait. QTL analysis refers to the detection of an association between trait phenotype and genotype of markers (Collard *et al.,* 2005). Markers are used to classify the mapping population (used for QTL identification) into different genotypic groups and determine whether significant differences exist between the groups with respect to the character being evaluated (Tanksley, 1993; Young, 1996). Usually, a significant difference among the different groups indicates the marker locus used to classify the population is linked to a QTL controlling that particular character. QTL detection is usually carried out through three methods: single marker analysis, simple interval mapping and composite interval mapping (Liu, 1998; Tanksley, 1993).

Several factors influence the identification of QTLs in segregating populations (Asins, 2002; Tanksley, 1993). The major ones include the genetic properties of QTLs, population size, experimental error and environmental effects. Additionally, there are certain requirements also like suitable mapping population, phenotypic screening methods, linkage map, and statistical methods to analyse the genotypic/ phenotypic information. The detection of QTL is crucial for marker assisted breeding while restricting linkage drag and fast recovery of recurrent genome. Recent advances in genomics have lead to availability of markers even in polyploid species and the option of whole-genome sequencing for crop species have aided in identifying the factors underlying a trait(s) and better understanding the trait(s) of interest (Varshney *et al.*, 2009).

Marker-Assisted Selection (MAS)

Selection of plants with correct combination of genes/traits is a very important component of plant breeding (Ribaut and Betran, 1999; Weeden *et al.*, 1994). Plant selection is very crucial, plant breeders usually work with thousands of plants (Witcombe and Virk, 2001) and it consumes lot of resources. Therefore, compared to conventional breeding, MAS increases both theeffectiveness and efficiency in plant selection and breeding. MAS can be defined as selection for a character based on genotype rather than the phenotype using molecular markers (Foolad and Sharma, 2005). Once the specific markers linked to particular QTL governing a specific trait are identified, they can be utilized as a diagnostic tool to identify plants carrying that QTL (Michelmore, 1995; Ribaut *et al.*, 1997). There are several advantages with MAS: (i) particular genotype can be selected at seedling stage (ii) simultaneous gene pyramiding is possible(iii) it is quite useful in cases where phenotypic evaluation is not feasible (iv) elimination/ minimizing of unnecessary phenotypic evaluation in the field (v) single plant based selections can be made and poor heritability is not a problem (vi) markers-based selection does not require progeny test in each generation to maintain recessive genes (vii) markers can aid in choice of parents in breeding programs (viii) most importantly MAS saves time. Though MAS is very powerful technique with substantial benefits, its usage in crop improvement has been limited. But, the adoption of MAS by researchers has increased substantially in the past decade as demonstrated in Table 1.

Table 1. Application of MAS in crop improvement

Crop	Examples of MAS	Reference
Barley	'Tango' variety with two QTL for adult resistance to stripe rust	Toojinda*et al.*, 1998
	'Sloope type' variety with cereal cyst nematode resistance	CRCMPB, 2001
	Introduction of *yd2* gene for yellow dwarf virus resistance	Jefferies *et al.*, 2003
Maize	Development of QPM maize through opaque2 gene transfer	Dreher*et al.*, 2000
	Use of yield-related QTLs	Koebner, 2004
Pearl millet	'HHB 67 Improved' resistance to downy mildew	Varshney*et al.*, 2009a
Sorghum	Pyramiding of stay green QTLs into drought susceptible lines	Varshney*et al.*, 2009a
Rice	Pyramiding of resistance genes against blight, and blast	Sanchez *et al.*, 2000, Singh *et al.*, 2001; He *et al.*, 2004
Wheat	Tolerance to high soil boron, cyst nematode, and several other traits	CRCMPB, 2001; Langridge, 2005

Association Mapping

Association mapping (AM) is an approach based on linkage disequilibrium (LD) to identify molecular markers and the genes responsible for the variation in a quantitative trait (Flint-Garcia *et al.*, 2003). Since AM refers to LD, AM is also known as "LD mapping". Unlike regular mapping population (recombinant inbred lines, double haploids or F_2), AM utilizes natural populations that have higher potential for QTL identification due to several cycles of recombination events will probably remove association between QTL and not so tightly linked marker (Varshney *et al.*, 2009a). Basically, AM refers to "identifying population wide marker-phenotype associations on the basis of previously unrecorded sources of disequilibrium" (Flint-Garcia *et al.*, 2003). The number of markers to be used (to scan the genome) and the resolution of association analysis depend upon the LD around a locus (Rafalski and Morgante, 2004). But, the linkage between and the trait/gene varies due to the uneven genetic recombination events. Pritchard *et al.* (2007) proposed population-based method that aids in large assessment of gene/allele-trait association in structured populations. This approach has been used to demonstrate AM in:

(i) Dwarf gene involved in flowering time and yellow endosperm colour in maize (ii) yield and yield stability in barley (iii) kernel morphology and milling quality in wheat (Varshney *et al.,* 2009a).AM coupled with advance statistical methods will enable better understanding of the trait variation and identify suitable candidates.

Whole-Genome and EST Sequencing

Whole genome sequencing provides the complete DNA-makeup and enabling us to understand the genetic variationsamong and between species/genus. The substantial increase in the fully sequenced plant genomes is due to exponential decrease in the cost of sequencing (Schatz *et al.,* 2012). Genome sequencing of several important crops including rice, sorghum, maize, soybean, barley, wheat, pigeonpea and chickpea has been completed (Table 2). Since genome sequencing provides information regarding gene(s) and genetic architecture the information could be used to manipulate agronomically important traits. Genome sequencing provides lot of important information but there is still a lot to be learned about the genetic organization, gene functions, and understanding the role of non-coding regions. Sequenced genome of a specific crop/plant identifies not only genes specific to that particular species but also a vast amount of non-coding space that needs further functional annotation.

Table 2. Published plant genome sequences of important food crops*

Scientific Name	Common Name	Year	Size (Mb)†	Journal	PMID
Oryzasativa	Rice	2005	389	Nature	16100779
Sorghum bicolour	Sorghum	2009	818	Nature	19189423
Zea mays	Maize	2009	2300	Science	19965430
Glycine max	Soybean	2010	1115	Nature	20075913
Solanumtuberosum	Potato	2011	844	Nature	21743474
Cajanuscajan	Pigeon pea	2011	833	Nature Biotech.	21623354
Setariaitalica	Setaria	2012	490	Nature Biotech.	22580950
Solanumly-copersicum	Tomato	2012	900	Nature	22660326
Hordeumvulgare	Barley	2012	5100	Nature	23075845
Triticumaestivum	Wheat	2012	17000	Nature	23192148
Cicerarietinum	Chickpea	2013	738	Nature Biotech.	23354103
Aegilopstauschii	Wheat DD	2013	4360	Nature	23535592

*Adapted from Michael and Jackson, 2013; †Mb – mega bases

In addition to whole genomes, large numbers of expressed sequence tags (ESTs) are now available from EST-sequencing projects. This information is useful for understanding transcriptome of target species (ex. EST sequencing in barley has revealed that large percent [~41%] of genes belong to multigene families and 4% undergo alternative splicing (Zhang *et al.*, 2004). EST resources have also been used to develop molecular markers (SSRs, SNPs and COS) and in functional genomics (Varshney *et al.*, 2009a).

Map-Based Cloning

Map-based cloning (MBC) involves the use of DNA-based molecular markers to construct genetic maps, physical maps and finally to isolate the gene of interest (Varshney *et al.*, 2009a) Due to the availability of large amount of resources it should be possible to isolate gene/QTL in short amount of time (ex. using rice genome sequence data ~ 1.75 and 0.5 million SNPs and InDels were identified respectively (Shen *et al.*, 2004). The candidates identified could further be utilized for allele mining to identify superior haplotypes and produce transgenic plants. Based on this approach a strategy called EcoTILLING was developed to identify several polymorphisms in germplasm line (Comai *et al.*, 2004). EcoTILLING enables the SNP discovery and haplotyping at much lower cost compared to SNP genotyping or haplotyping methods which require extensive sequencing. Thus utilizing genomic resources and novel approaches like AM coupled with MBC would help in gene isolation. Some of the achievements utilizing MBC in are listed in Table 3.

Table 3. Genes identified in various crops through MBC

Crop	Genes Isolated/Identified	Reference
Barley	Powdery mildew resistance genes –*Mlo, Mla,* and *Rar 1*	Buschges *et al.*, 1997; Wei *et al.*, 1999; Shirasu *et al.*, 1999
	Stem rust resistance – *Rpg1*	Brueggeman *et al.*, 2002
Maize	Flowering time QTL – *Vgt1*	Salvi *et al.*, 2002
	Root abcissic acid QTL – *ABA1*	Varshney *et al.*, 2009a
Rice	Plant architecture gene – *Dwarf1*	Ashikari *et al.*, 1999
	QTL grain production – *Gn1a*	Ashikari *et al.*, 2005
	QTL for salt tolerance	Ren *et al.*, 2005
Sorghum	A major aluminium tolerance gene – *AltSB*	Magalhaes *et al.*, 2006

contd...

Wheat	Vernalization genes - *VRN1, VRN2*	Yan *et al.*, 2003; Yan *et al.*, 2004
	Domestication gene *Q*	Simon *et al.*, 2006
	Leaf rust resistance gene - *Lr10, Lr21*	Feuillet *et al.*, 2003; Huang *et al.*, 2003

Functional Genomics

Functional genomics identifies the function of genes or sequence changes (allelic differences) that confers an improved phenotype. In case of allelic differences the major objective is to identify the changes in the sequence rendering the improved phenotype so that it could potentially be used as marker for that specific allele (Varshney *et al.*, 2009a). Therefore, functional genomics can very well be used in crop improvement programmes. Presently, simultaneous identification of large numbers of genes is possible through several platforms like serial analysis of gene expression (SAGE), massively parallel signature sequencing (MPSS) and microarrays (Varshney*et al.*, 2009a). Microarrays have been extensively used in cereal crops due to their cost advantage and high-throughput outputs. In case of rice, GeneChip arrays, full-length cDNA arrays and tilling microarrays are available; similarly AffymetrixGeneChip arrays have been developed (Close *et al.*, 2004).

The array technology has been used to understand developmental processes, environmental-stress responses and for genotyping of mutations (Potokina *et al.*, 2004; Milligan *et al.*, 2004; Aharoni and Vorst, 2002). These technologies have also been used for applied aspects of plant breeding, ex. Potokina *et al.*, (2004) has identified candidate genes for six malting parameters using cDNA array platform.Further, five out of the eight candidate genes were linked to known QTLs for malting quality demonstrates the applicability of functional genomics approach in molecular breeding. Thus functional genomics helps in identifying a candidate gene that can be used to develop perfect marker for use in MAS and applied breeding.

Comparative Genomics

Besides some cereal (rice, maize, wheat), legume (pigeonpea, chickpea) and oilseed crops (groundnut, soybean) many other crops like sorghum, pearl millet, greengram, black gram and small millets are also important for nutrition and economic purpose but at the local or regional level.Due to low economic returns these 'orphan crops' have not received adequate funding for research or in some cases the researchers themselves are reluctant to work with some of these crops. Since there is some level of genetic colinearity among different related species, bioinformatics tools provide an opportunity for transfer of information from model crop species to other 'minor/orphan' crops. There are several direct/indirect benefits from this technology transfer like: (i) crop diversity analysis and identification of variability or variants (ii) application of MAS to transfer desired alleles (iii) inter-taxa cloning and transfer of important alleles (Nelson *et al.,* 2004). Genomic research on minor or orphan crops could also identify some attributes that could be of use in major crops, ex. drought tolerance alleles found in pearl millet used in wheat and rice (Varshney*et al.,* 2009a).

Transgenics

Plant and animal genomes have been altered routinely by traditional breeding techniques for hundreds of years. Artificial selection of desirable traits has resulted in developing a variety of crops and crop cultivars, but the selection was restricted to naturally occurring variations. However, advances in different fields (as stated above) and genetic engineering now allow us to have a precise control over the changes to be made in an organism. Genetic engineering is the process in which recombinant DNA technology is used to alter/change the genetic makeup of an organism. New or alien genes can be transferred from species to a completely unrelated species through recombinant DNA technology or better known as "genetic engineering or transgenics", ex. transfer of Bt ä-endotoxin gene. Apart from crop species several animals and bacteria have been subject to genetic engineering. This technology could specifically be exploited to address major environmental constraints for crop production like drought, salinity and heat or cold stress.

Agricultural crop plants are the most cited examples of genetically modified organisms (GMOs). Some important benefits of genetic engineering include increased crop yields, reduced need for insecticides/fungicides, resistance to pests and disease, and greater food security. Early maturing crops that tolerate nutrient deficiencies (aluminium and boron), salinity and environmental stresses (drought, frost, and others) have allowed their cultivation in non-traditional areas (Takeda and Matsuoka, 2008). Despite the debate over transgenic crops the less developed or developing countries that face severe food shortages coupled with malnutrition, genetically modified crops with higher yields and better nutritional quality (developed via genetic engineering) could play a key role in alleviating these two major problems. There have been some crop cultivars that have been developed using this technology which are listed in Table 4.

Table 4. Development of transgenic crop cultivars

Trait	Crop	Genetic Change
Pest resistance	Maize	Resistance to European corn borer through expression of Cry1Ab protein from *Bacillus thuringiensis*
Herbicide tolerance	Soybean	Glyphosate tolerance conferred by expression of 5-enolpyruvyl shikimate-3-phosphate synthase from *Agrobacterium tumefaciens*, strain CP4
Virus resistance	Plum	Resistance to plum pox virus conferred by transferring coat protein gene from the virus
Fatty acid deposition	Canola	High laurate levels achieved through inserting ACP thioesterase gene from *Umbellulariacalifornica*

Integration of the above described approaches coupled with other areas like proteomics, metabolomics, transcriptomics and bioinformatics could effectively aid in combating the impacts of climate change on global agricultural production. However, there are several challenges like low heritability of traits, precise phenotyping, epistatic interactions, DNA methylation, RNAiand epigenetics that are to be addressed. In addition to these, technical and cost (economic) issuesare other major constraints.But, application of these robust and efficient 'omics' technologies to plant breeding will result in increased efficiency, reduced costs and increase the throughput of the assays. The newly developed genetic and genomics tools will definitely enhance the prediction of phenotype, they will not completely replace the conventional breeding process, rather they would complement the conventional plant breeding.

References

Aharoni, A. and Vorst, O. 2002. DNA microarrays for functional plant genomics. *Plant Molecular Biology* 48: 99–118.

Ashikari, M.,Wu, J.,Yano, M.,Sasaki, T. and Yoshimura, A. 1999. A rice gibberellin-insensitive dwarf mutant gene Dwarf 1 encodes the alpha subunit of GTP binding protein. *Proceedings of National Academy of Sciences U.S.A.* 96: 10284–10289.

Ashikari, M., Sakakibara, H., Lin, S., Yamamoto, T., Takashi, T., Nishimura, A., Angeles, E.R.,Qian, Q., Kitano, H.and Matsuoka, M. 2005. Cytokinin oxidase regulates rice grain production. *Science* 309: 741–745.

Asins, M. 2002. Present and future of quantitative trait locus analysis in plant breeding. *Plant Breeding* 121: 281–291.

Brueggeman, R., Rostoks, N., Kudrna, D., Kilian, A., Han, F., Chen, J., Druka, A.,Steffenson, B.and Kleinhofs, A. 2002. The barley stem rust resistance gene Rpg1 is a novel disease resistance gene with homology to receptor kinases. *Proceedings of National Academy of Sciences U.S.A.* 99: 9328–9333.

Büschges, R., Hollricher, K., Panstruga, R., Simons, G.,Wolter, M., Frijters, A., van Daelen, R., van der Lee, T., Diergaarde, P., Groenendijk, J., Töpsch, S., Vos, P., Salamini, F., Schulze-Lefert, P. 1997. The barley mlo gene: a novel control element of plant pathogen resistance. *Cell* 88: 695–705.

Close, T.J., Wanamaker, S.I., Caldo, R.A., Turner, S.M., Ashlock, D.A., Dickerson, J.A., Wing, R.A., Muehlbauer, G.J.,Kleinhofs, A. and Wise, R.P. 2004. A new resource for cereal genomics: 22k barley genechip comes of age. *Plant Physiology* 134: 960–968.

Collard B.C.Y, Jahufer,M.Z.Z.,Brouwer, J.B. and Pang, E.C.K. 2005.An introduction to markers, quantitative trait loci (QTL) mapping and marker-assisted selection for crop improvement: The basic concepts. *Euphytica* 142: 169–196.

Comai, L., Young, K., Till, B.J., Reynolds, S.H., Greene, E.A.,Codomo, C.A., Enns, L.C., Johnson, J.E.,Burtner, C.,Odden, A.R. and Henikoff, S. 2004. Efficient discovery of DNA polymorphisms in natural populations by EcoTILLING. *Plant Journal* 37: 778–786.

Cooperative Research Centre for Molecular Plant Breeding (CRCMPB, 2001). *Putting plant breeding into fast-forward. Annual Report 2001/2002*, p. 51 (www.molecularplantbreeding.com/ pdfs/ CRCMPB-Annual-Report-2001-02.pdf)

Dreher, K., Morris, M., Khairallah, M., Ribaut, J. M., Pandey, S. and Srinivasan, G. 2000. Is marker assisted selection cost-effective compared to conventional plant breeding methods? The case of quality protein maize. *Proceedings of Fourth Annual Conference ofInternational Consortium on Agricultural Biotechnology Research (ICABR), "The Economics of Agricultural Biotechnology,* Ravello, Italy.

Feuillet, C., Travella, S., Stein, N., Albar, L., Nublat, A. andKeller, B. 2003. Map-based isolation of the leaf rust disease resistance gene Lr10 from the hexaploid wheat (*Triticumaestivum* L.) genome. *Proceedings of National Academy of Sciences U.S.A.*100: 15253–15258.

Flint-Garcia, S.A., Thornsberry, J.M. and Buckler IV, E.S.2003. Structure of linkage disequilbrium in plants. *Annual Reviews in Plant Biology* 54: 357–374.

Ganal, M. W., Altmann, T. and Röder, M. S. 2009. SNP identification in crop plants. *Current Opinion in Plant Biology* 12: 211-217.

Gupta, P.K. and Varshney, R.K. 2000. The development and use of microsatellite markers forgenetic analysis and plant breeding with emphasis on bread wheat. *Euphytica* 113: 163–185.

He, Y., Xin, Li., Zhang, J., Jiang, G., Liu, S., Chen, S.,Tu, J.,Xu, C. and Zhang, Q. 2004. Gene pyramiding to improve hybrid rice by molecular-marker techniques. *New Directions for a Diverse Planet: Proceedingsof Fourth International Crop Science Congress,* Brisbane, Australia.

Huang, L., Brooks, S.A., Li, W., Fellers, J.P., Trick, H.N. and Gill, B.S. 2003. Map-based cloning of leaf rust resistance gene Lr21 from the large and polyploid genome of bread wheat. *Genetics* 164: 655–664.

Jefferies, S. P., King, B. J., Barr, A. R., Warner, P., Logue, S. J. and Langridge, P. 2003. Marker-assisted backcross introgression of the yd2 gene conferring resistance to barley yellow dwarf virus in barley. *Plant Breeding* 122: 52–56.

Koebner, R.M.D. 2004. Marker-assisted selection in the cereals: the dream and the reality. In *Cereal Genomics* (Eds. P.K. Gupta and R.K. Varshney), pp. 199–252, Kluwer Academic Publishers, Dordrecht, the Netherlands.

Korzun, V. 2002. Use of molecular markers in cereal breeding. *Cellular and Molecular Biology Letters* 7:811–820.

Langridge, P. 2005. Molecular breeding of wheat and barley. In *In the Wake of Double Helix: From the Green Revolution to the Gene Revolution* (Eds. R. Tuberosa, R.L. Phillips and M. Gale), pp. 279–286, Avenue Media, Bologna, Italy.

Liu, B. 1998. *Statistical Genomics: Linkage, Mapping and QTL Analysis* CRC Press, Boca Raton.

Magalhaes, J.V.,Kochian, L. and Outros. 2006. High-resolution mapping and cloning of AltSB, a major aluminium-tolerance gene in sorghum. *Plant and Animal Genome Conference XIV*: 61, San Diego, USA.

Michael, T.P. and Jackson, S. 2013. The first 50 plant genomes. *The Plant Genome* 6(2): 1-7.

Michelmore, R. 1995. Molecular approaches to manipulation of disease resistance genes. *Annual Reviews in Phytopathology* 33: 393–427.

Milligan, A.S., Lopato, S. andLangridge, P. 2004. Functional genomics studies of seed development in cereals. In *Cereal Genomics* (Eds. P.K. Gupta and R.K. Varshney), pp. 447-482, Kluwer Academic Publishers, Dordrecht, the Netherlands.

Nelson RJ, Naylor, R.L. and Jahn, M.M. 2004. The role of genomics research in improvement of 'orphan' crops. *Crop Science* 44: 1901–1904.

Potokina, E., Caspers, M., Prasad, M., Kota, R., Zhang, H., Sreenivasulu, N., Wang, M. andGraner, A. 2004. Functional association between malting quality trait components and cDNA array-based expression patterns in barley (*Hordeumvulgare* L.). *Molecular Breeding* 14: 153–170.

Pritchard, J.K., Stephens, M., Rosenberg, N.A. and Donnelly, P. 2000. Association mapping in structured populations. *American Journal of Human Genetics* 67: 170–181.

Rafalski, A. and Morgante, M, 2004. Corn and humans: recombination and linkage disequilibrium in two genomes of similar size. *Trends in Genetics* 20: 103–111.

Ren, Z.H., Gao, J.P., Li, L.G., Cai, X.L., Huang, W., Chao, D.Y., Zhu, M.Z., Wang, Z.Y., Luan, S. andLin, H.X.2005. A rice quantitative trait locus for salt tolerance encodes a sodium transporter. *Nature Genetics* 37: 1141–1146.

Reynolds, M.P. and Ortiz, R.2010. Adapting crops to climate change: a summary. In *Climate change and crop production* (Ed. M.P.Reynolds), pp. 1-8, CAB international.

Ribaut, J.M. and Betran, J. 1999. Single large-scale marker-assisted selection (SLS-MAS). *Molecular Breeding* 5: 531–541.

Ribaut, J-M., Hu, X.,Hoisington, D.and González-de-León, D.1997. Use of STSs and SSRs as rapid and reliable preselection tools in marker-assisted selection backcross scheme. *Plant Molecular Biology Report* 15: 156–164.

Salvi, S., Tuberosa, R., Chiapparino, E., Maccaferri, M., Veillet, S., van Beuningen, L., Isaac, P., Edwards, K. andPhillips, R.L. 2002. Toward positional cloning of Vgt1, a QTL controlling the transition from the vegetative to the reproductive phase in maize. *Plant Molecular Biology* 48: 601–613.

Sanchez, A.C., Brar, D. S.,Huang, N.,Li, Z. andKhush, G. S. 2000. Sequence-tagged site marker-assisted selection for three bacterial blight resistance genes in rice. *Crop Science* 40: 792–797.

Schatz, M.C., Witkowski, J. andMcCombie, W.R. 2012. Current challenges in de novo plant genome sequencing and assembly. *Genome Biology* 13(4):243.

Shariflou, M.R., Hassani, M.E. and Sharp, P.J. 2001. A PCR-based DNA marker for detection of mutant and normal alleles of the *Wx-D1* gene of wheat. *Plant Breeding* 120:121–124.

Shen, Y.J., Jiang, H., Jin, J-P., Zhang, Z-B., Xi, B., He, Y-Y., Wang, G., Wang, C., Qian, L., Li, X., Yu, Q-B., Liu, H-J., Chen, D-H., Gao, J-H., Huang, H., Shi, T-L.andYang, Z-N. 2004. Development of genome-wide DNA polymorphism database for map-based cloning of rice genes. *Plant Physiology* 135: 1198–1205.

Shirasu, K., Lahaye, T., Tan, M.W., Zhou, F., Azevedo, C. andSchulze-Lefert, P. 1999. A novel class of eukaryotic zinc-binding proteins is required for disease resistance signaling in barley and development in *C. elegans*. *Cell* 99: 355–366.

Simons, K.J., Fellers, J.P., Trick, H.N., Zhang, Z., Tai, Y.S., Gill, B.S. andFaris, J.D. 2006. Molecular characterization of the major wheat domestication gene Q. *Genetics* 172: 547–555.

Singh, S., Sidhu, J.S., Huang, N., Vikal, Y., Li, Z., Brar, D.S., Dhaliwal, H.S. and Khush, G.S. 2001. Pyramiding three bacterial blight resistance genes (*xa5*, *xa13* and *Xa21*) using marker assisted selection into indica rice cultivar PR106. *Theoretical and Applied Genetics* 102: 1011–1015.

Takeda, S. and Matsuoka, M. 2008. Genetic approaches to crop improvement: responding to environmental and population changes. *Nature Review in Genetics* 9(6): 444-457.

Tanksley, S.D. 1993. Mapping polygenes. *Annual Reviewin Genetics* 27: 205–233.

Toojinda, T., Baird, E., Booth, A., Broers, L., Hayes, P., Powell, W., Thomas, W. and Vivar, H. 1998. Introgression of quantitative trait loci (QTLs) determining stripe rust resistance in barley: an example of marker assisted line development. *Theoretical and Applied Genetics* 96: 123–131.

Varshney, R.K., Nayak, S.N., May, G.D. and Jackson, S.A. 2009. Next generation sequencing technologies and their implications for crop genetics and breeding. *Trends in Biotechnology* 27:522–530

Varshney, R.K., Hoisington, D.A. and Tyagi, A. 2009a. Advances in cereal genomics and applications in crop breeding. *Trends in Biotechnology* 24(11): 490-499.

Varshney, R.K., Thudi M, Aggarwal, R.K. and Börner A. 2007. Genic molecular markers in plants: development and applications. In *Genomics-Assisted Crop Improvement: Vol. 1: Genomics Approaches and Platforms* (Eds.R.K. Varshney and R. Tuberosa), pp. 13–29.

Weeden, N., Timmerman, G.M. and Lu, J. 1994. Identifying and mapping genes of economic significance. *Euphytica* 73: 191–198.

Wei, F., Gobelman-Werner, K., Morroll, S.M., Kurth, J., Mao, L., Wing, R., Leister, D., Schulze-Lefert, P. andWise, R.P. 1999. The *Mla* (powdery mildew) resistance cluster is associated with three NBS–LRR gene families and suppressed recombination within a 240 kb DNA interval on chromosome 5S (1HS) of barley. *Genetics* 153: 1929–1948.

Witcombe, J.R. and Virk, D.S. 2001. Number of crosses and population size for participatory and classical plant breeding. *Euphytica* 122: 451–462.

Yan, L., Loukoianov, A.,Tranquilli, G.,Helguera, M.,Fahima, T. and Dubcovsky, J. 2003. Positional cloning of the wheat vernalization gene VRN1. *Proceedingsin National Academyof Sciences U.S.A.* 100: 6263–6268.

Yan, L., Loukoianov, A., Blechl, A., Tranquilli, G., Ramakrishna, W., SanMiguel, P., Bennetzen, J.L., Echenique, V. andDubcovsky, J. 2004. The wheat VRN2 gene is a flowering repressor down-regulated by vernalization. *Science* 303: 1640–1644.

Young, N.D. 1996. QTLmapping and quantitative disease resistance in plants. *Annual Reviewsin Phytopathology* 34: 479–501.

Zhang, H., Sreenivasulu, N., Weschke, W., Stein, N., Rudd, S., Radchuk, V., Potokina, E., Scholz, U., Schweizer, P., Zierold, U., Langridge, P., Varshney, R.K., Wobus, U. andGraner, A. 2004. Large-scale analysis of the barley transcriptome based on expressed sequence tags. *Plant Journal* 40: 276–290.

□□□

Chapter 18

Molecular Approaches for Understanding Drought Tolerance in Moth Bean: Our Experiences

*RAMAVTAR SHARMA, S. KUMAR[1], H.R. MAHLA, R.K. BHATT AND R.K. KALIA

ICAR-Central Arid Zone Research Institute, Jodhpur-342003, Rajasthan, [1]AAU, Anand, Gujarat, * Email: ras_rau@rediffmail.com

Introduction

Moth bean [*Vigna acounitifolia* (Jacq.) Marechal] belonging to family leguminoceae and sub family papilionideae (Verdcourt, 1970) with 22 somatic chromosomes, occupies a position of considerable importance among pulses for its ability to grow with less water and inputs. A native of Indian subcontinent, it is grown in arid and semi arid areas throughout the tropics, sub-tropics and even warm temperate areas of different countries between 30°N and 30°S. Traditionally, it is cultivated in drier areas of India comprising of states of Rajasthan, Gujarat and Maharashtra, on marginal lands as a rainfed crop. It occupies 1.62 Mha area with production of 0.79 MT and productivity 486 kg ha^{-1}. Rajasthan ranks first in area and production contributing 98 per cent both in area (1.59 Mha) and production (0.79 MT) at the country level. The productivity in Rajasthan (486 kg ha^{-1}) was comparable to national average during the year 2010-11; however, during drought years it remains relatively low. Completely dependent on rainfall, production of moth bean is drastically influenced by amount and spread of precipitation received during crop growth. With changing climate fewer rain events are experienced with long dry spells affecting crop growth and production adversely. Natural evolution under such hostile climatic and growing conditions endowed this plant with capability to tolerate extreme drought and heat. Prolonged cultivation in the same adverse conditions without proper management and irrigation further maintained the intrinsic or genetic tolerance mechanisms developed.

After initial selection from natural diversity in germplasm mutation breeding have been successful in developing early dwarf and compact plant types (Sharma and Kakani, 2002). Recently bred short duration varieties have been helpful in coping with the changed climatic pattern capable of harvesting fewer rain events (Henry *et al.*, 1993; Kumar, 1996). Though, moth bean genotypes and varieties seem to grow well under water and heat stress they differ for their ability to stand water scarcity (Soni *et al.*, 2014a and b) and heat stress (Sharma *et al.*, 2014).

Despite of being a drought tolerant species, moth bean has not been properly investigated to understand drought tolerance mechanism at molecular level. In last few years we made some efforts to understand adaptive mechanism at morpho-physiological and molecular level involving a number of *in vitro* and field experiments (Soni *et al.*, 2011, 2012, 2014a and 2014b). We have tried to compile our experiences along with related review to understand drought tolerance in moth bean starting from screening and identification of tolerant types to gene expression level.

Morpho-physiological adaptation to drought tolerance

Fast root growth of short duration moth bean varieties probably is important trait to support early growth. Drought tolerance has been associated with root length, proliferation and dynamics since long (Johansen *et al.*, 1994, Kashiwagi *et al.*, 2005), though, drought tends to decrease the root growth resulting in reduced length and biomass (Muthuchelian *et al.*, 1986, Chartzoulakis *et al.*, 2002, Kusaka *et al.*, 2005, Nayar and Gupta, 2006). According to the Imanparast and Hassanpanah, (2009) genotypes that had good coleoptile length had excess germination percentage too with good root growth resulting in better settlement and higher yield under inhospitable environmental condition. The level of reduction in stem and leaf growth may depend on species and experimental conditions (Zhang *et al.*, 2006). The leaf growth was more sensitive to water stress in wheat, compared to maize (Nayyar and Gupta, 2006). Water stress usually has pronounced effect at the initial growth and establishment stage. The stress experienced during seedling growth may lead to reduced plant height as in perennial species like Citrus (Wu *et al.*, 2006). Phenological and leaf area plasticity is an important survival strategy of crops under water stress (Blum and Arkin, 1984; Blum, 2005), though, reduced leaf area leads to reduced crop yield through reduction in photosynthesis (Kramer, 1983). Prolonged water stress may also be associated with leaf senescence

(Bhatt and Srinivasa, 2005), diverting assimilates towards sink in case stress is experienced during reproductive stage.

Growth parameters like fresh and dry plant weights have a profound effect in water limited conditions. A usual adverse effect of water stress on crop plants is the decline in fresh biomass production (Ashraf and O'Leary, 1996). Drought stress decreases the mean plant biomass and it increases both the relative variation in plant biomass and the concentration of mass (Duan and Zhao, 1996; Xin *et al.*, 1998; Wu and Wang, 1999). Polyethylene glycol (PEG; 6%) induced dehydration increased dry matter (DM) percentage in soybean compared to the control (Sakthivelu *et al.*, 2008). Ashraf and Mehmood (1990) reported that a higher degree of drought resistance was associated with higher protein contents. However, the nature of plant species (Terri *et al.*, 1986) and the types of tissue modulate the concentration of soluble proteins under water stress (Irigoyen *et al.*, 1992).

Initially, screening of twelve genotypes of moth bean was taken up, for PEG (PEG-6000) desiccation tolerance in laboratory (RMO-423, RMO-435, RMO-40, RMO-225, RMO-257, RMM-12-Single, RMM-12-Zero, RMM-12–Poly, RMM-12-Double, CZM-105, CZM-18, and CZM-04-01). Germination in all twelve varieties of moth bean was completely inhibited at a concentration of 10% of PEG. However a differential response was visible at 8% of PEG. Whereas, the seedlings of the twelve varieties of moth bean sustained water stress induced by PEG up to 10% concentration. Incubation on 20% PEG of seven days old seedlings exhibited inhibited growth, chlorosis, wilting and blackening of roots as well as leaves in all twelve varieties though after varying periods of treatment. Differential visible effect induced at 15% PEG was considered suitable for further growth and metabolic studies. Leaf shrinking, leaf curling, yellowing of root tips, diffused growth of roots were the main stress induced symptoms observed after seven days of treatment. Genotypes were screened and classified as tolerant, moderately tolerant and sensitive based on germination inhibition at 8% PEG and number of seedling showing wilting symptoms at 15% PEG (Table 1).

Desiccation induced by PEG inhibited the germination and caused wilting in seedlings. Germination was, however, more sensitive to water stress as germination inhibited at much lower concentration of PEG (8%) compared to the concentration that induced wilting (15%) in seedlings. This could have evolved as a survival strategy of the species under stress conditions where germination may be restricted till sufficient availability for further growth. Low water uptake and restricted metabolic activities probably lead to inhibition of the germination (Singh *et al.*, 1994). Consequently, genotypes showing higher level of tolerance at seedling stage (RMO-435, CZM-105, RMM-12 (single), RMM-12 (poly)] had more and faster germination under stressed condition. Higher germination rate and proportion was reported for tolerant genotypes in *Onobrychis* and have been considered one of the indexes to evaluate for drought tolerance (Imanparast and Hassanpanah, 2009). The two traits seems to have co-evolved for better survival under stress conditions. Proper establishment of resource poor crops like moth bean may be ensured by their improvement for drought tolerance coupled with fast and uniform germination.

Table 1. Number of wilted plants (out of 36 plants) and effect on germination inhibition of *Vigna aconitifolia* after seven days of 15 % PEG treatment

Genotypes	Wilted seedlings	Germination inhibition	Remarks
RMO-435	5 (13.9 %)	6 (16.7%)	Tolerant
RMM-12 (Single)	7 (19.4%)	5 (13.9%)	Tolerant
CZM-105	6 (16.7%)	5 (13.9%)	Tolerant
RMM-12 (Poly)	6 (16.7%)	7 (19.4%)	Tolerant
RMO-225	30 (83.3%)	25 (69.4%)	Sensitive
RMO-40	28 (77.8%)	20 (55.6%)	Sensitive
RMO-257	34 (94.4%)	22 (61.1%)	Sensitive
CZM-04-01	26 (72.2%)	32 (88.9%)	Sensitive

Reduction in growth was observed for all the genotypes after seven days of seedling growth on PEG. However, genotypes having longer root and coleoptile showed comparatively lesser reduction and exhibited better tolerance level. Similarly, both root and shoot fresh weight was reduced under stress in all the genotypes. However, tolerant ones accumulated more fresh mass and experienced lesser reduction on treatment than susceptible types. It was noticeable that most of the tolerant genotypes accumulated more dry matter under control also and that too experienced increase on exposer to PEG stress. However, either lesser increase or even decrease in dry weight of roots and shoots was recorded in moderately tolerant or susceptible genotypes. Moreover, the proportion of dry matter also increased in most genotypes with much higher increase in tolerant genotypes both for root and shoot. This could be important in maintaining osmoticum of whole plant (Soni *et al.*, 2014a).

The genotypes were also evaluated in polythene tubes of 50 cm length filled with soil and vermi-compost in 1:1 ratio. The same amount of soil mix taken in the tubes was initially saturated with 1/4th MS salts. Subsequently, 40, 20 and 10 ml water was supplemented daily. The tubes were kept in uncontrolled field conditions under open sun and plants were allowed to grow for 30 days. The results have shown wide variation in traits like root length, root fresh weight, shoot fresh weight; and root and shoot dry weights while differences for coleoptiles and shoot length observed with PEG induced stress, disappeared (Soni *et al*., 2011, 2014a & b), probably because of the high light intensity. It is observed that moderate stress induced, root growth and more accumulation of fresh and dry weight in root and shoot that was reduced with further increase in the water stress. The genotypes classified as tolerant on the basis of PEG test had longer roots while accumulating more fresh and dry weight under both normal and stress conditions (Table 2).

Table 2. Response of moth bean genotypes, classified as tolerant, moderately tolerant and susceptible on the basis of germination and seedling growth on PEG, to various water levels (experiment conducted in 50 cm long polythene tubes)

Reaction type/traits		RL	CL	SL	RW	SW	RDW	SDW
Average of tolerant	40 ml	39.33	7.43	9.44	0.83	1.31	0.13	0.24
	20 ml	55.23	7.96	10.18	2.90	1.97	0.95	0.41
	10 ml	53.13	8.68	10.53	1.34	1.09	0.46	0.24
Average of susceptible	40 ml	31.99	7.74	8.89	0.41	0.52	0.06	0.10
	20 ml	49.67	7.90	10.47	1.49	1.90	0.41	0.37
	10 ml	44.88	8.20	10.02	1.09	0.87	0.34	0.18
Overall Average	40 ml	34.66	7.63	9.09	0.56	0.81	0.09	0.15
	20 ml	51.69	7.92	10.36	2.00	1.93	0.60	0.38
	10 ml	47.88	8.38	10.21	1.18	0.95	0.38	0.20

RL- root length, CL- coleoptile length, SL- shooth length, RW- fresh root weight, SW- Shoot weight, RDW- root dry weight, SDW- shooth dry weight

Molecular studies on drought tolerance

Maintenance of normal physiological functioning seems to contribute towards genotypic tolerance to water stress. Altered molecular interaction brought about by desiccation disrupts normal cellular function including photosynthesis. Plants respond to these changes by adjusting growth, physiology and molecular functioning brought about by altered gene expression. Drought-inducible genes from drought recognition to altered metabolism fall in two major categories those which are directly involved in metabolic processes leading to stress tolerance (functional genes) and those which are involved in regulation of gene expression and signal transduction in stress response (regulatory genes) (Ingram and Bartels,1996; Bray, 1997; Shinozaki and Yamaguchi-Shinozaki, 2003; Yamaguchi-Shinozaki 2005 and 2006; Seki *et al.,* 2002a and 2002b, Pinto *et al.,* 2010). Drought recognition and early responsive genes have more likely involvement in the signal transduction pathway and have a regulatory role over down- stream responsive genes.

The functional proteins directly participate in metabolic processes to protect cellular structures from damage and maintain water potential of cell. The late embryogenesis abundant proteins (LEA), chaperones and mRNA binding proteins protect cellular macromolecules like enzymes, lipids and mRNA from dehydration, whereas, detoxifying enzymes, such as glutathione S-transferase, superoxide dismutase and a soluble epoxide hydrolase protect cells from active oxygen species. Reactive free radicals produced during drought stress may directly react with cellular macromolecules causing inactivation of enzymes, lipid peroxidation, protein degradation and destruction of DNA strands (Forman, Maiorino and Ursini 2010, Jaspers and Kangasjärvi, 2010, Friso *et al.,* 2004; Gechev *et al.,* 2006; Baruah *et al.,* 2009). However, regulated degradation of damaged proteins involving proteinases and proteinase inhibitors is also important to maintain cell function.

Osmotic pressure is maintained by two classes of proteins that are involved in production of compatible solutes like proline, glycine betain and sugars which function as osmolytes and protect cells from dehydration and channels and transporter proteins like water channel proteins, sugar transporters and proline transporters. Transporters function in transport of water, sugars and proline through plasma membranes and tonoplast to adjust the osmotic pressure under stress conditions. The regulation of transduction and gene expression in stress response might include protein kinases, transcription factors and enzymes in phospholipids metabolism (Yamaguchi-Shinozaki and Shinozaki, 2003, Atkinson *et al.,* 2013, Lindemose *et al.,* 2013).

Stress enzymes like catalase, glutathione peroxidase and sugar content were higher in tolerant moth bean genotypes also which increased in most of the cases on treatment. However, some of the genotypes though recorded a reduction in these chemicals the level remained high in tolerant types (Soni, 2012, and Soni *et al.,* 2011).

Differential gene expression was assumed for the conspicuous distinction for desiccation tolerance in moth bean genotypes. In order to find out variations in gene expression among tolerant and susceptible types, total leaf proteins of the seedlings were resolved on SDS PAGE. Significant variations were observed among control and treated (15 % PEG) samples in banding pattern. While some genes were expressed only under stressed conditions others either showed quantitative variation or remained unchanged. Up-regulation of proteins of 97.4 KD and 18.5 KD size on treatment was noticeable in all the varieties. Whereas, two of the proteins of 54.5 KD and 17.0 KD size were induced only in tolerant genotypes. An 18.9 KD protein was down-regulated under stress (Fig.1). This might explain the differences observed among tolerant and sensitive in morpho-physiological traits.

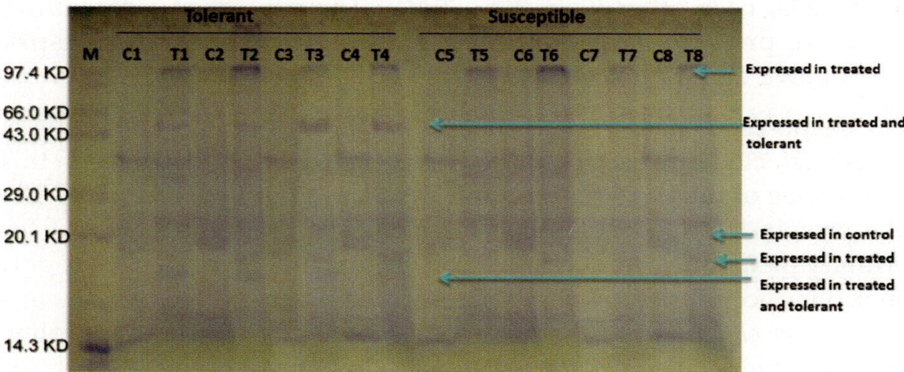

Fig. 1: Protein profile of 8 cultivars of *Vigna aconitifolia* seedlings extracted after 7 days of PEG (15%) induced water stress treatment. C-Control: Seedlings grown in absence of PEG, M-Marker, T-Treated, 1-RMM-12 (single), 2-RMM-12 (poly), 3-CZM-105, 4-RMO-435, 5-RMO-40, 6-RMO-225,7-RMO-257, 8-CZM-04-01

Though protein profiling provides some idea about quantitative change in gene expression it doesn't tell about genes involved. Hence, in order to have insight about stress induced genes a modified differential display of moth bean (using tolerant variety RMO-435) RNA was carried out. The standard procedure making using PAGE for resolution and radioactive material for visualization of amplified fragments was modified to use agarose gel by increasing the amplification cycles. Of 27 primer combinations (3 anchored and 9 arbitrary), 24 primer combinations amplified a total of 95 bands with an average of 4.04 amplicons per primer combination. Most amplicons generated were below 1 kb, but some of them approached 1.5 kb size.

A total of 40 up-regulated or newly expressed bands were recovered from stressed samples and sequences were generated for 35 ESTs. Out of 35, only 29 high quality ESTs (sequences having length >100bp) were obtained which represented 23 unigenes (4 contigs and 19 singletons). Amongst EST sequences large proportion was of chaperons viz. HSP 70, HSP101 and DnaJ like protein along with stress responsive and nucleotide binding regulatory ESTs.

Chaperons involved in the folding of proteins and the assembly of protein complexes within sub-compartments of the eukaryotic cell are represented by different members of the Hsp70 protein family. Whereas, the DnaJ-like protein (JK263604) family regulates the chaperone function of Hsp70 proteins by direct interaction of different Hsp70 and DnaJ-like protein pairs that appears to be specifically adapted to each other (Cyr *et al.*, 1994). The conserved J-domain of DnaJ-like proteins is responsible for their interactions with Hsp70 (Nishikawa and Endo, 1997).

Osmo-regulation and protection of photosynthetic apparatus is mainly affected by a number of stress regulated genes identified in this study. One of the ESTs represented 70kDa peptidyl-prolyl isomerase-like protein involved in proline biosynthesis. Proline is essential for primary metabolism and regulate cell osmoticum. Since the first report on proline accumulation in wilting perennial rye grass (*Lolium perenne*) (Kemble *et al.*, 1954), numerous studies have shown that the proline content in higher plants increases under different environmental stresses (Singh *et al.*, 2014, Djibril *et al.*, 2005). Hu *et al.* (1992) reported the cloning of the first gene for proline synthesis from moth bean (*Vigna aconitifolia*), ä1-pyrroline-5-caboxylate synthetase (P5CS) which is a bifunctional enzyme catalyzing first the activation of glutamate by phosphorilation and second the reduction of the labile intermediate ã-glutamyl phosphate into glutamate semialdehyde (GSA).

The availability of genotypes with sharp distinction for drought tolerance identified with simple screening techniques has been shown to have physiological and associated molecular basis. It is important that while considerable natural genetic variability for drought tolerance provides opportunity to understand and improve the trait it can be induced through mutagenesis increasing the scope of study in this crop.

The efficiency of simple techniques demonstrated for moth bean for identification of tolerant types may be extended for other crops providing the study material for better equipped laboratories for further basic studies.

References

Ashraf, M. and O'Leary, J.W. 1996. Effect of drought stress on growth, water relations and gas exchange of two lines of sunflower differing in degree of salt tolerance. *International Journal of Plant Sciences* 157: 729-732.

Ashraf, M. and Mehmood, S. 1990. Response of four *Brassica* species to drought stress. *Environmental and Experimental Botany* 30 (1): 93–100.

Atkinson, N.J., Lilley, C.J. and Urwin, P.E. 2013. Identification of genes involved in the response of Arabidopsis to simultaneous biotic and abiotic stresses. *Plant Physiol*ogy 162: 2028-2041.

Baruah, A., Simkova, K., Apel, K. and Laloi, C. 2009. Arabidopsis mutants reveal multiple singlet oxygen signaling pathways involved in stress response and development. *Plant Molecular Biology* 70: 547-563.

Bhatt, R.M. and Srinivasa, R.N.K. 2005. Influence of pod load on response of okra to water stress. *Indian Journal of Pant Phy*siology 10: 54-59.

Blum, A. and Arkin, G.F. 1984. Sorghum root growth and water-use as affected by water supply and growth duration. *Field Crops Research* 9: 131–142.

Blum, A. 2005. Drought resistance, water-use efficiency, and yield potential-are they compatible, dissonant, or mutually exclusive? *Australian Journal of Agricultural Research* 56: 1159-1168.

Bray, E.A. 1997. Plant responses to water deficit. *Trends in Plant Science* 2:48-54.

Chartzoulakis, K., Patakas, A., Kofidis, G., Bosabalidis, A. and Nastou, A. 2002. Water stress affects leaf anatomy, gas exchange, water relations and growth of two avocado cultivars. *Scientia Horticulturae* 95: 39–50

Cyr. D.M., Langer, T. and Douglas, M.G. 1994. DnaJ-like proteins: molecular chaperones and specific regulators of Hsp70 *Trends in Biochemical Scences* 19(4):176–181.

Djibril, S., Mohamed, O.K., Diaga, D., Diégane, D., Abaye, B.F. Maurice S. and Alain, B. 2005. Growth and development of date palm (*Phoenix dactylifera* L.) seedlings under drought and salinity stresses. *African Journal of Biotechnology* 4: 968–972.

Duan, S.S. and Zhao, S.L. 1996. Study on propagula hierarchy and sibling competition in spring wheat population under semi-arid conditions. In

Zhao, S.L. (Ed), *Arid Physioecology of Spring Wheat*, Shanxi Scientific and Technology Publishing House. Xian, pp 179-272 (in Chinese)

Forman, H.J., Maiorino, M. and Ursini, F. 2010. Signaling functions of reactive oxygen species *Biochemistry* 49(5): 835-42.

Friso, G., Giacomelli, L., Ytterberg, A.J., Peltier, J.B., Rudella, A., Sun, Q. and Van Wijk, K.J. 2004. In-depth analysis of the thylakoid membrane proteome of *Arabidopsis thaliana* chloroplasts: new proteins, new functions, and a plastid proteome data base. *The Plant Cell* 16: 478–499.

Gechev, T.S., Van Breusegem, F., Stone, J.M., Denev, I., and Laloi, C. 2006. Reactive oxygen species as signals that modulate plant stress responses and programmed cell death. *Bioessays* 28: 1091–1101.

Henry, A., Daulay, H.S. and Bhati, T.K. 1993. A promising moth bean variety for arid and semi arid regions. *Indian Farming* 43: 21-23.

Hu, C.A., Delauney, A.J. and Verma, O.P. 1992. A bifuctional enzyme (delta1-pyrroline-5 carboxylate synthetase) catalyzes the first two steps of proline biosynthesis in plants. *Proceedings of the National Academy of Sciences USA.* 89: 9354-9358.

Imanparast, L. and Hassanpanah, D. 2009. Response of Onobrychis genotypes to PEG 10000 induced osmotic stress. *Biotechnology* 8 : 365-369.

Ingram, J. and Bartels, D. 1996. The molecular basis of dehydration tolerance in plants. *Annual Review of Plant Physiology and Plant Molecular Biology* 47: 377-403.

Irigoyen, J. J., Emerich ,D. W., and Sánchez-Díaz, M. 1992. Alfalfa leaf senescence induced by drought stress: photosynthesis, hydrogen peroxide metabolism, lipid peroxidation and ethylene evolution. *Physiologia Plantarum* 84 (1): 67–72.

Jaspers, P., Kangasjärvi, J. 2010. Reactive oxygen species in abiotic stress signaling. *Physiologia Plantarum* 138: 405

Johansen, C., Krishnamurthy, L., Saxena, N.P. and Sethi, S.C., 1994. Genotypic variation in moisture response of chickpea grown under line-source sprinklers in a semi-arid tropical environment. *Field Crops Research* 37: 103-112

Kashiwagi, J., Krishnamurthy, L., Upadhyaya, H.D., Hari Krishna, Chandra, S., Vadez,V., and Serraj, R. 2005. Genetic variability of drought-avoidance root traits in the mini-core germplasm collection of chickpea (*Cicer arietinum* L.). *Euphytica* 146: 213-222.

Kemble, A.R. and MacPherson, H.T. 1954. Liberation of amino acids in perennial rye grass during wilting. *Biochemical Journal* 58: 46-59.

Kramer, P.J. 1980. Drought, stress and the origin of adaptations *In: Adaptation of plant to water and high temperature stress*. Wiley, New York

Kumar, D. 1996. Variability studies in induced mutants of moth bean on rainfed arid lands. *Annals of Arid Zone* 35: 125-128.

Kusaka, M., Lalusin, A. G. and Fujimura, T. 2005. The maintenance of growth and turgor in pearl millet (*Pennisetum glaucum* (L.) Leeke) cultivars with different root structures and osmo-regulation under drought stress. *Plant Science* 168: 1–14.

Lindemose, S., O'Shea, C., Jensen, M.K. and Skriver, K. 2013. Structure function and networks of transcription factors involved in abiotic stress responses. *International Journal of Molecular Sciences* 14: 5842-5878.

Muthuchelian, K., Krishna Rao, R., Paliwal, K. and Gnanam, A. 1986. Physiological responses of a tree legume to leaf water deficits. Physiological and biochemical and genetic aspects of crop plants in relation to environmental stress, 30–31. *National symposium* at Haryana Agricultural University, Hisar.

Nayar, H. and Gupta, D. 2006. Differential sensitivity of C3 and C4 plants to water deficit stress: association with oxidative stress and antioxidants. *Environmental and Experimental Botany* 58: 106–13.

Nishikawa, S.I. and Endo, T. J. 1997. The yeast JEM1p is a DnaJ-like protein of the endoplasmic reticulum membrane required for nuclear fusion. *Biological Chemistry* 16:12889-12892.

Pinto, R.S., Reynolds, M.P., Methews, K.I., McIntyre, C.I., Olivares-Villagas, J.J. and Chapman, S.C. 2010. Heat and drought adaptive QTL in a wheat population designed to minimize confounding agronomic effects. *Theoretical and Applied Genetics* 121: 1001-1021.

Sakthivelu, G., Akitha Devi, M. K., Giridhar, P., Rajasekaran, T., Ravishankar, G. A., Nedev T., Kosturkova, G. 2008. Drought-induced alterations in growth, osmotic potential and in vitro regeneration of soybean cultivars Drought-induced alterations in soybean cultivars. *General and Applied Plant Physiology* 34 (1-2): 103-112.

Seki, M., Ishida, J., Narusaka, M., et al. 2002a. Monitoring the expression pattern of ca. 7000 Arabidopsis genes under ABA treatments using a full-length cDNA microarray. *Functional and Integrative Genomics* 2: 282–291.

Seki, M., Narusaka, M., Ishida, J. et al. 2002b. Monitoring the expression profiles of ca. 7000 Arabidopsis genes under drought, cold, and high-salinity stresses using a full-length cDNA microarray. *The Plant Journal* 31: 279–292.

Sharma, R., Jain, M. and Kumar, S. 2014. Evaluation of differences among *Vigna aconitifolia* varieties for acquired thermotolerance. *Agricultural Research* 3: 104-112.

Sharma, R.C. and Kakani, R. 2002. *In Moth bean in India* (Eds. D. Kumar and N. B. Singh). Scientific Publishers (India), Jodhpur: 9-30.

Shinozaki, K., Yamaguchi-Shinozaki, K. and Seki, M. 2003. Regulatory network of gene expression in the drought and cold stress responses. *Current Opinion in Plant Biology* 6: 410–417.

Singh, A., Singh. A.K., Singh, V., Singh, N., Singh, V.N., Shamim, M.d., Vikram, P. and Singh, S. 2014. Genetic variability among traits associated with grain yield of rice (*Oryza sativa* L.) exposed to drought at flowering stage. *African Journal of Agricultural Research* 9 (16): 1252-1264.

Singh., B.B., Ambawatia, G.A. and Maharaj, M.S.S. 1994. Stress studies in lentil (*Lens esculanta* Moench). IV. Effect of removal of water stress on recovery of germination percentage in lentil. *Legume Research* 17: 8-12.

Soni, P., Rizwan, M., Bhatt, K.V., Mohapatra, T. and Singh, G. 2011. In-vitro response of *Vigna aconitifolia* to drought stress induced by PEG-6000. *Journal of Stress Physiology & Biochemistry* 7 (3): 109-121.

Soni, P., Sharma, R., Rizwan, M. and Singh, G. 2014a. Optimization of screening protocol for evaluation of drought tolerance in *Vigna aconotifolia* using hydroponic system. *Indian Research Journal of Genetics and Biotechnology* 6(4): 617-628.

Soni, P., Sharma, R., Rizwan, M. and Singh, G. 2014b. Combined effect of osmotic stress induced by polyethylene glycol-6000 and alcohol on in vitro growth of *Vigna aconotifolia*. *Annals of Agricultural Research New Series* 35 (1):1-6.

Soni, P., Mohapatra, T., Bhatt, K V., Singh, G., Rizwan, M., and Sharma, R. 2013. Cloning of drought related genes in *Vigna aconitifolia* through modified differential display. *Journal of Cell & Tissue Research* 13(2): 3701.

Soni, P. 2012. Identification of drought related genes in moth bean through differential display. Thesis: SKRAU, Bikaner.

Terri, J.A., Turner, M. and Gurevitch, J. 1986. The response of leaf water potential and Crassulacean Acid metabolism to prolonged drought in *Sedum rubrotinctum*. *Plant Physiology* 81: 678–680.

Wu, D.X. and Wang, G.X. 1999. The dynamics of size inequality in spring wheat population under semi-arid conditions and its physioecological basis. *Acta Ecologica Sinica* 19: 254-258 (in Chinese).

Wu, Q.S., Xia, R.X. and Zou, Y.N. 2006. Reactive oxygen metabolism in mycorrhizal and non-mycorrhizal citrus (*Poncirus trifoliata*) seedlings subjected to water stress. *Journal of Plant Physiology* 163: 1101–1110.

Xin, X.P., Wabng, G. and Zhao, S.L. 1998. Size hierarchy and its genetic analysis in spring wheat population under different water conditions. *Acta Phytoecolo. Sinica* 22: 157-163.

Yamaguchi-Shinozaki, K. and Shinozaki, K. 2005. Organization of cisacting regulatory elements in osmotic- and cold-stress-responsive promoters. *Trends in Plant Science* 10: 88–94.

Yamaguchi-Shinozaki, K. and Shinozaki, K. 2006. Transcriptional regulatory networks in cellular responses and tolerance to dehydration and cold stresses. *Annual Review of Plant Biology* 57: 781–803.

Zhang, J., Jia, W., Yang, J. and Ismail, A.M. 2006. Role of ABA in integrating plant responses to drought and salt stresses. *Field Crops Research* 97: 111–119.

□□□

Chapter 19

Role of Soil Microorganisms in Carbon Sequestration

N.R. PANWAR AND R.C. KASANA

ICAR-Central Arid Zone Research Institute, Jodhpur-342003, Rajasthan,
E- mail: npawar_soil@yahoo.com

Introduction

Human activities, especially the burning of fossil fuels such as coal, oil, and gas, have caused a substantial increase in the concentration of carbon dioxide (CO_2) in the atmosphere. This increase in atmospheric CO_2 —from about 280 to more than 380 parts per million (ppm) over the last 250 years—is causing measurable global warming (Sundquist *et al.,* 2008). Potential adverse impacts include sea-level rise, increased frequency and intensity of wildfires, floods, droughts, and tropical storms; changes in the amount, timing, and distribution of rain, snow, and runoff; and disturbance of coastal marine and other ecosystems. There is an urgent need of scientific and economically viable strategies to mitigate the consequences of adverse effect of increased atmospheric CO_2.

Carbon (C) is the major building block for life on earth and found in all living organisms. In the environment, carbon exists in many forms - predominately as plant biomass, soil organic matter, geologic deposits, and as the CO_2 in the atmosphere and dissolved in water bodies (mainly oceans). Carbon sequestration is the long-term storage of carbon in oceans, soils, vegetation (especially forests), and geologic formations. High levels of fossil fuel combustion and deforestation have transformed large pools of carbon from fossils (oil and coal deposits) and forests into atmospheric carbon dioxide. Carbon sequestration and reductions in greenhouse gas emissions can occur through a variety of agriculture practices.Along with anthropogenically enhanced fluxes of other greenhouse gases (e.g. CH_4 and NO_x), these changes have altered climate on a global scale, possibly moving Earth towards a 'tipping point'. Although the oceans will ultimately absorb much of the carbon mobilized by humans, the century to millennial time scales involved are too slow to prevent potentially dramatic changes (Schlesinger, 2003).

Soils play significant roles in global carbon cycle. It was estimated that soils have contributed as much as 55 to 878 billion tons (GT) of carbon to the total atmospheric CO_2 (Kimble *et al.*, 2002). The total soil carbon consists of the soil organic and inorganic carbon, estimated to be approximately over 2250 GT in the top 1 meter depth (Batjes, 1996).

Carbon sequestration - Meaning

Soil carbon sequestration is the process of transferring carbon dioxide from the atmosphere into the soil through crop residues and other organic solids, in a form that is not immediately reemitted. This transfer or "sequestering" of carbon helps off-set emissions from fossil fuel combustion and other carbon-emitting activities while enhancing soil quality and long-term agronomic productivity. Soil carbon sequestration can be accomplished by management systems that add high amounts of biomass to the soil, cause minimal soil disturbance, conserve soil and water, improve soil structure, and enhance soil fauna activity. Continuous no-till crop production is a prime example.

Carbon sequestion in soil - Process

Through the process of photosynthesis, plants assimilate carbon and return some of it to the atmosphere through respiration. The carbon that remains as plant tissue is then consumed by animals or added to the soil as litter. Primarily the carbon is stored in the soil as soil organic matter (SOM). SOM is a complex mixture of carbon compounds, consisting of decomposing plant and animal tissue, microbes (protozoa, fungi, and bacteria), and humus (Fig. 1). Carbon can either remain stored in soils for millennia, or it can be quickly released back into the atmosphere through respiration by soil microbes. Climatic conditions, natural vegetation, soil texture, drainage, and land use all affect the amount and length of time carbon is stored in soil.

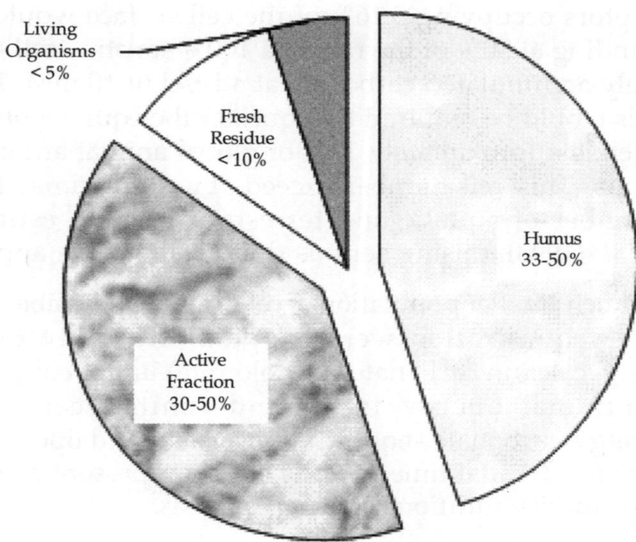

Fig. 1. Soil organic matter components. (Adapted from the Ecological Society of America, Ecoed.net, "Carbon Sequestration" http://esa.sdsc.edu)

Multiple options for reducing greenhouse gas impacts have been proposed. Some, including those known as geo-engineering, would sequester carbon in the geosphere, oceans and biosphere (Lal, 2008). These can be subdivided into engineered or abiological (deep ocean injection, oil well injection, coal bed etc.) and managed biological approaches (Biofuel production, biologically enhanced carbonate formation, dolomite precipitation, sequestration in biomass and soil organic matter, etc.), each of which has distinct advantages and disadvantages (Lal, 2008; King, 2011).

Several options for sequestering carbon that involve or exploit various properties of bacteria have been proposed. In addition to manipulating bacterial activity to increase soil carbon storage, these include bioengineering bacteria to trap CO_2 using novel complexing agents and manipulating microbial activity in certain hypersaline environments to enhance dolomite (a calcium–magnesium carbonate) formation. Tossel (2009) has speculated that small macrocyclic carbonate receptors similar to those described by Brooks *et al.* (2006) might be developed for the membranes of living organisms.

Assuming as an example a receptor size of 1 nm and a spherical bacterial cell 0.5 μm in diameter with a volume of 5.24×10^{-13} cm^3, a total of 1570 receptors occupying 0.16% of the cell surface would support carbonate binding at 50% of the rate of a 100% sorptive cell surface. If they passively accumulated carbonate at a level of 10 mM, 1.59 x 10^{31} bacterial cells would be required to sequester the equivalent of 1 Gt of carbon, which is approximately 12% of global annual anthropogenic CO_2 emissions. This cell number exceeds by approximately 50-fold the total number of aquatic and terrestrial bacteria estimated by Whitman *et al.* (1998), raising serious doubts about the approach.

A very much smaller population of cells might be feasible, however, if carbonate sequestration were coupled efficiently to mineral formation (e.g. calcium carbonates or dolomite) under conditions that required minimal energy input and limited cell turnover. Bioengineering a carbonate sequestration system and operating it at a suitable scale for decadal time periods clearly represent a substantial challenge but merits additional crucial analysis.

Both marine and terrestrial ecosystems have been proposed for managed carbon sequestration. Organic carbon production and storage has dominated the research agenda for these systems, but biologically enhanced carbonate deposition could offer an alternative approach suitable for some situations. Manipulation of the marine biosphere has been highly controversial and is not strongly supported (Glibert *et al.*, 2008). By contrast, terrestrial biospheric carbon sequestration might add as much as 100 Pg C (1 Pg = 10^{15} g) to existing soil carbon stocks (Houghton, 2007), although the amounts would vary considerably on local to regional scales (Canadell *et al.*, 2007). The 100 Pg C represents only approximately 12 years of current gross total anthropogenic CO_2 emissions (approx. 8.6 Pg C annually) (Houghton, 2007), it represents approximately 60 years of emissions from the transportation sector (Agrawal, 2007). This suggests that terrestrial carbon sequestration can meaningfully contribute to a portfolio of mitigation approaches and potentially offset a significant fraction of diffuse CO_2 sources for which direct capture is not yet feasible.

The potential for carbon storage in soils depends on numerous variables, including intrinsic storage capacities that reflect physical and chemical properties (Lal, 2003). Agricultural and land use practices affect plant and bacterial community composition, pools of organic matter, soil aggregate abundance and stability and pyrolysed biomass (biochar) formation from plant biomass represents a potentially significant mechanism for increasing sequestered carbon (Fig. 2). The microbial (bacterial and AMF) activities not only positively affect plant communities and primary production, but also contribute to carbon storage through organic matter turnover and through production of soil proteins that affect aggregation and formation of degradation resistant carbon pools (Six *et al.*, 2006). Fungi are important for aggregate formation and stabilization (Bossuyt *et al.*, 2001; Gupta and Germida, 1988), and aggregatesphysically protect plant-derived SOM and microbially derived organic matter (MOM) (Beare *et al.*, 1994; Six *et al.*, 2000). In addition to glomalin, other proteins, such as the hydrophobins and chaplins, might also be important for microaggregate formation (Rillig, 2005). In contrast to glomalin, hydrophobins appear much more widespread in their distribution among fungi, occurring in numerous basidiomycetes and ascomycetes (Claessen *et al.*, 2006; Linder, 2009). Hydrophobins not only playimportant roles in mycelia formation, but they also alter the hydrophobicity of spores, other cell surfaces and presumably various abiological surfaces. Chaplins are produced by streptomycetes (domain Bacteria, phylum Actinobacteria), however details of chaplin interactions with organic matter and microaggregates have not yet been explored nor is it known how their properties vary among streptomycetes (Gebbink *et al.*, 2005). In support of these concepts, Cleveland *et al.* (2007) have reported that particular members of the - *Proteobacteria* and *Firmicutes* in rain forest soils rapidly respond to additions of dissolved organic matter derived from plant litter leachate and probably account for changes in respiration.By contrast, changes in storage primarily reflect losses due to physical processes (e.g. erosion and transport) and differences between rates of plant inputs and organic matter mineralization, both of which are complex dynamic processes controlled by many factors. Although the possibility of managing soils for carbon sequestration on decadal time scales has been documented in several studies (Feller and Bernouxm, 2008), sequestration on the order of centuries has not been shown, and this is essential for managing CO_2- related climate change. Long-term carbon sequestration hinges on the lability and long-term accessibility of stored carbon to microbes and on carbon turnover rates. Clearly, a greater understanding of microbiology and ecology can contribute to the success of carbon management.

Carbon stored in soils worldwide represents the 3rd largest sink in existence, after oceans and geologic sinks. There is 2-4 times as much C stored in soils as there is in the atmosphere and approximately 4 times the C stored in vegetative material (i.e. plants). It is therefore understandable that the soil C sink is being viewed as one that could potentially have a significant impact on sequestering CO_2 emissions. However, before we consider the feasibility to store extra C in soil, it is essential to understand the forms in which soil C exists.

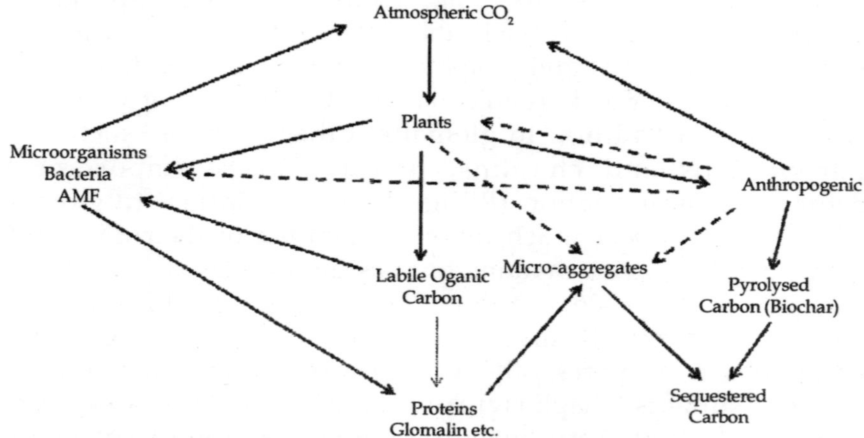

Fig. 2. Overview of carbon flows and selected interactions important for determining the extent and direction of soil carbon sequestration. Soil components subject to human disturbances are indicated with dashed arrows

Soil C is found as either inorganic (i.e. mineral) or organic materials. Inorganic soil C is generally found as carbonates of calcium ($CaCO_3$, or limestone) and magnesium ($MgCO_3$). Excluding concentrated deposits of these materials that arose from deposition of the shells of aquatic invertebrates, the majority of these inorganic forms of C are found in alkaline soils. $CaCO_3$ is effectively insoluble in water at pH > 8.3, so when these extreme conditions are met (typically in subsoils) of the black and grey cracking clays, $CaCO_3$ precipitates out of the soil solution to form whitish nodules embedded in the clay soil. Whilst these inorganic forms of C can represent a significant amount of the C stored in the profile of these particular soils in some areas, they are a significantly smaller store of C than that found in the organic form.

The organic forms of C in soil are a very diverse group of materials that can be defined as "everything in or on the soil that is of biological origin, whether it's alive or dead". It therefore includes live plant roots and litter (not shoots), humus, charcoal and other recalcitrant residues of organic matter decomposition. It also includes the organisms that live in the soil that are collectively called the soil biota (e.g., fungi, bacteria, mites, earthworms, ants and centipedes).

Fate of C and nutrients during transformations

Interestingly, as the microbial decomposition process occurs, C and some nutrients are liberated. The C is released as carbon dioxide (or methane under certain conditions) due to microbial respiration and surplus nutrients are released (mineralised) in inorganic forms suitable for use by other microbes and plants. However, soils are generally nutrient-poor environments so as the decomposition process occurs and the organic materials age in the soil, there is generally more C released as CO_2 than there are surplus nutrients released (i.e. a fungus needs 8 C atoms for every N atom to grow more hyphal threads, but can digest poor quality crop residue with a C:N ratio that starts at 100:1). As a result, surplus C is respired while the N is conserved, and through the aging process in soil the organic materials become increasingly nutrient rich. This enrichment of nutrients in humus and more recalcitrant, charcoal-like materials occurs particularly with N and sulphur (S). The humus that eventually forms from decomposition of, say, cereal straw will have C:N ratio of approximately 12:1, rather than 100:1 in its original form. After so many cycles of digestion and excretion these materials are less readily decomposed than when they entered the soil as plant residue, but the nutrients they contain ensure that they remain a valuable source of nutrients contributing to soil fertility. Bacteria or fungi growing on plant residues are predated by a number of more complex organisms like free living nematodes, which typically have higher C:N ratios than the organisms on which they feed. The excess N in this case will be released in an inorganic form (either ammonium-N or nitrate-N) for use by plants or other organisms. This inorganic N can build up as decomposition continues (if it is surplus to requirements of the microbial community) & will form the basis of the N supplied to the next crop.

Relative sizes of the different C pools

The size of the over all C pool in soils can be misleading as an indicator of how that soil will behave, unless we understand something about the relative proportions of the different 'pools' in which that C can be found. For example, soils that have differing proportions of soil as particulate organic matter (relatively young, labile material), humus and recalcitrant compounds like charcoal will behave very differently – both in terms of properties and the microbial communities (and resulting functions) they can support. Soils supporting open grasslands will have proportionally more char-like material in their native condition, from regular natural burning events, than similar soil types under a rainforest, for example. Similarly, soils in the sugar industry from areas where regular trash burning occurs can also have high (up to 50%) of soil C as charcoal, compared to much lower proportions with a long term history of trash blanketing (Bell and Lawrence, 2009).Consider the soils shown in the example (Fig. 3), where soil is examined at two times in its management history - once during the continuous cropping phase and a second occasion after further cropping and then an 8-10 year period under pasture. If simply analysed for total organic C, both soils would appear very similar. The amount of resistant/char like material is unchanged. However, while there has been a large increase inthe particulate/labilefraction in response to the pasture, the amount of humus (humus takes decades to centuries to form depending on base material) is much lower and still reflective of the end of the cropping rundown period. This means that (i) the long term soil nutrient stores have not recovered, and (ii) that this soil. organic matter will decline much more rapidly if the land is returned to cropping. In fact, the original rate of soil organic matter decline would effectively double in the next cropping phase and much more frequent pasture phases would be needed to maintain soil productivity.

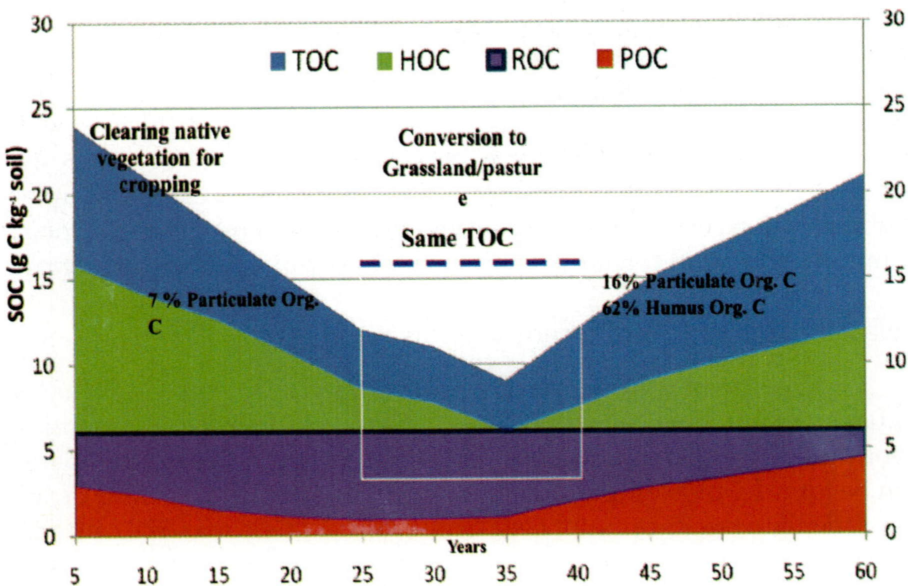

Fig. 3. Relative size of different carbon pools in soil (adapted from Bell and Lawrence, 2009 based on hypothetical data)

Soil C levels are fundamentally determined by the balance between organic matter inputs, primarily as plant residues, roots and root exudates, and organic matter losses due to decomposition, erosion, and leaching. Bacteria and fungi generally comprise >90% of the total soil microbial biomass, and they are responsible for the majority of SOM decomposition. Since soil microbial communities are key regulators of SOM dynamics and nutrient availability, shifts in microbial community composition and function (e.g., substrate utilization) in response to different agricultural management practices may play an important role in determining rates of C loss from the soil. The ratio of fungal:bacterial biomass has been shown to be particularly sensitive to soil disturbance, with lower ratios associated with increased intensity of cultivation (Bailey *et al.,* 2002), increased grazing pressure (Bardgett *et al.,* 1996) and increased N fertilization inputs (Frey *et al.,* 2004). In addition, fungal:bacterial biomass ratios were found to increase with successional age in a semi arid grassland community (Klein *et al.,* 1996).

Organic C taken up by the microbial biomass is partitioned between microbial cell biomass production, metabolite excretion and respiration. The degree to which MOM accumulates in soil depends on a balance between production and decomposition of microbial products, that is: (1) the microbial growth efficiency (MGE), the efficiency with which substrates are incorporated into bacterial and fungal biomass and byproducts, (2) the degree of protection of microbial biomass in the soil structure, and (3) the rate at which bacterial and fungal byproducts are decomposed by other microorganisms. The proportion of substrate C retained as biomass versus respired as CO_2 depends on MGE and the degree of protection of microbial biomass; the lower the MGE or the less protected the biomass, the more MOM-C is lost as CO_2.

Microbial contributions to carbon storage

Many variables affect soil carbon storage, but microaggregate formation substantially limits decomposition losses (McCarthy, 2008). At large scales (e.g. field to ecosystem), microaggregate formation depends on numerous factors, including organic matter loading rates, mechanical disturbances (e.g. tilling),temperature and water availability. Nonetheless, mycorrhizal fungi (MF), and possibly some bacteria, play additional key roles (Bedini *et al.,* 2009). Of course, MF also respire organic matter and thus participate in several aspects of belowground carbon cycling, the overall balance of which is uncertain.

Mycorrhiza / Fungi, *glomalin and microaggregates*

The details of contributions to microaggregate formation by MF at molecular scales are unclear but probably involve interactions among organic matter, metals, mineral surfaces and specific proteins. The most important of these proteins appears to be glomalin - a glycoprotein. Glomalin, which has been identified as a heat shock protein (Purin and Rillig, 2008), occurs in a phylogenetically narrow range of arbuscular my corrhizal fungi (AMF) within the order Glomales (Bedini *et al.*, 2010).

The effects of glomalin on microaggregates could be due to its impacts on hydration (water penetration and hydrophobicity) and the resulting accessibility of hydrolytic enzymes to bulk organic matter (Rillig, 2005). Soil structure and the pore-size distribution within soil aggregates indirectly influence microbial dynamics by restricting organism movement and thereby modifying the interactions between organisms. The diameters of bacteria, fungal hyphae, protozoa, and nematodes are: 0.15 to 4 (although typically 0.5–1 μm), 3 to 8, 10 to 100, and <50 μm, respectively (Coyne, 1999). The larger size of fungi relative to bacteria means that fungi are rarely found in micropores. However, small pores provide refuges for bacteria against attack from larger predators (i.e., protozoa and bacterivorous nematodes) that are typically unable to enter smaller pores. Spatial differentiation of bacterial community composition has been observed between water stable micro aggregate interiors and macroaggregates (Mummey and Stahl, 2004).

The location of bacteria and fungi within the pore network is a key factor in their survival and activity, and the larger size of fungi may make them more vulnerable to predation. Bacterial populations are consistently high in small pores, but highly variable in large pores where they are vulnerable to being consumed (Vargas and Hattori, 1986).

The amount of substrate C that is incorporated into microbial biomass and eventually transferred into the MOM pool is determined by the size of the microbial biomass and its MGE. The amount of microbial biomass is not only directly related to the C input level, but is also influenced by other factors including the availability of microbial habitat and consequently, the level of protection provided for the microbial biomass. Microbial biomass can be divided into unprotected and protected pools. Van Veen *et al.* (1985) split microbial biomass into a protected pool with a death rate of 0.5% per day and an unprotected pool with a death rate of 70% per day.

Soil microbial diversity and carbon sequestration

Soil microbial diversity, defined here as community composition (or taxon richness) and structure (or taxon abundance), has been the subject of numerous analyses (Fierer and Jackson, 2006). The first molecular ecological studies, whichwere based on small clone libraries and denaturing gelgradient electrophoresis, identified the phylogenetic affiliations and relative abundance of only the dominant community members (Campbell *et al.*, 2008). At present it is uncertain which is the most important for carbon sequestration: the limited genetic potential of a small number of common phylotypes or the greater genetic diversity of individually rare but collectively abundant phylotypes. Yuan *et al.* (2012) reported that microbial autotrophy plays a significant role in the sequestration of soil carbon in terrestrial C sequestration in between 2–3 Pg C yr^{-1}. Phylogenetic analysis showed that the dominant *cbbL*-containing bacteria were *Azospirillum lipoferum*, *Rhodopseudomon-as palustris*, *Bradyrhizobium-japonicum*, *Ralstonia eutropha*, and *cbbL*-containing chromophyticalgae of the genera *Xanthophyta*and *Bacillariophyta*. Differences in *cbbL*gene diversity were shown to be correlated with differences in SOC. Bacterial *cbbL* abundance was shown to be positively correlated with RubisCO activity (r=0.853, $P < 0.05$) and both *cbbL* abundance and RubisCO activity were significantly related to the synthesis rates of$_{14}$C-SOC (r=0.967 and 0.946, respectively, $P < 0.01$).

Although diversity indices, such as the Shannon index, might not be useful in predicting or understanding carbon sequestration, the specific composition of microbial communitie scould prove much more important. Much of the diversity revealed by molecular ecology approaches is not represented in cultures and often cannot be mapped unambiguously to taxonomic levels below families or genera. As a result, complementary approaches must be pursued, including an emphasison cultivating the uncultivated.

Impacts of key environmental variables on organic matter stabilization

Temperature effects

Numerous factors affect organic matter decomposition, but temperature is often regarded as a master variable (Davidson and

Janssens 2006). The effects of rising global temperatures have long been a focus of concern, because they could increase respiration and mobilize soil organic matter resulting in a positive feedback for climate change (Singh *et al.*, 2010). Hawkes *et al.* (2008) argue, for example, that temperature affects carbon allocation from plant to fungus and allocation within AMF networks, with a possible increase in respiratory carbon losses in a warmer climate that are not offset by increased AMF growth. Nonetheless, much remains unknown about interactions between temperature and managed soil carbon sequestration. Crucial research gaps include the responses of microaggregate formation and destabilization to both rising temperature and increasingly variable climate regimes, that is changes in the frequency of extreme events (e.g. high and low temperatures, precipitationregimes and snowfall).

Organic matter loading

Compared with conventional practices, organic farming practices have been shown to promote higher microbial biomass (Lundquist *et al.*, 1999; Petersen *et al.*, 1997) and to alter microbial community composition (Bossio *et al.*, 1998; Petersen *et al.*, 1997). Increased carbon storage in soils implies either relative or absolute reductions in respiration, increase in organic matter loading or some combination of both. Here, we consider possible impacts of increased loading. Organic matter loading rates have been reported to affect the temperature sensitivity of respiration (Gaumont-Guay *et al.*, 2006). This could occur if, for example diffusion limits substrate availability for respiration and increased loading reduces or eliminates that limitation. In this case, warming would lead to increased carbon turnover until a diffusion limitation was reestablished. Increased availability of labile low molecular weight organics can also affect turnover of bulk organic matter pools through a process referred to as priming (Blagodatskaya and Kuzyakov, 2008). Priming can potentially limit net carbon sequestration by accelerating respiration of bulk organic matter, thus decreasing total soil organic matter concentrations.

Responses to elevated CO_2

Responses of both plants and microbes to elevated CO_2 further constrain predictions of soil organic matter dynamics and carbon storage. These responses typically depend on interactions between CO_2

and numerous other variables (Billings and Zeigler, 2008). Nitrogen limitation in some mature, undisturbed soils can significantly diminish the ability of microbes to respond to changes in carbon availability or other environmental variables (Niklaus, 1998), but Barron-Gafford *et al.* (2005) have observed increased rates of soil respiration in experimental stands of eastern cottonwoods (*Populus deltoides*) exposed to elevated CO_2 and growing on nitrogen-replete soils. Increased microbial respiration occurs when plants are incubated with elevated CO_2 not only due to increased belowground carbon allocation but also as a result of specific interactions between plants, rhizosphere bacteria and MF (Oliver *et al.*, 2008). These interactions are dynamic, varying among plant hosts and as a function of plant physiological status. They include changes in rhizodeposition, microbial growth rates and community structure in response to passive and active chemical signals originating from roots, bacteria and fungi (Barron-Gafford *et al.*, 2005).

Results from a Mojave Desert study (Jin and Evans, 2007) illustrate the complexity of plant–microbe–CO_2 interactions and the need for improved biogeochemical models. Differential impacts were reported for elevated CO_2 on microbial substrate use, nitrogen cycling rates and exoenzymatic activities for soils beneath a dominant shrub and unvegetated interspace soils. Differences were attributed to variations in microbial biomass and turnover times between the two soils. The response of desert soils to climate change and elevated CO_2, and the ability of these soils to contribute to carbon sequestration, were found to sensitively depend on the dynamics of plant communities and specific plant–microbe interactions, which at present cannot be sufficiently well modeled to reliably predict outcomes.

Montealegre *et al.* (2000) observed a shift in the community composition of *R. leguminosarum* bv. *trifolii* occurred as a result of an increased atmospheric CO_2 concentration, and that elevated atmospheric CO_2 affects the competitive ability of root nodule symbionts, most likely leading to a selection of these particular strains to nodulate white clover.

According to the Intergovernmental Panel on Climate Change Fourth Assessment Report of 2007, sequestration and reduction of emissions over the next two to three decades will potentially have a substantial impact on long-term opportunities to stabilize levels of atmospheric CO_2 and mitigate impacts of climate change (Pachauri and Reisinger 2007).

Agricultural practices that increase soil carbon sequestration

Agricultural management practices differ in the type and intensity of disturbance; therefore, the impact on microbial biomass and MOM varies with the management practice employed. In agricultural systems, the amount and length of time carbon is stored in the soil is largely determined by how the soil resource is managed using variety of agricultural practices. The benefits of these practices as well as their potential hidden costs must be considered when management decisions are made. Removing CO_2 from the atmosphere is only one significant benefit of enhanced carbon storage in soils. Improved soil and water quality, decreased nutrient loss, reduced soil erosion, increased water conservation, and greater crop production may result from increasing the amount of carbon stored in agricultural soils.

Strategies to increase the soil carbon pool include soil restoration and woodland regeneration, no-till farming, cover crops, nutrient management, manuring and sludge application, improved grazing, water conservation and harvesting, efficient irrigation, agroforestry practices, and growing energy crops on spare lands. An increase of 1 ton of soil carbon pool of degraded cropland soils may increase crop yield by 20 to 40 kilograms per hectare (kg ha^{-1}) for wheat, 10 to 20 kg ha^{-1} for maize, and 0.5 to 1 kg ha^{-1} for cowpeas. As well as enhancing food security, carbon sequestration has the potential to offset fossil fuel emissions by 0.4 to 1.2 GT of carbon per year, or 5 to 15% of the global fossil-fuel emissions (Lal, 2004). Management techniques that sequester carbon in soils include:

Conservation tillage

No-tillage (NT) and minimum tillage (MT) systems often exhibit increased C storage compared with conventional tillage (CT) (West and Post, 2002); however, this difference disappears with the use of fallow rotations (Peterson *et al.*, 1998), demonstrating the importance of using multiple management practices to enhance soil C storage. Minimizes or eliminates plowing of soil for crop production through practices such as no-till (no plow) farming or mulch tillage. These procedures generally reduce soil erosion, improve water use efficiency and increase carbon concentrations in the topsoil leading to a significant increase in soil CO_2 sequestration. Conservation tillage can also lower the amount of fossil fuel consumed since it reduces the operation of farm machinery.Differences in tillage intensity also impact microbial community composition (Jackson *et al.*, 2003; Neely *et al.*, 1991). Frey *et al.* (1999) observed that a shift toward fungal dominated microbial communities under NT will be most important for residue decomposition and nutrient cycling processes near the soil surface. However, another study found no difference in fungal:bacterial ratio in surface NT and CT soils and a 10% lower fungal:bacterial ratio in NT soil at 6 to 12 cm. Less intensively managed agroecosystems (e.g., NT) bear the closest resemblance to natural ecosystems, which are fungal-dominated (Bardgett, 1996). These fungal-dominated agroecosystems require fewer inputs to sustain organic matter decomposition and nutrient cycling, that is, these systems show greater self-regulation. However, Allison *et al.* (2005) concluded that an improved metabolic efficiency due to increased relative abundance of AMF and saprophytic fungi does not promote the stabilization of C on cessation of tillage-based agriculture.

Organic Farming and Cover Crops

Compared with conventional practices, organic farming practices have been shown to promote higher microbial biomass and to alter microbial community composition (Bossio*et al.*, 1998). Cover crops improve carbon sequestration by enhancing soil structure and adding organic matter to the soil (Table 1). Organic farming practices frequently employ cover crops, which can change the soil microbial community and have a variable effect on MOM. The more N supply than C supply controls the effect of cover crops on soil carbohydrate content. More research is needed to investigate relationships between the use of organic farming practices (including cover crops) and soil microbial communities. Organic management appears to increase fungal biomass, which would favor increased soil C sequestration.

Crop rotation

Crop rotation as a management practice may increase soil C sequestration in comparison with continuous crop management or rotations that include fallow periods and even location specific farming systems (Table 1). More intensive cropping rotations not only increase soil C input but also increase microbial activity and biomass and alter microbial community composition, increasing levels of soil fungi. Effectiveness of crop rotating varies by region, crop type, and crop rotation timing.The hot water extractable carbohydrates are more influenced by crop rotations than are acid hydrolyzable carbohydrates. Since hot water extractable carbohydrates are considered more microbially derived than acid hydrolyzable carbohydrates, the microbially derived carbohydrates seem to be more affected by crop rotations than plant-derived carbohydrates (Six *et al.*, 2006).Varying the type of crops grown can increase the level of soil organic matter.

Table 1. Agricultural practices for enhancing productivity and increasing the amount of carbon in soils

Traditional practices	Recommended
Plough till	Conservation till or no-till
Residue removal or burning	Residue return as mulch
Low off-farm input	Judicious use of fertilizers and integrated nutrient management
Regular fertilizer use	Soil-site specific management
No water control	Water management/conservation, irrigation, water table management
Fence-to-fence cultivation	Conversion of marginal lands to nature conservation
Conventional practices	Organic farming (site specific) and cover crops
Land use along poverty lines and political boundaries	Integrated watershed management
Monoculture	Improved farming systems with several crop rotations

Epilogue

The potential for biospheric carbon sequestration could offset CO_2 emissions from the transportation sector for decades or even longer as biofuels, hydrogen and methane replace petroleum-based fuels. Realizing this potential, however, will require a much more detailed understanding of the specific organisms and mechanisms that can contribute to carbon storage, as well as continued improvement in understanding the responses of soil microbial communities to carbon loading and organic composition, and to changes in land use, temperature and hydrologic regimes, among othervariables. Microbial biomass and community composition as well as MOM concentrations are affected by crop rotations, tillage, organic farming practices and cover crops. All of these management regimes appear to induce shifts in overall microbial community composition and fungal:bacterial biomass ratios. Under NT, especially, higher fungal biomass levels are correlated with quantitative and qualitative improvements in SOM. Shifts in microbial community structure due to soil disturbance have important implications for ecosystem scale processes because bacteria and fungi function differently in the incorporation and turnover of C and in their respective effects on soil physical properties. Once sequestered, carbon remains in the soil as long as restorative land use, continuous no-till, andother Best Management Practices are followed. It is a win-win option.

There are a lot of conceptual advances in recent years about soil C sequestration mechanisms that need to be matched by methodological advances. Ways of using field sampling, lab analysis or modeling to understand how soil C sequestration changes, specifically in vulnerable to stable soil carbon pools. And it will be important as ever to integrate data that have already been collected into databases where people can use that collected information to understand things at a bigger scale about soil C sequestration. Numerous basic challenges remain to be addressed, beginning with an exploration of the extent to which soil organic concentrations represent localized optima. Some of the important challenges are as follows:

1. Develop a specific and predictive understanding of the mechanisms that determine soil carbon concentrations.
2. Identify the most important microbial groups that control soil carbon storage and mobilization.
3. Determine management strategies that can be successfully implemented at suitable spatial and temporal scales to promote contributions of microbes to soil carbon storage by using omics datasets.

To answer these challenges and implement successful long-term carbon management, it will be necessaryto couple approaches ranging from molecular to ecosystem scales in scope using observation and experimentation to address spatial and temporal issues of bacterial and fungal metabolic activity, hydrolytic enzyme distribution and microbial community composition and structure.

References

Agrawal, R. 2007. Sustainable fuel for the transportation sector. *Proceedings of National Academy of Sciences of the United States of America* 104: 4828–4833.

Allison, V.J., Miller, R.M., Jastrow, J.D., Matamala, R. and Zak, D.R. 2005. Changes in soil microbial community structure in a tall grass prairie chronosequence. *Soil Science Society of America Journal* 69: 1412–1421.

Bailey, V.L., Smith, J.L. and Bolton, H.Jr. 2002. Fungal-to-bacterial ratios in soils investigated for enhanced C sequestration. *Soil Biology and Biochemistry* 34: 997–1007.

Bardgett, R.D. 1996. Potential effects on the soil mycoflora of changes in the UK agricultural policy for upland grasslands. p. 163–183. In *Fungi and environmental change* (Ed. J.C. Frankland, N. Magan and G.M. Gadd). Cambridge Univ. Press, Cambridge.

Bardgett, R.D., Hobbs, P.J. and Frostegard, A. 1996. Changes in soil fungal: bacterial biomass ratios following reductions in the intensity of management of an upland grassland. *Biology and Fertility of Soils* 22: 261–264.

Barron-Gafford, G., Martens, D., Grieve, K., Biel, K., Kudeyarov, V., McLain, J.E.T., Lipson, D. and Murthy, R. 2005. Growth of eastern cottonwoods (*Populus deltoides*) in elevated CO_2 stimulates stand-level respiration and rhizodeposition of carbohydrates, accelerates soil nutrient depletion, yet stimulates above- and belowground biomass production. *Global Change Biology* 11: 1220–1233.

Batjes, N.H. 1996. Total carbon and nitrogen in the soils of the world. *European Journal of Soil Science* 47: 151–163.

Beare, M.H., Cabrera, M.L., Hendrix, P.F. and Coleman, D.C. 1994. Aggregate-protected and unprotected organic matter pools in conventional- and no-tillage soils. *Soil Science Society of America Journal* 58: 787–795.

Bedini, S., Pellegrino, E., Avio, L., Pellegrini, S., Bazzoffi, P., Argese, E. and Giovannetti, M. 2009. Changes in soil aggregation and glomalin related soil protein content as affected by the arbuscular mycorrhizal fungal species Glomusmosseae and Glomusintraradices. *Soil Biology and Biochemistry* 41: 1491–1496.

Bedini, S., Turrini, A., Rigo, C., Argese, E. and Giovannetti, M. 2010. Molecular characterization and glomalin production of arbuscularmycorrhizal fungi colonizing a heavy metal polluted ash disposal island, downtown Venice. *Soil Biology and Biochemistry* 42: 758–765.

Bell, M. And Lawrence, D. 2009. *Soil carbon sequestration - myths and mysteries.* The Department of Primary Industries and Fisheries.1-9.

Billings, S.A. and Zeigler, S.E. 2008. Altered patterns of soil carbon substrate usage and heterotrophic respiration in a pine forest with elevated CO_2 and N fertilization. *Global Change Biology* 14: 1025–1036.

Blagodatskaya, E. and Kuzyakov, Y. 2008. Mechanisms of real and apparent priming effects and their dependence on soil microbial biomass and community structure: critical review. *Biology and Fertility of Soils* 45: 115–131.

Bossio, D.A., Scow, K.M., Gunapala, N. and Graham, K.J. 1998. Determinants of soil microbial communities: Effects of agricultural management, season, and soil type on phospholipid fatty acid profiles. *Microbial Ecology* 36:0 1–12.

Bossuyt, H., Denef, K., Six, J., Frey, S.D., Merckx, R. and Paustian, K. 2001. Influence of microbial populations and residue quality on aggregate stability. *Applied Soil Ecology* 16: 195–208.

Brooks, S.J., Gale, P.A. and Light, M.E. 2006. Anion binding modes in macrocyclicamidourea. *Chemical Communications* 41: 4344–4346.

Campbell, J.H., Clark, J.S. and Zak, J.C. 2008. PCR-DGGE comparison of bacterial community structure in fresh and archived soils sampled along a Chihuahuan Desert elevational gradient. *Microbial Ecology* 57: 261–266.

Canadell, J.G., Pataki, D.E., Gifford, R., Houghton, R.A., Luo, Y., Raupach, M.R., Smith, P. and Steffen, W. 2007. Saturation of the terrestrial carbon sink. In *Terrestrial Ecosystems in a Changing World* (Eds. Canadell, J.G. *et al.*), pp. 59–78, Springer-Verlag

Claessen, D., Jong, W., Dijkhuizen, L. and Wosten, H.A.B. 2006. Regulation of Streptomyces development: reach for the sky! *Trends in Microbiology* 14: 313–319.

Cleveland , C.C., Nemergut, D.R., Schmidt, S.K. and Townsend, A.R. 2007. Increases in soil respiration following labile carbon additions linked to rapid shifts in soil microbial community composition. *Biogeochemistry* 82: 229–240.

Coyne, M.S. 1999. *Soil microbiology: An exploratory approach.* Delmar Publishers, Albany, NY.

Davidson, E.A. and Janssens, I.A. 2006. Temperature sensitivity of soil carbon decomposition and feedbacks to climate change. *Nature (Lond.)* 440: 165–173.

Feller, C. and Bernouxm, M. 2008. Historical advances in the study of global terrestrial soil organic carbon sequestration. *Waste Management* 28: 734–740.

Fierer, N. and Jackson, R.B. 2006. The diversity and biogeography of soil bacterial communities. *Proceedings of National Academy of Sciences of the United States of America* 103: 626–631.

Frey, S.D., Elliott, E.T. and Paustian, K. 1999. Bacterial and fungal abundance and biomass in conventional and no-tillage agroecosystems along two climatic gradients. *Soil Biology and Biochemistry* 31:573–585.

Frey, S.D., Knorr, M., Parrent, J. and Simpson, R.T. 2004. Chronic nitrogen enrichment affects the structure and function of the soil microbial community in a forest ecosystem. *Forest Ecology and Management* 196: 159–171.

Gaumont-Guay, D., Black, T.A., Griffis, T.J., Barr, A.G., Jassal, R.S. and Nesic, Z. 2006. Interpreting the dependence of soil respiration on soil temperature and water content in a boreal aspen stand. *Agricultural and Forest Meteorology* 140: 220–235.

Gebbink, M.F.B.G., Claessen, D., Bouma, B., Dijkhuizen, L. and Wosten, H.A.B. 2005. Amyloids - a functional coat for microorganisms. *Nature Reviews Microbiology* 3: 333–341.

Glibert, P.M., 2008. Ocean urea fertilization for carbon credits poses high ecological risks. *Marine Pollution Bulletin* 56: 1049–1056.

Gupta, V.V.S.R. and Germida, J.J. 1988. Distribution of microbial biomass and its activity in different soil aggregate size classes as affected by cultivation.*Soil Biology and Biochemistry* 20: 777–786.

Hawkes, C.V., Hartley, I.P., Ineson, P. and Fitter, A.H. 2008. Soil temperature affects carbon allocation within arbuscularmycorrhizal networks and carbon transport from plant to fungus. *Global Change Biology* 14: 1181–1190.

Houghton, R.A. 2007. Balancing the global carbon budget.*Annual Review of Earth and Planetary Sciences* 35: 313–347

Jackson, L.E., Calderon, F.J., Steenwerth, K.L., Scow, K.M., and Rolston, D.E. 2003. Responses of soil microbial processes and community structure to tillage events and implications for soil quality. *Geoderma* 114: 305–317.

Jin, V.L. and Evans, R.D. 2007. Elevated CO_2 increases microbial carbon substrate use and nitrogen cycling in Mojave Desert soils. *Global Change Biology* 13: 452–465.

Kimble, J. M., Lal, R. and Follett, R.R. 2002. Agricultural Practices and policy options for carbon sequestration: what we know and where we need to go. In *Agricultural practices and policies for carbon sequestration in soil* (EdsKimbel, J.M., R. Lal, and R.F. Follett), New York, Lewis Publishers, p 512.

King, G.M. 2011. Enhancing soil carbon storage for carbon remediation: potential contributions and constraints by microbes.*Trends in Microbiology* 19: 75-84.

Klein, D.A., McLendon, T., Paschke, M.W. and Redente, E.F. 1996.Nitrogen availability and fungal-bacterial responses in successional semi-arid steppe soils.*Arid Soil Research and Rehabilitation* 10: 321–332.

Lal, R. 2003. Global potential of soil carbon to mitigate the greenhouse effect.*Critical Reviews in Plant Sciences* 22: 151–184.

Lal, R. 2004. Soil carbon sequestration impacts on global climate change and food security. *Science* 304: 1623-1627.

Lal, R. 2008. Carbon sequestration.*Philosophical Transactions of the Royal Society B: Biological Science* 363: 815–830.

Linder, M.B. (2009) Hydrophobins: proteins that self assemble at interfaces. *Current Opinion in Colloid & Interface Science* 14: 356–363.

Lundquist, E.J., Jackson, L.E., Scow, K.M. and Hsu, C. 1999. Changes in microbial biomass and community composition, and soil carbon and nitrogen pools after incorporation of rye into three California agricultural soils.*Soil Biology and Biochemistry* 31: 221–236.

McCarthy, J.F. 2008. Protection of organic carbon in soil microaggregates via restructuring of aggregate porosity and filling of pores with accumulating organic matter.*Geochimicaet Cosmochimica Acta* 72: 4725–4744.

Montealegre, C.M., Kessel, C.V., Blumenthal, J.M., Hur, H., Hartwig, U.A. and Sadowsky, M.J. 2000. Elevated atmospheric CO_2 alters microbial population structure in a pasture ecosystem. *Global Change Biology* 6: 475-482.

Mummey, D.L. and Stahl, P.D. 2004. Analysis of soil whole- and inner-microaggregate bacterial communities. *Microbial Ecology* 48: 41–50.

Neely, C.L., Beare, M.H., Hargrove, W.L. and Coleman, D.C. 1991. Relationships between fungal and bacterial substrate-induced respiration, biomass, and plant residue decomposition.*Soil Biology and Biochemistry* 23: 947–954.

Niklaus, P.A. 1998. Effects of elevated atmospheric CO_2 on soil microbiota in calcerous grassland.*Global Change Biology* 4: 451–458.

Oliver, K.L., Hamelin, R.C. and Hintz, W.E. 2008.Effects of transgenic hybrid aspen overexpressing polyphenol oxidase on rhizosphere diversity.*Applied and Environmental Microbiology* 74: 5340–5348.

Pachauri, R.K. and Reisinger, A. 2007.*Climate Change 2007: Synthesis report*. IPCC, Geneva, Switzerland.pp 104.

Petersen, S.O., Debosz, K., Schjønning, P., Christensen, B.T. and Elmholt, S. 1997. Phospholipid fatty acid profiles and C availability in wet-stable macro-aggregates from conventionally and organically farmed soils. *Geoderma* 78: 181–196.

Peterson, G.A., Halvorson, A.D., Havlin, J.L., Jones, O.R., Lyon, D.J. and Tanaka, D.L. 1998. Reduced tillage and increasing cropping intensity in the Great Plains conserves soil C. *Soil & Tillage Research* 47: 207–218.

Purin, S. and Rillig, M.C. 2008.Immuno-cytolocalization of glomalin in the mycelium of the mycorrhizal fungus Glomusintraradices.*Soil Biology and Biochemistry* 40: 1000–1003.

Rillig, M.C. 2005. A connection between fungal hydrophobins and soil water repellency? *Pedobiologia* 49: 395–399.

Schlesinger, W. 2003.*Biogeochemistry, An Analysis of Global Change*. Academic Press.

Singh, B.K., Bardgett, R.D. Smith, P. and Reay, D.S. 2010. Microorganisms and climate change: terrestrial feedbacks and mitigation options. *Nature Review Microbiology* 2: 779–790.

Six, J., Elliott, E.T. and Paustian, K. 2000. Soil macroaggregate turnover and microaggregate formation: A mechanism for C sequestration under no-tillage agriculture. *Soil Biology and Biochemistry* 32: 2099–2103.

Six, J., Frey, S.D., Thiet, R.K. and Batten, K.M. 2006.Bacterial and fungal contributions to carbon sequestration in agroecosystems.*Soil Science Society of America Journal* 70: 555–569.

Sundquist, E., Burruss, R., Faulkner, S., Gleason, R., Harden, J., Kharaka, Y., Tieszen, L. and Waldrop, M. 2008.*Carbon sequestration to mitigate climate change.*U.S. Geological Survey.*http: //www.usgs.gov/.*

Tossel, J.A. 2009. Catching CO_2 in a bowl. *Inorganic Chemistry* 48: 7105–7110.

Van Veen, A.J., Ladd, J.N. and Amato, M. 1985. Turnover of carbon and nitrogen through the microbial biomass in a sandy loam and a clay soil incubated with 14C (U) glucose and $15N(NH_4)_2SO_4$ under different moisture regimes. *Soil Biology and Biochemistry* 17: 747–756.

Vargas, R. and T. Hattori. 1986. Protozoan predation of bacterial cells in soil aggregates. *FEMS Microbiology Ecology* 38: 233–242.

West, T.O. and Post, W.M. 2002. Soil organic carbon sequestration rates by tillage and crop rotation: A global data analysis. *Soil Science Society of America Journal* 66: 1930–1946.

Whitman, W.B., Coleman, D.C. and Wiebe, W.J. 1998. Prokaryotes: the unseen majority. *Proceedings of National Academy of Sciences of the United States of America* 95: 6578–6583.

Yuan, H., Ge, T., Chen, C., O'Donnell, A.G. and Wu, J. 2012. Microbial autotrophy plays a significant role in the sequestration of soil carbon. *Applied Environmental Microbiology* 78: 2328-2336.

❑❑❑

Chapter 20

Water Productivity Under Climate Change Scenario

ANURAG SAXENA AND R.K. GOYAL

ICAR-Central Arid Zone Research Institute, Jodhpur-342 003, Rajasthan

E-mail: Anurag.Saxena@icar.gov.in; drasaxena@gmail.com

Efficient use of water, a key input in agriculture, is a central issue throughout the world. Water, a finite source, is essential for both the human society and the ecological systems that humans rely on. With the population growth and economic development, water has become increasingly scarce in a growing number of countries and regions in the world. In spite of having the largest irrigated area in the world, India too has started facing severe water scarcity in different regions. Owing to various reasons the demand for water for different purposes has been continuously increasing in India, but the potential water available for future use has been declining at a faster rate. The agricultural sector (irrigation), which currently consumes over 80 percent of the available water in India, continues to be the major water-consuming sector due to the intensification of agriculture. Though India has the largest irrigated area in the world, the coverage of irrigation is only about 40% of the gross cropped area as of today. One of the main reasons for the low coverage of irrigation is the predominant use of flood (conventional) method of irrigation, where water use efficiency (WUE) is very low due to various reasons. Available estimates indicate that water use efficiency under flood method of irrigation is only about 35 to 40% because of huge conveyance and distribution losses. This requires an increase in water productivity (WP) and water use efficiency. It is much more pertinent for a vast tropical country like India, which experiences extreme variation in climate and rainfall. India, though achieved self-sufficiency in food with the help of efficient management of her natural resources leading to green revolution but arid and semi-arid regions are still lagging behind. Therefore, it is necessary to economize the use of water and at the same time increase the productivity per unit area and per unit quantity of water.

Arid regions are encountered by various problems. Harsh climatic conditions in arid regions allow the farmers to grow only one crop during rainy season that, too, depend on the rainfall. The average annual rainfall of western arid region is 317 mm. The rainfall is highly variable and erratic with frequent drought spells. The number of rainy days varies from 10 to 25. Groundwater is very deep, saline at many places and expensive to use. Indiscriminate use of water on undulating highly permeable sandy soils through conventional irrigation resulted in fall of groundwater by 0.6 – 1.0 m annually. The situation of over exploitation of groundwater is more serious in the region where out of 11 districts, 6 are in category of over exploited and remaining 5 are in category of semi critical zone. Rainfed farming is adversely affected by low and erratic rainfall coupled with high evaporative demand and low moisture retention by light textured soils. On the other hand indiscriminate use of scarce water through conventional irrigation management practices led to exhaustion of ground water resources and development of water logging in canal command area. Hence, conservation and efficient management of limited water is the need of the hour for achieving sustainable production for longer period on light textured soils of arid and semi-arid regions. Under climate threat water demand will be more.

Global warming due to enhanced greenhouse effect is expected to cause major changes in various climatic variables viz., absolute humidity, precipitation and net terrestrial and global solar radiation. Atmospheric temperature is the most widely used indicator of climatic change. Climate change may have significant impact on hydrological parameters viz. runoff, evapotranspiration (ET), soil moisture, ground water etc. ET is a major component of hydrological cycle after precipitation and determines the crop water requirement. The principle factors that influences the ET depends upon rainfall, temperature, humidity, sunshine hours etc. Any change in climatic parameters due to global warming will also affect ET. So, future crop planning and management of water resources will depend on expected change in ET.

Water productivity term plays a crucial role in modern agriculture which aims to increase yield production per hectare per unit of water used, both under rainfed and irrigated conditions. This objective can be pursued by:

1. Increasing the marketable yield of the crops for each unit of water transpired,
2. Reducing the outflows and the atmospheric water depletion,
3. Enhancing the effective use of rainfall, water stored and marginal quality water.

The first option concerns to the need for improving crop yield; the second one intends to increase the transpiration of water supply against the evaporation; the third aims to utilize efficiently the water resources (reservoirs, streams or groundwater sources). All these lead to the improvement in management practices which use less water for irrigation, decrease evaporation losses, optimize fertilizer supply, pest control, energy consumption, soil conditions, etc. This has more importance in arid and semi-arid regions with limited water supply, where the farmers are frequently constrained to apply deficit irrigation strategies and manage water supply in accordance with the sensitivity of crop's growing stages to water stress.

The basic intrinsic characteristic of water dynamics in soil cannot be altered to a large degree but availability can be modified enormously by water conservation and water management practices. During the complete life cycle of a plant the most dynamic property is water content in soil. Since water that is conserved or supplemented becomes available to the plant through the soil, monitoring of water content under field condition is of paramount importance. The whole life cycle of plant revolves around extraction of water from the soil through its root network and transpiring through the leaves in response to evaporation demand set up by the atmosphere. Any shortage of water in the soil is ultimately reflected by the growth of the plants in some form or another. Increasing water supply available to plants or enhancing water use relative to other losses and making efficient use of limited water are the three major concepts in optimizing water use in agriculture.

Increasing water availability to crop plants

Since rainfall is low and erratic, efforts are needed to enhance the total water supply available to crops. Some of these include tillage, mulching, soil moisture barrier, contour bunding and *in-situ* water harvesting. Mechanical structures like contour bunding, terracing are the pre-requisite for water conservation on sloppy agricultural lands. Off-season tillage on alfisols helped in controlling weeds and conserving water. However, such practice on arid soils may lead to soil erosion.

Hence, minimum tillage was found to control 73% weeds and result in 43% increase in yield. Use of weed mulch increased the water availability to crops by restricting evaporation directly from soil surface.

Crop planning based on rainfall and moisture availability period

In dryland areas crop production depends on amount and distribution of rainfall. Most of the farmers still grow long duration traditional varieties under conventional practices. The productivity of these crops varieties is affected due to uncertain rainfall. Hence, there is need to develop short duration varieties, crop plan and agronomic practices matching to rainfall pattern. On the basis of climatic data all over India, suitable crops and varieties for early, normal and late onset of monsoon, improved inter cropping, time of sowing and agronomic practices have been identified and developed. In arid part of Rajasthan short duration varieties of pearl millet, clusterbean, moth bean, mung bean and sesame have been developed to minimize the risk of crop failure. With the early onset of monsoon, pearl millet and sesame get preference. While with the late onset of monsoon clusterbean, mung bean and moth bean get preference. Since the moisture evaporates very fast, it is suggested to complete the sowing immediately after rains.

All operations carried out in the field, from land preparation to crop harvesting, with the aim of increasing the crop yield are included under agronomical practices. Certain simple agronomical practices like optimum tillage, administration of organic manure and suitable cropping pattern are effective in retaining soil fertility as well as giving satisfactory crop yield.

Tillage Operation

Tillage makes soil loose and hence prone to erosion. Timing and depth of tillage are the two important factors which need special attention. Tillage should be done immediately before the crop season to take advantage of one or two early showers for land preparation. In arid region, land tilled into ridges and furrows across the wind direction has been found to reduce the effects of wind erosion during the summer months. Deep tillage improved the soil moisture storage, water use efficiency and grain yield of pearl millet (Saxena *et al.*, 1997).

Land Leveling

Farm lands should be leveled for efficient management of water. Level lands allow more infiltration, thus increasing soil moisture and leaching. This in turn reduces run-off and hence soil erosion. Leveling of irregular land is done by the cut and fill method. Soil from the elevated portions is removed and placed in the depressed portions to obtain a level land.

Mulching

Mulching of open land surface is achieved by spreading stubble, trash or any other vegetation. The objectives of mulching are to minimize splash influence of rain drops on base surface, reduce evaporation, increase absorption of the rainfall, obstruct surface flow thereby retarding erosion and allow microbiological changes to occur at optimum temperature. Top surface of loamy sand soils acts as self-mulch after drying. In such conditions, mulch may not be useful practice in arid region, however, many a times droughts occur early in the vegetative stage. In such situation use of mulch immediately after sowing delay the process of top soil surface drying which helps in promoting better plant establishment and results in higher yield. Polyethylene mulch was highly effective in controlling evaporation losses (Gupta, 1978, 1980). Daulay *et al.* (1979) also reported beneficial effects of mulch in pearl millet. Application of grass mulch (6 t ha^{-1}) resulted in 200 per cent increase in the yield of green gram, dew gram and clusterbean (Gupta and Gupta, 1983). Use of mulch on no-tillage plots increased the yield of cowpea by 50-150 per cent over no mulch in different years (Gupta, 1986).) High yield of tomatoes and lady finger with grass waste mulch was also recorded (Gupta 1985). Increase in yield was owing to reduction in soil thermal regime evaporation and suppression of weeds (Gupta and Gupta, 1985).

Increasing water use relative to other losses

Evaporation of water directly from soil surface is an important component of the total water use of crops. Evaporation constitutes 19-46% of the total water on vertisols, 21% on alfisols and 12-18% on loamy sand soils. The best opportunity for water saving lies in methods increasing transpiration relative to other losses from the soil surface due to canopy shading. The elimination of weeds, optimum plant population and spacing and application of fertilizer/manure helped in better canopy growth and higher WUE of crops. Generally close row spacing and higher plant population is recommended in high rainfall areas or irrigated condition, but in arid region wide row spacing and low population is recommended. Paired row planting system helps in better canopy growth and hence evaporation from soil surface is restricted. Weeding improved the pearl millet yield by 44% over weedy condition (Table 1) in low rainfall season (Singh, 1988).

Table 1. Weeding enhances WUE (kg ha^{-1} mm^{-1}) and yield (kg ha^{-1}) of pearl millet

Treatment	Good rainfall (548 mm)		Low rainfall (252mm)	
	Yield	WUE	Yield	WUE
Weedy	1672	3.1	1053	4.2
Weed free	1868	3.4	1521	6.0

Management of limited irrigation water

In arid region water is limited and land is vast, hence water management should aim to maximize production per unit of water rather per unit of land. Some of the technologies like extensive, deficit and pressurized irrigation have been developed to maximize the production in arid region.

Extensive Irrigation

Extensive irrigation approach seeks to apply a small quantity of water over a large area rather than large quantity of water over a small area. Production per unit land may decrease but production per unit water may increase. Wheat and mustard required 840 and 250 mm water per hectare to produce maximum yield. When the same water applied optimally in 3, 1.5 and 4.0 ha land in wheat and mustard, respectively gave less production per unit of land but total productivity per unit of water was more by bringing larger area under irrigation (Table 2). Singh (1997) observed that under given water supply the area brought under irrigation in pearl millet was more in subnormal rainfall years than low rainfall years, however, the production enhanced in both the situation by bringing larger area under irrigation in pearl millet (Table 3).

Table 2. Extensive irrigation maximise production for given water supply

Crop	Water use (mm)	Area(ha)	Yield(kg ha^{-1})	WUE(kg ha^{-1}mm^{-1})
Wheat	840	1	5500	6.5
		3	9100	10.8
Mustard	250	1	1100	4.4
		1.5	2000	8.0

Table3. Increased area under S-IRR enhances pearl millet production

Good rainfall(331 mm)			Medium rainfall(266 mm)			Low rainfall(173 mm)		
S-IRR	Area	Yield	S-IRR	Area	Yield	S-IRR	Area	Yield
190	1.0	3660	292	1	3620	341	1.0	3540
76	2.5	7225	145	2	5380	209	1.6	4528
38	5.0	12950	73	4	7560	117	2.9	5829
0	1.0	2280	0	1	1400	0	1.0	1080

S-IRR = Supplemental irrigation (mm ha^{-1}), Area in ha, Yield in kg ha^{-1}

Deficit Irrigation

Deficit irrigation approach seeks to avoid irrigation at less critical stages of crop growth and apply less water at the end so as to eliminate water stored at harvest. One has to be well versed with the crop growth stages that are less sensitive to moisture stress and there is need for proper balancing between water given and ET demand of crop.

Rainwater management

In arid and semi-arid parts experiencing rainfall from 150-350 mm and 350-550 mm respectively, rainfall is not sufficient to ensure a good crop. Crop production is adversely affected due to low and erratic distribution. In dryland farming, the solution to soil moisture problem lies in the storage of rainfall in the potential root zone of the soil by the water harvesting methods. Therefore, water harvesting would be an alternative to leave land fallow for increasing the available water supply for plant growth. In last two decades, researchers and planners have given a lot of attention to dryland farming in such areas and for various regions specific agro-techniques have been evolved.

In-situ Rainwater conservation: Water stored in Soil profile Itself

Inter-plot water harvesting: In many arid areas, water harvesting is used to increase the total water supply available to crops. It is practiced in variety of ways depending upon topography, soil and rainfall. Inter-plot water harvesting that uses a portion of land with 5% slope as catchment to generate runoff and divert it to adjacent area was found highly beneficial to improve the yield of many crops. CAZRI has developed water harvesting technique with 3 m cropped area and 1.5 m catchment area (catchment to crop area ratio = 0.5) with 5% slope on both sides (Fig. 1). Response of pearl millet to water harvesting system was studied at CAZRI (Singh and Singh, 1988). Crop water supply, over the check, ranged from 23 to 60 mm in the extremely dry season and 108 to 286 mm in the season with a good rainfall. As a result, in water harvesting plot a yield of 3717 kg ha^{-1} was obtained with 69% of normal rainfall. This ratio (0.5) was found optimum for the type of soil, climate, and surface sealing pond sediment.

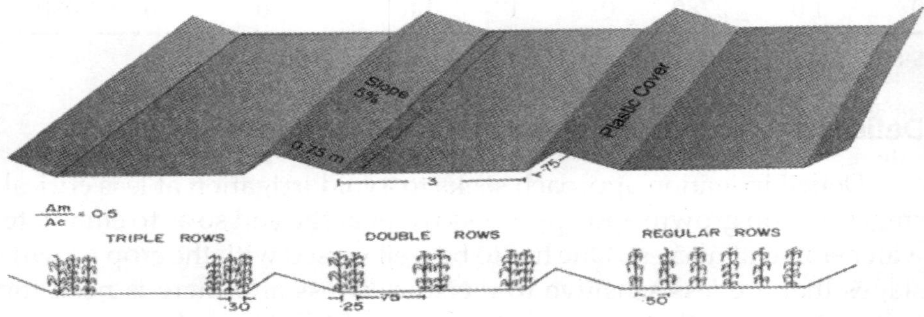

Fig. 1: *In situ* water harvesting system (Singh, 1988)

Integration of regular row (RR) and double row (DR) plant geometries into water harvesting system on yield and water use efficiency of pearl millet best suited to runoff farming (Fig. 1). In case of prolonged drought, 22% increase in yield with DR over RR was recorded (Singh, 1988). This increase was due to suppression of early plant growth, a mild crop canopy, micro-climate and conservation of water and nutrient by suppression of weed in DR system. Higher water storage in the 120 cm profile supported the pearl millet crop during ear emergence and grain development period that significantly increased the yield attributing characters. In the seasons with above normal, well-distributed rainfall, yield was similar for both RR and DR plant arrangements.

Water harvesting proved beneficial particularly in low rainfall years. The yield of pearl millet was almost 3 times with water harvesting compared to conventional sowing (Table 4). In low rainfall situation *in-situ* water harvesting helped in making efficient use of limited rainfall and N-fertilizer in pearl millet (Singh 1985). Application of 40 kg N ha^{-1} with *in-situ* water harvesting improved the pearl millet by 27% with 69 mm of additional runoff rainwater (Table 5).

Table 4. Water use, yield and WUE of pearl millet by water harvesting

System	Good rainfall			Low rainfall		
	Water use (mm)	Yield (kg ha^{-1})	WUE (kg ha^{-1} mm^{-1})	Water use (mm)	Yield (kg ha^{-1})	WUE (kg ha^{-1} mm^{-1})
Conventional	440	2320	5.3	143	400	2.9
In-situ water harvesting	630	2425	3.9	212	1240	5.9

Table 5. *In-situ* water harvesting makes efficient use of limited rainfall and N-fertilizer in pearl millet in low rainfall season (143 mm)

System	Added water (mm)	N- levels (kg ha^{-1})	Yield of pearl millet (kg ha^{-1})
		0	40
Rainfed	0	210	400
In-situ water harvesting	69	660	840

Micro catchment technique: Micro-catchment is one of the major forms of direct water conservation systems. Micro-catchment systems are relatively effective for use in growing trees and shrubs. In micro-catchment based cropping, rainwater is concentrated in a small portion of the cultivated area. The tree crops are deep rooted and can utilize the moisture stored in the sub-stratum and hence form a better option for micro-catchment based farming in sandy soil situations (Sharma *et al.*, 1986).

Arid horticultural plants like pomegranate, ber, and several others can be successfully grown with appropriate micro-catchments in the water scarce regions. In micro-catchment systems, the donor to collector area ratios may range from 1 : 1 to 20 : 1. The various sizes of micro-catchments have been used for different purposes. The sizes may be up to 5 ha for annual crops, 31 -144 m^{2}for *jujube* orchards in India (Yadav *et al.*, 1980), 0.35 ha for fuel wood plantation in Israel (Zohar *et al.*, 1988), 144-289 m^2 for fig, olive, pistachio plantation in Syria (Ibrahim, 1994). A number of catchment cropped area ratios and degree of slopes have been tried at CAZRI, Jodhpur. For *ber,* circular catchments of 1.5 m diameter and 5% inward slope with 54 m^2 of catchment has been found appropriate for conservation and proper utilization of rainwater (Sharma *et al.*, 1986). The circular micro-catchment and trenching techniques were compared for soil profile moisture storage in the arid region. The study revealed that the 20% increase in soil moisture was observed due to circular catchment or trenching than with any soil treatment (Ojasvi *et al.,* 1999). For further improvement in water use efficiency, these circular micro-catchments can be covered with plastic sheet (LDPE).

Ex-situ Rainwater conservation: Water stored in reservoir or pond for recycling

Ex-situ rainwater harvesting is a promising technology for enhancing the availability of water in arid areas. Ex-situ water harvesting involves collecting runoff originating from rainfall over a surface away from the field and storing it in surface storage systems for later use in the field. This type of rainwater harvesting provides supplemental or protective irrigation during dry periods of the cropping season.

Pressurised irrigation

Irrigation is a major problem in the arid regions which are characterized as having low rainfall, highly permeable soils and poor quality of irrigation water. The conventional surface or aerial (sprinkler) irrigation system can solve these problems only partially.

Though flood method of irrigation has been followed predominantly all over the world for cultivating crops, but it is no longer desirable for countries like India mainly due to limited availability of water resources and growing demand for water for irrigation and other purposes. Therefore, for achieving sustainable agricultural development it is essential to increase the existing water use efficiency. Further, in view of the limited water resources, water is to be efficiently and economically used in arid and semi-arid conditions, for which pressurized irrigation systems can be one of the viable options.

The pressurized irrigation system has the capability of applying desired quantity of water both accurately and uniformly. Drip and sprinkler are the two examples of pressurized irrigation and both are able to save water as water is applied under pressure through a network of closed pipes, sprinkler nozzles and emitters. The energy required is imparted to the water by a pumping unit, which in turn receives energy from either an electric motor or internal combustion engine.

Advantages of Pressurized Irrigation

- They provide a high degree of water application uniformly, often the highest of all irrigation systems in use.
- They allow excellent control of the amount and timing of irrigation. Small and frequent irrigations can be applied to match the trees' water needs. Runoff is minimized because of the low application rates, and deep percolation losses can also be minimized if the correct amount of water is applied. The frequent irrigation provides an excellent soil water condition for optimal tree performance.
- They can easily irrigate irregular terrain.
- Weed growth is minimized since only desired portion is wetted.

Disadvantages of Pressurized Irrigation systems

- High initial cost of the systems.
- Excellent management is needed to maintain the system since clogging of the emitters by physical particles, organic materials, and/or chemical precipitates may occur.
- The irrigation water must be pressurized, resulting in energy costs.
- Cover crops cannot be grown throughout year due to the localized nature of the water applications.

Drip irrigation system

The most befitting technique for efficient utilization of costly irrigation water in these areas is the 'Drip' method of irrigation. "Drip irrigation" is a method of watering plants at a rate equivalent to (and not more than) its consumptive use so that the plants would not experience any moisture stress throughout their life cycle. In this method of irrigation water is conveyed from the source (i.e., tube well, dug-well or pond) through a network of pipe-lines (known as drip system) and finally released near the plant base. The primary objective of this type of irrigation is to provide optimum quantity of water to the crop for the maximum yield and simultaneously saving of valuable water from wastage; thereby increasing the water use efficiently.

Irrigation efficiencies under different methods of irrigation have been shown in Table 6. Research suggests that drip method of irrigation (DMI) is not only suitable for those areas that are presently under cultivation but it can also be operated efficiently in undulating terrain, rolling topography, hilly areas, barren land and areas which have shallow soils. Reduction in water consumption due to drip method of irrigation over the surface method of irrigation varies from 30 to 70 percent for different crops. According to data available from research stations, productivity gain due to drip method of irrigation is estimated to be in the range of 20 to 90% for different crops and could be used in various crops as mentioned below.

Table 6. Irrigation efficiency (%) under different methods of irrigation

Irrigation efficiencies	Methods of irrigation		
	Surface	Sprinkler	Drip
Conveyance efficiency	40-50 (canal) 60-70 (well)	100	100
Application efficiency	60-70	70-80	90
Surface water moisture evaporation	30-40	30-40	20-25
Overall efficiency	30-35	50-60	80-90

Source: (Narayanamoorthy 2009).

- **Orchard crops**: Grapes, citrus, apples, pomegranate, pears, delicious fruits (peaches, apricots, plums etc.), nuts (almonds, pistachios), bananas, dates, olives, mangoes, guavas etc.
- **Vegetables**: Tomato, green pepper, cucumber, lettuce, green pea, cauliflower, okra etc.
- **Row and field crops**: Cotton, sugarcane, corn, groundnut and onion.
- **Others:** Berries, melons, alfalfa, flowers (carnations, gladioli and roses) and other ornamental plants.

Planting technique

Planting of seed or seedlings is done along the length of the lateral on one side or either side of it, irrespective of the position of the emitters on it. The emitters placed at a distance of 50 cm apart on a lateral line discharge the water at such rate that a continuous strip of about 40 cm width of soil remains wet throughout the length of the lateral. Different planting configurations (i) a single row (rectangular planting), (ii) double rows (square or equilateral planting) or (iii) triple row (hexagonal planting) were tried at CAZRI. The major objective of such different types of planting is to reduce the number of laterals, thereby reducing the cost of the system without reducing the plant population in the field. The results of the study revealed that equilateral planting with the side of 25 cm to 35 cm of a triangle performed the best in almost all the crops tried (Table 7). Therefore, equilateral planting is advocated for vegetable crops like tomato, cabbage, cauliflower, turnip etc.

Paired row planting with single lateral of drip combined with mulch resulted 40-50% saving in water and higher production of tomatoes. Drip irrigation at ET 100 saved 35% water and produced 40% higher yield over conventional irrigation. Drip ET 50 though yielded at par with conventional irrigation but saved 50% water (Singh 1978).

Table 7. Effect of planting configurations on crop yields

Plant arrangement	Yield (metric t ha^{-1}) of			
	Cabbage	Tomato	Turnip	Cauliflower
Rectangular	35.5	68.4	19.5	20.0
Square	33.2	73.1	20.3	23.8
Hexagonal	16.9	55.1	11.8	13.2
Equilateral	33.0	78.4	22.9	26.0

Irrigation scheduling and frequency

The main objective of the drip method of irrigation is to provide each plant with a continuous supply of soil moisture which is just sufficient to meet the evapo-transpiration demand. Thus it is desirable to irrigate the field daily at the end of the day break. The unique feature in drip method of irrigation is that since the water loss of the day is duly compensated every evening, the root soil always remaining nearly at the field capacity and consequently the crop never suffers from moisture stresses. This is the main reason for higher yield of crops as compared to other method of irrigations.

The potential evapo-transpiration as obtained from USDA class A pan may be multiplied with crop factor (approximately 0.7) to get daily consumptive use of crop. For a layman this approximate quantity or irrigation requirement may be translated into time period (in minutes) for which the system may be kept opened in a usual sunny or cloudy day of the *rabi* season (Shankarnarayan *et al.*, 1984). However, it has been observed by the scientists that by increasing the quantity of water in irrigation by 30 above the consumptive use through drip system the subsequent irrigation can be delayed by 2 days without much adverse effect.

Drip irrigation in cotton at 0.8 ETc gave 27.5% higher seed cotton yield compared to yield with furrow irrigation (1801 kg ha^{-1}) at CAZRI RRS Pali. Decrease in seed yield of cotton was 6.3% at 0.8 ETc and 25.5% at 0.6 ETc irrigation levels compared to full irrigation (1.0 ETc) (CAZRI Annual Report 2010-11).

Drip Irrigation for Horticultural crops

Drip method of irrigation is more economical for orchard crops than that for the vegetables. In orchards of ber, lemon, date-palm, pomegranate, etc. the plants as well as rows are kept almost 5 to 8 metre apart which necessitates where the laterals are spaced nearly 1 m to 1.5 m apart. The drippers for fruit trees are different from that of vegetable plants. Moreover, less number of drippers are used in each of the lateral lines. For example, in a lateral line of 30 m length laid out for tomato crops nearly 60 drippers are fixed on each lateral while for the same length of lateral to be used in ber orchard of grown up trees only 25 to 30 drippers will be necessary depending upon the quantity of water required. On the other hand it has been found that in arid areas these fruit trees when irrigated produce almost two times the fruit as compared to that in unirrigated conditions.

Use of Drip Irrigation for Saline water irrigation

In desert, over 60% of the total area has the problem of groundwater salinity. Conventional irrigation with saline water caused considerable reduction in yield and deteriorated the soil health. Drip method of irrigation has its adaptability to saline water irrigation. Since the frequency of irrigation in this method is quite high (i.e. every day) the plant base always remains wet which keeps the salt concentration in the plant root zone below the critical level that may otherwise be hazardous for the plant growth. The maximum salt concentration occurs around the periphery of the wet zone. Therefore, studies were conducted using drip irrigation to evaluate the utility of drip system in applying saline water. Study revealed that yield of potatoes irrigated by drip irrigation with water with a conductivity of 3 dS m^{-1} was 31% higher than that of furrow irrigation using good quality water (Table 8). In case of tomato, the yield obtaining using water of conductivity of 10 dS m^{-1} was nearly 14% less than that of sweet water. Daily drip irrigation provided gainful use of poor quality water in potato and tomato maintaining higher water content in the rooting volume and forcing the salts to the sides and below the root zone. Hence, drip system has shown the potential to use substandard water in our desert (Singh *et al.*, 1978).

Drip irrigation in pomegranate with saline water (EC 9.5 dS m^{-1}) resulted in better growth, canopy and fruit yield (2.6 kg per plant) of pomegranate over ring basin irrigation system (Research Highlights, CAZRI, RRS, Pali, 1959-2009).

Table 8. Gainful use of saline water in potato and tomato under drip irrigation system

System	Salinity (dS m^{-1})	Potato (t ha^{-1})	Tomato (t ha^{-1})
Drip	3	26	50
Drip	10	-	44
Conventional	0.08	20	50

Fertigation-Combining fertilization with Drip irrigation

During the irrigation process the plant roots take up nutrients (nitrogen, potassium) at a rate of up to 4-6 kg ha^{-1} day^{-1}. As in drip irrigation the volume of the root system is relatively small and the amount of nutrients available in the soil is quite limited. Furthermore, the water flowing through the zone of the root system washes away dissolved salts, including nutrients. Therefore, it is essential to resupply nutrients to the root system, particularly nitrogen, which is highly soluble in water and is rapidly depleted from the root zone. Although potassium and phosphorus are less soluble and will be washed downward at a slower rate, there is a need to replenish them also, especially in sandy soil.

Several methods have been developed to provide the plants during the growing season with required nutrients as often as necessary via the irrigation system. In fertigation the water discharged by the drippers actually constitutes nutrient solution. The composition of the nutrients in the solution and the concentration of each component can be adjusted in accordance with the specific needs of each crop at various times during the season. Each of the fertilizers used in fertigation needs to be highly soluble in water and free of undissolved residues. In the case of water rich in calcium (hard water) sedimentation may occur when phosphates are introduced to the water. To overcome this problem it is recommended to use either phosphoric acid or a compounded acidic fertilizer. Several devices are available for delivery of fertilizers via the irrigation system including a metal container for delivering fertilizers, pumps operating a venturi tube, and a wide range of pumps with different mechanisms and control devices.

Investigations were carried out at CAZRI, Jodhpur to economize and improve nutrients' efficiency with drip irrigation for tomato crop. Higher moisture content maintained with drip irrigation make the nutrients in available form to the crop (Singh *et al.*, 1989; Singh *et al.*, 1999). The fertilizer (one fourth of total fertilizer) applied on planted area basis with drip provided higher yield than broadcast application in check basin (Table 9). This shows a saving of 75% fertilizer with drip irrigation over conventional irrigation.

Table 9. Effect of fertility level on tomato crop

Method of fertilization	Nitrogen (kg ha^{-1})	Phosphorus (kg ha^{-1})	Potash (kg ha^{-1})	Tomato yield (t ha^{-1})
Surface (B)	224	88	168	50
Drip	56	22	42	60

Drip irrigation system was demonstrated in a farmer's field for tomato cultivation by CAZRI for six years in 0.4 ha land for ascertaining the advantages derived from drip irrigation system. In flood irrigation system only 0.24 ha of land was irrigated as compared to 0.4 ha of land under the drip irrigation system with the same quantity of water. As such an increase of 66.6% land was brought under irrigation. The tomato yield was obtained to the tune of 60 q ha^{-1} (Chauhan *et al.*, 1988).

The drip irrigation is neither affected by winds nor by saline water as it applies water directly in the rooting volume of crops. Studies conducted at CAZRI Jodhpur revealed that drip saves 30-50% water and provide 2-3 times higher yield than conventional irrigation (Singh and Saxena, 2001). Experiment conducted at CAZRI, Jodhpur with drip irrigation system and furrow irrigation system of surface irrigation method showed significant amount of water saving (50%) for potato crop without losing much yield (Table10). Adoption of drip irrigation in chillies resulted in 71% increase in yields over surface furrow irrigation (Singh *et al.*, 1999) at CAZRI Jodhpur.

Table 10. Saving of water in potato crop under drip irrigation system

Irrigation method	Water use (cm)	Yield (t ha^{-1})
Drip– ET 50	16	18
Furrow–ET 100	35	19

Yield of muskmelon, ladyfinger and tomato, drip system gave higher yields of 17.7, 37.8 and 22.6% compared to the yield under check-basin irrigation (Table 11) (Saxena *et al.*, 2014).

Higher water use efficiency and water productivity under drip irrigation over conventional method of irrigation (check basin) of crop sequences is presented in (Table 12). Lady finger – tomato – Melons crop sequence for a year with drip irrigation gave maximum water productivity (Saxena *et al.*, 2014).

Table 11. Yield of crops in cropping system (t ha^{-1})

Manure (2.5 t ha^{-1})	Ladyfinger		Tomato		Muskmelon	
	Drip	Check basin	Drip	Check basin	Drip	Check basin
Vermi-compost + Inorganic Fert.	5.8	3.9	55.0	44.0	45.3	36.6
FYM + Inorganic Fert.	5.4	3.6	49.6	40.3	43.4	35.0
N_{80}	6.4	4.4	57.4	48.5	46.9	39.4
N_{60}	4.6	3.7	51.1	41.6	40.3	36.1
N_{40}	3.2	2.9	46.2	37.1	36.4	33.2

Source: Saxena et al. (2014)

Application of micronutrients through drip irrigation in groundnut increased pod and haulm yields, shelling out-turn, seed mass and sound mature seeds (SMS) and nutrient use efficiency over their soil and foliar applications. The drip application of Fe, Zn and B increased pod yield by 31-36, 21-23 and 15-21%, respectively over control (Singh *et al.*, 2001). Further, it was recommended to apply micronutrients through drip irrigation for better nutrient utilization and high groundnut yield in semi-arid region.

Table 12. Water use efficiency and water productivity of different sequencesm

System	Water Use Efficiency (kg ha^{-1} mm^{-1})				Water productivity (kg m^{-3})
	Rainy	Winter	Summer	Mean	
Drip Irrigation					
Lady finger – Tomato – lady finger	20.1	10.7	14.1	14.9	1.49
Tomato – Gladiolus – Lady finger	10.5	–	14.1	–	–
Lady finger – tomato – Melons	20.1	10.7	21.2	17.3	1.73
Lady finger – Gladiolus – Melons	20.1	–	21.2	–	–
Check-basin					
Lady finger – Tomato – lady finger	6.5	8.0	6.1	6.9	0.69
Tomato – Gladiolus – Lady finger	7.3	–	6.1	–	–
Lady finger – tomato – Melons	6.5	8.0	15.5	10.0	1.00
Lady finger – Gladiolus – Melons	6.5	–	15.5	–	–

Source: Saxena et al. (2014)

Same amount of water was applied through conduit pipes in conventional, sprinkler and drip systems. The water applied was 69 cm in drip and 84 cm in sprinkler and furrow irrigation. The benefits of drip irrigation were not the same for all the crops. Increase in yield from drip irrigation over sprinkler and furrow irrigation was 44-47% in long gourd, 21-37% in round gourd, 9-22 % in watermelon and practically nil in ridge gourd (Table 13). Further, water use efficiencies were 8.1, 6.5 and 11.0 kg ha^{-1} m^{3} in long gourd, round gourd and water melon, respectively. Sprinkler system also showed superiority over conventional method (furrow) in terms of yield (Singh and Singh, 1978).

Table 13. Yield (t ha^{-1}) of various crops under different methods of irrigationm

Irrigation method	Long gourd	Ridge gourd	Watermelon	Round gourd
Drip	56	12	82	41
Sprinkler	39	10	75	34
Furrow	38	11	67	30

Under a demonstration of sprinkler irrigation system at farmer's field, yield of pearl millet increased by 64% and of wheat by 153% as compared to flood irrigation. Similarly the yield of mustard increased from 770 kg ha^{-1} to 1657 kg ha^{-1} compared to flood irrigation. It also saved 31 man days per hectare for preparation of plots which is to be done under the flood system of irrigation (Chauhan *et al.*, 1988).

Climate change due to global warming is likely to cause shortage of water particularly in arid and semi-arid parts of India. The problem of shortage of water to rainfed crops could be resolved either by increasing total available water to crop plants or restricting evaporation losses relative to transpiration by integrating various management options like soil moisture conservation practices, water harvesting, manipulation in planting arrangement, fertiliser application, eliminating weeds etc. In case of limited irrigation resources in arid regions, modest irrigation, deficit/extensive irrigation and pressurized irrigation exhibited the efficiency to improve water use and make efficient use of other resources like seed and fertilizer. For managing limited water supply some concepts like crop response factor as a tool to crop sensitivity rating, optimal leaf area concept, optimal yield concept, optimal irrigation scheduling, water use efficiency maximising concept (protected agriculture), partial root zone wetting concept may be adopted. Adoption of these approaches will certainly lead to efficient management of scarce water in arid and semi-arid region. The incentives to the farmers in terms of research support, training and extension can help the farmers of arid region to reap the benefits. Further, area specific researches on these aspects are needed to provide proper package of practices to the ultimate users.

References

Chauhan, K.N.K., Singh, R.N. and Kokate, K.D. 1988. Impact of transfer of agricultural technology in ORP villages: A case study. CAZRI Publication No. 34. Central Arid Zone Research Institute, Jodhpur. 35 p.

Daulay, H.S., Singh, H.P., Singh, R.P. and Singh, K.C. 1979. Effect of different mulches on yield and moisture use of pearl millet (*Pennisetum typhoides*). *Annals of Arid Zone* 18: 108-115.

Gupta, J.P. 1978. Evaporation from a sandy soil under mulches. *Annals of Arid Zone* 17: 287-290.

Gupta, J.P. 1986. Moisture and thermal regimes of the desert soil of Rajasthan, India and their management for high plant production. *Hydrological Sciences Journal* 31: 347-359.

Gupta, J.P. and Gupta, G.N. 1983. Effect of grass mulching on growth and yield of legumes. *Agricultural Water Management* 6: 375-383.

Gupta, J.P. and Gupta, G.N. 1985. Effect of mulches on hydrothermal environment of soil and crop production in arid western Rajasthan. *Annals of Arid Zone* 24: 131-142.

Gupta, JP. 1980. Use of an asphalt subsurface barrier for improving the productivity of desert sandy soils. *Journal Arid Environment* 3: 215-222.

Gupta. J.P. 1985. Improvement of soil physical environment for increased crop production. *Transactions of Indian society of Desert Technology and University Centre of Desert Studies* 10: 46-54.

Ibrahim, I.H. 1994. Rainwater harvesting project in Dier-Atye: A case study for Syria. In *Water Harvesting for Improved Agricultural Production*. FAO Water Reports 3, FAO, Rome, pp. 73-86.

Narayanamoorthy, A. 2009. Water Saving Technologies as a Demand Management Option: Potentials, Problems and Prospects. In: *Strategic Analyses of the National River Linking Project (NRLP) of India (Series 3): Promoting Irrigation Demand Management in India: Potentials, Problems and Prospects* (Ed. Rathinasamy Maria Saleth). pp. 93 – 126. IWMI.

Ojasvi, P.R., Goyal, R.K. and Gupta, J.P. 1999. The micro-catchment water harvesting technique for the plantation of jujube (*Ziziyphus mauritiana*) in an agroforestry system under arid conditions. *Agricultural Water Management* 41: 139-147.

Saxena Anurag, Goyal, R.K., Singh, R.K. and Roy, M.M. 2014.Water Management for Climate Resilience in Arid Region.Central Arid Zone Research Institute, Jodhpur, India, 52 p.

Saxena, A., Singh, D.V. and Joshi, N.L. 1997. Effect of tillage and cropping systems on soil moisture balance and pearl millet yield. *Journal of Agronomy and Crop Science* 178: 251-257.

Shankarnarayan, K.A., James, B.K. and Singh, S.D. 1984. Drip system of irrigation for arid land crop production. *Bhagirath* XXXI: 1-4.

Sharma, K.D., Pareek, O.P. and Singh, H.P. 1986. Micro-catchment water harvesting for raising jujube orchards in an arid climate. *Transactions of the American Society of Agricultural Engineers* 29(1): 112-118.

Singh, A.L., Ajay and Chaudhari, Vidya 2001. Drip irrigation- a potential system for micronutrient application in groundnut in semi-arid region. In *Microirrigation* (Eds. H.P. Singh, S.P. Kaushish, Ashwani Kumar, T.S. Murthy and Jose, C. Samuel), pp. 501-507, CBIP, New Delhi.

Singh, G.B. 1985. Prospects of irrigation in arid zone. In *Irrigated Agriculture in Arid Areas* (Ed. S.D. Singh), WAPCOS, New Delhi, pp. 490-498.

Singh, S. D., J. P. Gupta and Panjab Singh 1978. Water Economy and Saline Water Use by Drip Irrigation. *Agronomy Journal* 70:948-951.

Singh, S.D. 1978. Effects of Planting Configuration on Water Use and Economics of Drip Irrigation Systems. *Agronomy Journal* 70:951-954.

Singh, S.D. 1988. Water Harvesting in Arid Tropics. CAZRI Publication No. 37. Central Arid Zone Research Institute, Jodhpur. 68 p.

Singh, S.D. and Singh, P. 1978. Value of drip irrigation compared with conventional irrigation for vegetable production in a hot arid climate. *Agronomy Journal* 70: 945-947.

Singh, S.D., Singh, Y.V. and Bhandari, R.C. 1989. Tomato yield as related to drip lateral spacing and fertilizer application on total and wetted area basis. *Canadian Journal of Plant Science* 69: 991-99.

Singh, Y.V. 1997. Maximising water use efficiency for sustainable crop production in arid ecosystem. In *Recent Advances in Management of arid Ecosystems* (Eds. S. Kathju, R.K. Abhichandani, Amal Kar and N.L. Joshi), Jodhpur, pp. 135-136, AZRAI, Jodhpur.

Singh, Y.V. and Saxena, A. 2001. Chilli yield as related to water and nitrogen management under drip irrigation. In *Microirrigation* (Eds. H.P. Singh, S.P. Kaushish, Ashwani Kumar, T.S. Murthy and Jose, C. Samuel), pp. 467-471, CBIP, New Delhi.

Singh, Y.V., Joshi, N.L., Singh, D.V. and Saxena, A. 1999. Response of chilli to water and nitrogen under drip and check basin methods of irrigation. *Annals of Arid Zone* 38: 9-13.

Wetherald, R.T and Manabe, S. 1981. Influence of seasonal variation upon the sensitivity of a model climate. *Journal of Geophysical Research* 86 (C2): 1194-1204.

Yadav, R.C., Sharma, K.D., Singh, H.P. and Pareek, O.P. 1980. Some hydro-horticultural relations in a ber orchard (*Ziziphus mauritiana* Lamk.). *Annals of the Arid Zone* 19(3): 297-300.

Zohar, Y., Aronson, J.A. and Lo Evenstein, H., 1988. Cultivation of multipurpose trees in rain water harvesting system in the arid zone of Israel. *Commonwealth Forestry Review* 67: 339-349.

□□□

Chapter 21

Watershed Management in Arid Regions Under Climate Change Scenario

R.K. GOYAL AND ANURAG SAXENA

ICAR-Central Arid Zone Research Institute, Jodhpur-342003, Rajasthan,
E-mail: rkgoyal24@rediffmail.com

Introduction

Arid ecosystems are among the world's most fragile ecosystems, made more so by periodic droughts and over-exploitation of meager natural resources. Arid environments are extremely diverse in terms of their land forms, soils, fauna, flora, water, and human activities. Because of this diversity, no practical definition of arid environments can be derived. However, the one binding element to all arid regions is aridity. It is usually expressed as ratio of mean annual precipitation (P) to the mean annual potential evapotranspiration, where potential evapotranspiration is calculated by method of Penman taking into account atmospheric humidity, solar radiation, and wind speed. UNEP (1997) has recognized four main classes of aridity: hyper-arid ($P/EPT < 0.03$), arid ($0.03 < P/EPT < 0.20$), semi-arid ($0.20 < P/EPT < 0.50$), and dry sub-humid ($0.50 < P/EPT < 0.65$). Of the total land area of the world, the hyper-arid zone covers 4.2 percent, the arid zone 14.6 percent, and the semi-arid zone 12.2 percent (FAO, 1989.A large proportion of the inhabitants in arid regions are among the poorest in the world (Malagnoux *et al.,* 2007).

Soil and water are two essential basic natural resources for supporting the human and livestock population. Judicious management of these resources is extremely important for sustainable development. Since advent of human being on the earth soil and water resources are being used continuously for various purposes including agricultural production. In nature, soil and water resources are often in a fine balance. However, with unprecedented increase in human and livestock population demand for food and fodder has increased tremendously and these resources have been exploited indiscriminately. This has resulted in the imbalances that have led to reduction in land productivity and stress to sustain human development. In this paper, aspects related to soil and water management with an integrated outlook are presented in the context of arid regions of India.

Indian Arid Zone

The arid region of India is spread in 38.7 million hectare (Mha) area out of which 31.7 Mha is under hot arid zone and 7 Mha under cold arid zone (Fig.1). The hot arid region occupies major part of north-western India (28.57 Mha) between 22°30' and 32°05' N latitudes and from 68°05' to 75°45' E longitudes, covering western part of Rajasthan (19.6 Mha, 69%), north-western Gujarat (6.22 Mha, 21%) and 2.75 Mha in south-western part of Haryana and Punjab (Faroda *et al.*, 1999). The climate of Indian hot arid zone is characterized by an abundance of solar energy from cloud-less sky, high diurnal and seasonal temperature variations and annual and inter-annual irregular rainfall with long dry seasons associated with strong winds. Annual rainfall is approximately in the range of 100-500 mm, with a coefficient of variation ranging from 40 to 80% (Rao and Singh, 1998). The rainfall results largely from convective cloud mechanisms and is characterized by a relatively high intensity, short duration and limited aerial extent. The distribution of rainfall in space is very much influenced by the local terrain. With low ground level and little topographic relief, it is common to see rain evaporating before reaching the ground. The incoming radiation ranges from 15.12 - 26.50 MJ m^{-2} day^{-1} with very little cloud for most of the year. The annual potential evapotranspiration ranges from 1400 – 2000 mm $year^{-1}$ leading to a permanent negative water balance (Rao, 2009).

The soils in arid regions, as a result of these climatic conditions, are characterized by large variations in soil moisture, ground-water storage and high infiltration rates. A dominant feature of these soils is their susceptibility to erosion and degradation by wind, water and human activities. There are frequent dust and sand storms. The soils are dominantly sandy with 60-90 per cent fine sand and 2-10 per cent of silt-clay in the top soil. Soils are generally low in organic carbon, low to medium in available phosphorus and medium to high in available potassium (Joshi, 1993). The surface water resources are meager and directly resulted from rainfall, and therefore they dry up during the rainless season. Groundwater is deep and often brackish.

Last five decades in the arid region of India have witnessed unprecedented increase in human and livestock population. During the twentieth century, the region has experienced agricultural drought once in three years to every alternate year in one or the other part of the region. The overall probability of drought is 47 per cent. The weather condition, even in average years, for most part of the year remains too dry and inhospitable for successful growth of crops. Under such conditions of uncertainty, conventional cropping is risky and is essentially for sustenance only (Goyal and Vittal, 2008). Due to poor soil fertility and low water availability the average yield of principal cereal crop pearl millet (400 kg ha^{-1}) and grain legumes - moth bean (200 kg ha^{-1}), cluster bean (250 kg ha^{-1}) and cowpea (300 kg ha^{-1}) are low (Yadav *et al.*, 2009).

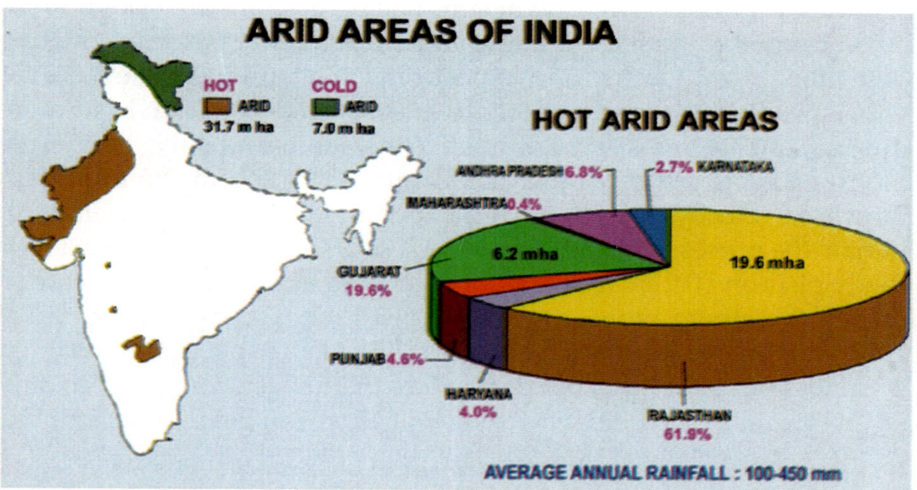

Fig. 1. Extent of Indian arid zone.(*Source:* Goyal, 2011)

The Challenges

The major physical manifestation of desertification in hot arid zone of India is wind erosion/deposition, followed by water erosion, as well as water logging and salinity (Ghose *et al.*, 1977). Wind erosion is a major problem in arid sandy tract of north-western India, especially in west Rajasthan. About 76 per cent area of western Rajasthan is affected by wind erosion. The other forms of degradation include vegetation degradation (3%), water erosion (2%), salinization (2%) and mine spoils (0.10%) (Kar *et al.*, 2009). Despite the present state of land degradation, Indian hot arid region supports 22 million human and 29 million livestock population over 19.6 Mha geographical area. The human: animal ratio ranges from 1:1.3 to 1:5 in arid regions against the national average of 1: 0.5 (Narain and Goyal, 2009). Heavy human and livestock pressure has put tremendous pressure on the fragile water and land resources of this region. Desertification, manifested in different forms of degradation is a real challenging issue in this region.

Wind Erosion

Wind erosion is the most dominant problem in arid India, especially the western Rajasthan. In this region the terrain is dominantly sandy with sparse natural vegetation cover. Wind erosion causes decrease in land productivity at both places i.e. the sites from where the finer particles are blown away and at sites where these are deposited. Wind erosion leads to the blowing away of the finer soil particles, causing depletion of soil fertility and exposure of roots (Fig. 2). The summer wind speed is high, reaching often a gust speed of 60-80 km h^{-1} or more during sandstorms, especially between March and July, before the monsoon rains set in. Gupta and Gupta (1981) found the minimum threshold velocity as 5 km h^{-1} at Bikaner and 10 km h^{-1} at Jaisalmer to initiate the process of wind erosion. A loss of 9.6 cm topsoil from a bare sandy plain and 0.2 cm from a stubble-covered sandy plain was recorded during the strong wind regime of April-June (Gupta and Aggarwal 1980). During the high wind regime at Bikaner, Gupta (1993) recorded a soil loss of 1449 t ha^{-1} from bare sandy plain when compared to only 22 t ha^{-1} from a crop field with a cover of 45 cm tall pearl millet stubble.

During sand storm with a mean wind speed of 20 km h^{-1}, the grass-cover of *Laisurus sindicus* near Jaisalmer reduced the erosion to 76.7 kg ha^{-1} day^{-1} when compared to barren land (273.7 kg ha^{-1} day^{-1}) (Gupta, 1993). Land use practices have also been found to significantly influence wind erosion. In the sandy plains, fields deep ploughed using tractors lost soil heavily during the 22-day long sandstorm of June (1985), while the soils under 8-12 per cent cover of natural vegetation remained protected (Dhir *et al.*, 1992).

Fig. 2. Severity of wind erosion. (*Source:* Goyal, 2011)

Water Erosion

The water erosion is more severe in most arid lands on account of rain's high erosive capacity and the poor vegetal cover. Since rainfall in this region is of convective nature and usually occurs at a very high intensity for shorter duration creating situation of flash flood particularly in urban areas where buildings and roads generate very high runoff even for little rainfall. During 1979 large parts in the state of Rajasthan witnessed flash flood due to very heavy downpour of more than 500 mm in just 3-5 days which cut off the state from rest of the country for several days. In the last 103 years western region witnessed nine moderate and 19 severe floods. Although, floods are considered a natural calamity, however, if the excess flood water or its part can be managed and utilized for rejuvenating the depleted aquifers as well as improving surface water resources the problem of water scarcity during droughts could be solved to a large extent.

Climate Change

Climate change is one of the greatest challenges of the present century. As per general circulation models (GCMs) with different scenarios of greenhouse gases emission, the globally averaged surface temperature is projected to increase by 1.4-5.8°C over the period of 1990 to 2100. The increase in temperature may have multiple effects on crop production on account of increased evapotranspiration (Irving, 1993). The overall changes in climate may invite new pests and diseases and some existing pests/diseases may become more resistant. Change in CO_2 concentration too can show uncertainty in crop yields. Simulations of the impacts of climate change using crop simulation models show that crop yield decreases due to climate change could have serious impacts on food security of this region (Shivakumar *et al.*, 2005).

Analysis of rainfall trend during the last 100 years by Indian Institute of Tropical Management (IITM) revealed that the summer monsoon rainfall, which contributes more than 85 per cent of the total annual rainfall in this zone, may decline by 10-15 per cent in arid western Rajasthan and the adjoining Punjab and Haryana states. A slight increase in winter rainfall with pronounced variability in rainfall and fewer rainy days is also predicted. Consequently, there will be higher incidence of droughts and floods in arid western India, affecting both *kharif* and *rabi* season crops (Joshi and Kar, 2009). Since a number of high-value crops are grown during *rabi,* with higher input cost and more temperature-sensitivity of *rabi* than the common *kharif* crops, the negative impact on farmers' economy would possibly be higher during *rabi*.

A recent simulation study by Kalra *et al.* (2008) has shown that a rise in temperature by 1°C can lead to a decline in wheat production by ~250 kg ha^{-1} in Rajasthan and by 400 kg ha^{-1} in Haryana. In mustard the decline may be ~100 kg ha^{-1} $degree^{-1}$ rise in temperature in Rajasthan, and by 200 kg ha^{-1} in Haryana. In chickpea the decline expected is ~200 kg ha^{-1} in Rajasthan and 500 kg ha^{-1} in Haryana. There are concomitant declines in biomass yield also.

Lal and Chander (1993) studied the effects of greenhouse warming on the water resources of the Indian subcontinent and suggested a rise in annual mean surface temperature of 2.0-3.5°C over the Indian subcontinent by the year 2090. According to them warming would be most pronounced over the northwestern India. As small as 1 per cent increase in temperature (0.42°C based on normal maximum temperature of Rajasthan) will enhance the evapotranspiration demand by 11.7 mm on annual basis (Goyal, 2004). It will cause an additional annual water demand of 718 Mm^3 and 2,245 Mm^3 for the whole state based on net irrigated area (61,345 km^2) and total cropped area (1,92,302 km^2), respectively.

In view of desertification due to wind and water erosions coupled with increased water demand and likely reduction in crop yield due to climate change, there is greater need for effective natural resources management for sustainability of production system of this region. More emphasis is needed to develop technologies for reducing water losses in reservoirs, conservation of rainwater and development of such crop varieties that require less water. Long-term strategic approaches to efficiently conserve and utilize rainwater on the one hand and in-season tactical approaches to mitigate the adverse effects of weather aberrations on the other are also needed. For resource management and utilization, conventional approach of soil and water conservation needs to be practiced within watershed boundary with sound watershed principles for effective resource base upgradation and utilization.

Watershed Approach

Initial approaches in conservation programs were based on particular land problem and individual holdings. In this approach often the upper reaches of the fields were left untreated and runoff from uplands used to damage the conservation work laid out in the individual holdings at downlands. The watershed approach helps in an integrated development of different parts of the watershed in accordance with their nature, and problems and potential with best quantum of interference. The integrated watershed management principles have accordingly been adopted for holistic development of drought prone and rainfed areas. Though no two watersheds are identical, their characteristics have common parameters, which can be judiciously used for obtaining the required results.

Objectives

The main objective of each watershed development project is to achieve sustainable production of biomass and restoration of ecological balance in the drought prone and rainfed areas improving thereby the standard of livelihood of people. The specific objectives are:

1. Conservation, upgradation and utilization of natural endowments namely land, water and vegetation in harmonious and integrated manner.

2. Generation of massive employment during the project period and regular employment after the project completion for enhancing the work opportunities in the backward rainfed areas with a view to ensure livelihood security, particularly for the underprivileged sections of the rural population.
3. Improvement of production environment and restoration of ecological balance through scientific management of land and rainwater, *in-situ* moisture conservation, introduction of scientific production system, creation of network of runoff management structures and devices for recharge of ground water and recycling for drinking purposes, domestic consumption and lifesaving irrigation.
4. Promotion of social forestry and horticulture activities in all suitable lands, and
5. More cash available with farmers and landless agricultural labourers through increased casual employment, and marketable surplus of agricultural and dairy produce.

Watershed Management Technology

The interaction between land and water determines the productivity of a given area, which can be made more productive by adequate understanding of physiography, climate, socio-economic conditions etc. and hydrological behavior of watershed. The PRA, land capability, choice of appropriate soil and water conservation measures, improved management practices, soil conservation technologies are unique features of detailed watershed planning. Optimization of different possible combinations and alternative uses of a specific resource should be tested. Priorities for the maximum benefits in the shortest possible time have to be selected. The inter-relationship between all factors has to be carefully studied and final decision must be made with local participation. Due to multidisciplinary nature of watershed development projects the various improvement activities need to be coordinated between farming community and government agencies. Therefore, success of such projects depends very much on the degree of co-ordination and co-operation between various organizations.

Economic consideration is a major factor in making decision on action and timing. However, these decisions have to be fit into the sequence of improvements required for effectiveness of all components in a plan. It is also important to note that the area development plan can be executed gradually, in sub-watershed from top to bottom. Thus, the development period can be spread over a number of years according to availability of finance and trained manpower. If and when all sub-watersheds have been treated and developed one by one, and main tributaries have also been improved, then the complete plan would have been implemented in a systematic manner and be based on sound technological and socio-economic principles. The subsequent maintenance is essential to ensure against potential natural hazards to the completed works and improvements.

Resource Conservation

Soil and water are the two basic resources, needed for crop production. Measures for soil and water conservation can be divided in two main categories i.e. physical/engineering and biological. While the physical measures are the first line of defence and are very necessary to prevent immediate damage, biological measures are the offensive attack made from a position of good defensive strength. Physical measures consist of construction of mechanical barriers across the direction of flow of water to retard/retain the runoff and thereby reduce soil and water losses. Contour bunds, contour cultivation, trenching, sub-surface barrier etc. are some of the physical measures, which are found to be useful in arid areas. Biological measures like vegetative barrier, windbreak/shelter belt, stubble mulching are very effective for moisture conservation and controlling soil erosion due to wind and water.

Resource Utilization

Efficient utilization of available resources is of paramount importance for successful crop production. Improved agronomic practices i.e. inter cropping, improved crops and seed varieties, crop rotation, minimum tillage etc. plays vital role in ensuring risk distribution. For aberrant weather situation contingency planning like life saving irrigation, integrated insect-pest management, differential fertilizer application etc. are crucial to minimize loss of seed, crop and fertilizer. The land which is unfit for conventional farming as per land capability classification may be used for alternative land uses, dairying, agro-forestry and agro-horticulture, cut flowers and fisheries, where water is stored for long-time. Value addition should be inbuilt with the program for income generation and livelihood.

Technological Integration

The success of watershed development projects greatly depend upon the availability of viable and acceptable location specific technology. Besides, availability of technological options, the success of the program also depends upon beneficiaries' resource base and level of income, timely availability of inputs and the skill of the farmers. As the people and location are the two most important factors, which are required for the selection of appropriate technology and for making suitable modification, it would be better to identify the proven traditional technology and skill, for adoption after suitable modifications. Some of these technologies are as follows:

Techniques for Enhancing Runoff from Catchments

Since primary source of water in arid zone of Rajasthan is rainwater, so for any improvement in availability of water, catchment conditions become crucial. Catchment is the base for harvesting rainwater in form of runoff. Runoff is highly dependent on catchment's shape, size, slope, and type etc. beside rainfall characteristics. Some general modifications can greatly help in enhancing the runoff percentage from catchments.

1. Simple earth smoothing and compaction helps increasing runoff from catchment areas. Success is generally greater on loam or clay loam soils. Care must be taken to reduce the slope and/or the length of slope to lessen runoff velocity and thereby reducing runoff.
2. Small amounts of sodium salts - particularly NaCl, $NaHCO_3$ applied to desert soils where vegetation has been removed- causes dispersion of the surface soil, reducing infiltration and increases runoff. However, this type of treatment to be effective requires some minimum amount of expanding clays in the soil.
3. Removal of stones and boulder and unproductive vegetation from catchment helps in uninterrupted flow, and enhances runoff to collection site.
4. Land shaping into roads and collection of water in channels.
5. Sandy soils have low water holding capacity. Spreading of clay blanket on the soil surface reduces infiltration and consequently accelerates runoff.

6. Application of wax, asphalt, bitumen and bentonite prevent downward movement of water, which augments runoff.

Technological Options for Non – Arable Lands

Contour furrowing

Contour furrowing is the most effective measure to reduce runoff and soil loss;it is commonly adopted in grasslands and forest lands. In very sandy soils or soils with heavy clay pan area, their benefit is, however, limited. Contour furrows varying from 30-60 cm wide and 10-25 cm deep can be used. The shape varies from "V" to square, rectangular, or parabolic. The cross section and depth of furrows mainly depend on soil and equipment used for making them. Furrows spaced 8-10 m apart give a better distribution of runoff water and higher yield of fodder. The effectiveness of contour furrows to hold water depends upon the degree of slope smoothness of the surface and accuracy in following contours; its life depends upon stability of soil.The water storage capacity of the furrow can be estimated by following equation.

$$Q = \frac{WxD}{100HI}$$

Where, Q = Depth of runoff water stored in cm from unit area

W = width of furrow, cm

D = depth of furrow, cm

HI = horizontal spacing, m

Construction of contour furrow is always started from the ridge and progressively extended towards the valley. In a study conducted in arid part of Iran it is found that contour furrow and pitting has significantly helped in controlling soil erosion, increasing water penetration and soil moisture content and promoted propagation of *Hammada saliconica* species, a desirable plant species for both soil conservation and livestock grazing in the region (Jahantigh and Pessarakli, 2009).

Contour Trenches

A contour trench is a useful practice in forestry areas. This practice can be adopted in area which is unsuitable for cultivation but suitable for forestry. Normal standard size of a trench is 60 cm x 30 cm x 60 cm with an unexcavated portion 1.5 m interval. Length, spacing or vertical interval depends on the slope of land. Spacing may vary from 30-60 m. After the trenches are excavated to correct size, they are refill partially, and stocking the remaining excavated material as a small bund on the downstream side. The storage capacity of trench is estimated by the following equation.

$$Q = \frac{WxD}{100HI\left(1 + \frac{X}{L}\right)}$$

Where, Q = depth of runoff from area in cm

W = width of trench in cm

d = depth of trench in cm

HI = horizontal interval in meters

X = gap between the trenches in m

L = length of trench in m

Mane *et al.*, (2009) reported effectiveness of continuous contour trench for runoff control and recommended as best soil conservation practice on area having 7 to 8 per cent slope in *Konkan* region of India.

Gradonies

Gradonies are steeply inward-sloping narrow bench terraces constructed on contours. Usually, gradonies are suitable for afforestation in uniformly steep sloping lands (Fig. 3). Based on the steepness of slope, vertical interval is kept from 1.0 to 1.5 m. The width of gradonies also varies from 1.0 to 1.5 m. The material dug from the inner side is heaped on the outer edge in order to make a berm of about 20 to 30 cm high with an inward slope of 7.5:1. In the middle of the gradonies, pits of 50 cm x 50 cm x 50 cm are made at spacing of about 3 m. On an average 1000 trees are planted within the gradonies and interspaces in one hectare area (Mahnot and Singh, 1993).

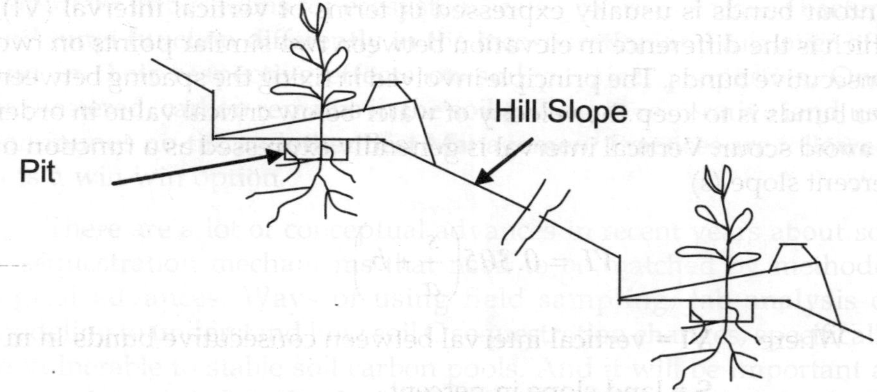

Fig. 3. Cross section of a gradoni. (*Source:* Goyal, 2011)

Technological Options for Arable Lands

Contour farming

Contour farming is beneficial on all slopes where line sowing is adopted. All ridges and rows of plants are placed across the slope to form continual series of miniature barriers to water and offer maximum opportunity for infiltration. Contour operations reduce the power of the water to erode, suspend and carry away soil particles and increase the moisture storage. Increased uniform moisture storage can boost up the yield above 10%. Contour farming alone cannot control runoff volume from higher sloppy lands which may need bunding and grass waterways in natural drainage. For small fields with uniform slopes, one contour guideline is enough and where slopes are irregular two to three guidelines may be enough for farming operations. Contour farming is most effective on moderate slope of 2 to 7%. Contour farming is reported to reduce soil loss upto 60 % in comparison to up and down farming on 1% slope (Smith and Wischmeier, 1962).

Contour bunds

Contour bunds are narrow base trapezoidal earthen embankment on contour, 1.5 to 2 m wide, constructed across the slope to act as barriers to runoff, to form water storage area on their upslope side and to break up a long slope into short segments. Contour bunds are recommended upto 6% slope and rainfall of upto 600 mm. Contour bunds are not suitable for shallow soils having depth less than 7.5 cm. Spacing between contour bunds is usually expressed in terms of vertical interval (VI), which is the difference in elevation between two similar points on two consecutive bunds. The principle involved in fixing the spacing between two bunds is to keep the velocity of water below critical value in order to avoid scour. Vertical interval is generally expressed as a function of percent slope (s)

$$VI = 0.305\left(\frac{s}{a} + b\right)$$

Where VI = vertical interval between consecutive bunds in m

S = land slope in percent

'a' and 'b' are constants specific to particular region

For soils having good infiltration rates, value of 'a' and 'b' are 3 and 2, respectively. For soils with low infiltration rate the value of 'a' and 'b' may be taken as 4 and 2, respectively. The horizontal interval (HI) is related to vertical interval (VI) by following equation

$$HI = \left(\frac{VI}{s} x100\right)$$

Depth of impounding is calculated on the basis of maximum rainfall of 24 hours for 10 years recurrence interval as follows :

Where 'h' is height of impounding (m) $h = \sqrt{\frac{\text{Re} \, x \, VI}{50}}$

'Re' is 24 hours maximum rainfall (cm) for 10 years recurrence interval. Total height of bund 'H' can be found out by adding 20- 25% freeboard to height of impoundment. For light red loam and sandy loam soils, the side slope of both the sides is taken as 1.5: 1, whereas for sandy soils it is taken as 2:1. The line of seepage should not cross the bottom of the bund while deciding the other dimension of the bund. On average, contour bunds had 27 percent higher soil moisture and 14 to 181 percent higher fodder yield than flat surfaces on grasslands of western Rajasthan (Wasi-Ullah *et al.*, 1972).

Graded bunds

Graded bunds are used to dispose off safely the excess water from the agricultural fields to avoid water stagnation. Graded bunds are suitable in areas where annual rainfall is 500 mm and if the soils are highly impermeable. Graded bunds usually have wide and shallow channels and earthen bund laid along a predetermined longitudinal slope. As graded bunds are essentially meant for the safe disposal of excess water from cropped lands, suitable outlets are required to be constructed on graded bund. Draining of excess water from one plot to another through outlets provided in the bund requires special attention since considerable amount of soil may be lost through these outlets. Provision should be made to arrest the silt and allow only clear water to flow away. The vegetated watercourse strengthens the system. In areas with high rainfall where the volume of water to be disposed from one plot to the next lower level plot is large and the vertical interval between the plots is reasonably high, a site specific outlet design like a pipe outlet, drop structure etc. must be provided for safe disposal of the excess water. Graded bund is reported to reduce the run-off from 20 to 4.8 per cent and soil loss from 24 to 4.12 t ha^{-1} yr^{-1}. Besides other benefits intercropping on contour resulted in 48 per cent higher grain yield (Singh *et al.*, 1997).

Grass waterways

Grass waterways are developed for safe disposal of excess water from agricultural fields. These may be natural or man-made courses protected against erosion by suitable grass cover. Grass waterways are also used for channelizing and regulating runoff flows for water harvesting purposes. The best location for waterways is a natural depression or along valley line. These may also be constructed along field boundaries for safe disposal of excess rainfall from agricultural fields. Vegetative waterways may be located in all classes of lands except hard rocks, where construction may be difficult. The cross section of waterways may be trapezoidal, triangular or parabolic with shallow depth and flat side slope to facilitate easy movement of man, animal and machinery. The depth of waterways may be kept within 20 to 50 cm and side slope more than 4:1. The channel should have free board of 15 cm. The channel cross-section and bed slope should be such that the computed velocity is within permissible limit (Singh *et al.*, 1990). Cost of construction of grass waterways depends upon type of soil, channel cross-section, length of channel and grass plantation technique.

Diversion Drains

Diversion drains are constructed to divert the runoff and to allow it to flow at a non-erosive velocity to a suitably protected outlet. They are particularly adapted to areas where already trees or grasses are in dominance. Usually diversion drains are constructed to protect conservation structure lower down in the cultivated areas by regulating flows from upland non-agricultural areas. Diversion drains are provided across the slopes on slight gradients. Their capacity is worked out and designed for a 10 year recurrence interval storms.

Contour vegetative barriers

Contour vegetative barriers are hedgerows of perennial grasses or shrubs planted at a regular interval on contours for conserving soil and water in sloping lands. Suitable grass species are grown along contours at suitable vertical interval to intercept part of runoff and to control erosion in agricultural fields having flat to slight undulating topography. The contour vegetative barrier moderates the velocity of overland flow and traps silt at low cost, and augment production of food, fuel and fodder or fiber from lands by growing suitable vegetation species. In recent years, contour vegetative barriers have found acceptability among the farmers as these are cheaper over mechanical measures and are protective while being also productive. Contour vegetative barriers can be easily established across a wide spectrum of soil-climatic conditions. Selection of species depends upon purpose of barrier, site-specific conditions, particularly soil and climatic variables. The spacing between plant-to-plant and row-to-row is governed by vegetation species to be planted as barrier. In general the plant-to-plant spacing of 20 to 30 cm at predetermined or 50 to 100 cm vertical interval between the barriers has been found effective for soil and water conservation. Generally, paired row of barrier planted in staggered form across the slope proves more effective.

Generally, dominant grass or shrub species of the region should be preferred for vegetative barrier. Among grasses *Cenchrus ciliaris, Cenchrus setigerus, Saccharum bengalense, Vetiveria zizanioides, Lasiurus sindicus, Panicum antidotale* and *Panicum turgidum* can be effectively used for soil and water conservation in arid areas. Shrubs like *Leptadenia pyrotechnica, Ipomoea carnea* and *Euphorbia antisyphylitica* can also provide good protection against water and wind erosion.Contour vegetative barriers of *Cymbopogan jwarancusa,Cenchrus ciliaris* and *Cenchrus setigerus* transplanted at 0.30 m apart on contours at 0.6 to 1.0 m vertical interval in sandy loam soil of Jodhpur (Rajasthan) have performed well and formed effective barriers in reducing soil erosion and increasing soil moisture storage. In a study conducted during 1992-1994 at 19 farmers' fields near Jodhpur rooted slips of local eight species of perennial grasses (*Cenchrus ciliaris, Cenchrus setigerus, Cymbopogon jwarancusa, Lasiurus sindicus, Panicum antidotale, Panicum turgidum, Saccharum bengalense* and *Vetiveria zizanioides*) and seedling of six species of shrubs (*Agave americana, Aloe barbadensis, Barleria prionitis, Euphorbia antisyphylitica, Ipomoea carnea* and *Leptadenia pyrotechnica*) were transplanted at 1 m vertical interval on contours across the slope. Result indicated that perennial grass species performed the best and formed effective barrier against soil erosion. Runoff volume and specific peak discharge were reduced by 28 to 97% and 22 to 96% respectively (Sharma *et al.*, 1999; Tiwari and Kurothe, 2006). In another study conducted at Kalyanpur (Distt. Barmer) during 1998, vegetative barrier of *Lasiurus sindicus, Saccharum munja* and *Cassia angustifolia* were established at horizontal interval of 30 m. The moisture data revealed 36.5%, 72% and 54.2% higher moisture storage as compared to control in *Cassia angustifolia, Lasiurus sindicus* and *Saccharum munja* respectively (Gupta and Rathore, 2002).

Khadin

Khadin is a unique practice of water harvesting, moisture conservation and moisture utilization in hyper arid region of Rajasthan. This system was designed & developed by the *Paliwal* Brahmins of Jaisalmer (Rajasthan) in the 15th century. This system has great similarity with the irrigation practices of the people of Iraq around 4500 BC and later of the Nabateans in the Middle East. A similar system is also reported to have been practiced 4,000 years ago in the Negev desert and in south-western Colorado 500 years ago. The main feature of *khadin* is a very long (100-300 m) earthen embankment built across the lower hill slopes lying below gravelly uplands. Sluices and spillways allow excess water to drain off. The *khadin* system is based on the principle of harvesting rainwater on farmland and subsequent use of this water-saturated land for crop production. The ratio of farmland and catchment areas is regulated to be about 1:10 so that a suitable moisture supply is uniformly maintained. It is suitable for deep soil surrounded by some natural rock outcrops constituting catchment area. CAZRI has developed *Khadin* of 20 ha areas in Baorali-Bambore watershed with surplussing arrangements (Fig. 4). Before construction of *Khadin*, uncontrolled runoff from upper catchment used to wash away seeds, fertilizers, and standing crops besides loss of valuable water. After construction of Khadin, farmer could take excellent *Kharif* and *Rabi* crops

(Narain and Goyal, 2005). Collecting water in a *khadin* aids the continuous recharge of groundwater aquifers. Studies of groundwater recharge through *khadins* in different morphological settings suggest that 11 to 48 per cent of the stored water contributed to groundwater in a single season. This replenishment of aquifers means that sub-surface water can be extracted through bore wells dug downstream from the *khadin*. The average water-level rise in wells bored into sandstone and deep alluvium was 0.8 metres and 2.2 metres, respectively (Khan, 1996).

Check dams are masonry overflow barriers (weirs) constructed across seasonal streams. A check dam as such has a relatively limited storage capacity but a large volume of water can still be pumped from such storage as the stream continues to flow and the check dam serves the purpose of an ideal intake structure. A check dam, by storing the base flow, maintains a supply of water for recharge as well as for direct use beyond the monsoon period (Goyal and Narain, 2006). It creates flooding of upstream area, which requires surplusing arrangements at suitable intervals to drain water. Check dam should be avoided in isolation. The number of check dams primarily depends upon the slope of the gully and the quantity of runoff. It may not be advisable to construct check dams on bigger streams with high gradient and where runoff is very high. The bigger streams should be treated with drainage line treatment like gabionic structures and boulder checks with masonry work to curtail the runoff.

Gabionic check dams

Gabionic check dams are useful in a locality where stones are readily available and their irregular shape makes them unsuitable for making loose stone check dams. If the expected water velocity is very high, gabion is recommended in place of loose rock dams. A gabion is rectangular shaped cage made of galvanized wire, which is filled with locally available boulders, rocks or stones. The gabion may be conveyed flat and are folded to shape at the construction site. Usually, gabions are 1 m wide and 0.75-12 m high with varying lengths ranging 2-10 m. The gabionic check dams are constructed by connecting several gabions in horizontal and vertical directions. The gabionic check dams are very stable and semi-permanent in nature. These structures are flexible; they may even change shape automatically according to the stream bed, even when the bed shape changes due to erosion, without losing stability.

Loose stone/dry stone masonry check dams (LSCD)

These structures are effective for checking runoff velocity in steep and broad gullies. These are suitable at upper reaches of the catchment. They have a relatively longer life and, usually require less maintenance. The bed of the gully is excavated to a uniform depth of about 0.3m. Stones are then hand packed from the foundation level. Flat stones of size 20-30 cm are the best for construction and laid in such a way that all the stones are keyed together. Large size stones are placed at the center of the dam and gaps between stones may be filled with small piece stones. The dam should go up to 0.3 to 0.6 meter into the stable portion of the sides of the gully to prevent end cutting. In the center of the dam, sufficient spillway is provided to allow maximum runoff to discharge. LSCD constructed at 1 m vertical interval in Jhanwar watershed area on 17 gullies proved to be very effective in controlling further extension of gullies (Goyal, *et al.*, 2007). Vangani *et al.* (1998) reported sediment deposition of 3.86 t ha^{-1} yr^{-1} against loose stone check dam in Osian-Bigmi watershed (Distt. Jodhpur).

Brushwood check dams

These check dams are constructed by using locally available brushwood and supported by wooden stakes and used in the small gully heads not deeper than 1 m. These check dam are of two types; single row post brush dam and double row post brush dam. Brushwood check dams are constructed in areas where wooden posts, brushwood etc., are available in plenty. These check dams can only be used in the small gully heads not deeper than 1m. Single row post brush dam are made of single row of wood stakes to which long branches of trees are tied length wise along gully with their butt ends facing upstream, while in double row post brush dam the straw and brushwood are laid across the gully between two rows of wooden posts the distance between the rows being not more than 0.9 m. The longest branches are laid at the bottom and the shorter length branches are laid above it till the required dam height (0.3 to 0.7 m) is obtained.

Earthen gully plugs

An Earthen gully plug is constructed out of local soil across the stream to check soil erosion and flow of water. It is suitable in the upper catchment areas having scope of water storage and where the soil for the embankment is close by. Depth of the gully should be less than 2 meter and gully bed slope is less than 10%. The site should have scope for side spillway. The bed of the gully is excavated to a uniform depth of about 0.3m. Foundation trench is filled with compacted earth and earthen embankment is constructed by using local soil.

Conclusions

A successful application of any soil and water conservation measures for improving production requires an integrated approach. In arid areas, conditions vary too much soil, climate, social factors and the list is endless so no two areas are identical, and therefore proper selection of technology for area specific is must. Watershed management techniques of rainwater harvesting and conservation will not only solve the problem of water but will also save valuable money, which can be utilized for other welfare schemes. All the technologies discussed above are essentially site specific and different components need to be integrated as a holistic approach to conserve rainwater to provide water for different needs on sustainable basis. These technologies are time tested and are of proven soundness for extreme conditions such as of arid. The modern scientific tools like remote sensing and GIS can be used to identify potential areas for watershed development. For implementation various government programs like MGNAREGA and other employment generation schemes can be coupled for various activities under watershed development program. This will not only help in proper utilization of valuable resources but will also create sense of belongingness among the farming community. The present day need is only to revive these technologies and apply them in the field on full scale.

References

Dhir, R.P., Kolarkar,A.S. and Singh, H.P. 1992. Soil resources of the Thar.*In*: *Perspectives on the Thar and the Karakum* (Eds. Amal Kar, R.K. Abichandani, K. Anantharam and D.C. Joshi), Department of Science and Technology, Govt. of India, New Delhi: 60-85.

FAO, 1989. Arid Zone Forestry: A Guide for Field Technicians. FAO Conservation Guide – 20.Food and Agricultural Organization, Rome, Italy.

Faroda, A.S., Joshi, D.C. and Balak Ram 1999.Agro-ecological Zones of North-western Hot Arid Region of India.Central Arid Zone Research Institute-Jodhpur, 24p.

Ghose, B., Singh, B. and Amal Kar 1977.Desertification around the Thar – A geomorphological interpretation.*Annals of Arid Zone* 16: 290-301.

Goyal, R. K. 2011. Soil and Water Conservation Measures for Management of Arid Lands. *In: Techniques of Water Conservation & Rainwater Harvesting for DroughtManagement* (eds.) Mishra, P.K., Osman, M., Satendra and Venkateswarlu, B. SAARC Training Program, 18-29 July, 2011, Central Research Institutefor Dryland Agriculture, Hyderabad, India, 714 p.

Goyal, R.K. 2004. Sensitivity of evapotranspiration to global warming: A case study of arid zone of Rajasthan (India). *Agricultural Water Management* 69: 1-11.

Goyal, R.K. and Vittal, K.P.R. 2008.Water resource management in hot arid zone of India.*Journal of Hydrological Research and Development* 23: 37-54.

Goyal, R.K. and Narain, P. 2006. Impact evaluation of watershed conservation measures in hot arid zone of India. In: *Conference on Natural resources management for sustainable development in Western India (NRMSD-2006)* Organized by Indian Association of Soil & Water Conservationist, Dehradun –Pune, 11-13 Oct. 2006, 77-79 p.

Goyal, R.K., Bhati,T.K. and Ojasvi,P.R. 2007. Performance evaluation of some soil and water conservation measures in hot arid zone of India. *Indian Journal of Soil Conservation*, 35 (1): 58-63.

Gupta, J.P. 1993. Wind erosion of soils in drought prone areas. *In: Desertification in Thar, Sahara and Sahel Regions* (Eds. A.K. Sen and Amal Kar), Scientific Publishers, Jodhpur, 91-105 p.

Gupta, J.P. and Gupta, J.N. 1981. A note on wind erosion from a cultivated field in western Rajasthan.*Journal of Indian Society of Soil Science* 29: 278-279.

Gupta, J.P. and Aggarwal, R.K. 1980. Sand movement studies under different land use conditions of Western Rajasthan. *In: Arid Zone Research and Development* (Ed. H.S. Mann), Scientific Publishers, Jodhpur, 109-114 p.

Gupta, J.P. and Rathore, S.S. 2002. *Biomass production and rehabilitation of degraded lands in arid zone*, 20p.

Irving M. Minitzer. 1993. *Confronting Climate Change: Risk, Implications and Responses*.Cambridge University press, London, 382p.

Jahantigh Mansour and Mohammad Pessarakli 2009. Utilization of Contour Furrow and Pitting Techniques on Desert Rangelands: Evaluation of Runoff, Sediment, Soil Water Content and Vegetation Cover. *Journal of Food, Agriculture & Environment 7: 736-739.*

Joshi, D.C. 1993. Soil resources of Rajasthan.*In: Natural and Human Resources of Rajasthan* (Ed. T.S. Chauhan), Scientific Publishers, Jodhpur, 77-92 p.

Joshi, N.L. and Amal Kar. 2009.Contingency crop planning for dryland areas in relation to climate change.*Indian Journal of Agronomy* 54: 237-243.

Kalra, N., Chakraborty, D., Sharma, A., Rai, H.K., Jolly, M., Chander, S., Ramesh Kumar P., Bhadraray, S., Barman, D., Mittal, R.B., Mohan Lal, and Sehgal, M. 2008. Effect of increasing temperature on yield of some winter crops in northwest India.*Current Science* 94: 82–88.

Kar, Amal, Moharana, P.C., Raina,P., Mahesh Kumar, Soni, M.L., Santra, P., Ajai, Arya, A.S. and Dhinwa, P.S.2009.Desertfication and its Control Measures.*In*: *Trends in Arid Zone Research in India* (Eds. Amal Kar, B.K.Garg, M.P.Singh and S.Kathju), Central Arid Zone Research Institute, Jodhpur, 1-47 p.

Khan, M.A. 1996. Inducement of groundwater recharge for sustainable development. In 28th Annual Convention, Indian Water Works Association, Jodhpur, India, 147-150 p.

Lal, M. and Chander, S. 1993.Potential impacts of greenhouse warming on the water resources of the Indian subcontinent. *Journal of Environmental Hydrology* 1: 3-13.

Mahnot, S.C. and Singh, P.K. 1993.Soil and Water Conservation. Inter-cooperation Coordination Office.Jaipur, 90p.

Malagnoux, M., Sene, E.H. and Atzmon, N. 2007. Forests, trees, and water in arid lands: a delicate balance. *Unasylva* 58: 24-29.

Mane, M.S., Mahadkar,U.V.,Ayare,B.L., and Thorat,T.N. 2009.Performance of Mechanical Soil Conservation Measures in Cashew Plantation Grown on Steep Slopes of Konkan. *Indian Journal of Soil Conservation* 37: 81-184.

Narain P. and Goyal, R.K. 2009. Arid zone management strategies in India with special reference to resource mobilization and outsourcing. *In*: *Converting Deserts into Oasis* (Eds. J.S.P.Yadav, R.K.Singh and V.P.Gupta). Stadium Press LLC, Texas USA, 176-185 p.

Narain, P. and Goyal, R.K. 2005. Lead paper on 'Rainwater harvesting for increasing productivity in arid zones'. National Symposium on "Efficient water management for eco-friendly sustainable and profitable agriculture".Symposium Abstract. Organized by Indian society of water management and Indian agriculture research institute from Dec. 1-3, 2005 at Water Technology Center, New Delhi , 141-142 p.

Rao, A.S. 2009.Climatic variability and crop production in arid western Rajasthan.*In*: *Trends in Arid Zone Research in India* (Eds. Amal Kar, B.K.Garg, M.P.Singh and S.Kathju). Central Arid Zone Research Institute-Jodhpur, 41-61 p.

Rao, A.S. and Singh, R.S. 1998. Climatic features and crop production. *In*: *Fifty Years of Arid Zone Research in India* (Eds. A.S. Faroda and Manjit Singh). Central Arid Zone Research Institute- Jodhpur, 17-38 p.

Sharma, K.D., Joshi, N.L., Singh, H.P., Bohra, D.N., Kalla, A.K. and Joshi, P.K. 1999. Study on the performance of contour vegetative barriers in an arid region using numerical models. *Agricultural Water Management*, 41: 41-56.

Singh, A.K., Kumar,A.K., Katiyar,V.S., Singh,K.D., and Singh,U.S. 1997.Soil and Water Conservation Measures in Semi-arid Regions of South-Eastern Rajasthan. *Indian Journal of Soil Conservation*, 25 (3): 186-189.

Singh, G., Venkatramanan, C., Sastry, G. and Joshi, B.P. 1990. Manual of Soil and Water Conservation Practices.Oxford and IBH Publishing Co. Pvt. Ltd.New Delhi, 385p.

Sivakumar, M.V.K., Das, H.P. and Brunini, O. 2005. Impacts of present and future climate variability and change on agriculture and forestry in the arid and semi-Arid tropics.*Climatic Change* 70: 31-72.

Smith, D.D. and Wischmeier,W.H. 1962. *Rainfall erosion. Advances in Agronomy*, 14: 109-148.

Tiwari, S.P. and Kurothe, R.S. 2006. Effect of vegetative barriers on soil and nutrient losses at 2% slope on agricultural lands of reclaimed Mahi ravines. *Indian Journal of Soil Conservation*. 34(1): 37-41.

UNEP 1997. United Nations Environment Program, World atlas of desertification 2ED. UNEP, London.

Vangani, N.S., Surendra Singh and Sharma,R.P. 1998.Index catchment - a new concept for sustainable integrated development. *Annals of Arid Zone*, 37 (2): 133-137.

Wasi-Ullah, Chakravarty, A.K., Mathur, C.P., and Vangani, N.S. 1972. Effect of contour furrows and contour bunds on water conservation in grasslands of western Rajasthan. *Annals of Arid Zone*, 11(3/4): 170-182.

Yadav, O.P., Khairwal, I.S., Rai, K.N., Kumar, D., Henary, A., Mahala, H.R., Singh, B.B., Dixit, G.P., Tikka, S.B.S., Acharya, S., Patel, J.B., Solanki, Z.S., Kumhar, S.R., Choudhary, B.R., Joshi, P., Dwivedi, N.K., Gopala Krishnan, S., Anwar, M.M., Kakani, R.K., Rajora, M.P., Jindal, S.K., Meghwal, P.R., More, T.A. and Singh, R.S. 2009. Genetic resources and improvement of crops, seed spices, grasses, trees and horticultural crops for arid zone.*In*: *Trends in Arid Zone Research in India* (Eds. Amal Kar, B.K.Garg, M.P.Singh and S.Kathju), Central Arid Zone Research Institute, Jodhpur, 238-277 p.

❑❑❑

Chapter 22

Potential of Utilizing Solar Energy for Reducing Carbon Emission

*P. C. PANDE, PRIYABRATA SANTRA AND A. K. SINGH

ICAR-Central Arid Zone Research Institute, Jodhpur-342 003, Rajasthan

*E-mail: pcpande52@gmail.com

Introduction

In rural areas of India most of inhabitants have agriculture as the main occupation. Since electricity is essential to meet the basic needs of villagers for lighting, water pumping, running of TV, telecommunication, medical centres and cottage industries, besides carrying out other agricultural tasks, it is necessary to ensure its availability for keeping the pace of development. Although there is several fold enhancement of electricity production from 1360 MW at the time of independence to more than 260, 000 MW, still there is shortage of electricity, more often reflected in power cuts. About 60% of electricity production is by coal based power plants, which emit about 800-960 g CO_2/kWh of electricity production having implications as observed on climatic aberrations. On the other hand, the general practice in the absence of electricity is to use kerosene oil for lighting and diesel for operating pumps and agricultural machineries. But, the fossil fuel is fast depleting and its burning has created a serious concern for the environment due to greenhouse gas emission. Moreover, people burn firewood, agricultural waste and cow dung cake for cooking food and water heating causing irreparable damage to the fragile eco-system of arid zone. Due to lack of energy resources, the villagers are not able to process their agricultural produce. In this connection the most preferable option is to harness free and abundant solar energy. The average irradiance on horizontal surface in India and Jodhpur are 5.6 and 6.0 kWh $m^{-2}day^{-1}$, respectively (Mani and Rangarajan, 1982). It is therefore aptly envisaged in National Solar Mission of the country to produce 100000 MW grid connected power through the use of solar energy by 2022 and in addition 2000 MW through decentralized systems.

Solar energy could be used both through thermal route by using flat plate collectors or concentrators or through photovoltaic cells to convert it directly to electricity. Concerted efforts have been made at CAZRI for more than four decades on developing different solar thermal and photovoltaic devices and reviewed (Pande et al., 1998; Pande et al., 2009). This included different types of solar cookers, dryers, stills, polish making machine, dual and multipurpose solar devices, PV pump based drip system, PV sprayer/duster, PV winnower cum dryer, PV mobile unit and integrated solar systems. Some of the basics and solar technologies are discussed for projecting their role in arresting the green gas emissions.

Photovoltaic cell

A solar cell technically known as photovoltaic (PV) cell is a device, which converts part of solar energy directly to electricity. In photovoltaic devices charge carriers generated by the incoming irradiance (energy greater than band gap) are separated and swept away by an internal electric field, which is created by either making metal semiconductors contact (Schottky barrier) or p-n junction or Metal Insulator Semiconductor (MIS) or Semiconductor Insulator Semiconductor) (SIS) devices. The height of potential barrier at the junction determines the voltage and the flowing charge carriers contribute to the current when connected externally. The choice of material depends on its band gap, absorption coefficient, diffusion length, mobility of charge carriers, ease in contact formation, type of junction with thermal and electronic compatibility, availability of material, toxicity and several other fabricational, opto-electronic and practical aspects (Chopra and Das, 1983; Fahrenbruch and Bube, 1983; Green *et al.*, 1994; Pande, 1984; Pande *et al.*, 1984; Pande *et al.*, 1996). PV cells based on single crystal and polycrystalline silicon, amorphous silicon, CdS with CdTe, CuInGaSe2 as absorber, InP, GaAs etc. have been much studied. PV modules of efficiency above 14-18 % on silicon and 7-10 % on CdTe, CuInGaSe2 and amorphous silicon are commercially available. Dye sensitized TiO_2 devices have also been receiving more attention. The total global production of PV has increased to some 3800 MW in 2007 to more than 233,000 MW till 2015. The cost of solar cells has come down from \$35/Wp in 1970s to less than 0.5 \$/Wp. Solar lantern, home lighting systems, pumps are commercially available. Several solar power plants have been installed under National Solar Mission.

Solar cells are modular, can be used right at the place of utility without extending long transmission lines, have long life of more than

25 years and are ecofriendly. The design of a PV system starts from an analysis of the load which is evaluated by considering the annual energy requirements, the seasonal and hourly energy diagrams, the peak power requirements, the type of supply i.e. DC or AC, hybridization with auxiliary source or suitable storage etc. (Parker, 1991). Generally the electric equipment are designed for their working by the utility network. Therefore, an inverter is required in order to feed the load with AC voltage at 50 Hz. However, in some cases it can be useful to supply the load by DC power and thus the inverter is eliminated. This feature has the advantages on the cost and efficiency of the system. For example, this can be case in solar pump when the load is an electric motor. Research work has been carried out at CAZRI on the performance evaluation of PV modules under harsh arid conditions with high ambient temperature and dust, both factors affecting the performance adversely. The effect of dust was found to reduce current output of PV panels in open field to more than 30 % if panels were not cleaned. Interestingly, the effect of dust was found more to reduce the PV generated current in winter months than in summer (Pande, 1992). Reflectors of extended length were found to enhance PV output by 17-20% with additional cost of 2-3 % of reflector (Pande and Dave, 2007). These studies are useful in designing PV systems. Design, development and experiences on a few PV and thermal integrated systems for arid agriculture are enumerated.

Solar PV generator for pumping and allied tasks

Drip irrigation system is required to economize the use of water while eliminating practically all the problems that are associated with flood irrigation such as water logging and salinity. However, it requires energy to run the system. In this context a solar Photovoltaic (PV) pump operated drip irrigation system is quite useful. For designing the system, the pressure discharge characteristics of the pump with changing irradiance are matched with the water requirements. This can be estimated by considering the pan evaporation rate, crop coefficient and wetted area. The size of the field, main and sub mains are also worked out for maximizing the field area with the available energy to reduce the cost of operation. The energy requirements and type of drippers are also investigated for ensuring uniform application of water in the field. Such a solar PV pump operated drip irrigation system was developed and its performance was evaluated (Fig.1). The system comprises 900Wp PV array with 800 W DC motor-pump mono- block and three openable low pressure compensating (OLPC) drippers on each plant in one ha pomegranate and ber orchard. Performance revealed that the system can command 4-5 ha pomegranate orchard with benefit cost ratio of more than 2 (Pande *et al.*, 2003).

Fig. 1. Pomegranate orchard irrigated with PV pump operated drip system

However, as irrigation is not required throughout the year, the system remains idle for many days making no use of the costly PV electricity throughout. In view of this a compatible PV generator was developed for contemplating some post- harvest and allied tasks (Fig. 2).

Fig. 2. PV generator

In the PV generator, a DC-DC voltage converter with air cooled fins is incorporated for charging 24 V battery bank, especially designed two tier chambers for encasing DC-DC converter and inverter at the top and keeping batteries at the lower end. The system can be used either for running the pump or operating small machines, illuminating the farm building and carrying out other agricultural activities. The pump is installed at a cistern (Tanka), which stores roof harvested water and thus such PV systems become useful round the year (Pande et al., 2008). Such a system would save about 35 tonnes of CO_2 emission during its life cycle. Now days, both AC and DC solar PV pumps of different capacities are commercially available. More than 19500 solar pumps were installed in our country till 2014. Several demonstrations of PV pump based drip system are in operation. There have been incentives for mass scale adoption. Replacement of diesel operated pumps with PV pumps indicate proportionately more mitigation of CO_2.

PV systems for plant protection

In agriculture there is a need to protect the crop from insects and pests and in this connection small PV based equipment are of great utility. An especially developed system comprises a Photovoltaic panel carrier, storage battery and a compatible dusting unit in which the PV panel is carried over the head that provides shade to the worker and simultaneously charges battery to run a duster. When the equipment is switched on, the stirrer agitates the powder that falls on the fast moving impeller and blown out through the exit pipe. The average field capacity of PV duster comes to about 0.75 ha/hr (Pande, 1990; Pande, 1998). In an attempt to enhance the capacity of earlier developed PV duster, a larger system was designed and fabricated. The duster (44 cm length and 14 cm diameter) has two motors, one for agitating the powder at the top compartment and other at the bottom for operating an impeller. Arrangements for mixing of air with the falling powder are made in the middle compartment with openings for air inlet. Since the size of the duster is bigger and it is difficult to carry on a handle, the dusting unit was attached to the back of the PV panel carrier with detachable features. A battery is also stacked in a socket on the backside beneath the dusting unit. Slings and belts are being provided to carry the PV panel as headgear and the chute of the duster is attached to a flexible pipe for dusting the powder on crop with ease (Fig. 3). The same system can run ultra-low volume (ULV) sprayer also. In addition, it can be used to operate LED for illumination.

Fig. 3. Larger Solar PV duster in operation

The performance of larger PV duster was studied with variation in the size of the holes through alignment of two eccentrically pivoted plates, provided at the base of the hopper for fall of powder. The capacity of dusting varies from 30 to 120g/min depending on the size of the holes in the hopper. Such systems replace the kerosene lamps generally used for lighting and diesel based dusters and thus helps reduce the CO_2 emission.

PV Winnower cum dryer

Another important application in the field is winnowing of the threshed grain. Generally farmers depend on natural wind and if there is a lull the material remains on the field and there is chance of its spoilage. With this in view a solar PV winnower was developed (Fig. 4). It is a convenient device for winnowing grains, spices and other agricultural produce. Since the winnowing is only seasonal task, a mixed mode dryer has been developed, which utilizes the winnowers fan and thus it can be used for dehydrating fruit and vegetables under forced circulation of air (Pande, 2006; Pande *et al.*, 2010).

Fig. 4. Winnowing of cluster bean

The system comprises PV modules, especially designed compatible winnower and drying cabinet. As a winnower 35 to 50 kg grain could be separated with in 1 to 1.5 hours from threshed materials. The drying chamber comprises a cabinet, especially designed to take the advantage of the sun's position in different months for getting the required energy gain and simultaneously to load the drying material conveniently on twelve trays stacked one above another in two parallel compartments. Provisions are made to insert the PV run fan of the winnower on the west opening through a pre air heating tunnel (Fig. 5). As a solar PV dryer 40-50 kg fruit and vegetables e.g. *kachra* (*Cucumis melo*) slices, grated carrot, *ber* (*Ziziphus* sp), coriander leaves, spinach, mushroom, chilies etc. could be dehydrated in less than half of the time required in open sun drying while retaining its colour and aroma. The system can be used for charging a battery for illumination of house. Thus it become more useful for domestic lighting and for agricultural purposes such as winnowing and cleaning of grains and dehydrating fruit and vegetables enabling farmer to get more benefits from the same system.

Fig. 5. PV winnower cum dryer with pre air heating tunnel

A module of portable solar PV dryer having provision to load five drying trays, each of size (0.9 m x 0.6 m) was designed and fabricated with alternative materials such as slotted iron angle, fibre glass sheet etc., eliminating the need of welding required otherwise in making the frame and the related problems of carrying heavy structure to field. After initial testing, an up-scaled portable PV dryer having six such modules, detachable pre air heating tunnel with fan and compatible PV module (75 W_p) with storage battery back-up (40 Ah) was developed at CAZRI, Jodhpur and installed at Krishi Vigyan Kendra, Pali to dehydrate 100-150 kg fruit/vegetables (Fig. 6). With the use of the pre heating tunnel (3.7 m length) the inside thermal gradient of the drying chamber reduced from 5-6°C to 2-2.5°C leading to uniform solar drying. The battery back-up (40 Ah) helped continue drying even during cloudy conditions. Experiments revealed better regulation of the drying temperature due to constant speed of the fan leading to uniform rate of drying in different trays. The device has been tested for dehydrating fenugreek, mint leaves and *Prosopis cineraria* pods, etc. Retention of colour and aroma was observed in the dried product with more acceptability. In addition to drying, the device can also be used for winnowing and lighting during night. Such a device has a potential of saving more than 32,000 kg of emission of CO_2 in a year besides the portability and multi- functionality of the device that make it more useful for rural areas.

Fig. 6. An up-scaled portable PV dryer cum winnower

Self-propelled PV mobile unit

The multipurpose self- propelling PV mobile unit (Fig. 7) was designed and developed with a view to providing a complete self sustained mobile power unit for various domestic, small agricultural and other rural applications in isolated cluster of houses (*Dhanis*) of arid region (Pande, 2009; Pande *et al.*, 2011). It comprises an especially designed mobile structure in which arrangements have been made to keep two PV modules (70W_p each) facing due south at 26 degree from horizontal to receive optimum solar radiation and also to provide shade to the subcomponents as well as to user. The mobile unit structure is made of rectangular iron angle frame, which is provided with four wheels at the lower end, two on the rear side of the rectangular frame and the remaining two on a guiding trapezoidal frame attached to it in the front side with a flexible metallic pulling sling, having handle to move it with ease through a DC motor driven system.

Fig. 7. PV mobile unit developed at CAZRI, Jodhpur

The multi crystalline silicon PV array (140 W_p) output is fed to the battery (12V 120Ah) through a regulator and a maximum power tracker to charge the battery in an optimum way and then to derive power through an in- built inverter for the working devices that could be operated for illumination, running radio, a small TV in the domestic front. The unit can be used also for butter extraction through the use of a churner and for winnowing with a blower. The utility of the system can also be extended for other cottage scale applications through operation of both AC and DC devices.

PV mobile unit has been successfully used to operate different post-harvest processing machines e.g. aloe vera gel extractor, ber grader etc. The device can contribute saving about 4 tonne of CO_2 emission besides the ease of operation for various cottage industrial and agricultural applications.

Solar thermal devices

In solar thermal applications, solar cookers and water heater are most common devices for the domestic use, although the applications are extended to distillation, drying, processing through steam generation, power generation, community scale operations, heating of buildings etc.

Generally available solar cookers can be used for boiling rice, lentil, vegetables, baking and roasting vegetables and cooking local food and feed for animals. BIS standards are available for solar cookers. At CAZRI stationary solar cookers have been developed (Pande and Thanvi, 1987; Pande and Thanvi, 1988). With the use of such cookers one can save about 30-40% of fuel requirement. Large scale application of solar cooker based on concentrator has been implemented at several places in India e.g. Bramhakumari, Ashram, Mount Abu, Tirupati temple etc., which saves huge amount of CO_2 emission daily. Solar cooker can also be used for boiling feed for cattle and may reduce CO_2 emission caused by burning of fuel wood. Similarly substantial amount of fuel can be saved by using solar waters heaters for domestic purposes like bathing, washing of clothes and utensils and in industries like textiles and dairy. Each m^2 of flat plate collector area can save 2.5-3 kWh electricity per day.

Three in one solar integrated device

After working on the concept of stationary solar cooker and developing dual purpose solar devices an integrated three in one solar device was developed. The integrated solar device (Fig.8) comprises especially designed trapezoidal GI tank to hold water, appropriate geometry to use the device in stationary mode (Pande and Thanvi, 1991; Pande, 2011), double glazed windows with reflectors at the top and in the front side to utilize solar energy round the year, facility to operate the system as cooker on especially designed cooking cum insulting tray and four plastic pipe nipples with caps for facilitating the air circulation while using it as a dryer with built-in storage. Three models were developed. The first ID-1 was conceptual model. Then the second ID - 2 was compact version while the improved three in one integrated device ID-3 has 25% enhanced capacity of water and 13.4 % less area of window but having 35% more effective utilization of energy compared to earlier developed model ID-2. The device can be used as solar water heater during winter months, solar cooker during clear days and solar dryer on availability of fruit or vegetables.

Fig. 8. Integrated three in one solar device

With this novel device as a solar cooker food for a family can be boiled within 2-3 hours (loading time 10 A.M) and that too without sun tracking. It can produce about 50 L hot water of 50-60°C utilizing the low altitude position of sun during winter and thus having energy gain both from top and front windows. As a dryer, fruit and vegetables like *ber*, grated carrot, spinach, watermelon flakes, tomato slices etc. could be dehydrated efficiently with regulation of temperature during day time when water works as a sink and helps continuation of the drying process even in the night through the solar heated water. The device has been successfully used in making watermelon candies also. The device can save about 400 kWh of energy every year.

PV clad enclosure and solar farming

Protected cultivation is considered appropriate specifically in the context of climate change, however, it requires thermal regulation inside the structure for optimum crop growth. Keeping in view the requirement, a PV clad enclosure (ground area 15.3 m^2) was developed with fibre sheets, two interconnected PV arrays of amorphous silicon solar cells (60 W_p each) provided at the top on the slots of inclined green corrugated fibre glass sheet roof and earth embedded pipes with provisions to suck air by PV operated fan through pipe network to create controlled environment (Fig. 9a). The observed data during different seasons indicated reduction in air temperature at the exit of the fan to a range of 6-10°C in summer and 2-4°C rise in winter, close to predicted values through a developed mathematical model. Performance of the PV clad enclosure with forced circulation of air through earth embedded pipes and PV mister indicated reduction in the inside temperature by 4-6°C below the ambient temperature during extreme weather conditions in summer (Pande *et al.*, 2012). Tomato plants were grown successfully inside the PV clad structure (Fig. 9b), which revealed regulation of the inside temperature to desirable range. The developed PV clad enclosure can be used for nursery raising and allied agricultural applications.

In this connection of PV in agriculture, more recently the concept of solar farming is in rage. In larger solar power plants the inter space between two rows of PV array may be used for growing plants by using the water required for cleaning PV array. Such features would enable farmers to use their part of the land as an ensured source of earning by selling the generated electricity and simultaneously arresting the CO_2 emission as each square meter of PV panel saves about 1330 kg of CO_2 emission per year.

Fig. 9. PV clad structure for multipurpose agricultural applications

Solar PV device and climate change mitigation

Solar PV devices discussed here have a role in reducing CO_2 emission in atmosphere by replacing either the use of coal based electricity or fossil fuel burring for different agricultural and domestic activities. Reduction in CO_2 emission for each solar PV device can be calculated by considering the respective emission factor (EF) of fossil fuels, which are otherwise used. For example, direct use of electricity from thermal grids leads to an emission of 0.82 kg CO_2e per kWh (CEA, 2014). Similarly, if the solar PV devices replace fossil burning fuelwood e.g. cooking, water heating etc then it can reduce the CO_2 emission by a factor 1.644 kg CO_2e per kg of fuelwood [EF for fuelwood = net calorific value of fuelwood (0.015 MJ/kg) x CO_2 emission factor fuel wood (109.6 kg/MJ)]. The EF for diesel is 2.667 kg CO_2e per litre of diesel used. Therefore, the solar PV devices discussed in this paper have a great role in mitigating climate change. Corresponding CO_2 emission reduction potential have been highlighted under each solar PV device. Here, the calculation of CO_2 emission reduction potential is shown for 3 HP solar PV pumping system, which may replace either grid generated electricity operated pumps or diesel operated pumps. In first case, solar pump will save about 2685 kWh energy after considering its use of 200 days in a year for irrigation and 6 hours per day. Thus a single 3 HP solar pump will reduce 2201 kg CO_2e in a year. In the second case, 3 HP solar pump will replace 790 litre of diesel per year for similar use as mentioned above and will reduce emission by 2106 kg CO_2e per year.

Conclusion

Solar energy can provide the needful support for futuristic energy requirements of 2070 in the country (Sukhatme, 2012). Photovoltaic cells are appropriate devices for energizing remote rural areas. Integrated multipurpose systems based on available PV modules are quite useful and concerted efforts need to be made to disseminate these to rural areas Some of these PV systems and integrated solar devices should be considered for commercialization through SAREGA (Solar Appliances for Rural Energy Gurantee and Assurance) (Pande, 2011b). Young entrepreneurs need to come forward for undertaking this task and all the concerned agencies should work in tandem to bring much required energy revolution in India. In view of the greenhouse gas emission by burning fossil fuel for the energy needs and its implications on climate, it is incumbent to use solar technology to arrest it and to make the future more greener.

References

Chopra, K.L. & Das. S.R. 1983. Thin Film Solar Cells. Plenum Press, New York.

CEA 2014. CO_2 Baseline Database for the Indian Power Sector, User Guide, Version 10.0 December 2014http://cea.nic.in/ reports/others /thermal /tpece/cdm_co2/user_guide_ver10.pdf

Fahrenbruch, A.L. and Bube, R.H. 1983. Fundamentals of Solar Cells Photovoltaic Solar Energy Conversion. Academic Press, New York.

Green, M. A., Wenham, S.R., Honsberg, C.B. and Hogg. D. 1994. Transfer of buried contact cell laboratory sequences into commercial production. *Solar Energy Materials and Solar Cells* 34: 83-89.

Mani, A and Rangarajan, S. 1982. *Solar Radiation over India*. Allied publisher Pvt. Ltd., New Delhi.

Pande P.C., Singh A.K., Ansari S. and Dave B.K. 2008. PV generator for multiple applications in arid farming. In *Diversification of Arid Farming Systems* (Eds. Pratap Narain, M.P.singh, Amal Kar, S.Kathju and Prvaeen Kumar), pp. 273-278. AZRAI & Scientific Publishers (India) Jodhpur.

Pande P.C., Singh A.K., Ansari S., Vyas, S.K. and Dave B.K. 2003. Design development and testing of a solar PV pump based drip system for orchards, *Renewable Energy* 28: 385-396.

Pande, P.C. 1984. A study on various forms of CdS solar cells. Ph.D. Thesis. University of Durham, U.K.

Pande, P.C. 1990. Development of PV systems for arid region. In *Energy and the Environment into the 1990s, Proceedings 1st World Renewable Energy Congress* (Ed. A.A.M. Sayigh), pp. 314-318. Pergamon Press, Oxford.

Pande, P.C. 1992. Effect of dust on the performance of PV panel. In *Proceedings 6th International Conference on Photovoltaic Science & Engineering* (Eds. B.K.Das and S.N. Singh), pp.539-542. Oxford and IBH Publishing Co.,Ltd.

Pande, P.C. 1998. A novel solar device for dusting insecticide powder. In *Proceedings of National Solar Energy Convention,* pp 117-122. University of Roorkee, India.

Pande, P.C. 2006. Design and development of PV winnower-cum-dryer. In *Proceedings. International Congress on Renewable Energy (ICORE 2006)*, (Eds. E.V.R. Sastry and D.N.Reddy), pp.265-269. Allied Publishers Pvt. Ltd.

Pande, P.C. 2009. A PV mobile unit for multipurpose rural applications. In *Proceedings International 18th Photovoltaics Solar Engineering and Science Congress,* (Eds. Swati Ray and P. Chatterjee), McMilllan Press, India.

Pande, P.C. 2011a. Integrated solar devices for rural areas: Experiments on drying mushroom, In *Proceedings of ISES Solar World Congress,* pp. 150-155. Kassel, Germany.

Pande. P.C. 2011b. Integrated solar devices towards achieving the National Solar Mission Goals. *Akshaya Urja* 4(4): 30-34.

Pande, P.C. and Dave, B.K. 2007. Economical production of electricity from PV modules for application in post harvest operations. In *Advances in Energy Research. Porceedings 1st International Conference on Advances in Energy Research*, pp. 296-300. IIT Bombay. Macmillan India Ltd.

Pande, P.C. and Thanvi, K.P. 1987. Design and Development of a solar cooker for maximum energy capture in stationary mode. *Energy Conversion and Management* 27(1): 117-120.

Pande, P.C. and Thanvi, K.P. 1988. Design and development of solar cooker-cum-dryer. *International Journal of Energy* 12: 539-545.

Pande, P.C. and Thanvi, K.P. 1991. Design and development of solar dryer cum water heater. *Energy Conversion and Management* 31(5): 419-424.

Pande, P.C., Bocking, S., Miles, R.W., Carter, M.J., Latimer, J.D. and Hill, R. 1996. Recrystallization of electrophoretically deposited CdTe films. *Journal of Crystal Growth* 159: 930-934.

Pande, P.C., Nahar, N.M., Chaurasia, P.B.L., Mishra, D., Tewari, J.C. and Kushwaha, H.L. 2009. Renewable Energy Spectrum in Arid Region. In *Trends in Arid Region Research in India* (Eds. Amal Kar, B.K.Garg, M. P. Singh and S .Kathju), pp. 210-237. Central Arid Zone Research Institute, Jodhpur.

Pande, P.C., Russell G.J. and Woods, J. 1984. The properties of electrophoretically deposited layers of CdS. *Thin Solid Films* 121: 85-94.

Pande, P.C., Singh A.K., Ansari S., Dave, B.K. and Purohit, M.M. 2011. A self - propelled mobile unit for rural development. In *Proceedings, International Congress on Renewable Energy (ICORE-2011)*, (Eds. S.K. Samdarshi, S. Mahapatra and S. Paul), pp. 272-276. Tejpur University, Assam. Solar Energy Society of India.

Pande, P.C., Singh, A.K., Dave, B.K. and Purohit, M.M. 2010. A preheated Solar PV dryer for economic growth of farmers and entrepreneurs. *SESI Journal*, 20(1&2): 73-79.

Pande, P.C., Singh, A.K., Santra, P, Purohit, M.M. and Dave, B.K. 2012. PV mister based cooled enclosure for arid agriculture. In *Proceedings International Congress on Renewable Energy (ICORE) 2012*, (Eds. S.K. Samdarshi and Jagat S. Jawa), pp. 13-16. Excel India Publishers and SESI, New Delhi.

Pande, P.C., Singh, H.P. and Nahar, N.M. 1998. Energy Management by non-conventional energy sources. In *Fifty Years of Arid Zone Research in India* (Eds. A.S. Faroda and Manjeet Singh), pp. 427-446. Central Arid Zone Research Institute, Jodhpur.

Parker, B.F. 1991. Solar Energy in Agriculture. Elsevier Science Publishers, Amsterdam, Netherlands.

Sukhatme, S.P. 2012. Can India's future needs of electricity be met by renewable energy sources? A revised assessment. *Current Science* 103(10):1153-1161.

❏❏❏

Chapter 23

Dynamics of Surface Energy Fluxes over Hot Arid Grassland Ecosystem

P. RAJA†

ICAR-Central Arid Zone Research Institute, Jodhpur-342 003, Rajasthan
† Present Address: ICAR-Indian Institute of Soil and Water Conservation, Research Centre, Udhagamandalam, Tamil Nadu, India-643 004
E-mail: *rajacazri@gmail.com*

Introduction

Energy is exchanged between the atmosphere and Earth surface. Radiation from Sun and long wave radiation from atmosphere warm the surface and provide energy to drive weather and climate. Part of this energy is stored in the ground or oceans and partly returned to the atmosphere, warming the air. The rest is used to evaporate water. These surface energy fluxes mainly constitute the Earth's energy budget. At the global scale, Earth's energy budget shows that the atmosphere has a deficit of energy while the earth surface has a surplus. The heat fluxes arise as sensible heat (carried by wind) and latent heat (carried by moisture) carried away from the surface. Over the land, sensible and latent heat fluxes are important determinants of microclimates. Microclimates are the climate near the ground and represent local climates due to topography, vegetation, soils, landforms, and structures. This chapter introduces the energy fluxes needed to understand microclimates especially in arid grassland system.

Macroclimate is the large-scale climate over 2000 km or more resulting from geographic variation in net radiation, the resultant transport of heat by atmosphere and oceans, and high and low surface pressure belts. In contrast, microclimate is local climatic features typically smaller than 2 km scale. A forest has a different microclimate than an adjacent barren land. Microclimates may vary within few centimeters, for instance soil covered with litter has different temperature and moisture than an adjacent uncovered patch of bare soil. At the local scale, the landscape is a mosaic of patches created by spatial variation in topography, soils, and land cover. By altering the energy and water cycle between land and atmosphere, landscape heterogeneity can create unique microclimate that differs from the prevailing macroclimate (Rosenberg *et al.* 1983).

The regional impacts of climate change are largely driven by hydrological variations and associated energy fluxes (Rach *et al.*, 2014). Thus, the impact of any changes in energy fluxes and consequent hydrological changes would be severe in the water stressed arid ecosystem. The extreme events of high temperature, high wind velocity, dust storms, and drought conditions are common phenomena in the Great Indian Thar Desert. Besides these, very low and erratic distribution of rainfall is also common and has significant influence on regional hydrological and associated nutrient-carbon cycling. Therefore, it becomes necessary to understand land surface-atmosphere interactions and their implications on arid region.

Furthermore, arid and semi-arid regions of the world ($\approx$ 1/3 of the global land surface) play a significant role in global climate in relation to biogeochemical and hydrological processes (Schlesinger *et al.*, 1990; Vernekar *et al.*, 2003). In view of the extent and amount of heat exchange between arid surface and the overlying atmosphere, it is necessary to examine how the surface heat flux varies with time so that, its impact on atmospheric circulation and climate can be better understood. The data on surface heat fluxes are being used to test, parameterize, and elicit outputs of land surface schemes in climate models (Williams *et al.*, 2009). This can help to develop local, regional, and global maps of water vapour fluxes (Jung *et al.*, 2009). If the inter-annual magnitude of change in surface energy balance were large, it would significantly affect the atmospheric circulation and the rain bearing system (Lare and Nickolson, 1994).

Additionally, dust storm is a frequent phenomenon during summer in Thar region (Raja *et al.*, 2013). Dust and rainfall are negatively correlated and are presumed to contribute substantially to droughts in arid/semiarid regions by significantly inhibiting rainfall (Hui *et al.*, 2008). Therefore, it is important to acquire long-term information on energy and mass exchange data to better understand complex feedback mechanisms between the terrestrial biosphere and climate system, and to estimate land surface energy balance that controls the height of the planetary boundary layer and climate

(Ryu *et al.*, 2008). However, there is a gap in understanding the drivers of energy and mass exchange processes in relation to biophysical and environmental controls especially for arid regions with proximity to sea.

Micrometeorological stations provide a promising observation system to measure the net exchange of radiation, energy and water over land for a footprint varying from 0.5 to 1.0 km^2 (Bhattacharya *et al.*, 2009). Worldwide studies on surface energy, water vapour and CO_2 exchange processes over arid/semi-arid savannas, temperate, Mediterranean grasslands and agricultural ecosystems have indicated considerable inter-seasonal and inter-annual variations (Wever *et al.*, 2002; Burba and Verma, 2005; Li *et al.*, 2006, 2007; Hao *et al.*, 2007; Ryu *et al.*, 2008; Aires *et al.*, 2008; Chen *et al.*, 2009; Bowling *et al.*, 2010). In arid climatic region of the Indian subcontinent, very few micrometeorological experiments have been conducted. The Monsoon Trough Boundary Layer Experiment (MONTBLEX) was conducted in 1990 (Goel and Srivastava, 1990) along the region of the monsoon trough. Some small-scale experiments (Vernekar *et al.*, 1993; Pillai *et al.*, 1998; Sethuraman *et al.*, 1998) were conducted for a limited period. A major multi-site field campaign, LASPEX (Land Surface Processes Experiment) was conducted in the semi-arid region of Gujarat, India to parameterize land surface processes (Vernekar *et al.*, 1993).

The Thar Desert arid ecosystem is unique because it undergoes pronounced water limited and green growth phase with water availability hardly confined to 2 to 3 months (July to September). Completely dry, water limited period extends from October to March. The period between April to September experiences some pre-monsoon and monsoon rainfalls. Moreover, rainfall is characterized by highly variable but low precipitation (~100 to 400 mm). Consequently, the study of intra-seasonal and inter-annual variability of surface energy fluxes in relation to biophysical characteristics can help to understand which period influences energy fluxes and closure behaviour and how the residual energy behaves with the timing of seasonal transition. Moreover, this ecosystem experiences large variability in precipitation, and therefore, it is imperative to understand regulating mechanism of evapotranspiration.

The long-term data on energy and water balance in arid grassland ecosystem with distinct growth cycle are scanty, particularly, in India. Here, characteristics of seasonal and annual variability of energy-water flux components and their drivers over a dominant desert grass (*Lasiurus sindicus*) are discussed.

Characteristics of the study site

The Great Indian Desert is a vast tract covering over 4000 sq.km in the northwestern part of the Indian subcontinent and it stretches from the western fringes of Aravalli Mountains to the Indus River. Geographically, the Thar Desert in India is located in the Rajasthan state between the foothills of the Aravalli ranges in the east and the international border with Pakistan in the west. Further west in Pakistan, the desert extends up to the fertile alluvial plains of the Indus. The study site is located (Fig. 1) at the Central Arid Zone Research Institute (CAZRI) experimental area (26°50283[3] N, 71°18283[3] E) at Chandan grassland farm covering ~ 96 hectares consisting predominantly of *Lasiurus sindicus*.

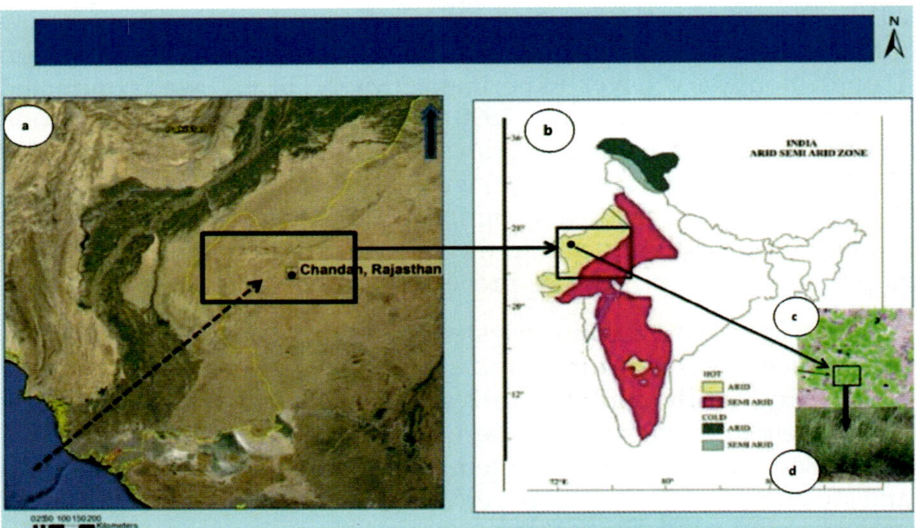

Fig.1. (a) Location of research site at the experimental grassland (small rectangular box denotes exact location of the experimental site) of Central Arid Zone Research Institute (CAZRI), Chandan, Rajasthan (dashed arrow indicates proximity with the Arabian sea) in the desert area, **(b)** location of arid region on the map of India,(modified after Dhir and Sighvi (2012), **(c)** true color composite from Landsat-7 (ETM+) over grassland at Chandan, **(d)** field photograph of grass growth pattern at Chandan.

The study was conducted during the year 2011 and 2012 using half-hourly micrometeorological data on radiation and energy balance components recorded through 10 m micrometeorological tower with multi-level sensors. The climatic situation of arid regions of Rajasthan state is characterized by highly variable but low precipitation (~100 to 400 mm), extremes of temperature (48°C to - 3°C) and high wind speed (~33 km h^{-1}). Mean monthly sunshine hours varied between 6.8 to 10.7 hrs day^{-1} and this leads to very high evaporative demand as well as generation of day-time convective turbulence. Moisture stress is an impediment to vegetation growth. Low humidity, combined with strong wind regime, leads to advection, a phenomenon that causes evaporation loss more than the energy actually available through solar radiation.

The soils of this region are predominantly Entisols (54.2%) followed by Aridisols (45.1%) and Alfisols (0.67%) (Shyampura, 2002). The soils are loose, sandy to sandy loam with very rapid permeability (> 25.4 cm hr^{-1}). Water retention at field capacity and permanent wilting point in Aridisols are comparatively higher than in Entisols and are 13% and 5%, respectively. Organic carbon content varied from 0.05 to 0.2 % (Kumar *et al.*, 2009). The monsoon season is very short (mid July to mid September) with only about 15-20 rainfall events and could supply moisture for not more than 90 - 110 days.

Micrometeorological observations

An INSAT (Indian National Satellite) linked micromete-orological tower (10 m) (Bhattacharya *et al.*, 2009; Bhattacharya *et al.*, 2013) was installed within arid grassland patch at Chandan (CAZRI), with a fetch ratio of approximately 1:50 (fetch from all direction was over 500 m). This recorded continuous and automated measurements of radiation and energy balance components from a set of sensors consisting of four-component net radiometer (at 4 m) (NR Lite, Kipp and Zonen, Delft, Netherlands), two-depths (at 0.1 and 0.2 m) soil heat flux plates (HFP01SC, Radiation & energy balance System Inc., USA), three-height air temperature-relative humidity probes with a shielded and aspirated sensor (HMP-45C, Campbell scientific, Logan, USA) and anemometer for wind speed and direction (KDS-131, KEPL, India) (at 1.25 m, 2.5 m and 7.5 m), three-depth soil thermometers (KDS-031, KEPL, India) (0.05 m, 0.2 m and 0.45 m) and rain-gauge of tipping bucket type (KDS-071, KEPL, India). In addition, monthly volumetric sampling of soil moisture was carried out from three depths (0.05 m, 0.2 m, 0.45 m) near the tower. A 25-channels data logger (DSAWSKM1-W, KEPL, India) logged data every second which were averaged over 30 minutes and stored in the data logger. In the present study, half-hourly averaged micro-meteorological data were analyzed for the two-year period from January 2011 to December 2012 covering two active growth cycles of arid grassland ecosystem.

Biophysical measurements

Systematic sampling plan was laid out for measuring leaf area index (LAI), plant height, and phenological observations. Twenty-five (5 × 5 m) randomly selected permanent quadrats were laid down for *Lasiurus sindicus* grass system. The LAI was measured at 10-day interval from the growth initiation (greening) phase from the first dekad of July (Jul 1d). The LAI measurements lasted up to senescence (browning) phase. The LAI measurements were taken using plant canopy analyzer (LAI-2000; Li-Cor, Inc., Lincoln, NE). In each quadrat, the mean LAI was obtained by taking average of eight readings (six below canopy and two above the canopy). Mean LAI of all the quadrats were computed to represent the temporal dynamics of LAI. Digital field photographs were taken from representative quadrats at weekly interval. These were analyzed to ascertain the transition dates from one growth stage to another.

Radiation balance and net available energy

The batch processing of micrometeorological data were carried out using 'Fluxsoft' software developed at Space Applications Centre (SAC-ISRO) in windows platform on Visual Studio using the protocol developed by Bhattacharya *et al.* (2009, 2013). An IDL (Interactive Data Language) routine was written to convert all half-hourly fluxes into daily averages. Radiation balance components and ground heat fluxes were averaged for daytime and for 24 hours period as well. Ten-day (dekadal) and monthly averages of fluxes were computed from daily averages. The data quality was screened to remove spurious spikes in the instantaneous and daily averages to have smooth temporal transition (Leuning *et al.*, 2012) and to obtain reliable dekadal averages.

The net radiation (Rn) was computed from the four components of radiation balance as given in equation (1).

$$R_n = RS_{net} + RL_{net} \quad (1)$$

where,

$$RS_{net} = RS \downarrow - RS \uparrow$$

$$RL_{net} = RL \downarrow - RL \uparrow$$

Where, RS and RL are shortwave and longwave radiation and the symbols '↓' and '↑' represent incoming and outgoing fluxes, respectively. The RS_{net} and RL_{net} are net shortwave and net longwave radiation fluxes at the surface.

The net radiation (Rn) is partitioned into sensible heat flux (SH), latent heat flux (LE), ground heat flux (G) and metabolic and storage energy (ÄH) in the vegetation canopy.

$$R_n = SH + LE + G + \Delta H \quad (2)$$

The storage heat of the canopy is being ignored in diurnal and seasonal cycle, with the assumption that the amount of energy stored during the heating cycle is being released during the cooling cycle in the short heighted, open-canopied, sparsely vegetated grassland ecosystem (Leuning *et al.,* 2012). The amount of metabolic heat is usually negligible as compared to other components.

The net available energy (Q) was computed as:

$$Q = Rn - G \quad (3)$$

10-days' day-time average latent heat flux from Bowen Ratio Energy balance (BREB) was computed as:

$$LE = \frac{Rn-G}{1+\beta} \quad (4)$$

Where: $\beta = \frac{SH}{LE} = \gamma \frac{dT_a}{de_a}$

The air temperature ($\frac{dT_a}{dZ}$) and vapour pressure ($\frac{de_a}{dZ}$) gradients can be determined through multi-height measurements within the surface layer (or laminar boundary layer) embedded in atmospheric boundary layer (ABL). λ is the psychrometric constant (0.66 mbar $^{o}C^{-1}$).

Sensible heat (SH) and latent heat (LE) fluxes

Turbulent heat fluxes were also computed from independent method following Monin–Obukhov Similarity Theory (MOST). The MOST (Monin and Obukhov, 1954) allows a valid parameterization of the surface atmospheric boundary layer. Following this theory, the turbulent fluxes of scalars within the surface layer can be related to the vertical gradients of their mean concentrations by an eddy diffusivity or turbulent transfer coefficient (Garrat, 1994). The eddy diffusivity caused by turbulent mixing is related to the size of the turbulent eddies and is therefore, proportional to the distance from the underlying surface or effective surface. The effective surface in case of vegetation canopy of height '*h*' is defined at a height '*d*' above the ground surface where, '*d*' is called zero plane displacement. Under neutral stability condition the wind profile equation above a homogeneous vegetation canopy is thus, expressed as follows (Thom, 1975):

$$u(z) = \frac{u_*}{k} \ln\left(\frac{z}{z_0}\right) \tag{5}$$

$$\frac{du}{dz} = \frac{u_*}{k\,(z-d)} \tag{6}$$

Here, z is the height of wind measurement above the zero plane displacement, z_0 is roughness length, u_* is frictional velocity and k is Von-Karman constant (= 0.4). The parameters, d and z_0 depend upon the aerodynamic roughness of the surface over which measurements are made. The zero plane displacement (d) and roughness length (z_0) were computed from plant height (h) using simple empirical relations: $d = 0.65 \times h$ and $z_0 = 0.1 \times h$ given by Campbell and Norman (1998). The eddy viscosity was estimated from wind speed and roughness parameters using the logarithmic wind profile equation. Solution of logarithmic equation is simple under neutral stability condition. However, absolute neutral stability cannot always have assumed value because very often, there is a thermal gradient in either direction in the vertical profile within the lower boundary layer. The atmosphere normally becomes stable during night time and unstable during day time because of substantial heating during daylight hours. Under unstable or stable conditions, the stability parameter, sensible heat flux and eddy viscosity become auto-correlated. In the present study, stability corrected aerodynamic resistance was determined through iterative method based on Monin-Obukov's Similarity Theory (MOST).

According to the Monin-Obukhov Similarity Theory (MOST) (Businger *et al.*, 1971; Xu and Qiu, 1997; De Bruin *et al.*, 2000) the vertical profiles of wind, temperature, and specific humidity for turbulent flows above a vegetation surface can be expressed as:

$$u_2 - u_1 = \frac{u_*}{k}\left[\ln\left(\frac{z_2}{z_1}\right) - \Psi_M\left(\frac{z_2}{L}\right) + \Psi_M\left(\frac{z_1}{L}\right)\right] \tag{7}$$

$$\theta_2 - \theta_1 = \frac{\theta_*}{k}\left[\ln\left(\frac{z_2}{z_1}\right) - \Psi_H\left(\frac{z_2}{L}\right) + \Psi_H\left(\frac{z_1}{L}\right)\right] \tag{8}$$

$$q_2 - q_1 = \frac{q_*}{k}\left[ln\left(\frac{z_2}{z_1}\right) - \Psi_Q\left(\frac{z_2}{L}\right) + \Psi_Q\left(\frac{z_1}{L}\right)\right] \quad (9)$$

In the above expressions, u = wind speed (ms^{-1}), è = potential temperature (□K), q = specific humidity (g kg^{-1}), z = height of measurement (m) above the zero plane displacement [(2.5 m - d) and (7.5 m - d) in the present study]. The subscripts, 1 and 2, represent lower and upper heights for the respective parameters in the profile measurement; è$_*$ = temperature scale, q_* = humidity scale and u_* = frictional velocity (wind scale); Ø = universal stability function for momentum (subscript M), heat (subscript H) and vapour (subscript Q); L = Monin-bukhov Length is expressed as follows:

$$L = \frac{Tu_*^3}{gk\theta_*} \quad (10)$$

The eddy sensible and latent heat fluxes are expressed as:

$$SH = -\rho C_p \theta_* u_* \quad (11)$$

$$LE = -\rho \lambda q_* u_* \quad (12)$$

where, ñ = air density, T = mean absolute temperature, *Cp* = specific heat of air, and g = acceleration due to gravity, ë = latent heat of vaporization of water.

For computation of the stability functions (Ø) for momentum (M), heat (H) and water vapour (Q) in this study, we adopted for unstable conditions the version of the Businger - Dyer flux-relationships proposed by Dyer (1974), which read in integrated form as:

$$\Psi_M = 2ln\left(\frac{1+x}{2}\right) + ln\left(\frac{1+x^2}{2}\right) - 2tan^{-1}(x) + \frac{\pi}{2} \quad (13)$$

$$\Psi_H = \Psi_Q = 2ln\left(\frac{1+x^2}{2}\right) \quad (14)$$

Where, $x = \left(1 - 16\frac{Z}{L}\right)^{\frac{1}{4}}$

For stable conditions, we used the equations proposed by Beljaars and Holtslag (1991):

$$-\Psi_M = \frac{aZ}{L} + b\left(\frac{Z}{L} - \frac{c}{d}\right)exp\left(-\frac{dZ}{L}\right) + \frac{bc}{d} \quad (15)$$

$$-\Psi_H = \left(1 + \frac{2}{3}\frac{aZ}{L}\right)^{\frac{3}{2}} + b\left(\frac{Z}{L} - \frac{c}{d}\right) exp\left(-\frac{dZ}{L}\right) + \left(\frac{bc}{d} - 1\right) \quad (16)$$

in which, $a = 1$, $b = 0.667$, $c = 5$ and $d = 0.35$ are constants taken as per Beljaars and Holtslag (1991).

The profile measurement of temperature, humidity, wind speed at the heights 2.5 and 7.5 m above ground surface were used for the computation process. The vapour pressure, specific humidity, and air density required in MOST were computed from the tower measurements of temperature, relative humidity, and air pressure using standard psychrometric equations. The standard values for specific heat of air (Cp = 1000 J kg^{-1} K), and acceleration due to gravity (g = 9.81 m s^{-2}), latent heat of vaporization of water ($\ddot{e}$ = 2.45×10^6 J Kg^{-1}) were used. The mean plant height was ~ 1.0 m. The zero plane displacement was computed from plant height using the empirical relation described earlier. The lower and upper heights of humidity, temperature, and wind speed measurements were 2.5 and 7.5 m, respectively. The iterative technique was applied by initializing stability factor Ø = 1 and L = 0 (neutral stability condition). The u_* and $\grave{e}_*$ were computed following the equations (7) and (8), respectively. Then L was computed from equation (10). Based on the temperature gradient (positive or negative), the iteration was started with stepwise increase or decrease of L following the stable or unstable loop, respectively. The iteration was repeated till the assumed L becomes nearly equal to the computed L (at 0.001 level of precision). The final value of L was used for the computation of SH and LE flux.

Residual energy (dE)

The residual energy for energy balance closure was computed as the difference between net available energy (Q) at surface and sum of sensible and latent heat fluxes from surface computed through MOST approach.

$$Residual\ energy\ (dE) = Q - (SH + LE) \quad (17)$$

Salient observations

Characteristics of Sewan grassland

The grass *Lasiurus sindicus*, locally known as *'Sewan'*, is an important nutritive fodder having several agronomic values in this desert. *Lasiurus sindicus* is a perennial C4 grass, having a number of genetic and epi-genetic strategies including physiological and phenological behaviour to survive the climatic extremes such as heat waves and drought. It is a unique plant system that efficiently manages to get green even at a meager precipitation and produces more biomass in comparison to other grass species. Plant height at maturity reaches up to ~ 1.0 m. The mature plant has 15-20 numbers of tillers with 25-30 leaves, up to 10 nodes and 4-6 reproductive branches on the main tillers. These structures contribute to biomass and LAI of the plant.

The phenology of the grass is such that it exhibits short green and prolonged brown growth stages in the annual phenological cycle with growth phase confined to 2 to 3 months only (July to September). The start of greening was found to vary and dependent upon the temporal variation of the monsoon rainfall. It gradually greened up within 2 to 3 days of start of rainfall (< 10.0 mm). The time-period from July to September exhibited the green stage characterized by green canopy (Fig.2 a). From the first week of October, traces of leaf browning began to appear. Plants progressively entered into physiologically dormant stage with the onset of dry winter and the rate of browning increased rapidly. By the first week of November canopy turned brown (Fig.2 b).

Fig.2. Pre - (a) and post-monsoon (b) phenology of grass species in CAZRI's experimental area, Chandan (Also seen :Agro Met Station (ISRO) tower.

Ground heat - moisture fluxes and their drivers

The annual behaviour of day-time averaged net radiation (Rn) showed a bimodal behavior. Maintenance of nearly constant inter-annual sensible heat fluxes is characteristic of the arid ecosystem. This characteristic may chiefly be attributed to little inter-annual variation in air temperature (AT) (mean of 3-height). The variability in inter-annual sensible heat fluxes may also be related to wind speed and wind direction. Advection of heat and moisture by winds from surrounding regions may alter temperature and humidity distribution at the experimental location (Leuning *et al.,* 2012). It has been observed that the sensible heat fluxes gradually increased from February to July followed by a usual dip in monsoon, coincident with vegetation growth and increase in soil moisture content and a secondary peak in post-monsoon before coming to minima in winter. The sensible heat flux was about three times higher than the latent heat flux in both the years indicating the energy transport to atmosphere occurred mainly by dry processes rather than through moisture transport. This could be attributed to the characteristics of arid atmosphere associated with poor rainfall and poor soil moisture quantities to support only sensible heat transfer.

The seasonal variations in sensible heat fluxes are related to those in vertical air temperature difference (VATD) (difference between temperature at heights 1.25 and 7.5 m) and surface wind speed. These changes are linked with atmospheric circulation, solar radiation, and land surface conditions. Sensible heat flux is dependent on the VATD as well as the turbulent transfer coefficient in atmospheric surface layer. The data indicated relatively less observations with large heat fluxes (< 400 Wm^{-2}) and VATD ^{3} 0.65 °C mainly distributed in summer season probably due to large surface heating and instability of the atmosphere. The sensible heat flux was also seen to increase linearly with wind speed *i.e.,* higher winds producing stronger heat fluxes and indicates the role of aerodynamic and advection effects. An examination of the annual rainfall pattern indicated close association between ten days mean latent heat fluxes and the rainfall. The LE was almost constant during dry period with no rainfall but increased linearly with increase in rainfall during the wet season. At lower soil water content (SWC > 2.5 m^3 m^{-3}), LE remained constant around 50-55 Wm^{-2}; however, increase in SWC during monsoon led to increase in LE. This behaviour probably indicated tight stomatal control on LE fluxes (Law *et al.,* 2002). The relatively higher LE in monsoon driven growing season suggested that it was controlled by the availability of soil moisture and hence the rainfall in these periods. Variation of LE with dekadal rainfall also defends the lagged response of vegetation growth to soil moisture vis-a-vis rainfall.

The relationship between Rn-VPD and LE helped to interpret how evapotranspiration was modulated. The relationship of evapotranspiration (ET) in terms of LE against net radiation and vapour pressure deficit (VPD) indicated direct dependence of ET on available energy and evaporative demand represented as VPD. During dry winter period (October to March) LE was nearly constant (~50 Wm^{-2}) and was much influenced by Rn. During wet period (March to October) LE increased with net radiation which is the principal source of energy for daytime evaporation when occasional rainfall occurs throughout the period. Our study indicated contribution of other influential environmental variables and vegetation in regulating VPD in addition to Rn and substantiates the argument that VPD is not a good representation of water availability at least in arid ecosystems.

During monsoon period when soil water content was high and VPD was low with high Rn, this led to export of energy which became close to zero later in the post-monsoon before becoming positive with the onset of winter. Generally, it has been found that the residual energy was of positive order especially during continuous rain spells. During the rest of the period, it was either negative or near zero. In a given vegetative patch, the rain splash creates relatively fast changes in surface (soil and canopy) wetness resulting into differential vegetative growths as compared to rate of changes in the atmospheric temperature and humidity regime at a given vegetated location. The faster changes in the canopy will have relatively slower impact on the exchange processes within surface layer. This could bring quicker and differential responses of outgoing shortwave and longwave radiation components than canopy-atmosphere exchange of heat fluxes. Therefore, any quicker and repeated disturbances like rain splashing is expected to disrupt equilibrium as well as energy balance. Coherence existed between precipitation and residual energy (dE) with some lag or lead time. It may be noted that rainfall is zero or tends to become zero when residual energy (dE) equals or more than zero.

Conclusions

Arid grasslands are unique ecosystems where precipitation and vegetation dynamics play an important role in the evolution of surface energy fluxes including evapotranspiration. In this ecosystems energy transport to atmosphere occurred mainly by dry processes rather than through moisture transport. The diurnal and seasonal dependence of sensible heat flux variations related to diurnal cycle of mean vertical air temperature differences, wind speed and direction indicated the roles of convective turbulence, wind shear and advection in energy transport. Frequent occurrence of dust storms leads to energy imbalance in arid grassland ecosystem. Sensible heat fluxes increase with the increase in vertical air temperature difference and wind speed that causes frequent dust storm events during summer. Close association was found among net available energy, rainfall and dry period. The way forward are: i. close monitoring of dust storm events with vertical air temperature difference (VATD) and wind speed to evolve possible model for dust storm occurrence. ii.Inter-annual comparison of net available energy, rainfall and dry period from two years of extremities (*i.e* excess rainfall and drought years) to further understand the implications of energy fluxes on monsoon. iii. Installation of more number of Agromet stations in whole of Arid Western Rajasthan to understand the dynamics of energy fluxes in entire Thar desert region.

Acknowledgements

The micrometeorological observations used in the study are generated at Central Arid Zone Research Institute (CAZRI) under the project titled "Energy and Mass Exchange in Arid Grassland System" which forms a part of SAC's (ISRO) national programme "Energy and Mass Exchange in Vegetative Systems (EME-VS)". Author is thankful to the Director, Space Applications Centre (ISRO) for funding this project and Director, ICAR-CAZRI for providing facilities to carry out this work.

References

Aires, L.M., Pio, C.A. and Pereira, J.S. 2008. The effect of drought on energy and water vapor exchange above a Mediterranean C3/C4 grassland in Southern Portugal. *Agricultural and Forest Meteorology* 148 (4):565–579.

Beljaars, A.C.M. and Holstag, A.A.M. 1991. Flux parameterization over land surface for atmospheric models; *Journal of Applied Meteorology* 30: 327–341.

Bhattacharya, B. K., Gunjal, K., Nanda, M. and Panigrahy, S. 2009. Protocol Development of Energy-Water Flux Computation from AMS and Model Development. SAC/RESA/ARG/EMEVS/SR/02/2009, chapter 3:34-53.

Bhattacharya, B.K., Singh, Nilendu., Bera, N., Nanda, M. K., Bairagi, G. D., Raja, P. , Bal, S.K., Vellu Murugan., Kandpal, B. K. , Patel, B. H. Jain, A. and Parihar J. S. 2013. Canopy-scale Dynamics of Radiation and Energy Balance over Short Vegetative Systems. *Scientific report*. SAC/EPSA/ ABHG/IGBP/ EME-VS/SR/ 02/ 2013, 97 p.

Bowling, D.R., Bethers-Marchetti, S., Lunch, C.K., Grote, E.E. and Belnap, J. 2010. Carbon, water, and energy fluxes in a semiarid cold desert grassland during and following multiyear drought. *Journal of Geophysical Research* 115, G04026, doi:10.1029/2010JG001322.

Burba, G.G. and Verma, S.B. 2005. Seasonal and interannual variability in evapotranspiration of native tall grass prairie and cultivated wheat ecosystems. *Agricultural and Forest Meteorology* 135 (1-4):190–201.

Businger J. A., Wyngaard J. C., Izumi I. and Bradley, E. 1971. Flux-Profile Relationships in the Atmospheric Surface Layer. *Journal Atmospheric Science,* 28: 181–189.

Campbell, G. S. and Norman, J. M. 1998. Introduction to Environmental Biophysics (*publ.*) Springer Science + Business Media, Inc, NY, USA, 71p.

Chen, S., Chen, J., Lin, G., Zhang, W., Miao, H., Wei, L., Huang, J. and Han, X. 2009. Energy balance and partition in Inner Mongolia steppe ecosystems with different land use types. *Agricultural and Forest Meteorology* 149 (11):1800–1809.

De Bruin, H. A. R., Ronda, R. J., Van, De. and Weil, B. J. H., 2000. Approximate solution for the Obukhov length and surface fluxes in terms of bulk Richardson numbers. *Boundary Layer Meteorology* 95: 145–157.

Dyer, A. J. 1974. A review of flux-profile relationships. *Boundary Layer Meteorology* 7: 363–372.

Dhir, R. P. and Singhvi A. K. 2012. The Thar Desert and its antiquity, *Current Science* 102, (7): 1001-1008.

Goel, M. and Srivastava, H.N. 1990. Monsoon Trough Boundary Layer Experiment (MONTBLEX). *Bulletin American Meteorological Society* 71: 1594–1600.

Garratt, J.R. 1994. The atmospheric boundary layer. Cambridge University Press, U.K., 316 p.

Hao, Y., Wang, Y., Huang, X., Cui, X., Zhou, X., Wang, S., Niu, H. and Jiang, G. 2007. Seasonal and interannual variation in water vapour and energy exchange over a typical steppe in Inner Mongolia. China. *Agricultural and Forest Meteorology* 146: 57–69.

Hui, J. W. C., Cook, B. I., Ravi, S., D'Odorico, P. and Fuentes, J. D. 2008. Dust-rainfall feedbacks in West African Sahel. *Water Resources Research* 44: W05202, doi:10.1029/2008WR006885.

Jung, M., Reichstein, M. and Bondeau. A. 2009. Towards global empirical upscaling of FLUXNET eddy covariance observations: Validation of a model tree ensemble approach using a biosphere model, *Biogeosciences* 6: 2001–2013, doi:10.5194/bg-6-2001-2009.

Kumar, Praveen., Tarafdar, J.C., Painuli, D.K., Raina, P., Singh, M.P., Beniwal, R.K., Soni, M.L., Mahesh Kumar, Santra, P. and Shamsuddin. M. 2009. Variability in Arid Soil Characteristics. In *Trends in Arid Zone Research in India* (*Eds.*) Amal Kar, B.K. Garg, M.P. Singh and S. Kathju, Central Arid Zone Research Institute, Jodhpur, 78-112 p.

Lare, A.R. and Nicholson, S.E. 1994. Contrasting conditions of surface water balance in wet years and dry years as a possible land surface–atmosphere feedback mechanism in the West African Sahel. *Journal of Climate* 7(5): 653–668.

Law, B.E., Falge, E., Gu, L., Baldocchi, D.D., Bakwin, P., Berbigier, P., Davis, K., Dolman, A.J., Falk, M., Fuentes, J.D,, Goldstein, A., Granier, A., Grelle, A., Hollinger, D., Janssens, I.A., Jarvis, P., Jensen, N.O., Katul, G., Mahli, K., Matteucci, G., Meyers, T., Monson, R., Munger, W., Oechel, W., Olson, R., Pilegaard, K., Paw U, K.T., Thorgeirsson, H., Valentini, R., Verma, S., Vesala, T., Wilson, K. and Wofsy, S. 2002. Environmental controls over carbon dioxide and water vapour exchange of terrestrial vegetation. *Agricultural and Forest Meteorology* 113: 97–120.

Leuning, R., Gorsela, E. V., Massmanb, W. J. and Isaacc, P. R. 2012. Reflections on the surface energy imbalance problem. *Agricultural and Forest Meteorology* 156: 65–74.

Li, S.G., Eugster, W., Asanuma, J., Kotani, A., Davaa, G., Oyunbaatar, D. and Sugita, M. 2006. Energy partitioning and its biophysical controls above a grazing steppe in central Mongolia. *Agricultural and Forest Meteorology* 137 (1–2): 89–106.

Li, S.G., Asanuma, J., Kotani, A., Davaa, G. and Oyunbaatar, D. 2007. Evapotranspiration from a Mongolian steppe under grazing and its environmental constraints. *Journal of Hydrology* 333 (1): 133–143.

Monin, A.S. and Obukhov, A.M. 1954. Basic laws of turbulent mixing in the ground layer of the atmosphere. *Transactions Geophysics*, Inst. Akad., Nauk USSR 151: 163-167.

Pillai, J.S., Saxena, S. and Vernekar, K.G. 1998. Diurnal Variation of Meteorological Parameters in the Land Surface Interface. *Boundary Layer Meteorology* 89: 197–209.

Rach, O. A., Brauer, H., Wilkes and D., Sachse. 2014. Delayed hydrological response to Greenland cooling at the onset of the Younger Dryas in Western Europe. *Nature Geoscience.*, DOI: 10.1038/ngeo2053.

Raja, P. Bhattacharya, B. K., Nilendu Singh, Sinha, N.K., Singh, J.P., Pandey, C.B., Parihar, J.S. and Roy, M.M. 2013. Surface Energy Balance and its

Closure in Arid Grassland Ecosystem: A case study over Thar Desert. *Journal of Agrometeorology*, spl.issue, v.1:94-99.

Rosenberg, N.J., B.L. Blad, and S.B. Verma. 1983. Microclimate, the biological environment. John Wiley & Sons, New York.

Ryu, Y., Baldocchi, D.D., Ma, S. and Hehn, T. 2008. Interannual variability of evapotranspiration and energy exchange over an annual grassland in California, *Journal of Geophysical Research*, 113, D09104, doi: 10.1029/2007JD009263.

Schlesinger, W.H. , Reynolds, J. F., Cunningham, G. L., Huenneke, L.F., Jarrell, W.M., Virginia, R.A. and Whitford ,W.G. 1990. Biological Feedbacks in Global Desertification. *Science* 247(4946): 1043-1048; DOI: 10.1126/science.247.4946.1043.

Sethuraman, A., Akin, D. E. and Eriksson, K. E. 1998. Plant - cell - wall - degrading enzymes produced by the white - rot fungus *Ceriporiopsis subvermispora. Biotechnology and Applied Biochemistry* 27: 37–47.

Shyampura, R.L., Singh, S.K., Singh, R.S., Jain, B.L. and Gajbhiye, K.S. 2002. "Soil series of Rajasthan", (NBSS & LUP publication. 95, Nagpur), 364 p.

Thom, A. S. 1975. Momentum, mass and heat exchange of plant community. In *Vegetation and the atmosphere* (*ed*) Monteith, J.L. (*publ.*) Academic Press, 57-109 p.

Vernekar, A.D., Thapliyal, V., Kripalani, R.H., Singh, S.V. and Kirtman, B. 1993. Global structure of the Madden-Julian oscillations during two contrasting summer monsoon seasons over India. *Meteorology and Atmospheric Physics* 52: 37-47.

Vernekar, K. G., Sinha, S., Sadani, L. K., Sivaramakrishnan, S., Parasnis, S., Brij Mohan, S., Saxena, S., Dharmaraj, T., Patil, M. N., Pillai, J. S., Murthy, B. S., Debaje, S. B. and Bagavathsingh, A. 2003. An overview of the land surface processes experiment (LASPEX) over a semi-arid region of India, *Boundary Layer Meteorology* 106: 561–572.

Wever, L.A., Flanagan, L.B., Carlson, P.J. 2002. Seasonal and interannual variation in evapotranspiration, energy balance and surface conductance in a northern temperate grassland. *Agricultural and Forest Meteorology* 112 (1): 31–49.

Williams, I.N., Raymond, T., Pierrehumbert and Huber, M. 2009. Global warming, convective threshold and false thermostats. *Geophysical Research Letters* 36 (21) L21805, doi:10.1029/2009GL039849.

Xu, Q. and Qiu, C. J. 1997. A variational method for computing surface heat fluxes from ARM surface energy and radiation balance systems. *Journal of Applied Meteorology* 36:4-11.

❑❑❑

Chapter 24

Farmers Coping Strategies to Climate Change and Contingency Planning in Drylands of India

A.K. MISRA AND R.N. KUMAWAT
ICAR-Central Arid Zone Research Institute, Jodhpur-342 003, Rajasthan
E-mail: mishraak17@yahoo.com

Introduction

Drylands occupy 68 % of India's cultivated area and supports 40% of the human and 65% of the livestock population (Misra*et al.*, 2009). It produces nearly 44 % of the total food grains and will continue to play a critical role in Indian agriculture (Singh *et al.*, 2004, Anonymous, 2011). The farming systems in dryland are full of diversity and are characterised with a variety of crops and cropping systems, agroforestry and livestock production. The changes in the quantity of rainfall received and its distribution pattern leads to intermittent drought during the crop period resulting in crop failures leading to debts and migration. Most of the rain in these areas is received during the south-west monsoon and quantity and distribution of rainfall remain uncertain during the season. Occurrence of drought once in 3 to 5 years is a common phenomenon either due to a deficit in seasonal rainfall or from inadequate soil moisture availability during the prolonged dry spells between successive rainfall events (Narain and Kar, 2005). As per the projections by the Intergovernmental Panel on Climate Change (IPCC, 2007), many arid and semi-arid areas are particularly exposed to the impacts of climate change and are projected to suffer a decrease of water resources (high confidence). According to Rajasthan's State Action Plan on Climate Change (SAPCC) some of the projected climate risks in the state (RPCB, 2010) are:

1. Slight decrease in mean annual rainfall combined with increase in frequency and intensity of extreme rainfall events.

2. Increase in annual mean surface temperature of 2 to 4 ^{0}C, which may increase flash floods and droughts.

According to the farmers of Jodhpur, the weather has become even more unreliable in the past two-three decades. They witness more erratic and untimely rainfall as well as less rainy day. Farmers depend solely on rainfall for growing of arable crops during *kharif* season and are having little contingency measures at their disposal, and are struggling to adapt to the additional stresses. Sometimes, the only coping strategy they perceive is to temporarily abandon fields and migrate for employment in nearby cities. The situation in fragile and marginal regions represented by arid and semi-arid tropical areas in India are likely to have increased risks and vulnerabilities due to climate change (IPCC, 2007; The World Bank, 2008). People follow subsistence farming that is typically characterized by degraded soils, small farm sizes, and low agriculture outputs. Climate change threatens agricultural production through extremes of temperatures, changes in precipitation patterns and increased occurrences of droughts and floods. Climate change in Indian sub-continent is already impacting negatively on rain-fed agriculture and livestock production systems. Since drylands are more vulnerable to climate change, the hazards of higher temperature and erratic precipitation caused significant reduction in productivity of rain-fed crops through reduced soil moisture, higher evaporative load, higher oxidation of soil organic carbon, terminal stress and forced maturity, low nutrient use efficiency etc. Similarly climate change affected animal productivity through shortage of feed and forage, heat stress, change in oestrus cycle etc. and also through more widespread distribution of important human, livestock and crop diseases.

Furthermore, our focus in this article is on agriculture; broadly defend to cover all land based activities including annual and perennial cropping, livestock based mixed farming systems involving farm-forestry, and their resource base. Agriculture not only constitutes one of the largest areas exposed to climate changes but sustains a majority of the rural population in these areas. Furthermore, the agriculture-dependent people are likely to be worse off in the face of climate change, due to their present marginality and vulnerability to risks as well as the limited resources and capacities to withstand global warming led crisis (Jodha, 1996; Kates 1985). At the same time there is increasing concern that advocates the traditional systems with visible potential to address the above mentioned issues along with the problems of food scarcity and poverty caused not only by climate change but other emerging changes that are marginalizing agriculture (Green Peace International, 2009; Oxfam International, 2007). The specific features of drylands such as high degree of fragility, marginality, diversity and limited accessibility generate the circumstances that keep them poor and contribute to their low productivity, associated with the low attention of mainstream policy makers to their problems (Jodha, 2005). However, having lived with limited and low productivity options for generations, the communities in these regions have evolved their own short and long term coping strategies to deal with the imperatives of their largely nature generated specificities (Narain and Kar, 2005; Jodha *et al.,* 2012).

The primary focus of this paper is on farmers' coping strategies against climatic variability in dry regions of India. The farmers' perceptions and coping strategies are largely governed by village level variables and weather conditions. The paper is based on the synthesis of key observations from the framers of drylands and review of studies that have dealt with farmers' adaptations or adjustments against climatic variability and associated risks.

The common strategies that have been recommended by different institutions include natural resource management, crop production techniques, livestock interventions and institutional arrangements to offset the impact of climate change. The farmers of the drylands have been living since centuries under conditions of variable temperature and moisture stress and have evolved many techniques for crop and livestock production suited to their sustenance farming.Two-way adaptation processes have been evolved by the farmers through trial and error, over generations. These involve adapting production and resource use systems to the agro-biological-climatic conditions on the one hand and adapting the natural resource base to sustenance and development needs of the community on the other. Keeping the land fallow for a certain period, livestock farming, mixed farming, mixed cropping, agro-forestry, traditional water harvesting in ponds, covered tankas etc., provision for grain and fodder storage, and animal migration are some of the mechanisms through which desert dwellers used to mellow the effects of drought and at the same time ensure long-term sustainability of the resources. Some examples of the coping mechanisms and strategies used by smallholders/farming communities to enhance resiliency against climatic variability in arid zone are mentioned here.

Rainwater harvesting

Rainwater is the key input in dryland agriculture. The rain water management assumes vital importance in cutting down risks and stabilizing crop production in dry areas. When rains are received with an intensity far reaching infiltration rate, runoff is inevitable. It varies from 10 to 40% of total rainfall depending upon the topography. Of this, at least 30% can be harvested into water storage structures. For coping drought, desert dwellers have a wealth of traditional wisdom to store, which was developed through ingenuity and experience to save, conserve and use each drop of water. Drylands farmer have developed a diversity of local water harvesting and management regimes that still continue to survive (Agarwal and Narain, 1997). Western Rajasthan is known for its traditional water harvesting systems and techniques. They have a history of continuous practice for over 8,000 years and are still functional. In India alone, there are over 35 traditional rainwater harvesting systems. These systems have been pivotal to the emergence and diversification of food production. Many of these rainwater harvesting systems require scooping of the earth and putting up embankments or erecting elongated soil heaps along farm boundaries. The earliest examples of these rainwater harvesting systems include the Jabalpur bandh and bandhulia of Satna, virda of Gujarat, ahar-pynes of Bijar, eri of Tamil Nadu, and khadins of Rajasthan. Khadin, which is also called a dhora, is an indigenous construction designed to harvest surface runoff water for agricultural production. The main feature of this system is a very long (100-300 m) earthen embankment built across the lower hill slopes lying below gravelly uplands. Sluices and spillways allow excess water to drain off. The system is based on the principle of harvesting rainwater on farmland and subsequent use of this water saturated land for crop production. It was designed by the Paliwal Brahmins of Jaisalmer, western Rajasthan in the 15th century. In this system runoff water from large rocky catchment is stored against an embankment and crops are grown from ridge to valley on receding soil moisture. Khandin system is very valuable for recharging groundwater and wells in their zones of influence.

Similarly, structures such as Talai, Tal, Nada, Nadi and Talab are village ponds/tanks of different capacities, which are used for drinking water by animals, humans, and sometimes for supplemental irrigation of vegetables and horticultural crops. In the olden days many landlords and other influential people used to construct tanks at grand scale for drinking water as well as recreation purposes. The ponds are constructed in low- lying areas, and earthen embankment called 'Agore' is constructed out of the dugout soil, with escape arrangement for surplus water. Existence of a murram/hard rock layer at the bottom of the pond reduces seepage naturally; otherwise the seepage is reduced by depositing silt at the bottom. Besides being a source of water, these community structures are excellent sites for groundwater recharge. Their maintenance and de-silting operation before the monsoon is taken care of by the community.

Field bunding is most popular in the country. It consists of trapezoidal bunds to impound runoff water behind them so that it can gradually infiltrate into the soil for crop use. In Nagaur district of Rajasthan, field is divided into small check basins to collect rainwater both for reclamation of salt affected soils and also to conserve rain water in to the field. In central Rajasthan, some parts of the field is kept fallow during rainy season which is used for animal grazing up to August and then subjected for repeated ploughing and planking to store more water into the deeper layers of soil received in the post rainy season. Some of the water management techniques having wider adaptations in the dry regions are presented in Table 1.

Table 1. Selected water management measures and their impacts

Technical measures	Impact
Digging of feeder channels to collect and divert additional rainwater into the village pond	Year round availability of water in the pond that can be used for irrigation, to keep fish and to grow water chestnut
Construction of gully plugs to slow down water runoff and increase water up-take of the soil	Less soil erosion Formation of natural ponds behind the plugs, serving as additional drinking sources for cattle
Construction of bunds around fields to preserve water and topsoil	Increase of soil moisture and prevention of erosion
Deepening of wells for allocation of drinking water	Guarantees drinking water in time of drought

Source: Sustainet

Conservation of soil moisture

In Indian arid zone precipitation is often erratic, both within growing seasons and between years. The coefficient of variation of annual rainfall varies from 37 per cent to more than 59 per cent (Narain and Kar, 2005). The high inter-annual variability of rainfall is the major single factor influencing crop and pasture yields. *In-situ* rain water harvesting through inter-row water harvesting and paired row water harvesting are common techniques adopted by the farmers of Pali-Marwar region to store more rain water in the root zone for efficient utilization of rain water.

High evaporative loss coupled with water stress at critical stages lead to severe reduction in crop yields. Use of mulches could form a viable practice to minimise evaporative losses from the agricultural fields. The beneficial effects of different mulches in conservation of soil moisture have been widely reported. Many practices are in vogue in different parts of the dry region. Farmers carried out different soil works in the standing crops to remove weeds and to create soil/dust mulch. Soil mulch of surface 5-8 cm dry soil effectively reduces the evaporation losses by obstructing the rise of soil moisture through capillary action. The soil mulch also prevents deep cracks in soils (especially black soils) by reducing the direct action of atmosphere and hence evaporation is also reduced. The repeated inter-cultivations done in *rabi* crops even in the absence of weeds help in reducing evaporation losses. Among the different mulches soil mulch is the cheapest one.

In fruit orchards and vegetables, farmers of Gujarat are using straw (residues of castor, cotton stalk etc.) on soil surface for moisture conservation. Straw mulches reduce both the amount of energy absorbed by the soil and its upward movement and thus reduces evaporation. Farmers of the Shekhawati region of Rajasthan and Gujarat have started using plastic mulch in vegetable cultivation both for checking the growth of weeds and to check evaporative loss of water from the planted rows in which water is supplied through drip irrigation (Fig. 1). Similarly, mulching with pebbles in the rings of fruit trees and use of *murram*(a kind of product found in the hills of red colour) in groundnut cultivation is common in the Saurashtra region of Gujarat for soil moisture conservation.The pebbles placed in the basins of trees not only reduce evaporation but also facilitate infiltration of rain water into the basin.

In the desert districts, shelter belt plantation of the indigenous trees on the field boundaries is a common practice to minimize the abrasive action of sand laden winds, desiccating (evaporative) effect of dry winds and effect of cold waves. Generally, shelterbelts give protection from desiccating winds to the extent of 5 to 10 times their height on windward side and up to 30 times on leeward side. Due to reduction in wind speed, evaporation losses are reduced and more water is available for plants. The beneficial effects of shelterbelts are seen more clearly in drought years. In the Indira Gandhi Canal Command area, sissam (*Dalbergia sissoo*) is used commonly on the field boundaries. The farmers of the tube well cultivation mostly used bordi (*Ziziphus rotundifolia*) and Isaraelibabool (*Acacia tortilis*) for shelter belts in the Jaisalmer (Fig. 2). Similarly, the farmers of the khadin areas used gonda (*Cordia myxa*) and desibabool (*Acacia nilotica*) around field boundaries to minimize the evaporative loss of water both from the khadin and soil surface. In other parts of Rajasthan where rainfall is less than 300 mm, trees such as rohida (*Tecomella undullata*) jharberi (*Ziziphus nummularia*), pilu (*Salvedora oleoides*), kummut (*Acacia senegal*) etc. serve the purpose. In semi-arid parts of the Rajasthan where rainfall is greater than 350 mm, farmers keep khejri (*Prosopis cineraria)* in the field for making the ambient temperature congenial for crops.

Fig 1. Plastic mulch in tomato at farmers' field of Jhunjhunu, Rajasthan

Fig 2. Use of *Acacia tortilis* at field boundaries by the farmers of Jaisalmer, Rajasthan

Biranibadi-farming of Watermelons and Cucurbits

Traditional cultivation of watermelons and cucurbits in small areas with limited water during summer, particularly in Bikaner Division, is called 'Biranibadi' which is practiced by the 'Mali' community to evade the effects of drought and to generate income up to Rs. 10,000 ha. The practice consists of making ring basins at 1m x1 m on the northeast aspect and erecting wind breaks of kheemp in between two rows (Kavia *et al.,* 2004). The system helps moisture conservation in post- rainy season, and if rain comes during winter one harrowing and planking is done. At Shivaratri (end of February or early March) each basin is given 4 litres of water in the centre and two pre-soaked seeds are pressed in a 15-cm deep pit and covered first with wet soil and then with dry soil. After 2 months, 5 litres of water is again given. Crop becomes ready after 2.5-3.0 months.

Crop diversification and mixed cropping

Farmers in western Rajasthan are adapting and coping with climate variability in various ways. These mechanisms are manifested by the diversity of resource management and cropping systems, which are based on indigenous knowledge of management of the fragile and variable environment, local genotypes of food crops, intercropping, and agroforestry systems. These coping mechanisms not only help meet the farmers' subsistence needs but also encourage biodiversity conservation. To offset crop failure arising from rainfall variability and unpredictability, farmers cultivate several hardier (or drought-tolerant) types of the same crop species.

To avoid total crop failure during drought, most promising strategies practiced by the farmers of western Rajasthan is mixed cropping, in which up to 20 different crops are grown in temporal and spatial interaction on the same field. In general, the traditional rainfed cropping practice during *kharif* season in the region is to mix the seeds of pearl millet, moth bean, mung bean, clusterbean, sesame and kachri (a local vegetable crop) for sowing (Faroda, 2002). Arid zone farmer has also coping system through some cash crops like isabgol, cumin, spices and condiments, as well as medicinal herbs and shrubs. Some of these crops are fairly drought-tolerant and provide livelihood during adverse time. The different components of the system complement one another, conserving soil fertility and moisture, making a fallow period dispensable. The system is designed to meet diverse household requirements over time, as all the crops are not harvested at the same time. On the one hand, the system produces a wide variety of different crops meeting most of the basic food requirements of a household. On the other hand, this mixed cropping scheme is more resilient to detrimental weather influences as its components are mutually reinforcing and the multitude of crops is diffusing the risk of complete crop failure.

The multipurpose trees are generally grown on cropland. Many fodder trees, shrubs and grasses area traditionally grown as border rows or live hedges. These prevent wind erosion and provide fodder and feed for livestock and humans and fuel for cooking. Sowing strips of grasses around the fields and pearl millet in the centre is a traditional practice for controlling wind erosion and producing fodder for livestock.

Exploitation of underutilised plants

In many parts of the drylands, farmers' still gather wild plants on a larger areas for their subsistence in and around crop fields. In many agro pastoral societies, collection of edible leaves, berries, roots, tubers, fruits, etc. in the bush lands surrounding the villages constitutes an important strategy for diversification of the food base. During droughts or other times of environmental stress, many plants are gathered and consumed. Besides, many shrubs like *Lawasonia alba, Capparis decidua, Cassia angustifolia, Commiphora wightii* and a variety of medicinal and aromatic plants are utilized by the people for income generation through their fruits, leaves exudates, etc. these are source of survival and for coping drought. The fruits of *Cordia gahraf, Z. mauritiana, P. cineraria Salvedora,* and seeds of *Citrullus, C. biflorus* and roots of *Boerhavia diffusa* used as food by the desert dwellers during drought years in western Rajasthan.

Agroforestry

Traditional agroforestry systems provide unique basis of life support in dry region. During droughts when crops fail, the much-needed fodder, fuel, timber, food (fruits/seeds/pods) and income from minor forest products, like gum and medicinal products are provided by multipurpose perennial trees, bushes and weeds. Trees like *Prosopis cineraia, Techomella undul*ate and *Salvadora oleoides* have been the life of the desert.

The presence of trees in agroforestry designs stands out as a key strategy for mitigation of microclimate variability in smallholder farming systems. Farmers moderate the microclimate by retaining and planting trees, which reduce temperature, wind velocity, evaporation and direct exposure to sunlight and intercept hail and rain. Many farmers also apply mulches of ground-covering plants or straw to reduce radiation and heat levels on newly planted surfaces, inhibit moisture losses, and absorb the kinetic energy of falling rain and hail. Agroforestry systems contribute simultaneously to buffering farmers against climate variability and changing climates, and to reducing atmospheric loads of greenhouse gases because of their high potential for sequestering carbon. As intensive agriculture is now widely practiced the traditional systems are on decline. With tractorization the regeneration of native vegetation is also declining.

Home gardens

The home garden is representative of a household's needs and interests, providing food, fodder, firewood, market products, construction materials, medicines, and ornamental plants for the household and local community. Now new species have also been incorporated, including papaya (*Carica papaya*), guava (*Psidium spp.*), banana(*Musa spp.*), lemon (*Citrus limon*) and orange (*Citrus aurantium*). Under the shade of trees, a series of both indigenous and exotic species of herbs, shrubs, vines and epiphytes may be grown. Seeds from useful wild species brought into the garden by the wind or animals are often not weeded out and are subsequently integrated into the home garden system.

System of cultivation

Traditionally, farmers of the region were using indigenous country plough or a double bottom plough, called 'durfan' drawn by camel or bullock for shallow ploughing. A funnel used to be attached as a sowing device. Since seed was placed at a shallow depth, covered by a thin layer of soil, chances of encrustation due to post sowing rains were few. The system encouraged minimum tillage with energy that caused least damage to soil structure. Farmers in arid zone get only 2-3 days' time for sowing due to fast drying of land surface. With traditional system, sowing of small acreage was possible, leaving large chunks of fallow lands with grasses and bushes to serve as grazing ground for animals. Regeneration of trees like khejri (*Prosopis cineraria),* as well as of bushes and grasses was possible, serving excellent fodder in different agroforestry systems. Similarly, the competition of weeds with crops was controlled generally through hand weeding and using the biomass for animals, or leaving it on land as mulch for moisture conservation.

Traditional fodder storage system/Fodder bank

During good rainfall years, excess fodder of pearl millet, wheat, barely or any other crop is stored for adverse years. The cereals are stored in huge structures called 'Karai' or 'Pachava', which are made-up of crop stalks/branches, either in conical shape with a circular base, or as a rectangle at base and top like a hut shape. Chaffed stover is tightly packed in such structures. The grasses grown during rainy season are harvested and sun dried for some days and stored as hay for feeding livestock during lean periods in winter commonly practiced by the farmers.

Pasture/Grazing Lands

Pasture lands are community lands which occupy nearly 4% of geographical area. These used to sustain the animal population in the region and through that the humans. Orans and Gochars (sacred forests) are also the village pasture lands allotted to deity. Their management rests with a trust or the village community. Conventionally felling of trees is prohibited in Orans and Gochar, but is now encroached upon. These were excellent traditional coping system during droughts, and were also beautiful reserves of arid biodiversity.

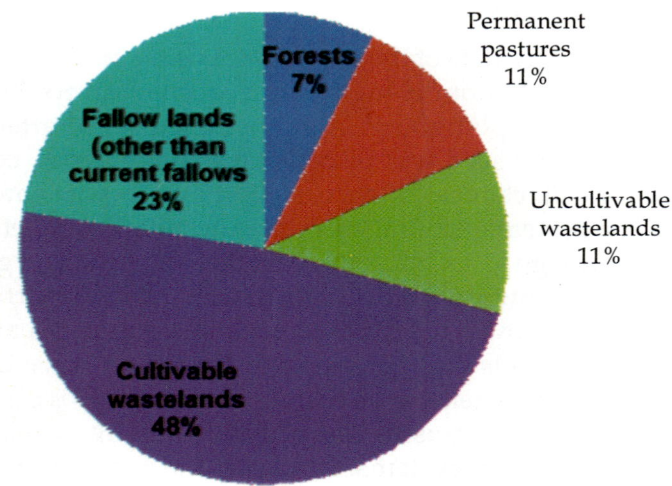

Fig. 3. Grazing resources of western Rajasthan (74,71014 ha)

Common pasture land tends to be a particularly important grazing resource during the rainy season. Since animals are excluded from farmers' fields during this season, many households (particularly those with little private grazing land of their own) depend almost entirely on Common property resources (CPRs) at this time of the year. As much as 30% of livestock forage in India as a whole comes from CPRs (both forests and non-forest CPRs).

CPRs are an important component of livelihood assets of communities in semi-arid areas of India and offer vital income and sustenance opportunities in the harsh agro-climatic conditions (Jodha, 1995). Before, independence, community pasture land management in Rajasthan was entrusted to '*thikanedars*' or '*caretakers*' who were appointed by the Princely States. These caretakers were responsible for ensuring the productivity of the pastures through soil and water conservation measures, controlled livestock rearing and preventing misuse and encroachment by the individuals (Misra, 2006). After independence, with the transfer of power from the Princely States to the State Government of Rajasthan, the management of common properties was handed over to government authorities at different levels. Today, common lands in Rajasthan fall under 3 categories: Village community pastures (*Charagah* or *Charnot*), which have been entrusted to the Gram Panchayats, revenue wastelands, which fall under the authority of the Revenue Department; and forests, which are under the authority of the Forest Department. The village community pastures are accessible to all groups and classes of the rural communities. Village commons in Rajasthan are the primary source of fodder for grazing animals, but also a source of fuel wood and many important non-timber forest products. In arid regions, common lands provide 66 % of the fodder requirement. Common lands are therefore a critical livelihood resource for landless, small and marginal farmers and there is priority need to protect and invest in both improving the quality of these lands as also enable community management and governance of these lands. The crop residues from the rainfed crops as well as from irrigated crops contributed 22% in the total fodder requirements in these regions. The remaining demand of fodder is met out from the purchased fodder brought from the neighbouring states such as Punjab and Haryana.

Traditional livestock production and feeding systems

The traditional livestock production systems were complex and based on tradition and socio-economic considerations mainly guided by available feed resources. In dry area, livestock and food production systems are closely integrated. Crops provide feed and fodder to the animals, while in return animals provide milk, meat and wool as a source of nutrition and cash income, supply draught power and manure for crop production. A close link exists between livestock and CPRs, as grazing is predominate feeding practice among the livestock keepers. Animals of the poor depend completely on CPRs during the monsoon season (6 months). This complex inter-relationship between CPRs, livestock and crops in rainfed farming systems has contributed to the sustainability of rainfed agriculture for generations. However, this relationship is now under increasing pressure from different sources (Misra, 2006).

According to farmers, straw, stovers and grasses makes the largest contribution towards livestock feed and comes largely from cereal crops. Preferential feeding was a common practice. Farmers tend to ration their home grown crop residues for optimum use: large ruminant receive priority for crop residues and milking animals in addition receive some supplementary feeding made up of a combination of home grown by-products like bran and purchased raw materials like oil cakes. Small ruminants particularly goats, and camel were given crop residues of legume crops and leaves of trees. Salt and mineral mixture are occasionally provided to animals. Animals eat only leafy portion and left dry stem which ultimately ends up as manure. This practice appears wasteful, but reflects the people's age old tradition. Farmers have their own ways to respond to the scarcity of fodder. There is very little cultivation of fodder crops in drylands on account of water scarcity even though farmers are aware of the importance of feeding green fodder and also the existing high yielding fodder varieties.

Livestock management

Due to low and erratic rainfall and frequent droughts, agriculture is not a dependable source of livelihood in western Rajasthan. Nearly 30.18 million livestock, including huge population of sheep and goat, provide sustainability in the region, and helps to cope with drought. As animals are the first casualty during drought, issues related to animal management are of paramount importance. Traditional coping system relied more on drought-hardy and well adapted breeds of animals like Tharparker, Rathi, Kankerj, Malvi, and Nagauri cows, Marwari, Magra, Chokla, Nali, Malpura and Sonadi sheep, and Marwari, Jakharana and Sirohi goats. People also relied on traditional system of medicines and management of native breeds. Unproductive animal are let loose or sent to Gaushalas as culling or slaughtering is a taboo.

Migration and Nomadism

In drylands of India, domestic animals are living assets contributing to household income, food security and health through dietary diversification. During period of scarcity of fodder, feed and water, migration of animals from resource- poor to resource-rich neighbouring states has been a well-established way of life and coping mechanism. The degree of mobility depends on flock/ herd size, location of the family or village, as well as on the amount of fodder produced in a given year. A study conducted by CAZRI (2014) revealed three kinds of livestock migration occurs.

1. *Local/temporary migration*: People with smaller flock (<50 head) size generally migrates for short periods. They move from their home villages to grazing areas in neighbouring villages within the district when the local pastures are exhausted. Migration takes place mainly between November and June.

2. *Semi-migration*: People with larger flock/herd sizes face difficulty in maintaining the animals near the native surroundings during the dry season (November to July/August). They typically migrate out of the district or state during the dry season, but return to their native area in the following monsoon season.

3. *Permanent migration*: In this kind of migration, large herds/flocks are permanently moving from the home location to other districts or states.

For long distance migration, about 10-20 persons join together and move simultaneously with their animals seeking safety and uniting against other management related problems and threats on the way. One experienced person within the group is appointed as leader who visits in advance different locations where animals are to be taken for grazing. Migration strategy of livestock owners revolves around availability of fodder and water resources in native region. In good rainfall years, they may delay their departure by 1-2 months. The departure of herds is linked with crop harvesting in other regions. During the migration animals graze on fallow harvested fields of farmers.The migration with livestock has definite routes to MP, UP, Haryana, Punjab and Gujarat. Migrating herds are allowed to camp in fields for small money to improve soil fertility. They return back to their native places with the onset of rainfall. In this system, they are facing difficulties due to shrinkage of common grazing land, and also dwindling in favour of sedentary way of life thus exerting huge pressure on limited resources of arid region.

The understandings generated by the above discussions indicate a number of potential approaches and options to facilitate coping strategies against climate change in the drylands of India. The primary issues, central to dryland agriculture, such as crop technologies, natural resource management and livestock production systems are focused. Over the centuries, generations of farmers and herders have developed complex, diverse and locally adapted agricultural systems, managed with time-tested, ingenious combinations of techniques and practices that lead to community food security and the conservation of natural resources and biodiversity. These farming systems are based on cultivation of a diversity of crops and varieties in time and space that have allowed traditional farmers to avert risks and maximize harvest security in uncertain and marginal environments, under low levels of technology and with limited environmental impact. The short and long term adaptation strategies should include:

1. Use of locally adapted varieties/species exhibiting more appropriate thermal time and vernalization requirements and/ or with increased resistance to heat shock and drought.
2. Enhancing organic content of soils through compost, green manures, cover crops, etc., thus increasing water holding capacity.
3. Wider use of local knowledge and practical means to "harvest" water and conserve soil moisture (e.g., crop residue retention and mulching), and more effective use of irrigation water.
4. Managing water to prevent waterlogging, erosion, and nutrient leaching where rainfall increases; Use of crop diversification strategies (intercropping, agroforestry, crop-sequencing, etc.) and integration with other farming activities such as livestock raising.
5. Preventing pest, disease, and weed infestations through management practices that enhance biological and other natural regulation mechanisms (antagonisms, allelopathy, etc.), and development and use of varieties and species resistant to pests and diseases; and
6. Using climate forecasting to reduce production risk.

References

Agarwal, A. and Narain, S.1997. *Making Water Management Everybody's Business: Water Harvesting and Rural Development in India.*No. 87 London, International Institute for Environment and Development.Gatekeeper Series.http://www.iied.org/NR/agbioliv/gatekeepers/documents/GK87.pdf

Anonymous. 2011. *Report of the XII Plan Working Group on Natural Resource Management and Rainfed Farming*. Planning Commission, Government of India. New Delhi.

CAZRI. 2014. CAZRI *Annual Report 2014*.CAZRI Jodhpur.

Faroda, A S. 2002. Some considerations for drought management in Indian arid zone. In: *Drought Management in Indian arid zone*.(Eds: P Naraian, D C Joshi, S Kathju and A kar), CAZRI, Jodhpur.

Green Peace International. 2009. *Agriculture at cross roads - Food for survival*. Amsterdam, The Netherlands.

IPCC. 2007. Summary for Policymakers. In *Climate Change 2007*: The Physical Science Basis. Contribution of Working Group I to the Fourth Assessment Report of the Intergovernmental Panel on Climate Change (Solomon S, Qin D, Manning M, Chen Z, Marquis M, Averyt KB, Tignor M and Miller HL, eds.). Cambridge, United Kingdom and New York, NY, USA: Cambridge University Press.

Jodha, N.S., Singh, N.P. and Bantilan, M.C.S. 2012. Enhancing Farmers' Adaptation to Climate Change in Arid and Semi-Arid Agriculture of India: Evidences from Indigenous practices: *Developing International Public Goods from Development-oriented Projects*. Working Paper Series no. 32.Patancheru 502 324, Andhra Pradesh, India: International Crops Research Institute for the Semi-Arid Tropics. 28 pp.

Jodha, N.S. 1996. Enhancing Food Security in More Crowded and Warmer World: Factor and Processes in Fragile Zones. In *Climate Change and World Food Security* (Eds. Downing TE, ed.).Berlin: Springer.

Jodha, N.S. 2005. Changing Ecosystem-Social System Links and Future of Drylands in India. *Annals of Arid Zone* 44 (3&4):387-402

Kates, R.W. 1985. Interaction of climate and society.In *Climate impact assessment* (Eds. Kates R.W., Ausubel, J.H. and Berberian, M.), New York: John Wiley & Sons pp 3-36.

Misra , A. K. 2006. Integration of livestock in land use diversification. In: *Land Use Diversification for sustainable rainfed areas agriculture*. (Eds Sharma K.D. and Soni B.), Atlantic Publishers and Distributor's, New Delhi pp 277-299.

Misra, A. K., Rama Rao, C.A., Saubramanyam, K.V. and Ramakrishna, Y.S. 2009.Improving dairy production in India, rainfed agroecosystem: constraints and strategies.*Outlook on Agricuture* 38:284-292.

Narain, P. and Kar, A. 2005. *Drought in Western Rajasthan: impact, coping mechanism and management strategies*. CAZRI, Jodhpur.

Oxfam International. 2007. Adapting to Climate Change: What is needed in poor countries and who should pay. *Oxfam Briefing Paper* 104. Oxford: Oxfam International.

RPCB (Rajasthan Pollution Control Board). 2010. *Climate Change Impacts, Mitigation and Adaptation Science for Generating Policy Options in Rajasthan*, Government of Rajasthan

Singh,H. P., Sharma, K. D., Reddy, G. S. and Sharma, K. L. 2004. Dryland agriculture in India. In *Challenges and strategies for dryland agriculture*.CSSA Special Publication No. 32. Madison, USA pp. 67-92

Sustainet. Adaptation of small-scale farmers to climatic risks in different states of India.(www.sustainet.org)

The World Bank. 2008. Environment Matters at the World Bank: Climate Change and adaptations. Washington DC: The World Bank.

□□□

Chapter 25

Conservation Agriculture - An Opportunity to Reduce Greenhouse Gases

D.K. PAINULI

ICAR - Central Arid Zone Research Institute, Jodhpur-342 003, Rajasthan
E-mail: painulidk@gmail.com

Introduction

Climate in a narrow sense is usually defined as the "average weather" or more rigorously, as the "statistical description in terms of the mean and variability of relevant quantities over a period of time ranging from months to thousands or millions of years". The **classical period**, as defined by the World Meteorological Organization (WMO), is 30 years. These quantities are most often surface variables such as temperature, precipitation, and wind. Thus, climate in a wider sense is the state, including a statistical description, of the climate system. **Climate change** on the other hand refers to "a statistically significant variation in either the mean state of the climate or in its variability, persisting for an extended period (typically decades or longer)". Thus, events that used to be rare occur more frequently. For example, summer time maximum air temperatures increasingly break records each year, or vice-versa duration and thickness of seasonal lake ice is decreasing with time. Occasionally, an event or sequence of events occurs that has never been witnessed before or recorded before, such as the exceptional hurricane season in the Atlantic in 2005. If such a season does not recur within the next 30 years, we would call it an exceptional year and not a precursor of change. Therefore, only a persistent series of unusual events taken in the context of regional climate parameters can suggest that a potential change in climate behaviour has occurred. The persistent series of extreme weather events of heat waves, droughts, fires, hurricanes, and rains that are happening world over convince scientists to agree that climate change is not a myth but reality.

India has in recent years experienced "extreme weather events" in regular intervals like sudden and extreme rainfall in Mumbai in 2005 causing flood, cloudburst in Leh in 2010, Uttarakhand tragedy in 2013, Jammu and Kashmir flood in 2014, Chennai rains in 2015 etc.

Reasons for Concern about Climate Change

The Intergovernmental Panel on Climate Change (IPCC) warned in 2007 that if global temperatures are allowed to increase at current rate then up to 30% of all species on Earth will be at increased risk of extinction; hundreds of millions of people will be exposed to greater water scarcity; there will be a significant increase in wild fire risk; most of the world's coral reefs will be bleached; there will be increased mortality from floods, heat waves and droughts; and there will be an increasing burden of malnutrition and infectious disease in many parts of the world. Similar, grave concerns have been expressed in 2015 Paris Climate Conference (COP 21).

Causes of Climate Change

Solar radiation at the frequencies of visible light largely passes through the atmosphere to warm the planetary (Earth) surface, which emits this energy at the lower frequencies of infrared thermal radiation. Before all this emitted infrared thermal radiation could escape to outer space part is absorbed by a range of anthropogenic gases present in the earth's atmosphere. Principal among them are methane (CH_4), carbon dioxide (CO_2) and nitrous oxide (N_2O). These gases in turn re-radiate back much of the energy which makes the atmosphere warmer. As the atmosphere gets warmer, it makes the Earth's surface warmer too. Due to this effect on temperature, experienced in "Green/ Glass House" also, these gases are known as "Green House Gases" (GHGs). Re-emission of heat by the GHGs performs a useful function of maintaining the earth's temperature as otherwise the heat energy would escape leaving the earth cold and unfit to support life.

The amounts of GHGs in atmosphere were fairly stable prior to industrial revolution but have since grown significantly by 41.2%, 153.0% and 20.4% for CO_2, CH_4 and N_2O, respectively (Table 1). This has lead to corresponding increase in radiative forcing, in said order for these gases, by 1.88, 0.49 and 0.17 W m^{-2}resulting in global warming or the rise in global temperature compared to that prevailing prior to industrial revolution. Global temperature rose by 0.6 ^{0}C during the 20th century (IPCC, 2001) and is projected to increase by 2 to 6 ^{0}C during the 21st century (IPCC, 2007). Rise in global temperature or global warming is an acknowledged factor responsible for climate change. Therefore, COP 21 has resolved to keep global temperature rise this century well below 2 ^{0}C and even further to 1.5 ^{0}C above pre-industrial levels by adopting various approaches of limiting emissions of GHGs. Scientists believe that 1.5 ^{0}C limit is a significantly safer defense line against the worst impacts of a changing climate.

Table 1. Pre-industrial revolution and recent tropospheric concentration of major greenhouse gases (adapted from Blasing, 2014)

Greenhouse gases	Pre-1750 tropospheric concentration[1]	Recent tropospheric concentration[2]
Carbon dioxide (ppm)	280	395.4[3]
Methane (ppb)	722	1893[5]/1762[4]
Nitrous oxide (ppb)	270	326[4]/324[4]

1. *Chapter 8.3.2 (IPCC, 2013).*
2. *Averages over a specific 12-month period for all gasesbecause atmospheric concentrations of gases tend to vary systematically over the course of a year.*
3. *National Oceanic and Atmospheric Administration Earth System Research Laboratory website.*
4. *The first value in a cell represents Mace Head, Ireland, a mid-latitude Northern-Hemisphere site, and the second value represents Cape Grim, Tasmania, a mid-latitude Southern-Hemisphere site. "Current" values given for these gases are annual arithmetic averages based on monthly background concentrations for year 2012. Source: Advanced Global Atmospheric Gases Experiment (AGAGE).*

Share of Agriculture in GHG Emissions

In 2010, CO_2 contributed 76% of global GHG emissions, CH_4 about 16% and N_2O about 6%. The contribution of agriculture was about 11% of global GHG emissions (7% CH_4 and 4% N_2O). On individual gas basis share of agriculture was 44% and 67% of global CH_4 and N_2O emissions, respectively (IEA, 2012). FAO estimates that agricultural emissions from crop and livestock production grew from 4.7 billion tonnes of carbon dioxide equivalents (CO_2 eq) in 2001 to over 5.3

billion tonnes in 2011, a 14% increase. The increase occurred mainly in developing countries, due to an expansion of total agricultural outputs. It is estimated that world would require at least 50% more food by 2050 than it does today and a "business as usual" approach to boosting global food production would have substantial negative consequences for climate change (McArthur, 2015). Therefore, there is a global consciousness to promote "climate-smart agriculture" (CSA) - meaning agriculture that uses inputs as efficiently as possible, is resilient to climate change and significantly decreases greenhouse gas emissions, while still meeting the demand for food. Literature suggests that despite large annual exchanges of CO_2 between the atmosphere and agricultural lands the net flux is approximately balanced-with CO_2 emissions around 0.04 Gt $CO_2\,yr^{-1}$ only. Thus, on the mitigation side agriculture's main GHGs are N_2O and CH_4 rather than CO_2. The Global Warming Potential (GWP), which provides a simple measure of the radiative effects of emissions integrated over a specified time horizon relative to an equal mass of CO_2 emissions, considered over 100-yr time horizon is 28 times and 265 times more for CH_4 and N_2O respectively, compared to CO_2. Thus, even a small change in the amount of N_2O and CH_4 will result in larger change in temperature.

Conservation Agriculture

Food and Agriculture Organization of the United Nations (FAO) defines conservation agriculture (CA), perhaps the most generic definition, as follows:

"CA is a concept for resource-saving agricultural crop production that strives to achieve acceptable profits together with high and sustained production levels while concurrently conserving the environment. CA is based on enhancing natural biological processes above and below the ground. Interventions such as mechanical soil tillage are reduced to an absolute minimum, and the use of external inputs such as agrochemicals and nutrients of mineral or organic origin are applied at an optimum level and in a way and quantity that does not interfere with, or disrupt, the biological processes. CA is characterized by three principles which are linked to each other, namely: (1) continuous minimum mechanical soil disturbance; (2) permanent organic soil cover; and (3) diversified crop rotations in the case of annual crops or plant associations in case of perennial crops."

FAO has also provided the following quantifying parameters to define CA -

1. *Minimal soil disturbance:* zero tillage is ideal, but the system may involve controlled tillage in which the disturbed area must be less than 15 cm wide or no more than 25% of the cropped area must be disturbed. Also, no periodic tillage disturbs an area greater than the aforementioned limits.
2. *Soil cover:* retention of crop residues or other organic soil surface cover on no less than 30 % of the area - if possible continuously throughout the year (the ideal level of soil cover is site-specific).
3. *Crop rotation:* the rotation should involve at least three different crops (where farmers do not have enough land to rotate crops, intercropping can be used).

Indeed, not disturbing the soil has a lot of positive outcomes: soil life can develop in a stable habitat in quantity and quality better than on tilled soils, the structural integrity of the soil is maintained so continuous vertical macro-pores are not destroyed and remain as drainage channels for rainwater into the soil, the weed seed bank in the soil does not receive the stimulation for germination.

A permanent organic soil cover ensures the protection of the soil surface against wind, rain, sun and from faster drying out. It provides a regular supply of organic matter also, which is a key feature for soil fertility. In addition, the mulch suppresses germination of the weeds, provides habitat for beneficial fauna, and feed for the soil organisms.

And a well designed crop rotation allows sustainable management of the system, because it aids in the suppression of weed and pest problems. Different crop species with different root systems explore different soil horizons and hence increase the efficiency of the use of soil nutrients. Legumes are recommended as rotational crops for their nitrogen-fixing functions.

The combination of minimum tillage, soil surface cover and crop rotation - the three essential components of the conservation agriculture - with their synergistic effects leads to a sustainable, resource-saving agriculture that at the same time is productive and profitable.

Impact of Conservation Agriculture on GHGs

Conservation agriculture can, as discussed in the following sections, lower GHGs in the atmosphere through their reduced production as well as sequestration of carbon.

a) Reduced emission

i) Reduced fossil fuel consumption

Conventional tillage based agriculture has many variants and the same applies to conservation agriculture. Tullberg (2007) compared fossil fuel/energy use and GHG emissions under Conventional tillage (CA), Reduced/ zero tillage (CA1) and Permanent bed controlled traffic minimum tillage (CA2) systems in Australia. Permanent wheel lanes are an essential component of CA2. For fuel consumption he took into account various pertinent points/steps involved such as energy value of the herbicide's constituents, energy value of the herbicide manufacturing/distribution process, overhead energy used in equipment manufacture and maintenance, etc. To illustrate consideration of the pertinent points: i) in CA2 system wheels operate on permanent compacted traffic lanes resulting in greater tractive efficiency and therefore less energy requirement in all the operations, and ii) because production of a given amount of food and fibre in CA1 and CA2 systems entails the use of less equipment, and that equipment is used for less hours per hectare, a reduction of at least the same magnitude might be expected in the energy requirements of equipment manufacture and maintenance. The data (Table 2) demonstrates that conservation agriculture could reduce sum of energy on field operation by machinery (Fuel) and herbicide by 26% and 67% for CA1 and CA2 respectively, compared to CA.

Table 2. Herbicide, machinery (Fuel) and total energy requirements for three tillage systems (adapted from Tullberg, 2007)

Representative tillage system	Energy (MJ ha^{-1})			Energy saving % CA
	Herbicide	Machinery	Total	
CA: Conventional till, no herbicide	0	1941	1941	/
CA1: Reduced/ zero till, < 1 tillage/crop	320	1116	1436	26
CA2:Permanent bed, minimum till	240	397	637	67

The net energy value of most petroleum fuels is in the range 40-45 MJ L^{-1} and the CO_2 and other GHG emissions resulting from the combustion of petroleum fuels is approximately 2.75 kg CO_2 per litre of fuel. From these two values and the energy used in the three tillage systems (Table 2) mean fossil impact on emission of GHGs was estimated. It clearly showed advantage of conservation agriculture in reducing the emissions of GHGs (Fig. 1). Emission of GHGs was 133.0, 98.7, and 43.8 kg CO_2 eq per crop in conventional tillage, reduced/zero tillage and permanent bed minimum tillage systems, respectively. Erenstein and Laxmi (2008) calculated that adoption of zero tillage for land preparation and crop establishment in rice-wheat system of the Indo-Gangetic Plains (IGP) could on an average save 36 L diesel ha^{-1}, translated in equivalent to a reduction in 93 kg CO_2 emission ha^{-1} yr^{-1}.

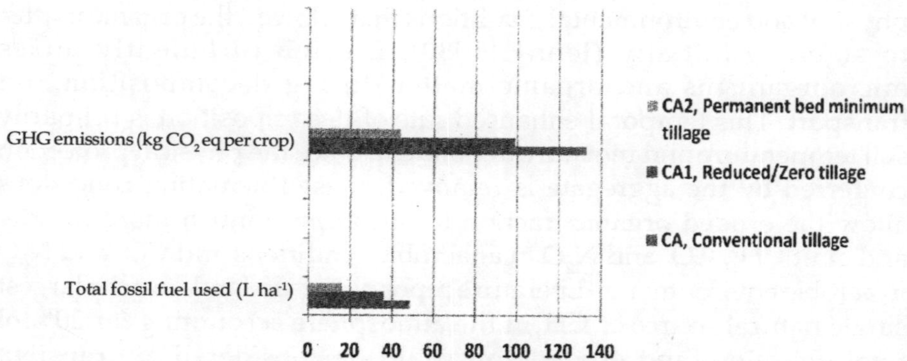

Fig. 1. Impact of fossil fuel consumed in three tillage systems on the emissions of GHGs (adapted fromTullberg, 2007)

In conservation agriculture diversification of species grown in rotation reduces the requirements for pesticides and other agrichemicals, and therefore there is less fossil fuel consumption required for their manufacture, transport, etc. Based on literature review Jensen *et al.* (2012) concluded that the ability of the legumes to fix their own nitrogen via a symbiosis with rhizobia bacteria reduces emissions of fossil energy-derived CO_2 and results in lower N_2O fluxes compared to cropping system that are fertilized with industrial nitrogen.

ii)Reduced soil erosion

Jenkinson and Rayner (1977) defined soil organic matter (SOM) by function in five categories namely decomposable plant material, resistant plant material, soil microbial biomass, chemically stabilized organic matter and physically stabilized organic matter. The last one, physically stabilized organic matter, is associated with sand, silt and clay, and in interior micro pores it is inaccessible to microbes. This portion of SOM is the primary reservoir of carbon (C) and nitrogen (N). It gets released through the mechanical action of water erosion and is subsequently exposed to environmental conditions favourable to microbial decomposition. Thus, stable SOM (humus) which can have a mean residence time of centuries becomes a "liable" source of GHGs through a series of biochemical transformations initiated by the physical process of erosion. Erosion enhances SOM decomposition at two locations: the eroded surface of the land and the eroded "in transit" soil. Erosion creates a new pool of mineralizable organic matter also that is different from the stable organic matter remaining. This soil organic fraction, that is carried away, is no longer under the same physical and environmental conditions that allowed the organic matter to stabilize initially (Jenny, 1980). Erosion turbulently mixes microorganisms and organic matter during decomposition and transport. This temporal enhanced rate of decomposition is primarily soil temperature and moisture dependent. Once the physical protection conferred by the aggregate is removed, these fluctuating conditions allow the eroded organic fraction to decompose much more rapidly and emit CH_4, CO_2 and N_2O in anaerobic conditions and CO_2 and N_2O in aerobic environments. Literature reports that wetlands are the largest single natural source of CH_4 in the atmosphere accounting for 20% of total emission, and eroded soil could be considered a "transient wetland" with its similar nutrient dynamics.

As discussed below in the forthcoming section iv) the aerobic nitrification process is a source of atmospheric N_2O. After organic nitrogen is metabolized to ammonium (NH_4^+), but before chemoautotrophic bacteria convert it to nitrate (NO_3^-), a fraction of the nitrogen is "leaked" as N_2O. The anaerobic denitrification process can also release N_2O when NO_3^- is anaerobically converted to nitrogen gas per dinitrogen (N_2). Both aerobic and anaerobic reactions can occur simultaneously in unsaturated soils. Aerobic reactions predominate in large well-aerated pores while anaerobic activities in the smaller pores that can be found inside the aggregates. Anaerobic reactions, such as methanogenesis and denitrification, increase with soil moisture content.

Tillage loosens soil and renders it vulnerable to erosion, while cover on soil surface protects surface soil against the detaching and eroding actions of rain (Fig. 2). Therefore, conservation agriculture will reduce both soil erosion and consequent, as is discussed above, emissions of the GHGs.

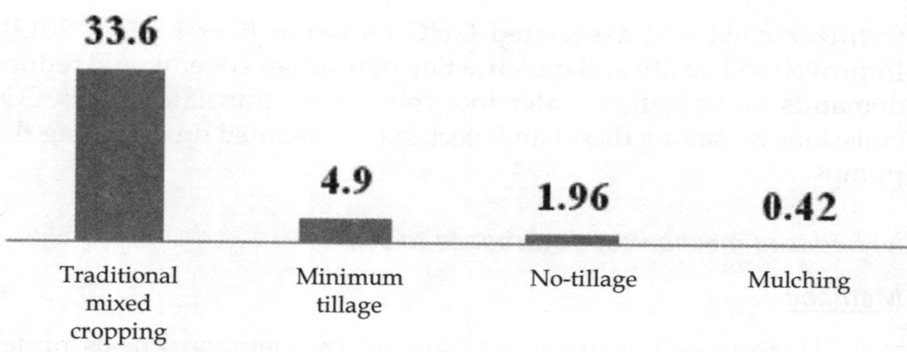

Fig. 2. Effects of tillage systems on soil loss at Kwadaso in Ghana, soil loss from bare fallow was 313 Mg $ha^{-1} yr^{-1}$ (adapted from Mensah-Bonsu and Obeng, 1979).

iii) No residue burning

In order to dispose of residues quickly, farmers prefer labour-saving residue burning method over residue incorporation or removal from the field. Residues burning however produce various GHGs including CO_2, CH_4 and N_2O. Emission of CO_2 is however considered neutral, as it is reabsorbed during the next growing season. Residues burning of five major crops; namely sugarcane, rice, wheat, maize and cotton in India during 2010-11 produced 301.60 Gg carbon equivalents CH_4 and 77.85 Gg carbon equivalents N_2O. The emissions attributed to these crops were 141.60, 114.81, 84.74, 22.24 and 16.07 Gg carbon equivalent respectively, with share of CH_4 and N_2O as reported in Table 3 (Lenka *et al.*, 2014).

Table 3. Effect of burning of different crop residues on methane and nitrous oxide emissions in India for the year 2010–11 (adapted from Lenka et al., 2014)

Crop	CH_4(Gg carbon equivalent)	N_2O (Gg carbon equivalent)
Sugarcane	115.12	26.48
Rice	82.37	32.44
Wheat	73.32	11.42
Maize	16.74	5.50
Cotton	14.06	2.01

Conservation agriculture, which advocates retaining residues in the fields, will not generate GHGs on this account. Further, retention of residues will improve soil health by adding to soil the carbon fixed in biomass thereby resulting over the time in reduced amount of fertilizer need and associated GHG emission (Corsi *et al.*, 2012). Improved soil health and mulch action of residues cover would reduce demands on irrigation water too; this would translate in less CO_2 emissions by saving diesel and electricity consumed on operating the pumps.

iv) Reduced mechanical disturbance to soil

Methane

CH_4 from soil is produced (source) by methanogenesis under anaerobic conditions and the major biological sink of atmospheric CH_4 is oxidation by methanotrophic bacteria in aerobic soils. Therefore, soils can be net source or sink for CH_4 depending on the land use and management practices including the complex interactions with the climatic parameters. Because tillage affects soil properties (Hati, 2014) therefore it may influence GHG emissions too.

Flux measurements from different ecosystems have indicated that CH_4 uptake is generally higher in forest soils than in regularly cultivated arable lands suggesting that disturbances associated with agriculture damage CH_4 oxidation sites thereby reducing CH_4 oxidation capacity of soils. Bronson and Mosier (1993) reported a 90% decline in CH_4 oxidation capacity of tilled and irrigated soils under wheat (*Triticum aestivum*) and corn (*Zea mays*) relative to those under native grassland. Similarly, Ball *et al.* (1999) reported higher CH_4 oxidation rate from soil under no-till relative to under moldboard tillage. In 43 years of continuous no-till practiced in United States Corn Belt region on an Alfisol (Crosby silt loam), Ussiri *et al.* (2009) reported that the global warming potential of measured CH_4 emissions were 84.6, 69.6 and -9.8 kg CO_2 equivalents ha^{-1} yr^{-1} for moldboard, chisel and no-till practices,respectively. Thus, suggesting that adoption of no-till in long term may contribute in mitigating global warming through reduced CH_4 emission.

The increase in CH_4 uptake under no-till could be attributed to greater pore continuity resulting in both the enhanced gas diffusivity and increased rate of atmospheric CH_4 supply, and presence of ecological niches for methanotrophic bacteria (Hutsch, 1998; Smith *et al.*, 2003) including higher soil organic carbon content.

Nitrous oxide

N_2O emission from the soil is governed mainly by denitrification (Eq 1) and nitrification (Eq 2) mechanisms.

NO_3^- <u>Nitrate reductase</u> → NO_2 <u>Nitrite reductase</u> → NO <u>Nitric oxide reductase</u> → N_2O <u>Nitrous oxide reductase</u> → N_2 (Eq 1)

NO NO

↑/ NO \ ↑

NH_4^- → NH_2OH → [HNO]NO_2 → NO_3^- (Eq 2)

\ [X] / ↓

N_2O

{NB: Eq 2 after Paul and Clark (1996) involves incompletely known processes and intermediates.}

Denitrification occurs when nitrate is present in anaerobic microsites developed wherever microbial demand for O_2 exceeds diffusion mediated O_2 supply. This may occur where O_2 diffusion is impeded by water, either at the centers of soil aggregates or in water saturated regions, or wherever local O_2 demand is exceptionally high. Denitrification in soils also consumes N_2O through the reduction of N_2O to N_2. Hence, this bacterial process may serve both as a source and sink of N_2O. Nitrification also contributes to N_2O emission during oxidation of $NH4^+$ or NH_2OH (Hydroxylamine) to NO_3^-. The relative importance of either of these two processes in N_2O emission that can occur simultaneously in the range of soil water contents normally encountered in agricultural soils is difficult to assess and is likely to vary appreciably with various factors affecting soil conditions viz. soil water content, aeration status, type of N fertilizer, organic matter content, climate etc. Also, the primary source of N_2O emission is often uncertain. It is generally assumed that a majority of N_2O production occurs in proximity to the soil surface but some reports suggest significant production in sub-surface.

Researches on the effect of tillage on N_2O emission have generated mixed results. Some studies have reported higher N_2O emission from no-tillage (Aulakh *et al.*, 1982; Linn and Doran, 1984; Palma *et al.*, 1997; Ball *et al.*, 1999; Baggs *et al.*, 2003) while others have reported lower N_2O emission from no-tillage (Kessavalou *et al.*, 1998a; Robertson *et al.*, 2000) and still others have reported no difference among tillage practices (Choudhary *et al.*, 2002; Elmi *et al.*, 2003). Six *et al.* (2004)

compiled and analyzed available data of soil derived GHG emissions from conventional tillage and no-tillage systems. They concluded that at least for the first decade or so switching from conventional to no-till system increased N_2O emissions by 2.5 ± 0.5 kg N ha^{-1} $yr^{-1}$for humid environments and 0.8 ± 1.0 kg N ha^{-1} yr^{-1} for dry environments, and afterwards emissions from no-tillage were lower than those from conventional tillage. Nitrogen fertilization and cultivation can both decrease CH_4 uptake and increase N_2O production, thereby contributing to the increasing atmospheric concentrations of these gases (Mosier *et al.*, 1991).

Tillage alters soil structure and thus physical, chemical and biological environments including retention and transmission behavior of water and air. Rudaz *et al.* (1999) observed largest rates of N_2O and NO_x emissions when water filled pore space (WFPS) was above 70-80% but tended to decline when WFPS approached 100%. Ball *et al.* (1999) observed that reduced gas diffusivity and air-filled porosity caused by heavy rainfall resulted in periods of high N_2O fluxes. As explained above, the N_2O is emitted by nitrification in aerobic soils and denitrification in anaerobic soils. Nitrification occurs under aerobic conditions and in presence of NH_4^+, while denitrification occurs under anaerobic conditions and in presence of simple C compounds. Stimulation of denitrification by dry-wet cycles is primarily due to reduced soil O_2 concentration for aerobic respiration rather than from more availability of substrates for denitrifying bacteria. Tilling soil disturbs structure and releases more substrate for decomposition. In drier soils where production of N_2O through denitrification is limited, the increased decomposition of organic matter stimulated by tillage may result into more nitrification and elevated N_2O emission compared to no-tillage.

The mechanism proposed to explain higher N_2O fluxes from no-tillage includes: i) litter carbon accumulation at the surface which causes increase in pool of substrate nitrogen in surface layer (Baggs *et al.*, 2003), ii) decreased O_2 availability caused by increased bulk density and water filled pore space (Ball *et al.*, 1999) and iii) decreased crop yields in recently established no-till soil which reduces N uptake resulting in increased residual N available for nitrification/ denitrification (Six *et al.*, 2004). However, lower bulk density and greater soil aggregation and mean weight diameter of soil aggregates observed under no-tilled in a long term study of 43 years continuous corn (Ussiri *et al.*, 2009) suggested that O_2 diffusion rates were higher under no-tilled than under tilled soils. In the reported study of Ussiri *et al.* (2009) overall annual N_2O flux was significantly more from chisel tilled (1.96 kg N_2O-N ha^{-1} yr^{-1}) and moldboard tilled (1.82 kg N_2O-N ha^{-1} yr^{-1}) soils than no-tilled (0.94 kg N_2O-N ha^{-1} yr^{-1}) soil. The N_2O emitted were equivalent to 1690, 1825 and 875 kg CO_2 equivalents ha^{-1} yr^{-1} for chisel tilled, moldboard tilled and no-tilled, respectively after 43 years of tillage and continuous corn practice on Crosby silt-loam. Net N_2O emission as well as GWP from no-till were therefore 48 and 52% lower than those from moldboard till and chisel till, respectively. Grant *et al.* (2004) predicted for no-tillage about 33% reduction in N_2O emission for western Canada due to less decomposition of soil organic matter, but a slight increase for eastern Canada mainly due to higher soil moisture resulting from higher precipitation,which increases denitrification rates. Their work has therefore also exhibited the influence of climate on tillage moderated N_2O emission.

Carbon dioxide

Emission from soil due to tillage practices is highly variable and complex (Kessavalou *et al.*, 1998b). Some researchers (Hendrix *et al.*, 1988; Franzluebbers *et al.*, 1995) reported higher CO_2 emissions for no-tillage soils than for the conventional tillage soil. However, some others (Fortin *et al.*, 1996; Aslam *et al.*, 2000) did not observe any significant differences in CO_2 emissions between conventional tillage and no-tillage treatments. Still others have observed significantly lower CO_2 emission from no-tillage than conventional tillage during the short period after tillage disturbance (Reicosky and Lindstrom, 1993; Dao, 1998; Al-Kaisi and Yin, 2005). In contrast, higher emissions were reported for some periods and lower for others under no-tillage relative to conventional tillage also (Ball *et al.*, 1999; Vinten *et al.*, 2002). Through life-cycle analysis of wheat production in Northwest India, Sapkota *et al.* (2014) reported 10 times higher global warming potential per unit of wheat yield in conventional tillage based production (~ 400 kg CO_2 equivalent per Mg wheat yield) than in zero tillage based production (~ 35 kg CO_2 equivalent per Mg wheat yield). For the entire one year monitoring period in a 43-year old experiment CO_2 fluxes under moldboard tillage, chisel tillage and no-tillage averaged 2.14, 2.07 and 1.61 g CO_2-C m^{-2} day^{-1}, respectively. Higher CO_2 fluxes from moldboard tillage and chisel tillage than no-tillage suggest accelerated decomposition of soil organic matter attributable to tillage.

Lower CO_2 emission from no-tillage than tillage could be attributed to: i) slower decomposition of residues placed on the surface in no-tillage than when they are incorporated with tillage, ii) more crop residues on soil surface under no-tillage could serve as barrier for CO_2 emissions from soil atmosphere, reduce soil temperature and minimize soil-residue contact, iii) long-term changes in physical, chemical and biological soil properties after many years of no-tillage. In contrast, higher CO_2 emission from no-tillage than tillage observed also in some studies was attributed to both higher soil water content and higher SOC content maintained at the soil surface under no-tillage system which favored increased microbial decomposition. Linn and Doran (1984) attributed 90% CO_2 emission variations between no-tillage and plowed soils to the differences in WFPS. Smith *et al.* (2003) reviewed effects of soil physical factors and biological processes on soil - atmosphere gas exchange. They concluded that CO_2 emissions by aerobic respiration was primarily temperature dependent, but became moisture dependent as soil dried.

To sum up, the differences in CO_2 emissions among tillage systems may not be the sole result of short-term effects of tillage, but instead both the combined short- and long- term effects. Short term effects of tillage result from physical soil disturbance occurring during plowing and the location of crop residues. While long term effects are attributed to changes in physical, chemical and biological properties of soil after several years of practicing a specific tillage.

v) Inclusion of legume crop

Inclusion of legume is emphasized while opting for diversification of crop species grown in rotation in conservation agriculture. Based on literature review Jensen *et al.* (2012) concluded that less quantitative data are available concerning N_2O losses from legume residues following a legume phase in a cropping sequence. But, from available information while the potential losses of N_2O can be large from leguminous residues containing high concentrations of N such as nodules or fresh foliage the contribution of N_2O emissions from senesced vegetative residues remaining after grain harvest of a crop legume could be small. Further work is therefore required to better understand how the management and quality of legume residues affects N_2O emissions in subsequent crops.

b) Enhanced carbon sequestration

Carbon sequestration refers to both natural and deliberate processes by which CO_2 is either removed from the atmosphere or diverted from emission sources and stored in the terrestrial environments (vegetation, soils, and sediments), ocean, and geologic formations. The idea is to: i) prevent carbon emissions produced by human activities from reaching the atmosphere by capturing and diverting them to secured storage (otherwise emitted CO_2 will remain in the atmosphere to add to global warming) or ii) remove carbon from the atmosphere by various means and store it.

Under conservation agriculture, soil turns into a net sink of carbon for 30 to 50 years. The decomposition of roots and crop residues increases the organic matter content of the soil at various depths, while the release of carbon to the atmosphere occurs slowly. In the total balance, carbon is sequestered in the soil. Lal (1999) reported that conservation agriculture can sequester 0.5 – 1.0 t C ha^{-1} yr^{-1} in humid temperate region, 0.2 – 0.5 in humid tropics and 0.1 – 0.2 in semi-arid zones. The increase in soil carbon stock was attributed to reduced CO_2 efflux as a result of no or reduced mechanical disturbance to soil and higher input of organic matter to soil. Depending on climatic conditions, in general, a minimum addition of carbon in the form of vegetative residues @ 4.2 t ha^{-1} yr^{-1} in cropping systems and 4.5 t $ha^{-1}yr^{-1}$ in mixed systems of pastures and crops is necessary for maintaining soil organic matter at stable levels. Below these values CO_2 emission will or can take place and the soil then becomes itself a source of carbon. Therefore, for carbon sequestration to be successful, it is necessary to include crops in the rotation that produce a large quantity of biomass. Powlson *et al.* (2011) based on literature review have, however, observed that sequestering carbon in soil as a means of climate change mitigation has limitations and its potential has been over-emphasized. The reasons given by them include the following constraints: i) the quantity of carbon stored in soil is finite, ii) the process is reversible, and iii) even if SOC is increased there may be changes in the fluxes of other greenhouse gases, especially the N_2O and methane.

Paustian *et al.* (1997) compared 39 paired tillage experiments ranging in duration from 5 to 20 years and estimated that no-till resulted in an average soil carbon increase of 285 g m^{-2} compared to conventional tillage. Using average experiment duration of 13 years implies an approximate carbon sequestration rate of 0.22 Mg ha^{-1} yr^{-1}. West and Post (2002) analyzed global data on a paired conventional-tillage and no-tillage plots and concluded that changing from moldboard plow tillage to no-tillage can sequester at an average rate of 0.57 Mg C ha^{-1} yr^{-1} reaching equilibrium in 15-20 years. However, conversion to no-tillage under wheat (*Triticum aestivum* L.) - fallow rotations showed no increase in soil organic carbon pool. Puget and Lal (2005) compared 56 different pairs of moldboard tillage and no-tillage and reported 42 pairs with an increase of SOC content under no-tillage and 11 with no response. Among the 42 pairs with SOC gain, statistically significant differences were only reported for 10 pairs. They concluded that changing from moldboard tillage to no-tillage sequestered 0.33 Mg C ha^{-1} yr^{-1}.No studies actually indicated a significant soil organic carbon decrease under no-tillage compared to moldboard tillage. Soils under long-term no-tillage generally contain a larger SOC pool in surface than conventionally tilled soils (Fig. 3). This presents a challenging limitation of SOC accumulation taking place mostly in the top 5 cm only in no-tillage/conservation agriculture systems. It is, therefore, said that over the years surface soil under no-tillage/conservation agriculture will exhaust its capacity to store carbon, which is a function of physicochemical characteristics inherent to soils (Six *et al.*, 2002), thereby rendering the objective of carbon sequestration irrelevant.

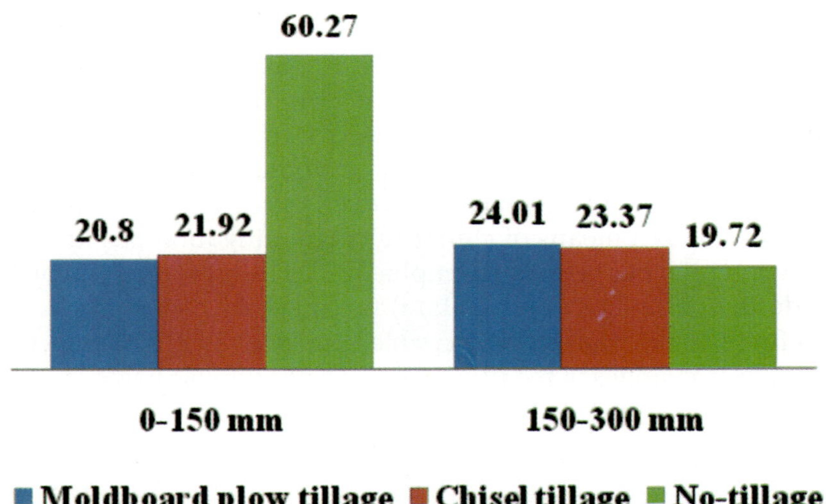

Fig. 3. Total soil organic carbon pool (Mg ha^{-1}) in 0 – 150 mm (LSD at 5% - 7.93) and in 150 - 300 mm (LSD at 5% - 6.61) after continuous 43 years of different tillage practices (adapted from Ussiri and Lal, 2009)

To address this limitation "Occasional or Strategic tillage" for no-tillage systems is proposed as a one time tillage, conducted once in 12 or more years as required, to invert the high SOC surface layer with less carbon containing deeper soil in an otherwise continuous no-till system. This will result in increased SOC in the profile in the long term. Quincke (2006) imparted occasional tillage to two experimental fields in Eastern Nebraska that were under long term no-tillage management for growing rainfed corn or grain sorghum rotated with soybeans. He observed that a one time tillage with moldboard plow was effective in reallocating SOC from the surface into deeper soil depths to 30 cm. Soil organic matter in the surface 2.5 cm was significantly lower after occasional tillage with moldboard plow when compared to continuous no-tillage. It had, however, increased the sequestration capacity/potential of the surface soil under continuous no-tillage.

Conclusion and Way Forward

As discussed in this chapter evidences are both for and against the impact of conservation agriculture on reducing emissions and enhancing sequestration of the greenhouse gases. However, considering all individual sources and sinks together, it appears that conservation agriculture will reduce greenhouse gases. Therefore, conservation agriculture should be adopted as a long term strategy for reducing greenhouse gases in the atmosphere. Further, instead of adopting piecemeal approach an integrated monitoring of all the parameters and their impact on greenhouse gases is required to have information on net emission, because they exhibit complex interactions. This may be done on agro-ecoregion basis and its dominant cropping systems to arrive at a clear understanding and make definitive recommendations.

The way forward for reducing greenhouse gases in the atmosphere is thus, adoption of conservation agriculture as a long term strategy vis - a - vis for unambiguous recommendations integrated monitoring of all the parameters and their impact on greenhouse gases on agro-ecoregion and dominant cropping systems basis.

References

Al-Kaisi, M. M. and Yin, X. H. 2005.Tillage and crop residue effects on soil carbon and carbon dioxide emission in corn-soybean rotations.*Journal of Environmental Quality* 34: 437-445.

Aslam, T., Choudhary, M. and Saggar, S. 2000. Influence of land-use management on CO_2 emissions from a silt loam soil in New Zealand. *Agriculture Ecosystems & Environment* 77: 257-262.

Aulakh, M. S., Rennie, D. A. and Paul, E. A. 1982.Gaseous nitrogen losses from cropped and summer-fallowed soils.*Canadian Journal of Soil Science* 62: 187-196.

Baggs, E. M., Stevenson, M., Pihlatie, M., Regar, A., Cook, H. and Cadisch, G. 2003. Nitrous oxide emissions following application of residues and fertilizer under zero and conventional tillage.*Plant and Soil* 254: 361-370.

Ball, B. C., Scott, A. and Parker, J. P. 1999.Field N_2O, CO_2 and CH_4 fluxes in relation to tillage, compaction and soil quality in Scotland.*Soil & Tillage Research* 53: 29-39.

Blasing, T. J. 2014. Recent greenhouse gas concentrations.*DOI: 10.3334/CDIAC/atg.032* (Accessed 19 January 2016).

Bronson, K. F. and Mosier, A. R. 1993. Nitrous oxide emissions and methane consumption in wheat and corn cropped systems. In: *Agriculture Ecosystem Effects on Trace Gases and Global Climate Change*. Harper, L A, Mosier, A R, Duxbury, J M, and Rolston, D E (Eds.). ASA-CSSA-SSSA, Madison, WI, pp133-144.

Choudhary, M. A., Akramkhanov, A. and Saggar, S. 2002. Nitrous oxide emissions from a New Zealand cropped soil: tillage effects, spatial and seasonal variability. *Agriculture Ecosystems & Environment* 93: 33-43.

Corsi, S., Friedrich, T., Kassam, A., Pisante, M. and Joao, de Moraes S. 2012.*Soil organic carbon accumulation and greenhouse gas emission reductions from conservation agriculture: A literature review*. Plant Production and Protection Division Food and Agriculture Organization of the United Nations Rome, 2012.www.fao.org/fileadmin/user_upload/agp/icm16.pdf (Accessed 19 January 2016).

Dao, T. H. 1998. Tillage and crop residue effects on carbon dioxide evolution and carbon storage in a Paleustoll.*Soil Science Society of America Journal* 62: 250-256.

Elmi, A. A., Madramootoo, C., Hamel, C. and Liu, A. 2003. Denitrification and nitrous oxide to nitrous oxide plus dinitrogen ratios in the soil profile under three tillage systems. *Biology and Fertility of Soils* 38: 340-348.

Erenstein, O. and Laxmi, V. 2008. Zero tillage impacts in India's rice-wheat systems: A review. *Soil Tillage Research* 100: 1–14.

Fortin, M. C., Rochette, P. and Pattey, E. 1996.Soil carbon dioxide fluxes from conventional and no tillage small grain cropping systems.*Soil Science Society of America Journal* 60: 1541-1547.

Franzluebbers, A. J., Hons, F. M. and Zuberer, D. A. 1995. Tillage induced seasonal changes in soil physical properties affecting soil CO_2 evolution under intensive cropping. *Soil Tillage Research* 34: 41-60.

Grant, B., Smith, W. N., Desjardins, R., Lemke, R. and Li, C. 2004. Estimated N_2O and CO_2 *emissions* as influenced by agricultural practices in Canada.*Climatic Change* 65: 315-332.

Hati, K. M. 2014. Effect of conservation tillage on soil physical properties.In *Conservation Agriculture for Carbon Sequestration and Sustaining Soil Health.* (Eds. J. Somasundaram, R. S. Chaudhary, A. Subba Rao, K. M. Hati, N. K. Sinha and M. Vassanda Coumar), pp. 51-60. New India Publishing Agency, New Delhi.

Hendrix, P. F., Han, C. R. and Groffman, P. M. 1988.Soil respiration in conventional and no-tillage agroecosystems under different winter cover crop rotations.*Soil Tillage Research* 12: 135-148.

Hutsch, B. W. 1998. Tillage and land use effects on methane oxidation rates and their vertical profiles in soil. *Biology and Fertility of Soils* 27: 284-292.

IEA 2012. CO_2 emissions from fuel combustion (*2012 Edition*) - III.1. *International Energy Agency. Part III*: *greenhouse-gas emissions.* edgar.jrc.ec.europa.eu/ docs/IEA_PARTIII.pdf (Accessed 19 January 2016).

IPCC 2001. Climate change 2001: The scientific basis. Cambridge University Press, Cambridge, UK.http://www.grida.no/publications/other/ ipcc_tar/?src=/climate/ipcc_tar/wg1/index.htm (Accessed 19 January 2016).

IPCC 2007.Summary for Policymakers. In *Climate Change 2007: The Physical Science Basis. Contribution of Working Group I to the Fourth Assessment Report of the Intergovernmental Panel on Climate Change.*(Eds. S. Solomon, D. Qin, M. Manning Z. Chen, M. Marquis, K. B. Averyt, M. Tignor *and* H. L. Miller).Cambridge University Press, Cambridge, United Kingdom and New York, NY, USA.https://www.ipcc.ch/pdf/assessment-report/ar4/wg1/ ar4-wg1-ts.pdf(Accessed 19 January 2016).

IPCC 2013. *Climate Change 2013: The Physical Science Basis. Contribution of Working Group I to the Fifth Assessment Report of the Intergovernmental Panel on Climate Change.* (Eds. T. F. Stocker, D. Qin, G. K. Plattner, M. Tignor, S. K. Allen, J. Boschung, A. Nauels, Y. Xia, V. Bex and P. M. Midgley).Cambridge University Press, Cambridge, United Kingdom and New York, NY, USA.http://www.climatechange2013.org/images/report/ WG1AR5_TS_FINAL.pdf (Accessed 19 January 2016).

Jenkinson, D. S. and Rayner, J. H. 1977.The turnover of soil organic matter in some of the Rothamsted classical experiments.*Soil Science* 123: 298-305.

Jenny, H. 1980. The Soil Resource: Origin and Behaviour. Ecological Studies 37.Springer-Verlag.XX, 377 p.

Jensen, E. S., Peoples, M. B., Boddey, R. M., Gresshoff, P. M., Hauggaard-Nielsen, H., Alves, B. J. R. and Morrison, M. J. 2012. Legumes for mitigation of climate change and the provision of feedstock for biofuels and biorefineries.A review. Agronomy for Sustainable Development 32, 329-364. DOI 10 1007/s13593-011-0056-7.(Accessed 19 January 2016).

Kessavalou, A., Doran, J. W., Mosier, A. R. and Drijber, R. A. 1998a. Greenhouse gas fluxes following tillage and wetting in a wheat-fallow cropping system. *Journal of Environmental Quality* 27: 1105-1116.

Kessavalou, A., Mosier, A. R., Doran, J. W., Drijber, R. A., Lyon, D. J. and Heinemeyer, O. 1998b. Fluxes of carbon dioxide, nitrous oxide, and methane in grass sod and winter wheat-fallow tillage management.*Journal of Environmental Quality* 27: 1094-1104.

Lal, R. 1999. Global carbon pools and fluxes and the impact of agricultural intensification and judicious land use. In *Prevention of land degradation, enhancement of carbon sequestration and conservation of biodiversity through land use change and sustainable land management with a focus on Latin America and the Caribbean*, pp. 45-52. World Soil Resources Report 86, FAO, Rome.

Lenka, S., Lenka, N. K., Singh, R. C., Subba Rao, A., Kundu, S., Raghuwansi, J. and Patidar, C. P. 2014.Greenhouse gas emission and soil properties as influenced by wheat biomass burning in Vertisols of central India. *Current Science* 107: 1150-1154.

Linn, D. M. and Doran, J. W. 1984.Effect of water-filled pore space on carbon dioxide and nitrous oxide production in tilled and nontilled soils.*Soil Science Society of America Journal* 48: 1267-1272.

McArthur, J. W. 2015. Agriculture in the COP 21 agenda. In: *COP21 at Paris: What to expect – The issues, the actors, and the road ahead on climate change.* (Ed. A. Bhattacharya), pp. 37-41. Brookings Global Economy and Development program. November 2015. www.brookings.edu/research/reports2/2015/11/17-paris-climate-talks(Accessed 19 January 2016).

Mensah-Bonsu and Obeng, H. G. 1979.Effects of cultural practices on soil erosion and maize production in the semi-deciduous rainforest - savanna transitional zones of Ghana.In *Soil Physical Properties and Crop Production in the Humid Tropics*. (Eds. D. J. Greenland and R. Lal), pp. 509-519. John Wiley, Chichester.

Mosier, A., Schimel, D., Valentine, D., Bronson, K. and Parton, W. 1991. Methane and nitrous oxide fluxes in native, fertilized and cultivated grasslands. *Nature* 350: 330-332. doi:10.1038/350330a0 (Accessed 19 January 2016).

Palma, R. M., Rimolo, M., Saubidet, M. I. and Conti, M. E. 1997. Influence of tillage system on denitrification in maize-cropped soils. *Biology and Fertility of Soils* 25: 142-146.

Paul, E. A. and Clark, F. E. 1996.Soil Microbiology and Biochemistry, Academic Press, California, USA.

Paustian, K., Andren, O., Janzen, H. H., Lal, R., Smith, P., Tian, G., Tiessen, H., VanNoordwijk, M. and Woomer, P. L. 1997. Agricultural soils as a sink to mitigate CO_2 emissions. *Soil Use and Management* 13: 230–244. DOI: 10.1111/j.1475-2743.1997.tb00594.x (Accessed 19 January 2016).

Powlson, D. S., Whitmore, A. P. and Goulding, K. W. T. 2011. Soil carbon sequestration to mitigate climate change: a critical re-examination to identify the true and the false. *European Journal of Soil Science*62: 42–55, doi: 10.1111/j.1365-2389.2010.01342.x (Accessed 19 January 2016).

Puget, P. and Lal, R. 2005. Soil organic carbon and nitrogen in a Mollisol in central Ohio as affected by tillage and land use.*Soil & Tillage Research* 80: 201–213.

Quincke, J. A.2006. *Occasional Tillage of No-till Systems to Improve Carbon Sequestration, and Soil Physical and Microbial Properties*.Ph D Dissertation, Faculty of The Graduate College, University of Nebraska Lincoln, 170 p. lacs.ipni.net/ipniweb/...nsf/.../Quincke%20PhD-Ocassional%20tillage.pdf (Accessed 19 January 2016).

Reicosky, D. C. and Lindstrom, M. J. 1993.The effect of fall tillage method on short-term carbon dioxide flux from soil.*Agronomy Journal* 85: 1237-1243.

Rudaz, A.O., Walti, E., Kyburz, G., Lehmann, P. and Fuhrer, J. 1999. Temporal variation in N_2O and N_2 fluxes from a permanent pasture in Switzerland in relation to management, soil water content and soil temperature. *Agriculture Ecosystems & Environment* 73: 83-91. http://dx.doi.org/10.1016/S0167-8809(99)00005-5 (Accessed 19 January 2016).

Six, J., Conant, R. T., Paul, E. A. and Paustian, K. 2002. Stabilization mechanisms of soil organic matter: Implications for C-saturation of soils. *Plant and Soil* 241: 155-176.

Six, J., Ogle, S. M., Breidt, F. J., Conant, R. T., Mosier, A. R. and Paustrian, K. 2004. The potential to mitigate global warming with no-tillage management is only realized when practiced in the long term. *Global Change Biology* 10: 155-160. doi: 10.1111/j.1529-8817.2003.00730.x (Accessed 19 January 2016).

Smith, K. A., Ball, T., Conen, F., Dobbie, K. E., Massheder, J. and Rey, A. 2003. Exchange of greenhouse gases between soil and atmosphere: interactions of soil physical factors and biological processes. *European Journal of Soil Science* 54: 779-791.

Sapkota, T. B., Majumdar, K., Jat, M. L., Kumar, A., Bishnoi, D. K., Modonald, A. J. and Pampolino, M. 2014. Precision nutrient management in conservation agriculture based wheat production of Northwest India: Profitability, nutrient use efficiency and environmental footprint. *Field Crops Research* 155: 233-244.

Tullberg, J. N. 2007. Agriculture engineering in support of the Kyoto protocol - The clean development mechanism for conservation agriculture. 22p. http://www.unapcaem.org/publication/ConservationAgri/CA.pdf (Accessed 19 January 2016).

Ussiri, D. A. N. and Lal, R. 2009. Long-term tillage effects on soil carbon storage and carbon dioxide emissions in continuous corn cropping system from an alfisol in Ohio. *Soil & Tillage Research* 104: 39-47.

Ussiri, D. A. N., Lal, R. and Jarecki, M. K. 2009.Nitrous oxide and methane emissions from long-term tillage under a continuous corn cropping system in Ohio.*Soil & Tillage Research* 104: 247-255.

Vinten, A. J. A., Ball, B. C., O'Sullivan, M. F., Henshall, J. K., Howard, P., Wright, F. and Ritchie, R. 2002. The effects of cultivation method and timing, previous sward and fertilizer level on subsequent crop yields and nitrate leaching following cultivation of long-term grazed grass and grass-clover swards. *Journal of Agricultural Science* 139: 245–256.

West, T. O. and Post, W. M. 2002. Soil organic carbon sequestration rates by tillage and crop rotation: a global data analysis. *Soil Science Society of America Journal* 66: 1930–1946.

❑❑❑

Chapter 26

Role of Organic Farming in Mitigation of Climate Change

ARUN K. SHARMA
ICAR - Central Arid Zone Research Institute, Jodhpur-342 003, Rajasthan
E-mail :arun.k_sharma@yahoo.co.in

Introduction

The monitored increases in the green house gases carbon dioxide, methane and nitrous oxide have been identified as most likely being anthropogenic. This has been clearly demonstrated the likelihood of changes in climate and the predictable impact on future life on earth in the fourth assessment report of the Intergovernmental Panel on Climate Change (IPCC, 2007). The global increase in atmospheric carbon dioxide concentration is primarily due to burning fossil fuel, deforestation and changes in land use pattern, methane and nitrous oxide emissions are primarily due to agriculture.

Predicted climate change will influence weather events globally through a) Increase in temperature (warmer and fewer cold days and nights,warmer and more frequent hot days and nights, heat waves); b) increase in extreme weather events; c) more drought frequency and d) Rising sea levels.

Agriculture is extremely vulnerable to climate change. Higher temperatures and frequent droughts eventually reduce yields of food crops while encouraging weed and pest proliferation. Changes in precipitation patterns with long spell of drought particularly in hot arid region and also declining in production increase the likelihood of short - run crop failures and long - run production declines. Therefore, it is becoming imperative to find out mitigative and adaptive measures in agriculture to climate change so that food production can be sustained for meeting the demand of population.

Conventional agriculture-substantially contributing to climate change

Manufacturing of fertilizers and pesticides require significant amounts of energy through which huge quantity and after application several greenhouse gases release in the atmosphere. By this way conventional chemical farming contributes significantly to the climate change. World wide 90 million tons of mineral oil or natural gas are processed to get Nitrogenous fertilizers every year. This generates 250 million tons of CO_2 emission. On the contrary, organic farms returns 575 to 700 kg CO_2 to the soil. Organic farming thus reduces CO_2 emission by eliminating synthetic fertilizers, and at the same time reduces atmospheric concentration of this gas by storing in the soil, a win-win system (Niggli, 2008). Further, soils with higher humus content can adapt to the adverse effect of climate change.Therefore, organic system can be a option for mitigating climate change effects on the agriculture.

Organic farming : A viable alternative

In last few decades awareness about social and environmental issues has been increased. Agriculture is one of basic enterprise that has major role in these two issues. Organic agriculture is one of the agriculture production system that not only supportive to the environment but also sensitive to the social issues like employment, health, migration etc. Definitions given by two international organizations also verify this concept. These definitions are :

1. Organic farming is a holistic management system, which enhances agro-ecosystem health, utilizing both traditional and scientific knowledge. Organic agriculture systems rely on ecosystem management rather than external agricultural inputs. (IFOAM, 2006)
2. Organic farming is an environmentally and socially sensitive food supply system. The primary goal of organic farming is to optimize the health and productivity of independent communities of soil life, plants, animals and people. (FAO, 2002)

Organic farming represents a broad set of practices that emphasize farming based on ecosystem management, integrated cropping and livestock systems, diversity of products, and reliance on natural pest and disease control without the use of synthetic inputs. The objectives of organic agriculture are to produce sustainable and healthy food through harnessing natural biological and ecological processes. The Codex Alimentarius Commission (2001) defines organic agriculture as "a holistic production management system which promotes and enhances agro-eco systems health including biodiversity, biological cycles, and soil biological activity. It emphasizes the use of management practices in preference to the use of off - farm inputs, taking into account that regional conditions require locally adapted systems. This is accomplished by using, where possible, cultural, biological, and mechanical methods, as opposed to using synthetic materials, to fulfill any specific function within the system". An organic agricultural system is therefore designed to rely on renewable energy sourced as much as possible from on - farm or natural local systems.

In simple words organic farming is the production system with the optimum utilization of local resources in such a way so that sustainability of production and wellness of the society & environment can be maintained for long time.

Objectives of organic farming

According to the International Federation of Organic Agriculture Movement (IFOAM) the objectives of organic farming are as follows:

- To produce food of high nutritional quality in sufficient quantities.
- To work in harmony with natural systems rather than seeking to dominate them.
- To encourage and enhance the biological cycles within farming system involving microorganisms, soil flora and fauna, plants and animals.
- To maintain and increase the long term fertility of soils.
- To use, as far as possible, renewable resources in locally organized agricultural systems.
- To work, as much as possible, within a closed system with regard to organic matter and nutrient elements.
- To give all livestock, conditions of life that allow them to perform all aspects of their innate behavior.
- To avoid all forms of pollution that may result from agricultural techniques.
- To maintain the generic diversity of the agricultural system and its surroundings, including the protection of plants and wild life habitat.
- To allow agricultural producers an adequate return and satisfaction from their work including a safe working environment.

- To consider the wider social and ecological impact of the farming system.

Essential characters of organic farming

The most important essential characters of organic farming are as follows:

- Maximal but sustainable use of local resources.
- Use of purchased inputs, only as complementary to local resources.
- Increasing crop and animal diversity in the form of poly-cultures, agroforestry, integrated crop/livestock systems, etc. to minimize risk and improve ecological balance and economic stability.
- Ensuring the basic biological functions of soil-water-nutrient-humus continuum.
- Creating an attractive overall landscape which gives satisfaction to the local people.

Thus, to sum up mixed farming, crop rotation and organic cycle optimization are the three key principles of organic agriculture.

Essential components of the organic system

There are three major components of organic system that creates core environment for organic production system. There may be some minor components of location specific.

Three major components are-

1. Maintaining agro-diversity through cropping system, integration of animals and other beneficial flora and fauna.
2. Conservation and efficient utilization of local resources with least dependency on external inputs.
3. Fairness in production and utilization so that every stakeholder of the system gets his due share.

Advantages of organic farming

Some important advantages of organic farming experiences worldwide are as follows:

1. It creates optimal conditions in soil for improved growth and physiological activities of plants and the higher crop yield.
2. It improves soil physical, chemical and biological conditions.
3. It improves soil carbon content (carbon sequestration).
4. Utilizing organic wastes as an input minimizes pollution.
5. It mitigates environmental degradation .
6. It is more resistant to diseases, insects and pests.
7. Organically grown produce are believed to be nutritionally superior and good for health.
8. Higher demand of organically produced food coupled with its premium price may improve economy of the farmer and the region.

Organic farming vs conventional (chemical) farming- ideological differences

Although organic farming is seems to be just the exclusion of synthetic external inputs but it is the ideological differences with conventional farming that makes organic farming, friendly to society and environment. These differences are-

Organic farming	Conventional (chemical) farming
Holistic approach	Reductionist approach.
Decentralize production	Centralize production
Harmony with NATURE (Harness the *benefit*)	Domination on NATURE (Exploit for *profit*)
Diversity	Specialization
Input optimization (save more)	Output maximization (spend more)

Organic farming : An energy efficient and ecofriendly system

Organic farming is energy efficient system as compared to conventional chemical farming. Also while conventional farming contribute to green house gases substantially, the OF kept this to minimum and therefore proved ecofriendly.

Energy use in crop production

When examining energy use in food systems, it is logical to begin with the activities involved in growing crops and raising livestock. Agricultural systems utilize both, direct energy for farm activities such as operating machinery and maintaining infrastructure, and indirect energy, used for the manufacturing of machines and inputs. Different production methods drastically alter the amount of energy needed to grow a particular crop or raise livestock. Nemecek *et al.,* (2005) found energy use in Swiss organic systems on a per hectare basis to be 46–49% lower than in a mineral fertilizer based system and 31–35% lower than in a conventional manure based system. On a crop unit basis the differences were 36–43% and 10–20% respectively. Green house warming potential in organic systems on a per hectare basis was 29–32% lower than in a mineral fertilizer system and 35–37% lower than in the conventional manure based system.

Energy use in fertilizer production

Nitrogen fertilizer is cited as the biggest energy sink in non-organic production. Not only is nitrogen fertilizer produced from the raw materials of fossil fuel but the conversion process to usable fertilizer is energy intensive. The production of one tone of nitrogen fertilizer utilizes one to one and half tones of equivalent petrol. In productions, like grains, in which high amounts of synthetic nitrogen fertilizers are applied, about half of the total energy (direct and indirect) needed, of the energy used for the manufacturing of the nitrogen fertilizers. Compared to such conventional production systems, the energy use in organic agriculture is therefore, about half. According to the Soil Association (2006), the largest portion of energy utilized in conventional agriculture on average,37 percent of the total energy is synthetic pesticides and mineral fertilizers, particularly nitrogen, and to a lesser extent, phosphorous, and potassium. Refsgaard *et al.,* (1998) found that energy consumption through the use of fertilizers accounted for any where from 25-68 percent of total energy use depending on the type of crop and growing conditions. Manufacture of synthetic nitrogen fertilizer is a very energy intensive process, and currently requires large amounts of fossil fuel energy. Natural gas is the main fuel and feed stock, which accounts for 62 percent of the energy used in synthetic N fertilizer production. Less efficient and more polluting fuels such as naphtha and fuel oil also represent a high share, 15 and 9 percent respectively, of the energy used in fertilizer manufacturing.

Organic agriculture utilizes manure, legumes, and other natural sources of nitrogen, which replace the fossil fuels to manufacture synthetic nitrogen fertilizer with natural biological processes. Legumes fix atmospheric nitrogen naturally in their root - nodules by the activity of micro-organisms. Water and other plant nutrients are supplied via the active soil biology of organic systems; soil microbes break down and cycle minerals from soil rock particles, decaying plant matter, manure, and compost to the plant roots. This natural symbiosis of soil microfauna and crops is suppressed in many non-organic systems with less biologically active soils.

Because of its reliance on natural fertilizers, organic farming often performs relatively better in terms of energy efficiency despite lower yields. In the majority of field trials and studies of operating farms, the increase in yield for conventional production over organic did not off set the energy used in the fertilizer to produce that gain (Stolze*etal*, 2000). In an analysis of conventional rice farming in the Philippines, concluded that increases in rice yield come with a 6 to 25 fold increase in energy consumption predominately because of fertilizer used to achieve incremental gains in crop yield (Pretty, 1995). In India synthetic fertilser production contributes about 6% of total greenhouse gas emission (Roy *et al.*, 2009).

In addition to emissions from manufacture, N fertilizers when applied to farm soils result in emissions of N_2O. The concern over N_2O emissions arises from its long atmospheric life (166 ± 16 years) and its higher global warming potential (296 times that of CO_2) (IPCC 2007). The amount of N_2O emitted from N fertilizers in soils depends largely on the amount of fertilizer applied, and to a lesser extent on temperatures, soil or crop type (Dobbie *et al.*, 1999; Dobbie and Smith, 2003).

Going by conservative estimates, greenhouse gas emissions (GHGs) from synthetic N fertilizer for India reached ~100 Mt of CO_2 -eq. in 2006-07 (Roy *et al.*, 2009). Together, both the manufacture and application of synthetic N are responsible for six per cent of India's total greenhouse gas emissions, comparable to sectors like cement or iron and steel industries, and to emissions from the entire road transport system.

Pesticide/agrochemicals production and consumption

In conventional chemical agriculture wide range of agrochemicals such as pesticides, herbicides, fungicides, and insecticides are used to enhance the production. The energy burden of these agrochemicals mainly comes from their manufacturing and transport, with a relatively small amount of energy used in their application. Because of the diversity of synthetic chemicals and annual variation in their application, an average energy burden of production and transport is generally used (Saunders *et al.*, 2006). Although the energy consumption of fertilizers typically outweighs that of agrochemicals in conventional systems, chemical pesticides are themselves a significant contributor to the energy inefficiency of many conventional operations.

Climate change mitigations options with organic farming

The most important factor to mitigate climate change is cutting emissions of CO_2 from fossil fuels i.e., prevention is better than cure. As organic farming does not use mineral fertilizers, which are produced using fossil fuel, it is a positive contributor even though it mostly leads to lower yields. Energy produced from perennial crops additionally sequesters carbon to the soil and if produced in a sustainable way these crops may serve as an infinite energy source, safe guarding the environment with minimal climatic impact.

Organic farming as a response measure: Organic agriculture is based on improving soil fertility, diversification and integration of crops and livestock at farm level. In this production system, organic matter will accumulate in the soil, while direct and indirect greenhouse gas emissions can be reduced substantially. The water balance is greatly improved and biodiversity on the farm increases rapidly. This form of agriculture compared to other has clear advantages with respect to climate change and the protection of the environment. Organic management practices save energy and build up fertility that may render the system less vulnerable, so payments and support for these benefits are justified and on these grounds farmers will be motivated for additional environmental services.

Soil carbon sequestration: Agriculture is unique in that as a system it not only produces greenhouse gases from energy consumption but also provides a sink for these emissions. Carbon, which is emitted in the form of carbondioxide as a result of fossil fuel consumption, can be sequestered in soil organic matter and plant biomass. Despite the limitations in both annual rates and attainable soil carbon levels, sequestration of carbon in soil organic matter can contribute to climate change mitigation. In addition to removing CO_2 from the atmosphere soil carbon accumulation has positive influences on soil quality and wider environmental benefits. In the event of severe weather changes, better soil quality can counteract the risk of losing soil fertility and stability. Evidence from farming system trials worldwide shows that organic farming enhances carbon stocks, while productivity is lower compared to systems using mineral fertilizers and pesticides. Manure based farming systems trial in Switzerland had upto 15% higher organic carbon (5.4 tons/ha) than the system with mineral fertilizer (Fliessbach *et al.*, 2007). It is interesting that with identical initial manure amounts the long-term use of composted manure showed 12% (4.4 tons/ha) higher values than the use of rotted or stacked manure. Higher soil organic carbon levels in the organic systems may have had implications for the documented increase in soil aggregate stability, with consequences for improved water infiltration and lower soil erosion often observable in the field.

Organic farming has a greater potential to sequester carbon in biomass and soil than most forms of conventional agriculture. Organic agriculture, especially on farms where cover cropping, grazing on pastures, and establishing permanent hedgerows and buffer zones are utilized, increases carbon sequestration. Application of manure, compost, and crop residues, which are vital to the maintenance of soil fertility in organic systems, has been proven to increase soil organic carbon in amounts unparalleled by conventional methods (SARE, 2001).

Organic farming stores carbon in soil and plant biomass by building organic matter, encouraging agro-forestry and forbidding the clearance of primary ecosystems. Organic crop rotations often include more mulching and green manure crops, which build soil fertility and increase the content of soil organic matter. This binds carbon and thus reduces the net CO_2 emissions from farming. In agroforestry systems where trees are combined with crop production the potential for carbon binding is even higher. Diverse agroforestry systems has a combination of annual and perennial crops could be a significant contribution to carbon sequestration and combined food and energy production. Marriott and Wander (2006) analyzed soil samples from nine farming system trials that were started in the USA between 1981 and 2000. The soil organic carbon concentrations were 14 percent higher in organic systems than in conventional ones.

The mitigation potential of globally applied Organic Agriculture is therefore highly significant especially in the next 20 to 40 years which is critical in policy terms for delivering major greenhouse gas reductions, transitioning to a low-carbon economy and above all avoiding catastrophic climate change.

Organic farming for mitigating climate change

Agriculture is highly vulnerable to climate change and our food supply relies on successful adaptation. Adaptation actions include those necessary to restore the resilience of eco-systems and their productivity to enable sustainable economic development. Organic Agriculture increases the ability of the farming system to continue functioning when faced with the adverse effects of climate change by increasing resilience within the agro-ecosystem.

No contribution in climate change : Organic farming reduces greenhouse gases, especially nitrous oxide, as no chemical nitrogen fertilizers are used and nutrient losses are minimized. Nitrogen-fixing legumes used as green manure can provide an alternative to all the synthetic fertilizer currently in use worldwide, without compromising food production. Additionally, recycled organic residues, like manure or compost, prove to be good substitutes to synthetic N. Doing away with the production itself, which accounts for high carbon emissions (48Mt of CO_2-eq. in 2006/07), we can halve the emissions to ~50Mt of CO_2-eq., and the contribution of fertilizers to the country's emissions would drop from 6 to 3 percent (Badgley *et al.,* 2007). Agriculture produces greenhouse gases through many practices including soil tillage, fertilizer application, and manure handling; utilization of fossil fuel energy in the manufacturing of synthetic inputs, however, contributes the most greenhouse gas equivalents in cropping operations and is the second largest greenhouse gas contributor besides methane emissions from animals on livestock farms (Ziesemer, 2007). Because organic agriculture prohibits the use of these chemical pesticides and fertilizers, the greenhouse gas emissions in organic systems are already significantly lower than in conventional systems (Kotschi *et al.,* 2004). Reduced off site production and transportation of concentrated feed for organic livestock systems also decreases carbon emissions for organic production when compared to conventional (Kotschi *et al*, 2004).

Minimizes energy consumption : Organic farming use energy by 30-70% per unit of land as compared to CF by eliminating the energy required to manufacture synthetic fertilizers, and by using internal farm inputs, thus reducing fuel used for transportation. Improving energy efficiency and self reliance is one of many principles and objectives in organic agriculture. Therefore, it should be balanced against other objectives such as improving nutrient recycling, soil fertility, animal welfare and biodiversity. In these aspects organic agriculture has an advantage and it is therefore a major challenge to improve energy use efficiency and self reliance without compromising the other objectives. More focus seems to be needed in order to reduce dependence on fossil energy sources and combine renewable energy production with food production in organic farming. Organic agriculture uses less energy per hectare mainly because of the absence of artificial fertiliser. But most farms still depend on fossil fuels. Thus, when yields are comparable to conventional or maximum 20-30 percent lower, the energy use per kg crop or milk is lower in organic farming.

Despite the complexities, uncertainties, and gaps in knowledge regarding energy consumption in both conventional and organic agriculture, a few general conclusions can be drawn. Typically, organic agriculture uses 30 to 50 percent less energy in production than comparable non-organic agriculture. Though organic agriculture on average uses energy more efficiently, it often requires an indirect trade-off of energy intensive inputs with additional hours of human labour approximately one third more than conventional agriculture. Most importantly, organic systems seek to establish closed (or atleast semi-closed) energy systems, based on energy use efficiency and biomass recycling.

Resilience with agro-diversity : The diversity of landscapes, farming activities, fields, and agro-diversity is greatly enhanced in organic agriculture (Niggli *et al.*, 2008), which makes these farms more resilient to unpredictable weather patterns that results from climate change. Organic agriculture systems build on a foundation of conserving and improving diversity by using diverse crops, rotations and mixed farm strategies. Enhanced biodiversity reduces pest outbreaks (Zehnder *et al.*, 2007). Similarly, diversified agro-ecosystems reduce the severity of plant and animal diseases, while improving utilization of soil nutrients and water.

Organic farming helps farmers adapt to climate change

1. **Prevention of losses of resources :** Prevents nutrient and water loss through high organic matter content and soil covers, thus making soils more resilient to floods, droughts and land degradation processes. Organic farming creates robust and environmentally benign farming systems that are resilient to temperature extremes, drought and which avoid soil erosion. Organic farming also promotes sustainable community based ecosystem management, conservation and restoration activities.
2. **Low financial liability** : The financial requirements of Organic Agriculture for adaptation are low. Additional costs mainly come from information provision, education and extension services. Organic Agriculture however offers innovative farmer based group systems that facilitate best practice knowledge exchange in a systematic and cost free manner. This is particularly important for the empowerment of vulnerable and poor people in rural populations that rely on agriculture for their livelihoods. Organic Agriculture fulfils many of the requirements identified for successful adaptation strategies.
3. **Protection shield of local seed and crop diversity** : Preserving seed and crop diversity, which increases crop resistance to pests and disease. Maintenance of diversity also helps farmers evolve new cropping systems to adapt to climatic changes.This minimizes risk as a result of stable agro-ecosystems and yields, and lower production costs.

Organic farming under hot arid environment

Considering the possibilities and to explore them through experimentation on organic production system for low rainfall areas, a 2.0 ha model organic farm (MOF) has been established in 2008, within the central research farm CAZRI, Jodhpur. The farm was registered with Rajasthan Organic Certification Agency, Jaipur for organic certification in August, 2008. The model organic farm declared "Certified" in august 2011 after completion of 3 years of registration.

A rotation of four crops including cluster bean and sesame in rainy season (July-October) and cumin and psyllium in winter (November- March) was selected for the study. Yields were found to be increasing significantly with the increase in use of organic manures from 2.5 to 7.5 tons/ha in all the crops. Legume cultivation in rainy season on an average contributed to 25-30% increase in yield in the subsequent crops of cumin and psyllium.

Increase in soil moisture retention with the use of organic manure was observed that helped in better growth and yield of crops. Similarly increase in soil organic carbon from 0.23% to 0.29% was recorded after three years application of compost @ 5.0 t/ha (D.K. Painuli and A.K. Sharma unpublished data). Crop resilience to climatic variability was enhanced with the use of organic manure.

There is a widely held view that organic systems give poor yields. However, the findings at the MOF show that, while during the initial developmental stage of an organic system there may be slightly lower yield than in a conventional one, after 2-3 years once the system is developed the yield levels are comparable to the conventional (chemical input based) system. In the fifth year yields of 917.5kg/ha for sesame, 1122.2 kg/ha for cluster bean, 830.9 kg/ha for cumin and 856.4 kg/ha for psyllium were recorded. This is comparable to the average yield of a conventional system. Since most of the inputs are being prepared on-farm the cost of production was reduced by 30-70%, depending upon the crop (Sharma,2013).

Future strategy

Supporting organic farming means supporting climate change mitigation and adaptation therefore, support of all the stakeholders are required in promotion of organic farming.

Government : Governments should acknowledge Organic farming as an effective strategy to reduce greenhouse gases and sequester carbon. They should help farmers adapt to climate change by promoting Organic farming through research and extension services.

Research organizations : Researchers should study and quantify the role of Organic farming in mitigating and adapting to climate change in order to improve farming techniques and disseminate findings.

Farmers : Farmers should grow organically to increase their farm's resilience to the effects of climate change. Incentives should be given to the farmers for promoting organic farming.

Consumers : Consumers should choose locally-grown, organic products to reduce their climate change impact.Infrastructure and market should be developed for marketing of organic produce.

Conclusion

Agriculture's role in both climate change and non-renewable resource consumption needs a more prominent position in the global discussion of curbing greenhouse gas emissions. Organic farming offers possibilities for carbon credits through sequestration, emissions reductions through better energy efficiency and develops resilience in crops to climatic adversities; therefore, world wide framework need to be developed either through international agreements or national policies to capture these opportunities. Given Organic farming's role as both a mitigative and ameliorative in climate change, policies need to be established for promotion of organic farming not only for human health but also health of the climate and our living planet as a whole.

References

Codex Alimentarius Commission. 2001. http://www.fao.org/fao-who-codexalimentarius/en/

Badgley, C., Moghtader, J.,Quintero, E.,Zakem, E.,Chappell, M.J.,Avilés-Vázquez,K., Samulon, A. and Perfecto, I.2007. Organic agriculture and the global food supply. *Renewable Agriculture and Food Systems* 22: 86-108.

Dobbie, K.E. and Smith, K.A.2003. Nitrous oxide emission factors for agriculturals oils in Great Britain: the impact of soil water-filled pores pace and other controlling variables.*Global Change Biology* 9:204-218.

Dobbie, K.E.,McTaggart, I.P. and Smith, K.A.1999. Nitrous oxide emissions from intensive agricultural systems: variations between crop sand seasons, key driving variables, and mean emission factors. *Journal of Geophysical Research* 26:891-899.

Fan S, Gloor M, Mahlman J, Pacala S, Sarmiento J, Takahashi T, Tan P.1998. A large terrestrial carbon sink in North America implied by atmospheric and oceanic carbon dioxide data and models. *Science* 282:442–446.

FAO. 2002. Organic Agriculture, environment and food security. FAO, Rome, Italy, 252p.

Fliessbach, A., Oberholzer, H.R., Gunst, L.andMäder, P.2007. Soil organic matter and biological soil quality indicators after 21 years of organic and conventional farming. *Agriculture, Ecosystems and Environment* 118: 273–284.

Ghoshal, S.2004. Unpublished PhD Thesis, Bidhan Chandra Krishi Viswa Vidyalaya, Mohanpur, W. Bengal (Referred by Roy *et al.*, 2009).

IFOAM. 2006. http://www.ifoam.org/growing_organic/definitions/doa/index.html

IPCC. 2007. Climate change 2007- synthesis report: http://www.ipcc.ch/pdf/assessment-report/ar4/syr/ar4_syr.pdf

Kotschi, J.,Müller-Sämann, K.2004. The Role of Organic Agriculture in Mitigating Climate Change : a Scoping Study. IFOAM, Bonn, Germany.http://infohub.ifoam.bio/sites/default/files/page/files/role_of_oa_migitating_climate_change.pdf

Marriott, E.E., Wander, M.M.2006. Total and labile soil organic matter in organic and conventional farming systems. *Soil Science Society of America Journal* 70: 950-959.

Nemecek, T., Huguenin-Elie, O., Dubois, D., Gaillard, G. 2005. Ökobilanzierung von Anbau system enim Schweizerischen Acker-und Futterbau Schriftenreihe der FAL 58. FAL Reckenholz, Zürich. (Referred by Niggli, 2008)

Niggli, U.2008. Organic farming and climate change-Monograph, FIBL, Switzerland, 29p.

Pretty, J.1995. Regenerating Agriculture: Policies and Practice for Sustainability and Self- Reliance. Earthscan Publications Limited, London.320p.

Refsgaard, K.,Halberg, N.,Kristensen, E.1998. Energy Utilization in Crop and Dairy Production in Organic and Conventional Livestock Production Systems. *Agriculture Systems* 57(4): 599-630.

Roy, B.C.,Chattopadhyay, G.N. and Tirado, R.2009. Subsidizing food crisis. Greenpeace India, Bangalore, 41p.

SARE.2001."The Transition from Conventional to Low-Input or Organic Farming Systems: Soil Biology, Soil Chemistry, Soil Physics, Energy Utilization, Economics and Risks."Project Report SW99-008. http://wsare.usu. edu/pro/pr2001/SW99-008.pdf.

Saunders, C., Barber, A.,Taylor, G. 2006. Food Miles—Comparative Energy/Emissions Performance of NewZealand's Agriculture Industry. *Research Report 285*, Agribusiness & Economics Research Unit, Lincoln University, Lincoln. http://ucanr.edu/datastoreFiles/608-324.pdf

Sharma, A.K. 2013. Organic system management. Indian Farming.63(4): 13-16.

Soil Association. 2006. Organic Works: Providing more jobs through organic farming and local food supply. Bristol House, Bristol.http://www.soilassociation.org/LinkClick.aspx?file ticket= 60CVIT1Nw0U%3D & tabid=387

Stolze, M., Piorr, A., Häring, A., Dabbert, S.2000. The Environmental impacts of organic farminginEurope. *Organic Farming in Europe: Economics and Policy, Volume 6.* University of Hohenheim Department of Farm Economics, Stuttgart.

The Codex Alimentarius Commission. 2001. Codex Alimentarius Commission. 2001. *Codex Alimentarius: Organically Produced Foods.* Food and Agriculture Organization of the United Nations and the World Health Organization. Rome, Italy.http://www.fao.org/docrep/005/y2772e/y2772e00.html.

Zehnder, G., Gurr, G.M., Kühne, S., Wade, M.R., Wratten, S.D. and Wyss, E. 2007. Arthropod pest management in organic crops. *Annual Review of Entomology* 52: 57-80.

Ziesemer, J.2007. Energy use in organic food system. FAO, Rome , Italy.http://www.fao.org/docs/eims/upload/233069/energy-use-oa.pdf

□□□

Chapter 27

Effect of Climate Change on Herbivore Insects

NISHA PATEL

ICAR -Central Arid Zone Research Institute , Jodhpur-342 003 , Rajasthan
E-mail: npatelcazri@gmail.com

Introduction

Although there is still a debate going on globally about the truth in the climate change predictions, most of them indicate new environmental conditions in the near future, viz., increased temperature and UV radiations, higher CO_2 and O_3 concentrations, changes in rainfall patterns and occurrence of extreme climatic events. Flora and fauna of any agroclimatic region is closely related with the climatic conditions as evidenced by the variations in biodiversity of different regions. Change in climatic conditions and extreme weather events whether natural or anthropogenic/manmade will certainly influence plant and animals greatly. Composition and richness of diverse species strongly influences functioning and stability of ecosystems, which ultimately are our life-support systems (McCann, 2000).

Insects form almost half of the total faunal biodiversity (Speight *et al.,* 1999) and provide valuable major ecosystem services such as pollination, being important links in the food chain, decomposition, recycling, herbivory etc. (Hillstrom and Lindroth, 2008). The crucial factors in environment that influence the physiology of insects are temperature and humidity, therefore any alteration in these will directly affect insects. Moreover insects have a close relationship with plants, so the modifications in their host plants due to climate change are also likely to influence them indirectly (Cornelissen, 2011).

Global climatic changes might affect insects directly, through changes in developmental rates, survival, fecundity, voltinism and dispersal of insect pest population (Karuppaiah and Sujayanad, 2012). Beside food, climatic conditions are basic factors that influence insect range (Jawaroski, 2013). The biology of insects is closely linked to environmental temperature, as they are ectothermic. Temperature determines insect survival, population dynamics and distribution. Similarly humidity also plays an important role in the life cycle of insects. As climate changes will impact diversity, distribution and abundance, phenology and quality of host plants, insects'dependant on them will also be affected indirectly. The alterations in activity of natural enemies viz; predators, parasitoids and entomopathogens with the changing climate, will also affect insect herbivores. Changes in temperature, humidity, O_3 and CO_2 levels, and extreme climatic events candirectly impact insect population dynamics, distribution and abundance of insect through changes in physiology, behavior and life history parameters of insects (Huey and Berrigan, 2001).

Direct effects

All climate change scenarios predict an increase in both global temperature means and variations in seasonal and diel temperatures. These changes lead to physiological, life history, and ecological consequences for ectothermic insects as temperature influences every stage of their life cycle, right from egg hatching to adult emergence (Colinet *et al.*, 2015). Temperature can alter insect life cycle duration, voltinism, population density, size, genetic composition, (Bale *et al.*, 2002). In temperate ecosystems, increased temperatures that do not exceed the optimum ranges generally improve insect metabolism, reproductive performance and development of insects. On the other hand, temperature changes that are outside the optimum range may increase mortality, reduce fecundity and decrease growth rates. With rise in each degree in the global temperature, the life cycle of all insects is expected to become shorter, which is expected to further lead to increased number of generations and higher populations of many insect pests. In temperate climates, increase in temperature will lead to early emergence and completion of life cycle leading to increase in number of generations. Milder and shorter winters will result in warm weather pests starting to breed sooner (Bale *et al.*, 2002). Lange *et al.* (2006) have predicted with the help of model simulations, that warmer summers can result in two generations of the most aggressive and serious insect pest of European spruce forests i.e, the spruce bark beetles per summer, instead of one. This will provide the bark beetles the opportunity to attack forests twice in a single year in Norway. Similarly Juszczak *et al.* (2013) have predicted that the important developmental stages of the codling moth may occur much earlier in the future climate conditions than currently in Poland and the fastest changes may be observed in the emergence dates of 95% of larvae of the second codling moth generation. This could increase the emergence probability of the pest third generation that has not currently occurred in Poland. Other degree-day or models have also predicted early emergence of insects, accelerated development of maggots and borers- possibly resulting in more generations (and more crop damage) per year.

The time period available to insects in temperate climates increases with rise in temperature which provides them more time to feed feeding and mating, locate proper place for laying eggs and more frequent oviposition, resulting in higher possibility of colonizing larger number of host plants (Moore and Allard, 2008; Nethererand Schopf, 2010).

The impact of enhanced temperatures on insects in tropical climate is however expected to be different from that of temperate climates. Studies on the direct effect of warming on insect fitness concluded that though warming in the tropics, will be of relatively small magnitude, its consequences will be more harmful because tropical insects are relatively more sensitive to temperature changeas these species are living in environments that are already close to their physiological optimum (Deutsch *et al.,* 2008). Studieson the brown plant hopper, *Nilaparvata lugens* (Stål), which is the most serious pest of rice across the world, revealed that climate warming in tropical regions and occasional extreme high temperature events are likely to become important limiting factors affecting the survival and distribution of *N. lugens* (Piyaphongkul, *et al.,* 2012). Small increases in temperature in tropical ecosystems can have large impacts as most of these species have small thermal safety margins, and thus even small amounts of warming will likely decrease performance. However, the thermal adaptation of a species may also be influenced by spatial temperature variations over its geographical range.

Indirect effects

As a result of changes in climatic parameters, plant diversity, abundance and richness will also change (Thuiller *et al.,* 2005),which will consequently affect insect diversity and abundance. Bellard *et al.*

(2012), reviewed different possible effects of climate change on individual, population, species, community and ecosystems and observed that the majority of models indicate alarming consequences for biodiversity, with the worst-case scenarios leading to extinction. Changes in host plant morphology, physiology and biochemistry will affect the herbivore insects indirectly (Morrison and Morecroft, 2006; Yuan *et al.*, 2009; Yadugiri, 2010).

Change in vegetation

Due to change in climate, the vegetation diversity and abundance of different regions will be affected (Kazakis *et al.*, 2007). Many plant species are adapted to particular temperature and moisture conditions. As these conditions change, habitats become unsuitable for these plants to grow. Rising concentrations of CO_2 in the atmosphere act as fertilizer to many plants (Reich *et al.*, 2014). These changes may stimulate growth in certain crops, trees, and weeds of cold climates, under moderate temperature increases. Many insects are host specific and changes in host abundance or quality will lead to increase or decrease in population of such insects. Some insect species face the threat of extinction due to their reliance on threatened plant species. Moir and Leng (2013), based on a study of 1019 insect species on 104 plant species, identified that 70 species were of immediate conservation concern as they rely on plant species which themselves are threatened. Further they have also predicted that sessile feeders and highly host specific groups such as whiteflies, scales and mealy bugs appear most prone to extinction. Species, which have small dispersal ranges have increased risk of extinction through small changes in habitat or environment due to climate change (Franco *et al.*, 2006). Species with small geographic ranges, specific habitat needs and difficult dispersing are likely to be out competed by faster species in the competition (Urban, 2012) However some species will expand their geographical range to places where climatic factors will be less harsh with global warming. This has been already documented for several butterflies, beetles, dragonflies and grasshoppers (Parmesan, 2006).

Effect on the host plant chemistry and physiology leading to changes in herbivore performance

Effects of elevated CO_2

Global CO_2 level is continuously increasing because of human activities. The concentration of CO_2 in the year of 2100 is expected to range from 540 to 970 ppm compared to about 280 ppm in the pre-industrial era (www. acer.acre). Increased atmospheric CO_2 influences plant physiology, biochemical composition resulting in changes in plant productivity and quality (Lindroth, 2010). The physiological changes caused by elevated CO_2 levels may lead to altered herbivory patterns insect diversity and abundance (Stiling *et al.*, 1999). Due to elevated CO_2 conditions plants may show enhanced photosynthetic activity, increased productivity and increased leaf area or biomass (Owensby *et al.*, 1999). Higher CO_2 levels may also change plant primary and secondary metabolism. Coviella *et al.* (2002) have predicted that for the future expected elevated CO_2 concentrations; plant allocation to defensive compounds will be affected enough to impact plant-herbivore interactions. The increased carbon availability for plant tissues, results in a "nitrogen dilution effect". The decreased concentration of nitrogen witha high C/N ratio has potential effects on plant secondary metabolism, a lowered concentration of leaf protein and therefore reduced nutritive value to herbivores. Elevated CO_2 conditions are also expected to cause changes in plant secondary chemistry due to an increased carbon supply and allocation to the production of carbon based secondary and structural compounds.

Some scientists have suggested that herbivores would respond to altered plant primary and secondary metabolism under elevated CO_2 by increasing food consumption to compensate for the plant lowered nutritional quality (Dermody *et al.*, 2008) and by reducing their growth rates and prolonging their development time (Goverde and Erhardt, 2003). These modifications in herbivore performance under CO_2 enrichment would make them more susceptible to natural enemies ultimately reducing herbivore abundance, richness and diversity (Stiling *et al.*, 2003). Studies on the effects of enhanced CO_2on insects with different feeding behaviors have indicated that as compared to miners and gall forming insects chewing insects will be significantly more affected (Stiling and Cornellison, 2007).

Effects of Increased O_3 in the troposphere

In coming years, all types of ecosystems are bound to be affected by tropospheric ozone, which is one of the most harmful and prevalent pollutant. Plants are likely to face the danger of increasing levels of ozone (O_3) concentration in the troposphere. The changes in plants will subsequently affect herbivores. Increase in O_3 levels can decrease photosynthesis even at relatively low concentration (Noormets*et al.*,

2001), consequently resulting in decreased growth and lowered leaf nutritional quality of plants. The decline in plant quality may lead to increase in herbivory levels due to over compensation for poor nutritional quality of food. Degree of insect herbivory shall depend on amount of change in host plant quality due to elevated O_3 levels.

Phenology

Phenology is principally concerned with the dates of first occurrence of biological events in plants and animals like germination, flowering, fruiting, host finding, mating, egg laying, and breaking of hibernation or diapause in their annual cycle. Most periodic biological events are affected by weather parameters. The timing of these biological events usually depends on rainfall, seasonal temperatures, etc. in both natural and agricultural systems. With change in temperatures and time of rainfall over the years, the timing of occurrence of phenological events has also been observed to change and found to be occurring earlier/ later in the season than their normal times. Life cycle of many insects is closely dependant/synchronized with interacting plants. Pollinators are present when plants are blooming, predators are abundant when host insects are available, host specific root, shoot or foliar eaters require the presence of that plant during a specific season. Therefore often a species may be strongly dependent on the presence of another. However with changes in climatic parameters this synchrony might be disturbed and these close interactions may become unsynchronized.

Ecosystem services provided by insects

Insects provide important ecosystem services, acting as pollinators, predators, parasitoids and herbivores and changes in their abundance and diversity have the potential to alter the services they provide (Hillstrom and Lindroth, 2008). Many of the ecosystem services provided by insects such as nutrient cycling, pest regulation and pollination – directly contribute to agricultural production (Losey and Vaughan, 2006). The healthy functioning of these ecosystem services ensures the sustainability of agriculture. Climate change has the potential to alter these services. Pollination service provided by insects is estimated to range from 1% to 16% of the market value of agricultural production (Hein, 2009). Pollinators and their host plants may respond differently to changes in temperature. Increased temperatures may postpone the flowering time of some plants while their pollinators might remain unaffected. High rainfall may limit reduce the foraging activity of pollinators. Decreased rainfall, leading to drought periods and water stress may decrease flower numbers and nectar production. (FAO 2011).

Activity and abundance of natural enemies might be affected due to change in status of individual pest species as a result of global warming. Changes in temperature might alter diurnal activity patterns of different groups of insects, and changes in interspecific interactions could also alter the effectiveness of natural enemies for pest management. Extended periods of drought followed by heavy rainfall are detrimental to the natural enemies causing increase in population of Oriental armyworm, *Mythimna separata* (Walk.) populations (Sharma *et al.*, 2002). Temperature not only affects the rate of insect development, but also has a profound effect on fecundity and sex ratio of parasitoids (Dhillon and Sharma, 2009).

Extreme events, Precipitation and drought

The risk of drought in some areas or that of extreme precipitation and flooding in others is increasing due to climate change. Changes in rainfall pattern can affect insect populations to a great extent. Warming would alter the timing of rains and snow melt, affecting the seasonal availability of water. Although many plants especially in arid zones are resilient to some degree of drought, increases in temperature could make future droughts more damaging than those experienced in the past. Drought stressed plants also have less resistance against feeding damage by insect. Lack of water in stressed plants concentrates the amino acids causing it to become more nutritious to pests like aphids. Therefore sucking pests viz. mites and jassids may flourish during short droughts as plant nutrients become more concentrated. Soil insects are likely to be highly vulnerable to low soil moisture, caused by a reduction in summer rainfall, which is predicted for some regions under current climate change scenarios. In studies on effect of drought on egg hatching of clover root weevil (*Sitona lepidus* Gyllenhal, Coleoptera: Curculionidae) it was found that egg viability was significantly reduced and viable eggs took more time in hatching (Johnson *et al.*, 2010).

Extreme events like sudden heavy downpours have an impact on pest incidence in the short term. Rapid changes in the climate caused by extreme events can lead to sudden pest outbreaks that can lead to yield losses. Very good rainfalls in the arid region in the past have caused sudden increase in population of defoliators such as *Achaea janata, Heliothis* sp, *Spodoptera* etc.

Survival of many natural enemies of insect pests also such as predators and parasites is affected by good precipitation or droughts. Entomopathogens such as insect-killing fungi which require humidity to germinate and are generally less effective during droughts.

Way forward

Interactions between climate, crops and biodiversity are complex and need further studies. Regular monitoring of insect pest in specific regions will provide the key to this change and its mitigation. The type and extent of insect pest problems, the effect on ecosystem services they provide will differ in specific regions and crops and hence these issues need to be addressed regionally for specific situations.

Better understanding of the interactive relationships between pests, their hosts and climate, is important to develop crop-pest models for each specific region. Understanding and maintaining local agro-ecological balance, biodiversity and ecosystem functions will be essential to ensure food security and enhancing productivity. Some management strategies which will help in maintaining the resilience of agro-ecosystems are-discouraging monocultures, increasing crop diversification and agricultural diversity through crop variety and germplasm adapted to shifting agro-ecologies, adoption of no-till or low-till plantings, uses of mulches for water retention and soil management practices that reduce fertilizer use like promotion of legumes in crop rotations.

References

Bale, J.S., Masters, G.J., Hodkinson, I.D., Awmack, C., Bezemer, T.M., Brown, V.K., Butterfield, J., Buse, A., Coulson, J.C., Farrar, J., Good, J.E.G., Harrington, R., Hartley, S., Jones, T.H., Lindroth, R.L., Press, M.C., Symrnioudis, I., Watt, A.D. and Whittaker, J.B. 2002. Herbivory in global climate change research: direct effects of rising temperature on insect herbivores. *Global Change Biology* 8:1-16.

Bellard, C., Bartelsmeier, C., Leadley, P., Thuiller, W. and Courchamp, F. 2012. Impacts of climate change on the future of biodiversity. *Ecology Letters*15(2): 365-377.

Colinet, H., Sinclair, B.J., Vernon, P. and Renault, D. 2015. Insects in fluctuating thermal environments.*Annual Review of Entomology* 60:123-40. doi: 10.1146/annurev-ento-010814-021017.

Cornelissen, T. 2011. Climate change and its effects on terrestrial insects and herbivory patterns. *Neotropical entomology* 40 (2): 155-163.

Coviella, C.E., Stipanovic, R. D.and Trumble, J.T. 2002. Plant allocation to defensive compounds: interactions between elevated CO_2 and nitrogen in transgenic cotton plants. *Journal of Experimental Botany* 53(367): 323-31.

Dermody, Orla., O'Neill Bridget, F., Zangerl, Arthur, R., May, R. and DeLuc, Evan. H. 2008. Effects of elevated CO_2 and O_3 on leaf damage and insect abundance in a soybean agroecosystem *Arthropod-Plant Interactions* 2: 125–135 DOI 10.1007/s11829-008-9045-4

Deutsch, C.A., Joshua, J. T., Raymond, B.H., Kimberly, S.S., Cameron, K.G., David, C.H. and Paul R.M., 2008.*Impacts of climate warming on terrestrial ectotherms across latitude.Proceedings of the National Academy of Sciences (USA).*105:6668–6672.

Dhillon, M.K. and Sharma, H.C. 2009. Temperature influences the performance and effectiveness of field and laboratory strains of the ichneumonid parasitoid, *Campoletischlorideae. BioControl* 54: 743-750.

FAO. 2011. *Potential effects of climate change on crop pollination. www.fao.org/3/a-i2242e.pdf*

Franco, A., Hill, J.K., Kitschke, C., Collingham, Y.C., Roy, D.B., Fox, R. and Huntley, B. 2006. Impacts of climate warming and habitat loss on extinctions at species' low-latitude range boundaries. *Global Change Biology* 12 (8): 1545-1553.

Goverde, M. and Erhardt, A. 2003. Effects of elevated CO_2 on development and larval food-preference in the butterfly *Coenonymphapamphius* (Lepidoptera, Satyridae). *Global Change Biology* 9: 74-83.

Hein, Lars. 2009. The Economic Value of the Pollination Service, a Review Across Scales. *The Open Ecology Journal.*2009(2): 74-82.

Hillstrom, M.L. and Lindroth, R.L. 2008. Elevated atmospheric carbon dioxide and ozone alter forest insect abundance and community composition. *Insect Conservationand Diversity* 1: 233-241.

Huey, R. B. and D,Berrigan. 2001. Temperature, demography, and ectotherm fitness. *American Naturalist*. 158: 204-210.

Jawaroski, Tomasz. 2013. The effect of temperature and humidity changes on insects development their impact on forest ecosystems in the expected climate change. *Lecne Prace Badawcze* (Forest Research Papers). 74 (4): 345–355.

Johnson, S.N., Gregory P.J., McNicolJ. W., Oodally,Y., Zhang X. and Murray, P.J. 2010. Effects of soil conditions and drought on egg hatching and larval survival of the clover root weevil (*Sitona lepidus*). *Applied Soil Ecology* 44(1): 75-79.

Juszczak, R., Kuchar, L., Lesny, J. and Olejnik, J. 2013. Climate change impact on development rates of the codling moth (*Cydia pomonella* L.) in the Wielkopolska region. *Poland International Journal of Biometeorology*. 57(1): 31–44.

Karuppaiah, V. and Sujayanad, G.K. 2012. Impact of Climate Change on Population Dynamics of Insect Pests.*World Journal of Agricultural Sciences* 8 (3): 240-246, 2012.

Kazakis. G., Ghosn, D., Vogiatzakis, N. and Papanastasis, V.P. 2007. Vascular plant diversity and climate change in the alpine zone of the Lefka Ori, Crete. *Biodiversity and Conservation*. 16: 1603-1615.

Lange, H., Økland, B. and Krokene, P. 2006. Thresholds in the lifecycle of spruce bark beetle. *Interjournal for Complex Systems* 1648 p.

Lindroth, R.L. 2010. Impacts of elevated CO_2 and O_3 on forests: phytochemistry, trophic interactions, and ecosystem dynamics. *Journal of Chemical Ecology* 36: 2-21.

Losey, J.E. and Vaughan, M. 2006. The economic value of ecological services provided by insects. *Bioscience*. 56 (4): 331–323.

McCann, K. S. 2000. The diversity–stability debate.*Nature*. 405: 228-233

Moir, M.L.and Leng, M.C. 2013. *Managing coextinction of insects in a changing climate: Developing management strategies to combat increased coextinction rates of plantdwelling insects through global climate change*. National Climate Change Adaptation Research Facility, Gold Coast, 104 pp.

Moore, B. A. and Allard, G. B. 2008. Climate change impacts on forest health. *Forest Health & Biosecurity Working Paper*s FBS/34E. Forest Resources Development Service, Forest Management Division, FAO, Rome.

Morrison,J.I. and Morecroft, M.D. 2006. *Plant growth and climate change*. New York, Wiley-Blackwell, 232p.

Netherer S. and Schopf A. 2010. Potential effects of climate change on insect herbivores in European forests – General aspects and the pine processionary moth as specific example. *Forest Ecology and Management*. 259: 831–838.

Noormets, A., So^ber, A., Pell, E.J., Dickson, R.E., Podila, G.K., So^ber, J., Isebrands, J.G. and Karnosky, D.F. 2001. Stomatal and non-stomatal limitation to photosynthesis in two trembling aspen (*Populus tremuloides* Michx.) clones exposed to elevated CO_2 and/or O_3. *Plant, Cell and Environment* 24: 327–336.

Owensby, C.E., Ham, J.M., Knapp, A.K. and Allen, L.M. 1999. Biomass production and species composition change in a tallgrass prairie ecosystem after long-term exposure to elevated atmospheric CO_2. *Global Change Biology* 5: 497-506.

Parmesan, C. 2006. Ecological and evolutionary responses to recent climate change. *Annual Reviewof Ecology, Evolution and Systematics* 37: 637-669.

Piyaphongkul, J., Pritchard, J. and Bale, J. 2012. Can Tropical Insects Stand the Heat? A Case Study with the Brown Planthopper. *Nilaparvata lugens* (Stål) *PLoS* 2012;7(1):e29409. doi: 10.1371/journal.pone.0029409. Epub 2012 Jan 12

Reich, P., Hobbie, Sarah. E., Lee, Tali, D. 2014. Plant growth enhancement by elevated CO_2 eliminated by joint water and nitrogen limitation . *Nature Geoscience* 7: 920–924.

Sharma, H.C., Sullivan, D.J., and Bhatnagar, V.S. 2002. Population dynamics of the Oriental armyworm, *Mythimna separate* (Walker) (Lepidoptera: Noctuidae) in South-Central India, *Crop Protection* 21: 721-732.

Speight, M.R., Hunter, M.D. and Watt, A.D. 1999.*Ecology of Insects – concepts and applications*. Oxford, Blackwell Science, 340p.

Stiling, P. and Cornelissen, T. 2007. How does elevated carbon dioxide (CO_2) affect plant-herbivore interactions? A field experiment and a meta-analysis of CO_2-mediated changes on plant chemistry and herbivore performance. *Global Change Biology* 13: 1823-1842.

Stiling, P., Hunter, M.D., Rossi, A.M., Hymus, G.J. and Drake, B.G. 2003. Elevated CO_2 lowers relative and absolute herbivore density across all species of a scrub-oak forest. *Oecologia* 134:82–87.

Stiling, P., Rossi, A.M., Hungate, B., Dijkstra, P., Hinkle, C.R., Knott, W.M. and Drake, B. 1999. Decreased leaf-miner abundance in elevated CO_2: reduced leaf quality and increased parasitoid attack. *Ecology Applications*9: 240-244.

Thuiller, W., Lavorel, S., Araujo, M.B., Sykes, M.T. and Prentice, C. 2005. Climate change threats to plant diversity in Europe. *Proceedings of the National Academy of Sciences* 102: 8245-8250.

Urban, Mark, C., Tewksbury, Joshua, J., Sheldon and Kimberly, S. 2012. On a collision course: competition and dispersal differences create no-analogue communities and cause extinctions during climate change. *Proceedings of the Royal Society B: Biological Sciences* DOI: 10.1098/rspb.2011.2367 Evans, Daniel M., Turley Nash E.,www. acer.acre.ca/resources/…./ global/….. projections- for carbon-dioxide

Yadugiri, V.T.2010.Climate change: the role of plant physiology. *Current Science* 99: 423-425.

Yuan, J.S., Himanen, S.J., Holopainen, J.K., Chen, F. and Stewart, C.N. 2009. Smelling global climate change: mitigation of function for plant volatile organic compounds.*Trends Ecology Evolution* 24:323–331.

❑❑❑

Chapter 02

Pollinator Dynamics under Changing Climatic Conditions: A Study from Arid Western Plains of India

SHARMILA ROY

ICAR-Central Arid Zone Research Institute, Jodhpur-342003, Rajasthan
E-mail: roysharmilaigfri@gmail.com

Introduction

The global climate is predicted to change considerably and that may result in major shifts in food production. Climate model projections of the fourth assessment report by the Intergovernmental Panel on Climate Change (IPCC) indicated an increment in the global surface mean temperature of 1.1–2.9°C for their lowest emission scenario and 2.4–6.4°C for the highest one by the end of the century. In regions where cold temperatures are one of the primary factors limiting crop production, temperature increases will be beneficial. The food production may increase at many places due to increases in rainfall and temperature while in other places it may decrease due to change in rainfall and its pattern (Jarvis *et al.*, 2010). South Asia is predicted to have the least increase (1.8-5°C), except for the Himalayas (IPCC 2007). Rainfall was projected to change between -5 to 20 per cent during the cold season and between -40 to 15 per cent in the warm season.

Pollination of flowers is an essential step in the sexual reproduction of plants. It is mutually beneficial to the pollinators for food in terms of yield increase in many crops, and indirectly by contributing to the healthy functioning of the terrestrial ecosystems (Klein *et al.*, 2007; Potts *et al.*, 2010). Major pollinator dependent crops are fruits, vegetables, spices, plantation crops, pulses and oilseeds. The area covered by pollinator-dependent crops has increased by more than 300 per cent during the past 50 years (Aizen and Harder, 2009). It has been estimated that the total annual economic value of crop pollination worldwide is about € 153 billion and the 87 of the world's leading food crops depend upon animal pollination, representing 35 percent of global food production (Gallai *et al.*, 2009).

Most flowering species rely on insects or other animals, than other methods for transfer of pollen. Globally, over 80 per cent of the pollination is performed by insects and bees (Robinson and Morse, 1989). Lack of sufficient number of suitable pollinators causes decline in fruit and seed production (Partap, 2001).

Plant and pollinator species are increasingly at risk of local and global extinction due to human activities, including habitat loss, altered land use, introduction of alien species and climate change. This global warming and climate change influences plant phenology and their interactions with other plants and animals. Climate warming may influence metabolism of poikilotherm organisms (insects, reptiles, etc.) affecting their population dynamics and distribution to different latitudes. Similarly, in plants onset of flowering and flowering *per se* may be changed due to warming, photoperiodicity, precipitation, soil humidity, etc. or combinations of various factors.

In view of climate change it is also possible that the fruit trees, due to their perennial nature and capability to sequester more carbon will be resilient to climate change and may be made liable to counter the impacts of climate change. But it is equally possible that like other plants climate change will have both positive and negative impacts on fruits in tropical regions. In regions where the prevailing temperatures are already high, further increases in temperature will adversely affect the yield and quality of fruits.

For the arid Western region of India, *Zizyphus mauritiana* ('ber') was recommended as an alternate crop in the early 1970 (Chopra and Srivastav, 1973). Since then the plant has naturalized itself and provided economic sustenance to the region and assurance against ecological degradation under the high temperature, scanty and erratic rainfall (100-400mm) and frequent droughts. The IPCC and PRECIS (Providing Regional Climates for Impact Studies) models have projected hotter days, warmer nights and decreased rainfall in this region by 21st century.

In this article a general discussion on the impact of climate change on plants and associated pollinators followed by a case study on *Zizyphus mauritiana* flower phenology and associated pollinator dynamics from Arid Western Plains of India is presented.

Pollination and pollinators

Pollination of flowers is an essential step for reproduction. It was first addressed in the 18th century by Christian Konrad Sprengel. Stone carvings and bricks from the palace of Assyrian kings as early as 800 B.C. depicted the significance of pollen and pollination of fruits to enhance quality and yield of seeds and fruits (www.wikipedia.org). It is the process by which pollen is transferred from the anther to the stigma of the flower. Thereafter pollen grain germinates and its pollen tube grows down the style to the ovary, one nucleus fuses with the polar bodies to produce the endosperm tissues, and the other with the ovule to produce the embryo. This process is thus termed as double fertilization.

Pollination can be accomplished by cross or self-pollination. Cross-pollination, also called allogamy, occurs only when pollen is delivered to a flower from a different plant. An estimated 48.7 per cent of plant species are self-incompatible and this necessitates cross pollination. It is also estimated that about 42 per cent of flowering plants exhibit a mixed mating system in nature. Cross pollination requires abiotic or biotic vectors. Plants adapted to cross pollination adopt various mechanisms to ensure spread of pollen to other plants' flowers (www.wikipedia.org). About 80 per cent of all plant pollination is through vectors, usually insects, but also reptiles, birds, mammals that may also carry pollen. Pollination occurs as a result of their activities when visiting plants for feeding, breeding or shelter.

The integrated sets of floral traits (morphology, colour, odour, size and phenology) associated with particular pollinator groups are known as floral or pollination syndromes (Faegri and Pijl, 1979). Co-evolution of flowering plants and their pollinators started about 225 million years ago. Darwin in 1862 considered that pollinator-mediated selection on floral traits (Table 1) has been an important evolutionary force that caused the diversification of flowering plants. Different pollinators may be responsible for floral traits divergence or same pollinators may bring trait convergence (Fenster*et al.*, 2004). However, the flowers of a particular species may be visited by several other secondary pollinators not strictly associated with floral traits hence the entire assemblage of flower visitors are important for pollination and play important role on plant reproduction especially in absence of primary pollinators and have potential for evolutionary shifts in pollination syndrome (Guerrero *et al.*, 2014).

Table 1. Floral traits and associated pollinators

Trait	Bat	Bee	Beetle	Bird	Butterfly	Fly	Moth	Wind
Color	White, green or purple	Bright white, yellow, blue, or UV	White or green	Scarlet, orange, red or white	Bright red and purple	Pale, or dark brown, purple	Pale red, purple, pink or white	Pale green, brown, or colorless
Nectar guides	None	Present	None	None	Present	None	None	None
Odor	Strong and musty; emitted at night	Fresh, mild, pleasant	None to strongly fruity or foul	None	Faint but fresh	Putrid	Strong sweet; emitted at night	None
Nectar	Abun-dant; some where at hidden	Usually present	Some times present	Ample; deeply hidden	Ample; deeply hidden	Usually absent	Ample; deeply hidden	None
Pollen	Ample	Limited; often sticky, scented	Ample	Limited	Limited	Limited	Limited	Abunda nt; small, smooth
Flower Shape	Bowl shaped; closed during day	Shallow; with landing platform; tubular	Large and bowl-shaped	Large, funnel - like; strong perch support	Narrow tube with spur; wide landing pad	Shallow; funnel-like or complex with trap	Regular; tubular without a lip	Regular and small

Source: www.pollinator.org

Effect of climate change on plants and pollinators

The climate change is expected to force species extinct by hampering individual survival and reproduction, by reducing the amount and accessibility of suitable habitat, or by eliminating other organisms that are essential to the survival of the species (Wall *et al.*, 2003). All living beings respond to changes in temperature and humidity by altering their activity and metabolism (Biesmeijer *et al.*, 2006). Temperature and photoperiods mainly affect development of plants.

The extreme climatic conditions like drought which are short-term events affect plants, particularly during anthesis.

The increase in average global temperature may be associated with advancement in the phenology and reproduction of plants and their pollinators. It is also possible that some species may not response to increased temperatures or climate variability in their phenology (Hegland *et al.*, 2009). In plants onset of flowering may be advanced due to warming, change in photoperiodicity, precipitation, soil humidity etc. and/or combinations of these various climatic factors. Under the climate warming conditions, plants and their interaction with other organisms' (herbivory, parasitism and pollination) is important (Fig.1).

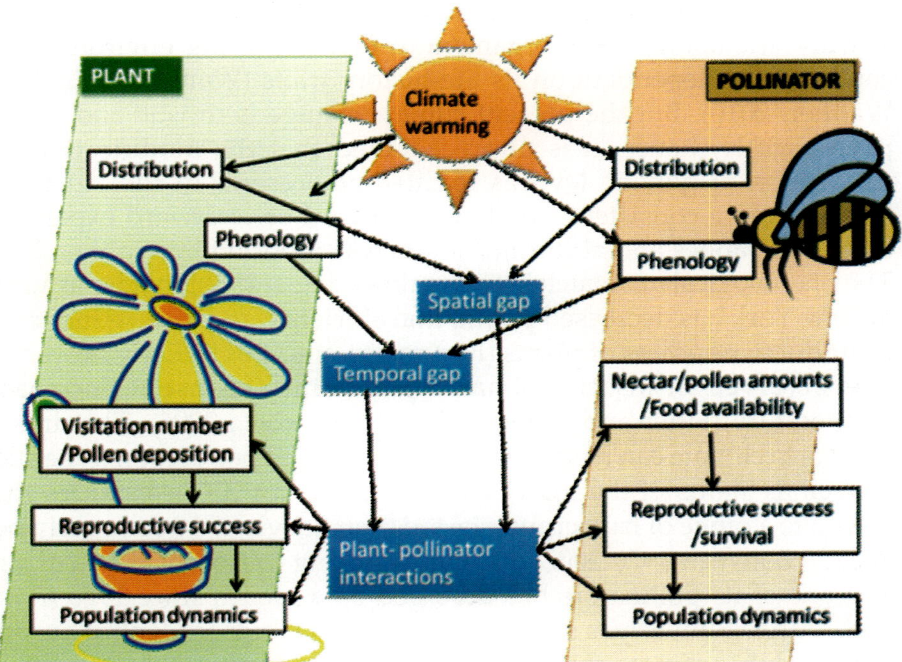

Fig. 1. Diagram showing probable effects of climate warming on phenology and distribution of plants-pollinators (modified from Hegland *et al.*, 2009).

In Western Illinois, USA observation on 14983 interactions between 1419 pollinator, 429 plant species were recorded in prairie forest transition zone (Robertson 1929). Memmott *et al.* (2007) in their simulation predicted that with doubling of atmospheric CO_2, phenological shift will reduce floral resources and the 17-50 per cent of all the pollinators will suffer disruption in food supply. The result also predicted extinction of plants, pollinators and their interactions.

Effect on plant phenology

Temperature and the rainfall are the most effective abiotic factors that guide flowering time and duration of flower in the plants. The climate change may disrupt the overlap in seasonal timing of flower production and of pollinator abundance, thus changing the opportunity for interaction between the plants and animals. Firstly, the impact of climate change on plant and pollinators may not be uniform globally as the amplitude of temperature change varies geographically. Secondly, the changes in climatic conditions are not always identical at different times of the year, even at the same location. For example, Ahas *et al.* (2002) and Parmesan (2006) reported that spring phases have been advanced in western and central Europe but delayed in Eastern Europe. Thirdly, the ability of an organism to follow shifting environmental conditions is dependent on its life-history traits (Walther *et al.*, 2002; Walther, 2010). But the present climate change is drastic and taking place swiftly, hence, it is unprecedented that the modification for adaptation at genetic level is feasible at present. In the Northern Hemisphere a consistent trend of northward or westward expansion of species ranges and altitudinal shifts was reported by Walther 2010. The phenological mismatch between flowering and pollinators in apple at New York has decreased pollination efficiency and thus reduced the reproductive success of plants (Bartomeus *et al.*, 2011). Early flowering was observed in insect pollinated plants in response to increased temperature (Fitter and Fitter, 2002; Menzel *et al.*, 2006). The drastic change in climate can have adverse effects on the population dynamics as well as status of the insects. For instance, six UK bee species that pollinate a range of fruit crops, started their activity 6-10 days early per decade between 1970-2007 and the flowering period of blackcurrant and one of its pollinators has diverged by a month since 1970 (Potts *et al.*, 2010). Similarly, a study on terrestrial orchids in Central Europe indicated that climate change has affected flowe[illegible]ime of 79 per cent of the orchids (31 taxa out of 39 studied) studied. [illegible]n average of all taxa studied, flowering time advanced by 3 days (3.8% of flowering period) during the last 50 years compared with the period before 1960 (Attila Molnár *et al.*, 2012).

Due to warming, flower numbers may or may not increase but mass flowering is generally induced due to increase in temperature in temperate climate. However, more number of flowers ensure food and population density of the pollinators. Similarly, flowering duration is another phenological response of plant and important for pollinators for their food. Prolonged flowering duration was observed in plant communities in Europe. Menzel *et al.* (2006) used systematic phenological network data set (1971–2000) of more than 125,000 observational series of 542 plant and 19 animal species in 21 European countries. The results showed that the average advance of spring was 2.5 days per decade in Europe and plants responded to this change. 78 per cent of all leafing, flowering and fruiting records advanced (30% significantly) and only 3 per cent were significantly delayed. In a warming experiment on subalpine meadows flowering, based on species, indication of prolonged flowering as well as and no change in flowering was observed (Dunne *et al.*, 2003).

Effect on pollinator activity

An increase in average air temperatures affects the population dynamics and distribution of insects whose internal body temperature depends on external environment (Willmer and Stone, 2004). The average increase in global temperature due to climate change may cause migration to suitable places for reproduction and alternate host selection for food and survival. The likely responses of pollinator species to climate change due to temperature and host phenology, are adaptation to the new environment, migration to other suitable areas and extinction. The first response is not expected as the climate change occurring too rapidly for populations to adapt by genetic change. Roy and Sparks (2000) showed that butterflies appear earlier and remain active for longer period during warmer period of Europe (1976-1998). More than 50 years data of Spain showed that appearance of bees is correlated with the spring temperature (Gordo and Sanz, 2006). Similarly, a long term study conducted at Spain showed a clear relationship with *Apis mellifera* appearance with early spring warming. Likewise bumble bees have advanced their spring flight times in UK by approximately 2 weeks from 2001 to 2007 (Spark and Collinson, 2007), probably because of higher soil temperature which affected queen bee's hibernation duration.

Climate plays strong role in geographical distribution of species. Hewitt (2000) discussed historical climate events like during major glaciations period species were concentrated to Equator region while during inter glacial period i.e. warmer period species migrated to higher latitudes and altitudes. For example, flies appear to become more abundant in colder and wetter areas whereas bees are abundant in warmer and drier habitats, suggesting that the composition and dominance of pollinator assemblages may change with climate warming (Hegland *et al.*, 2009). The climate warming impact is more prominent in temperate regions. Kelly and Goulden (2008) showed that nine dominant plant species and their Lepidoptera pollinators on a Southern California mountain shifted upward about 65° N in the last 30 years.

Effect of climate change on tropical plants and pollinators

Phenological responses of plants to rapid climate change is least understood in the tropics. The temperature, photoperiod and seasonally available rain are affected by global warming. This could potentially have serious consequences for plants and the organisms dependant on them. Tropical pollinators may respond to different temperature cues than pollinator species at higher latitudes. Many species move towards poles and higher altitudes with increase in temperatures (Coope, 1995). Climate change will have both positive and negative impacts on fruits in tropical regions. In these regions where the prevailing temperatures are already high, further increases in temperature will adversely affect the yield and quality of fruits. In regions where cold temperatures are one of the primary factors limiting crop production, temperature increase will be beneficial. Alternatively, the crops that require low temperature like apple adapt to altitudinal shift and reduction in onset of timely flowering in the mountain belt of the Hindu-Kush Himalayas (Basannagari and Kala, 2013).

In the sub-tropics, fruit tree development is mainly driven by atmospheric temperature. After cessation of growth in winter, phenophases usually appear early in spring with the onset of higher temperatures. Distinct changes in air temperature since the end of the 1980s led to clear responses in mango and guava phenology in many parts of India (Rajan 2009). Reddy *et al.* (2012) in their study noticed a decline in activity of *Apis florae* in mango at temperatures above 32°C. The pattern of diurnal variations indicated a peak of activity between 8.00-11.00 hrs. The number of out-going bees as well as the number of pollen collectors has declined with increase in temperature but showed a significant positive correlation with relative humidity. Also, areal time field observations and herbarium data study on changes on flowering phenology of *Rhododendron arboretum*, an important tree of Indian Central Himalayan region, predicted 88–97 days early flowering over the last 100 years (Gaira *et al.*, 2014).

Further, a pollination and seed production study of *Mikania micrantha* Kunth was undertaken in Viti Levu, Fiji to determine the impact of rainfall upon the reproductive potential of this weed (Macanawai *et al.*, 2011). Wet weather reduced the number of floral visits by all pollinator groups and thus influenced the production of filled seed. There was a greater proportion of filled seed produced by mikania plants per m^2 in the moderate rainfall region (61%) than in high rainfall region (51%).

Measures to moderate the impact of climate change on plants and pollinators

Effective pollination services

The key points of effective pollination services are (i) extension and consolidation of current knowledge Base (ii) adaptive management (iii) human capacity building (iv) mainstreaming. The Global Action provides guidance to member countries and relevant tools to use and conserve pollination services that sustain ecosystem functions, and to formulate policies that will ensure sustainability of these ecosystem services. The action plan for the International Pollinator's Initiative (IPI) included six major areas of concern; (1) reducing taxonomic obstacle on pollinators, (2) monitoring decline of pollinators, (3) identifying the causes of pollinator decline, (4) quantifying the economic value of pollinators to agriculture, (5) conservation of pollinator diversity, (6) sustainable use of pollinators.

Biodiversity management

Biodiversity ensures ecosystem functions by enhancing performance and buffering the temporal and spatial variations. Biodiversity can enhance mean levels of pollination due to functional complementarities among pollinator species and also lead to more stable pollination across time and space (Klein *et al.*, 2003). Bartomeus *et al.*, (2013) in their study on 46 years' time series data set on bloom phenology of apple and independent data set of wild bee, pollinator of apple collected at the same period indicated that high levels of biodiversity are needed to sustain ecosystem function in actual ecosystems. But long-term temporal effects, such as responses to climate warming, that is likely to play a critical role in maintaining everyday ecosystem functions need to be explored, to ensure biological interaction at spatial and temporal scales.

Study on *Ziziphus mauritiana* pollinators from arid western plains

Ziziphus mauritiana is an economically important tropical tree, which is grown all over Indian subcontinent for fruits. The plant is tolerant to wide climatic and soil conditions, can be effectively cultivated in the marginal lands with limited inputs and little attention. The tree propagates freely and withstands high temperature (about 40°C) and long periods of drought. In the arid Western region where rainfall is scanty the plant has naturalized itself and provided economic sustenance to the region and assurance against ecological degradation. It gives steady annual income of approximately Rs. 30000/ha under rain fed conditions and Rs. 60000/ha under irrigated conditions to the farmers from January to March (Meghwal *et al.*, 2007).

Since the 'ber' is basically insect pollinated plant and phenological responses of plants depend on climate change possibility exits that regional increase in temperature and change in rainfall pattern due to global warming may affecton flower phenology and the pollinator dynamics. In this respect a number of short term studies were conducted. However, to predict future impact of global warming long term uniform data sets are required. Such data sets are not available by and large. Gap in the available information was bridged by two approaches wherein firstly, the evidences were generated from real time field observations from three representative sites of Arid Western plains and secondly, information from published literature was tracked.

Climate change scenario of the zone

The impact of temperature change can be clearly seen from the fact that the northern parts of India are warmer than the southern parts, with a general increase of 3-6°C over the base-period (Lal *et al.*, 1996; Lonergan 1998). The arid Western region of our country is known for drastic climate, where rainfall is scanty (100-400mm) and erratic and frequent droughts are common. Further IPCC and PRECIS (Providing Regional Climates for Impact Studies) model have also projected hotter days, warmer nights and decreased rainfall in Thar region by 21st century.

The western part of India is primarily desert and broadly categorized by the Planning Commission as Western Dry Zone (# 14) commonly known as Thar Desert. On the basis of rainfall, this Zone is further divided into five sub zones (Table 2).

Table 2. Sub divisions of Western Dry Zone (#14)

Zone	Agro climatic Zone	Districts under zone	Rainfall (mm)
IA	Arid Western Plain	Barmer, Jodhpur	200-370
IB	Irrigated North Western Plain	Ganganagar, Hanumangarh	100-350
IC	Hyper Arid Western Plain	Bikaner, Churu, Jaisalmer	100-350
IIA	Transitional Plain of Inland Drainage	Jhunjhunu, Nagaur, Sikar	300-500
IIB	Transitional Plain of Luni Basin	Jalore, Pali, Sirohi	300-500

A long-term study on trends in mean air temperature (1971-2011) for Arid Western region by Poonia and Rao (2013) showed that if the present rate of warming continues there will be an increase in temperature by 3.3°C at Bikaner, 3.4°C at Jaisalmer, 2.9°C at Jodhpur and 2.5°C at Pali by the end of 21st century. This study also predicted that there will not be significant rise in rainfall (@ 0.56 mm/year) in Western arid zone. But there could be location specific variations, for example+100 mm at Bikaner, +124 mm at Jaisalmer, -40 mm at Jodhpur) and +21 mm at Pali. At the same time spatial and temporal variation in potential evapotranspiration will range from 2.1 mm/ day to 12.2 mm/day and on an annual basis may range between 1500 mm to 2220 mm. Further, the study showed that the projected air temperature may rise(upto 4°C) by 21st century that may increase evapotranspiration requirements by 9 to 23 per cent during monsoon period and 13-47 per cent during winter period and this is likely to increase water demand subsequently.

Floral biology of *Zizyphus mauritiana*

To understand the impact of climate change it becomes important to understand flower morphology and phenology. Flowers of Z. *mauritiana* are greenish-yellow, faintly amenoid fragrant, pedicels 3-8 mm long, calyx with 5 deltoid lobes, hairy outside, glabrous within, petals 5, subspathulate, concave and reflexed. Flower buds emerge in axillary position and inflorescence is cyme, consist of clusters (9-10, range 7-18, on tertiary branches) of 12-14 buds located at each node/ cluster of a branch. Cluster are 1.5 – 3.0 cm from neighboring clusters.

Flowers are protandrus (anthers release their pollen before the stigma of the same flower is receptive) hence; fruit set depends on cross pollination by insects attracted by the fragrance and nectar. *Z. mauritiana* floral traits partly imply for flies pollination syndrome. The swollen floral disc deceitfully create impression of nectar drop and provide sufficient stimulus to attract the wasps, bees, and flies.

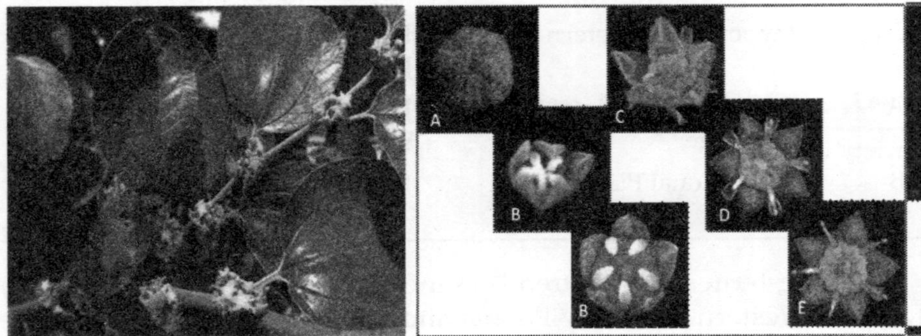

Fig.2. *Ziziphus mauritiana* inflorescence & stages of flower development (A- Bud, B-) A sexual phase: Flower partly opened; petals are upright; stamens covered by petals, C- Male phase: anthesis occurred; anthers dehisced and pollen present, D- Female phase: Stamens drooped onto petals; stigma developed and receptive, E-Brown stigma and no longer receptive).

Floral phenology of Zizyphus mauritiana

At Jodhpur (26°28′N-73°02′E) ten varieties (Gola, Umran, Mundia, Tikadi, Banarasikadaka, Illaichi, Dandan, Seb, ZG3 and Chhuhara) were observed for varietal variation in flower phenology. The flowering duration was from August to October (32nd - 44th Metrological weeks) depending upon the cultivars. Tikadi, Chuhara and Seb are early flowering varieties; flower bud initiation takes place during 32nd-33rd metrological week. In Ilaichi, Dandan, ZG-3, Mundia, Gola, Umran budding starts at about 34th -35th metrological week. BanarasiKadaka was late flowering plant; bud initiation takes place in the 37th metrological week. The flower opens between 12.00-14.00 hrs (temperature 25-27°C) in the Gola, Umran, Mundia, Tikadi, Banarasikadaka, ZG3 and Chhuhara varieties while in Illaichi, Dandan and Seb flower opening takes place in 7.00-9.00hrs (temperature 34-37°C). However, a deviation in flower opening time was found in case of ZG3 and Chhuhara in published literature. Pareek (2001) reported that in these two varieties anthesis occur during

7.00-8.30hrs. Some variations were also observed in the flowering duration and initiation of flowering time from the present study and the earlier works. It is reported that flowering initiation may be influenced by pruning and other agronomic practices (Kundu *et al.*, 1998). But no similar report was found by the authors regarding the flower duration. This kind of variation in flower phenology and their causes need to be thoroughly investigated to understand climate change impact on the ber in arid environment.

Table 3. Flower phenology of Z. *mauritiana* at Jodhpur climatic conditions

Variety	Flowering phenology of Z. *mauritiana*					
	Observed (2011-2013)			Reported by earlier workers		
	Initiation	Peak	End	Initiation	Peak	End
Gola	34	36	39	35	36	39
Mundia	35	37	41	-	-	-
Umran	34	36	41	35	39	43
Tikadi	32	34	39	36	40	43
BanarasiKadaka	37	39	42	35	36	40
Illaichi	34	36	41	35	37	43
Dandan	35	36	39	35	36	43
Seb	32	34	37	35	39	40
ZG-3	35	37	41	35	36	43
Chhuhara	33	34	38	35	36	43

* The reported information is a synthesis from the work of Pareek (2001)

The latitudinal variation in flower biology may also provide clues for the impact of predicted climate change on the variety. Four study sites were selected based on agroclimatic condition of arid Western India namely Jodhpur (26°28' N - 73°02' E), Pali (25°46'N - 73°25'E), Bikaner (28°01'N - 73°22'E) and Chandan (26°60'N - 71°21'E) near Jaisalmer. Authors have observed about 20 days early initiation of flowering (32nd met week) in Pali weather condition followed by Jodhpur (34th met week), Bikaner (36th met week) and Chandan (37th met week). However this observation needs further confirmation and being arid region correlation with abiotic factors like air temperature, rainfall, sunshine period, RH, wind speed and evapotranspiration rate is also important. Peak flowering was observed in the 38th metrological week at Pali and Jodhpur while this was achieved in 41st metrological week at Jaisalmer.

Variation in flowering period was noted from different parts of India also. In Southern India (9°55'N - 78°07' E) the flowering period was from 10th May to 20th September while in Arid western part (26°18'N - 73°01'E) it was from 25th August to 30th October and in North Western part (30°20'N - 76° 28'E) it was from 10th September -20th November (Pareek, 2001). Single flowering season was reported from Hyderabad but Devi *et al.* (1989) reported two blooms in a year during January – September at Vishakhpattanam (17°42'N - 82°18'E). In peninsular India (9°-20°N-and 74°-80°E) under tropical climate flowering duration is longer. In this region maximum and minimum temperature during flowering is 28.9-38.1°C and 19.7-26.3°C,respectively. In Northern Indian subtropical climate (maximum and minimum temperature during flowering are 28.2-35.7°C and 9.8-25.4°C respectively) flowering duration is shorter (August –November). Varietal variation in total flowering duration was reported by many workers. In Gola flowering duration was recorded as 71 days, in Seb 68 days, in Tikadi 47 days and in Umran 71 days (Pareek *et al.*, 2007).

Pollinator's activity

Temporal variation in pollinator's activities

More than thirty species of insects (6 Lepidoptrans, 12 Dipterans, 7 Hymenopterans, 3 coleopterans, 1 Orthoptera, and 2 Hemipterans) were found to be visiting the flowers of jujube (Table 4). Dipterans were the most abundant (52%) visitors followed by Hymenopterans (36%). However, intensity of visits varied yearly (Fig. 3). In year 2011 *Musca domestica* (1.6/flower/min) and *Polistes hebraeus* (1.3/flower/min) were most dominant while in year 2012 and 2013 *Apis indica* (0.9/flower/min) was dominant followed by *Musca domestica* (0.7/flower/min). The average stay of *Polistes hebraeus* on a flower was much longer (4.25sec/flower) than *Musca* spp. (3.33 sec/flower), *Api ssp.* (3sec/flower) and *Eristalinu* ssp. (2.87sec/flower).

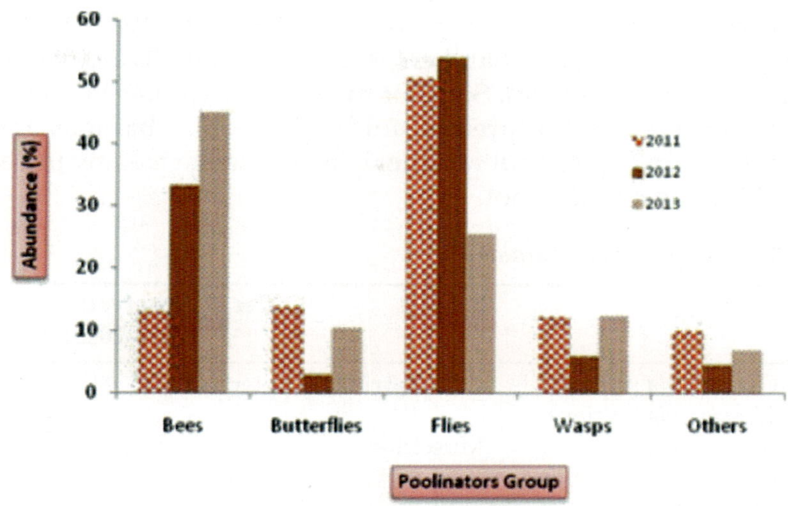

Fig. 3. Abundance of *Z. mauritiana* pollinators in Jodhpur climate during 2011-2013

Table 4. Flower visitors of *Z. mauritiana* at Jodhpur observed during 2011-2013

Order	Family	Identified species	Genera
Diptera	Calliphoridae	*Chrysomyia bezziana* Villeneuve, *Chrysomyia rufficalis*	2
	Muscidae	*Musca domestica, M. conducens, M. sorbens*	3
	Ottidae	*Physiophora aenea, P. Demandata*	3
	Sarcophagidae	*Sarcophaga dux* Thomson	1
	Syrphidae	*Eristalis obliquus* Wiedemann, *E. taphicus* Wiedemann	2
	Sepsidae	*Sepsis* sp.	1
Hymenoptera	Apidae	*Apis dorsata* F., *Apis florae* F, *Tetragonula laeviceps* (Smith)	4
	Vespidae	*Polistes hebraeus, Vespa orientalis*	2
	Formicidae	*Monomorium indicum* Forel, *Monomorium* sp.	2
Coleoptera	Scarabaeidae	Not identified	3
Hemiptera	Tessaratomidae	Not identified	2
Lepidoptera		*Eurema sp., Papilio sp., Tarucus sp., Hypolimnas sp., Danaus sp., Castalius sp.*	6

A comparison of this data of 2011-2013 with the 1980-1983 report by Singh (1984) revealed that there is adjustment in diversity of major pollinators' group (Fig. 4). Some of the taxa are similar but few were not encountered in the present study (Table 5). This may require elaborate study to point out whether climate change has any impact on the pollinators' guild or not.

Table 5. Pollinators of *Z. mauritiana*

No.	Genus	Family	Order	Observed (2011-2013)	Reported (1984)
1	*Chrysomyia rufficalis*	Calliphoridae	Diptera	+	+
2	*Chrysomyia bezziana*			+	
3	*Musca domestica*	Muscidae		+	+
4	*M. conducens*			+	
5	*M. sorbens*			+	
6	*Physiphora aenea*	Otitidae		+	+
7	*Physiophora demendata*			+	+
8	*Sarcophaga sp.*	Sarcophagidae			+
9	*Sarcophaga dux*			+	
10	*Eristalinus sp.*	Syrphidae			+
11	*Allobaccha saphirhina*				+
12	*Eristalis obliquus*			+	
13	*E. taphicus*			+	
14	*Apis indica*	Apidae	Hym-enoptera	+	+
15	*Apis sp.*			+	+
16	*Apis florae*			+	
17	*Tetragonula laeviceps*			+	
18	*Delta compasterenes esurience*	Fumenidae			+
19	*Nomioides varegatus*	Halictidae			+
20	*Chalybion bengalense*	Sphecidae			+
21	*Oxybelus lamellatus*				+
22	*Philanthus basalis*				+
23	*Sceliphron madraspa tanumpictum*				+
24	*Campsomeriella collaris*	Scolidae			+
25	*Iswara luteus*	Iiphidae			+
26	*Vespa orientalis*	Vespidae		+	+
27	*Vespa orientalis*			+	
28	*Monomorium indicum*	Formicidae		+	
29	*Monomorium* sp.			+	
30	*Sepsis sp.*	Sepsidae		+	

Further, there is temporal variation in pollinator visits on the blossom. More number of insects visited the flower during late morning hours followed by a decline in the noon (Fig. 5). Maximum activity of insects was recorded during 9.00-14.00hrs and this result is in corroboration with the studies of Singh (1984).

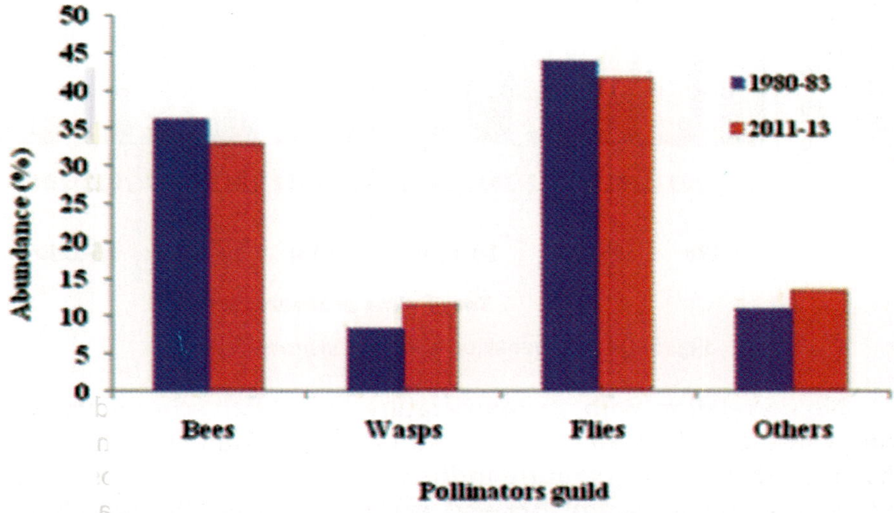

Fig.4. Abundance of major pollinator groups studied at different periods

The favorable temperature range of insect's activity was found between 28-32°C. However peak activity varied species wise. For example *Physiophora* spp. were more abundant up to 10.00 hrs in a day and are the first pollinators appear in orchard. Honey bees take over during peak flowering period i.e. during October. The activity of major pollinators viz., *Physiophora* spp., *Chrysomia* sp., *Sarcophaga* sp., *Eristalinus* sp and *Apis* spp., was observed to decrease with increasing air temperature. But the *Polisteshebraeus*, was found to be more active with increasing diurnal temperature.

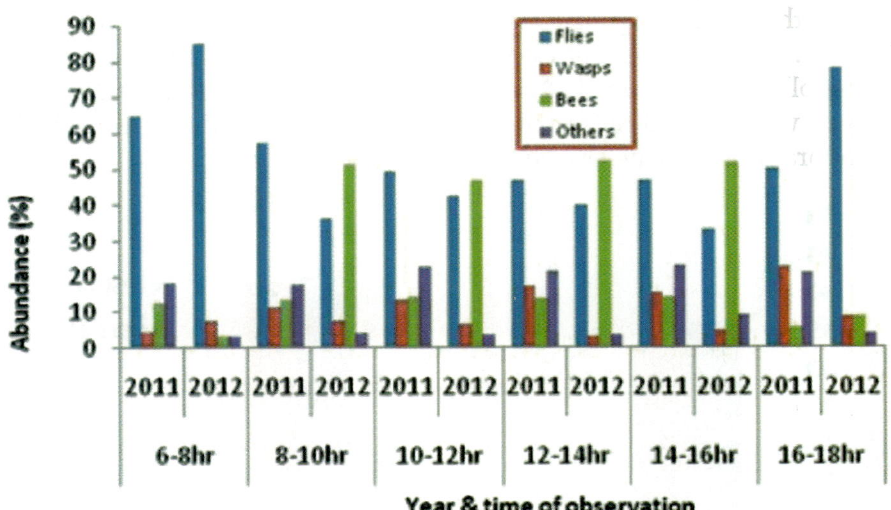

Fig. 5. Diurnal behavior of flower visitors in ber

No correlation with air temperature and relative humidity was found in the activity of *Musca* spp and wasps (Table 6). The study indicated that the temperature influences the activities of most of the pollinators, but impact of RH and rainfall could not be established. Since the activity of Muscidae and wasps was observed to be not affected by diurnal temperature, further studies are needed on their effective role as jujube pollinators under predicted warmer climatic conditions.

Table 6. Correlation of pollinator's activity with temperature and relative humidity

Insect Visitors	Temperature (°C)	Relative Humidity (%)
Physiophora sp.	-0.89**	0.78**
Chrysomia sp.	-0.87**	0.76**
Stomoxys sp.	-0.83**	0.74**
Musca sp.	0.01	-0.12
Papilio sp.	-0.59**	0.45**
Sarcophaga sp.	-0.81**	0.72**
Polistes hebraeus	0.74**	-0.74**
Apis florea	-0.52**	0.52**
Eristalinus sp.	-0.61**	0.51**

(N = 60; r (5%) = 0.26; r (1%) = 0.34)

Spatial variation in pollinator's activities

Information from pollination studies along altitudinal or latitudinal gradients (proxies for temperature influence) may help to infer how pollinator assemblages could be affected by climate warming. Three different sub agro-climatic zones of the Western Arid Plain viz., Pali (IIB), Chandan (IC), Bikaner (IC) and Jodhpur (IA) were observed for abundance/diversity of pollinators so as to understand the predicted impact of future climate change of the zone. Maximum number of pollinators was observed at Pali followed by Jodhpur location (Table 7). Metrological parameters of peak flowering period (35th to 44th metrological week) recorded from the nearby weather stations. The wind speed, maximum and minimum air temperature varied significantly during this period. The maximum and minimum air temperature and highest wind speed was recorded at Chandan (37.20°C, 20.31°C, 8.0km/hr. respectively). Comparison of insect abundance with climatic parameters suggested that the rainfall, temperature and wind speed may be affecting the activity of pollinators. However, correlation studies are needed.

Table 7. Pollinators abundance at different climatic zones of Arid Western Rajasthan (N =120, ^{abc} = Duncan sub set, Alpha = 0.05) (AT= Air temperature; AT & wind speed is mean of 37th - 44th met week)

Place	Bees	Wasps	Flies	Other	Total	Rain fall (mm)	AT (°C) max	AT (°C) min	Wind speed (km/hr.)
Jodhpur	2.90^{c}	0.80^{c}	1.63^{b}	1.11^{b}	6.44^{c}	145.10	35.08	23.68	3.62
Pali	3.35^{c}	0.45^{b}	4.83^{c}	1.21^{b}	9.83^{d}	103.40	34.97	22.78	4.67
Bikaner	0.52^{a}	0.07^{a}	0.93^{ab}	0.24^{a}	1.76^{a}	0.20	36.39	23.59	4.84
Chandan	1.53^{b}	0.63^{b}	1.39^{a}	0.30^{a}	3.86^{b}	0.00	37.20	20.31	7.96

Way Forward

In the earlier period, the diversified agriculture systems that have successfully integrated livestock and orchards were less vulnerable to sudden climatic changes (Gurungand Bhandari 2009). Hence for resilient agriculture greater plant and crop diversity as well integrated farming systems are the need of the day. Like natural ecosystem greater bio diversity in man made ecosystem may help to mitigate losses due to changed plant-animal interactions which in turn may also influence crop yield. Perennial tree species may be less affected by climate change than annual grain crops. But tropical fruit trees may have the opportunity to spread beyond the present latitudes to much higher latitudes and altitudes. Such changes may have both positive and negative impacts on the lives of local public. Hence the information on the plant physiology and phenology will be helpful for the people to manage their cropping system.

As seen from the case study, climate change has affected 'ber' flower phenology and pollinator abundance and diversity. To advocate conclusive measures simulation studies along with real time studies on flower phenology and pollinator's dynamics in relation to abiotic factors are required.

References

Ahas, R., Aasa, A., Menzel, A., Fedotova, V.G. and Scheifinger, H. 2002. Changes in European spring phenology. *International Journal of Climatology* 22: 1727–1738.

Bartomeus, I., Ascher, J.S., Wagner, D., Danforth, B.N., Colla, S., Kornbluth, S. *et al*. 2011. Climate-associated phenological advances in bee pollinators and bee-pollinated plants. *Proceeding of National Academy of Science* USA 108: 20645–20649.

Bartomeus, I., Park, M.G., Gibbs, J., Danforth, B.N., Alan N. Lakso, A.N. and Win free, R. 2013. Biodiversity ensures plant–pollinator phenological synchrony against climate change. *Ecology Letters* 16: 1331–1338.

Basannagari, B. and Kala, C.P. 2013. Climate Change and Apple Farming in Indian Himalayas: A Study of Local Perceptions and Responses.doi: 10.1371/journal.pone.0077976

Biesmeijer, J.C., Roberts, S.P.M., Reemer, M., Ohlemu¨ller, R., Edwards, M., Peeters, T. *et al*. 2006. Parallel declines in pollinators and insect-pollinated plants in Britain and the Netherlands. *Science* 313: 351–354.

Chopra, D.P. and Srivastava, G. 1973. Ber varieties for Western Rajasthan. *Indian Horticulture* 17 (4): 27-30.

Coope, G.R. 1995. Insect faunas in ice age environments: why so little extinction? *In* J. Lawton R. May, eds. *Extinction rates,* pp 55-74. Oxford, UK, Oxford Univ. Press.

Devi, K. R., Atluri, J.B. and Reddy, C.S. 1989. Pollination ecology of *Zizyphus mauritiana* (Rhamnaceae). *Indian Academy of Science (Plant Science)* 99: 223-239.

Dunne, J.A., Harte, J. and Taylor, K.J. 2003. Subalpine meadow flowering phenology responses to climate change: integrating experimental and gradient methods. *Ecology Monograph* 73: 69–86.

Faegri, K. and van der Pijl, L. 1979. *Principles of Pollination Ecology*. Pergamon Press, Oxford.

Fenster, C.B., Armbruster, W.S., Wilson, P., Dudash, M.R. and Thomson, J.D. 2004. Pollination syndromes and floral Specialization. *Annual Review of Ecology Evolution Systematic* 35:375–403.

Fitter, A.H. and Fitter, R.S.R. 2002. Rapid changes in flowering time in British plants. *Science* 296: 1689–1691.

Gallai, N, Salles, J, Settele, J andVaissière, B.E. 2009. Economic valuation of the vulnerability of world agriculture confronted with pollinator decline. *Ecological Economics* 68: 810-21.

Gordo, O. and Sanz, J.J. 2006. Temporal trends in phenology of the honey bee *Apis mellifera* (L.) and the small white *Pierisrapae* (L.) in the Iberian Peninsula (1952–2004). *Ecological Entomology* 31:261–268.

Gaira, K.S., Rawal, R.S., Rawat, B., Indra, D. and Bhatt, I.D. 2014.*Current Science* 106 (12):1735-1738.

Gurung, G.B. and Bhandari, D.N. 2009. An integrated approach to climate change adaptation. LEISA Magazine 24.

Guerrero, V.R., Aguilar, R., Rodrýguez, M.S., Ashworth, L., Mikel, M.L., Bastida, J. M. and Quesada, M. 2014. A quantitative review of pollination syndromes: do floral traits predict effective pollinators? *Ecology Letters* 17: 388–400.

Hegland, S.J., Nielsen, A., La´ zaro, A., Bjerknes, A. L. and Totland, Ø. 2009. How does climate warming affect plant-pollinator interactions? *Ecology Letters* 12: 184–195.

Hewitt, G.M. 2000. The genetic legacy of the quaternary ice ages. *Nature* 405: 907–913.

IPCC (Intergovernmental Panel on Climate Change). 2007. *IPCC Fourth Assessment Report: Climate Change (AR4)*.

Jarvis, A., Ramirez, J., Anderson, B., Leibing, C. and Aggarwal, P. 2010. Scenarios of climate change within the context of agriculture. In: *Climate Change and Crop Production* (Ed. M.P. Reynolds) CAB International. 633p.

Kelly, A.E. and Goulden, M.L. 2008. Rapid shifts in plant distribution with recent climate change. Proc. *National Academy of Science* USA, 105: 11823–11826.

Klein, A.M., Dewenter, I.S. And Tscharntke, T. 2003. Pollination of *Coffea canephora* in relation to local and regional agroforestry management. *Journal of Applied Ecology* 40: 837–845.

Klein, A.M., Vaissière, B.E., Cane, J.H., Steffan-Dewenter, I., Cunningham, S.A. and Kremen, C. *et al.* 2007. Importance of crop pollinators in changing landscapes for world crops. *Proceedings of the Royal Society of Biological Sciences*, 274: 303-13.

Kundu, S.S., Parrek, O.P. and Gupta, A.K. 1998. Effect of time and severity of pruning on growth, flowering and fruiting in ber cv. Umran. *Annals of Arid Zone* 33 (1): 49-52.

Lal, M., Srinivasan, G. and Cubasch, U., 1996. Implications of increasing greenhouse gases and aerosols on the diurnal temperature cycle of the Indian subcontinent. *Current Science* 71 (10): 746-752.

Lonergan, S. 1998. Climate warming and India. In: *Measuring the Impact of Climate Change on Indian Agriculture* (Eds. A. Dinar, *et al*). World Bank Technical Paper No. 402, pp 33–67. Washington DC.

Macanawai1, A.R. Day, M.D. Tumaneng-Diete, T. and Adkins, S.W. 2011. The impact of rainfall upon pollination and reproduction of *Mikania micranthain* Viti levu, Fiji. In: Proceedings of 23rd Asian-Pacific Weed Science Society Conference 316-321p.

Meghwal, P.R., Khan, M.A. and Tewari, J.C. 2007. *Growing ber (Zizyphus mauritiana lam) for sustainable income and employment in arid and semiarid regions*. CAZRI, Jodhpur India. 26pp.

Memmott, J., Craze, P.G., Waser, N.M. and Price, M.V. 2007. Global warming and the disruption of plant–pollinator interactions. *Ecology Letters* 10: 1–8.

Menzel, A., Sparks, T.H., Estrella, N. and Roy, D.B. 2006. Altered geographic and temporal variability in phenology in response to climate change. *Global Ecology and Biogeographer* 15: 498–504.

Molnár, A.V., Tökölyi, J., Végvári, Z., GáborSramkó, G., Sulyok, J. and Barta, Z. 2012. Pollination mode predicts phenological response to climate change in terrestrial orchids: a case study from central Europe. *Journal of Ecology* 100:1141–1152.

Pareek, O.P. 2001. *Ber*. International Centre for Underutilized Crop. Southampton, UK. 282pp.

Pareek, O.P., Mukherjee, S. and Paliwal, R. 2007. Floral biology of ber –A review. *Agriculture Review* 28(4): 277-282.

Parmesan, C. 2006. Ecological and evolutionary responses to recent climate change. *Annual Review of Ecology Evolution Systematic* 37: 637–669.

Partap, T. 2001. Mountain agriculture, marginal land and sustainable livelihoods: Challenges and opportunities. In: *Proceeding of International Symposium on Mountain Agriculture in HKH Region*. ICIMOD, Kathmandu, Nepal.

Poonia, S. and Rao, A.S. 2013. Climate Change and Its Impact on Thar Desert Ecosystem. *Journal of Agricultural Physics* 13 (1): 71-79 pp.

Potts, S.G., Biesmeijer, J.C., Kremen, C., Neumann, P. Oliver Schweiger, O. and William E. Kunin, W.E. 2010. Global pollinator declines: trends, impacts and drivers. doi:10.1016/j.tree.2010.01.007

Rajan, S. 2009. Impact assessment of climate change in mango and guava research. In: *Information Technology Applications in Horticultural Crops* (Eds. P.M. Govindakrshnan, J.P. Singh, S.S. Lal, V.K. Dua, S. Rawat, S.K. Pandey,) pp. 36-42. Central Potato Research Institute, Shimla, India.

Reddy, P.V.R., Verghese, A. and Rajan V.V. 2012. Potential impact of climate change on honeybees (*Apis* spp.) and their pollination services. *Pest Management in Horticultural Ecosystems* 18(2): 121-127.

Robertson, C. 1929. *Flowers and Insects: Lists of Visitors to Four Hundred and Fifty Three Flowers*. Carlinville, IL.

Robinson, W.E., and Morse, R.A. 1989. The value of honeybees as pollinators of US crops. *American Bee Journal* 129 (1): 477-487.

Roy, D.B. and Sparks, T.H. 2000. Phenology of British butterflies and climate change. *Global Change in Biology* 6: 407–416.

Singh, M.P. 1984. Studies on the activity of some insect pollinators on Jujube (*Zizyphusmauritiana* Lamak). *Entomon* 9 (3): 177-180.

Wall, M.A., Timmerman-Erskine, M. and Boyd, R.S. 2003. Conservation impact of climatic variability on pollination of the Federally endangered plant, *Clematis socialis* (Ranunculaceae). Southeast. *Nature* 2: 11–24.

Walther, G.R. 2010. Community and ecosystem response to recent climate change. *Philosophical Transactions of the Royal Society* 365: 2019-2014.

Walther, G.R., Post, E., Convey, P., Menzel, A., Parmesan, C., Beebee, T.J.C. *et al.* 2002. Ecological responses to recent climate change. *Nature* 416: 389–395.

Willmer, P.G. and Stone, G.N. 2004. Behavioral, ecological and physiological determinants of the activity patterns of bees. *Advances in the Study of Behavior* 34: 347-466.

❑❑❑

Chapter 29

Livelihood Security through Management of Small Ruminants in Hot Arid Zone

B.K.MATHUR*, J.P. SINGH, UBED ULLHA AND R.C. BOHRA

ICAR- Central Arid Zone Research Institute, Jodhpur--342 003, Rajasthan

*E-mail: mathurbasant@yahoo.co.in

Introduction

Indian hot arid zone which is about 12% of total geographical area of the landmass of 0.32 million km^2 has maximum covering in western Rajasthan i.e. 61% of the total area whereas the other areas in arid region are available in the states of Gujarat, Punjab, Haryana, Andhra Pradesh and Karnataka accounting for 20, 5, 4, 7 and 3% of hot arid area whereas the cold arid area of 8.4 million ha lies in the state of Jammu & Kashmir covering the Leh and Ladakh region.The hot arid area is characterized by frequent droughts of 47% of frequency in the last century of moderate to severe nature. Due to higher occurrence frequency of droughts in this region of state, the livestock assumes great importance as a drought management measure as agriculture is at the mercy of rainfall pattern which is very uncertain in amount as well as distribution coupled with poor soil condition, higher evapotranspiration and higher wind velocity causing the soil erosion. According to the latest (2012) livestock census, Rajasthan harbors 57.77 million heads of domestic animals while it was 57.89 million in 2007, which showed a little increase of 0.12 million heads. However the population was 54.35 and 49.1million in 1997 and 2003, thus during this period a decrease of 10.08% was observed mainly due to drought years. The hot arid zone of Rajasthan is comprised of twelve districts of the state lying in the western part and these are Barmer, Bikaner, Churu, Hanumangarh, Jaisalmer, Jalor, Jhunjhunu, Jodhpur, Nagaur, Pali, Sriganganagar and Sikar and this arid region of Rajasthan has livestock population of 30.18 million which is about

52.27% of the total population of the state (Census, 2012). Out of the total livestock species available in the state large number of sheep and goats 6.88(%) and 12.79(%) millions are present in arid zone of Rajasthan. Excepting the districts of Hanumangarh and Sriganga-nagar where the large ruminant production system of cattle and buffalo is dominant in all other districts of arid zone the small ruminants production system accounting 65.17% of total livestock population is prevalent.

Table 1. Trend of Livestock Population in Arid Zone of Rajasthan (Million)

Livestock	Livestock Census Year				% change		
species	1997	2003*	2007	2012	1997-12	2003-12	2007-12
Cattle	4.96	4.12	5.05	6.18	24.59	50.00	22.38
Buffaloes	3.16	3.2	3.45	3.95	25	23.44	14.49
Sheep	10.49	7.36	8.43	6.88	-34.41	-6.52	-18.39
Goat	9.53	8.37	11.7	12.79	34.21	52.81	9.32
Camel	0.53	0.39	0.35	0.28	-47.17	-28.21	-20.00
Others	0.22	0.18	0.13	0.11	100	-38.89	-15.38
Total	28.57	23.62	29.11	30.18	5.63	27.77	3.68

*Between 1997 -2003 drought years

The harsh climatic conditions prevailing in the arid region, e.g., erratic rains and frequent droughts would suggest that, it is not very suitable for crop farming. Livestock farming has some in built superiority over crop farming as far as growth; stability and resource conservation are concerned. On an average, the region experiences three years of drought in every 10 years. The natural forces constituting the soil-climatic complex, which conspire to reduce the crop productivity and cause instability in agricultural production, have much less impact on livestock farming. This is due to differences in the nutritive value of natural vegetation, which mainly sustains livestock.

The superiority of livestock farming for development of arid region is further highlighted by the fact that this region is endowed with some of the best breeds of livestock and drought-hardy perennial grasses. The local breeds of livestock have acquired certain characters to withstand the arid climate, and the characters have been transmitted through generations to make the present hardy breeds of animals. However, due to lack of proper nutrition, genetic potentiality of these animals has not been expressed to its maximum level.

The major livestock breeds in the region are:

Cattle Tharparkar, Rathi, Nagori and Kankrej.

Sheep Marwari, Jaisalmeri, Chokla, Nali, Magra, Pugal and Sonadi.

Goat Marwari, Kutchi and Parbatsari(Sirohi).

Camel Bikaneri and Jaisalmeri.

For arid region of Rajasthan the general climatic conditions, topographical features and biotic factors do not encourage agricultural operations in the absence of extractive industry the peasantry has to fall upon animal husbandry as their main occupation. Rearing some of the finest breed of cattle, camel, sheep and goats known for their endurance making much use of the meagre feed resources which are grasses ,herbs, shrubs, tree leaves and cultivated feed and fodder crops.

Table 2. Common available feeding material from natural rangelands in arid region with crude protein percentage

S.No.	Botanical name	Local name	C.P.%
	Grasses		
1.	*Aristida funiculate*	*Lapra*	4.38
2.	*Brachiaria ramosa*	*Murat*	11.37
3.	*Cenchrus biflorus*	*Bhurat*	15.31
4.	*Cenchrus ciliaris*	*Dhaman*	8-12
5.	*Cenchrus setigerus*	*Moda Dhaman*	8.77
6.	*Cynodon dactylon*	*Dob*	10.10
7.	*Dactyloctenium aegypticum*	*Ganthia*	10.10
8.	*Eluesine comprass*	*Tantiya*	13.99
9.	*Eragrostis tremula*	*Chidi grass*	6-10
10.	*Lasiurus sindicus*	*Sewan*	9.05
11.	*Panicum antidotale*	*Gramna*	7.1
12.	*Tetrapogon tenellus*	*Kagiyo*	13.36
13.	*Urochloa panicoides*	*Kuri*	12.02
	Weeds		
1	*Achryanthes aspera*	*Antipada*	19.9
2	*Amaranthus viridis*	*Chandelio*	4.9
3	*Celosia argentia*	*Makhamal*	12.5

contd...

4	*Cyperus rotundus*	*Motha*	9.1
5	*Digera muricata*	*Lolru*	12.07
6	*Indigofera cordifolia*	*Bekario*	18.0
7	*Trianthema portulacastrum*	*Sathi*	14.31
8	*Tribulus terristris*	*Kanti*	18.18
	Trees/shrubs		
1	*Acacia nilotica*	*Desi babul*	13.9
2	*Acacia senegal*	*Kumat*	17.48
3	*Acacia tortalis*	*Israili babool*	14.6
4	*Azadirachita indica*	*Neem*	15.8
5	*Calotropis procera*	*Aak*	13.5
6	*Capparis deciduas*	*Kair*	2.05
7	*Cordia gharaj*	*Gundi*	15.3
8	*Euphorbia caducifolia*	*Thor*	5.61
9	*Prosopis cineraria*	*Khejri*	14.63
10	*Tecomella indica*	*Rohida*	12.2
11	*Prosopis juliflora (Pods only)*	*Vilayati babul*	13.8
12	*Salvadora persica*	*Metha Jal*	10.5
13	*Ziziphus nummularia*	*Jharberi*	14.1

Source: Mathur B. K. 2008

Simultaneously, the grazing/browsing areas are progressively shrinking with encroachment by ever growing human population necessitating to evolve economical supplementary feeding (Mathur 1996). Cultivated feed & fodder crops grown are: *Pennisetum typhodes* (Bajra); *Sorghum bicolor* (Jowar); *Phaseolus aconitifolius* (Moth); *Cyamopsis tetragonoloba* (Guar); *Sesamum indicum* (Til); *Phaseolus aureus* (Moong).

Fodder resources in hot arid region

These resources include leaves of various fodder trees, shrubs and pods. A large number of trees and shrubs have been identified which are being lopped for fodder purpose. Important tree fodders and utilization of their nutritive value have been reviewed by many authors (Patnayak, *et al.*, 1995; Patnayak, 1992; Sharma *et al.*, 1992; Singh, 1981). Although a large number of fodder trees have been identified, only a very few species have been studied in detail and utilized in conventional feeding practice. There is need to identify and study in detail promising species of tree fodders in terms of forage production, animal acceptability and nutritive value. The production potential of livestock is not optimum due to scarcity of feed and fodder resources. The deficiency of green fodder, dry fodder and concentrate was estimated to 36, 61 and 79 per cent, respectively in western Rajasthan during normal years. This situation is further aggravated during drought and severe drought years.

Although tree fodders are considered as emergency fodders, they serve as potential feed resources for small ruminants and provide about 50% of the total dry matter consumption. They are utilized as freshly loped leaves or by conserving as dry fodders. Dry matter in most tree varies from 12 to 20% on dry basis. Very high calcium and low phosphorus are a common feature in almost all tree fodders which when fed solely leads to a very wide Ca:P ratio. Dry matter intake from most of the tree leaves varies from 2.5 to 3.5 kg/100 kg B W. Although they are rich in protein, the digestibility of crude protein is very low (25 to 30%) in many cases. Inspite of higher intake of nitrogen and calcium, retention of nutrients in general is poor and phosphorus balance is negative or marginally positive. Tree fodders, therefore, should be fed in mixtures with other feeds, specially containing high phosphorus, such as wheat bran, rice bran, grains etc.

Although tree fodders are consumed by all livestock, goats have special preference compared to others. The digestibility of nutrients in tree fodders in generally higher in goats with better growth rates than in sheep when fed dried tree leaves. (Bohra, 1980; Singh and Bhatia, 1982). Low digestibility of protein is anti-nutritional factors such as mimosne, cyanogens, saponins, oxalates etc. The deleterious effects of various anti-nutritional factors and their possible amelioration have been reviewed by Kumar (1991).

Proper methods of collection, conservation and processing are important aspects of utilization of tree fodders such as *Ailanthus excelsa, Z. nummularia* and *P.cineraria* were used in complete feeds for meat lambs. (Bhatia and Patnayak 1982). The proximate analysis of some common top feeds particularly of desert region is given in table 3.

Nutritive value of some most common top feed : Leaves of *A. excelsa, Z. nummularia* and *P. cineraria,* the three most important top feeds, are also high in the TDN (total digestible nutrients) scale (64%, 50% and

41.0%, respectively), while the DCP (digestible crude protein) value of *A. excelsa* leaves in quite high (16%), whereas, that of the more palatable *Z. mummularia* or *P. cineraria* leaves, are considerably low (5.6% and 4.5%).

Table 3. Proximate components (% DM basis) of some desert top feeds

Scientific name	Local name	Crude protein	Ether extract	Crude fibre	Nitro-gen free extract	Ash	Phosp-horus	Cal-cium	Magne-sium
Acacia nilotica	Babul	13.9	-	9.2	69.8	7.1	0.10	2.6	0.4
Acacia Senegal	Kumut	10.3	-	9.5	65.7	16.4	0.05	6.9	0.6
Ailanthus excelsa	Ardu	19.6	3.7	13.5	47.7	15.5	0.20	2.3	-
Albizia lebbeck	Siris	16.8	4.0	31.5	36.2	11.5	0.15	2.57	-
Azadirachta indica	Neem	12.4	3.5	11.4	66.6	9.4	0.10	3.4	0.6
Dalbergia sissoo	Shisham	16.2	4.2	24.6	46.2	9.2	0.21	2.3	-
Ficus bengalensis	Bargad	9.6	2.6	26.8	51.6	9.3	-	-	-
Ficus religiosa	Pipal	9.7	2.7	26.9	45.8	-	-	-	-
Prosopis cineraria	Khejri	13.9	1.9	20.3	59.2	6.5	0.20	1.9	0.5
Prosopis juliflora	Vilayati babul	21.4	-	20.8	50.0	7.7	0.20	1.5	0.5
Salvadora oleides	Pieujal	9.6	-	9.3	40.2	40.8	0.10	11.9	0.7
Salvadora persica	Khara jal	14.2	-	9.4	44.9	31.4	0.20	8.8	0.7
Tecomella undulata	Rohida	12.2	-	15.8	63.0	8.9	0.20	3.4	0.9
Ziziphus nummularia	Bordi	11.7	3.0	16.0	65.0	7.2	0.20	1.6	0.3

Sources: Ganguli *et al.* (1964); Acharya and Patnayak (1977)

In the by-products mentioned above the following characteristics are most important: protein concentration and its biological value quantitative and qualitative composition of amino acids, digestibility, level of energy, fats and carbohydrates, vitamin and mineral content and the amount of fibre and substances that might be hazardous or toxic to the organism. The quantity of by products used as feed depends on the country's resources and the technical equipment's used for its preparation preservation and improvement.

Table 4. Levels of different anti-nutritive factors in agro-industrial byproducts and methods for their removal

No.	Name of feeds	Anti nutritive factor	Level of anti nutritive factor	Method for removal/ destroying
	1	2	3	4
1.	Babul seeds (*A.nilotica*)	Tannins	5%	-
2.	Cotton seed cake (Deoiled)	Gossypol	22g/100g including 0.15g/100 in free form	Biling with water
3.	Castor bean-meal	Ricin	-	High pressure (20 psi for 60 min) Treatment with *Aspergillus niger*
4.	Groundnut-cake	Aflatoxin	0.12 to 1.0ppm	Ammonia treatment
5.	Guar-meal	Trypsin inhibitor and gum	-	-
6.	Kapok	Cyclopono penoic acid	-	-
7.	Karanj-cake (*Pongamia glabra*)	Karanjine Pongamol Glabarin	10.15 mg/100g	Solvent extraction
8.	Kosum-cake (*Seleichera oleosa*)	HCN	2.4 mg/100 g	-
9.	Linseed-meal	Aflatoxin HCN Linamarin	- 0.03% -	- - -
10.	Mahua-cake (*Madhuca indica*)	Saponins (Mowrin) Tannins Sapogenol	4.0-7.0% 1.47% 8.39%	Washing with water/Alkali treatment followed with HCl treatment
11.	Mango seed Kernal	Tannins	5-10%	-

contd...

12.	Neem-cake (*Azadirachta indica*)	Nimbin Salalnnin Azadirachtin Tannins	- -	1% NaOH treatment followed by water washing
13.	Nahar seed-cake	Unidentified	-	-
14.	Panewar seeds (*Cassia tora*) Acid	Crysophanic	0.083%	-
15.	Rapeseed/ Mustard-cake	Tannins Glucosinolates	- 5 to 10 mg/g	-
16.	Rubber seed-cake	HCN	9 mg/100g	-
17.	Soybean-cake	Trypsin inhibitor Antivitamin E Antvitamin D Aflatoxin	420 mg/g	Heat treatment at proper temperature moisture, duration and particle size. Steam treatment for 15 min Auto claving Ammonia treatment
18.	Sunflower	Aflatoxin Chlorogenic acid Tannins	- - 1.2%	- -
19.	Safflower-meal	Aflatoxin	-	-
20.	Sal seed-meal	Tannins	8.3-13.8%	0.1 N NaOH treatment
21.	Spent tea leaves	Tannins	12%	-
22.	Spent annatto seeds (*Bixa orellana*)	Tanic acid	10%	-
23.	Tumba seed-cake (*Citrullus colocynthis*)	Glucosinolates	-	Soaking in water
24.	Vilayati babul (*Prosopis juliflora*)	pods Tannins	1-1.5%	-
25.	Water hyacinth	Oxalic acid	2.4%	-

Source : National Dairy Research Institute, Karnal, Mathur BK 1997&2011,Ranjhan SK 1990

Utilization of Un-Conventional Feed Resources

Non conventional feed (NCF) or non-conventional feed resources (NCFR) or unconventional feed (UCF) are relative term and may differ from country to country and region to region and time to time in the same country. Non-conventional feeds may be grouped according to availability from agro industrial by products and classified as (a) Vegetable protein source (b) Animal protein sources (c) Energy sources and (d) other miscellaneous unconventional feeds. For this purpose, various non-conventional feeds need to be screened for their possible use as regular livestock feed to enhance the nutritional output per unit area. Additional, utilization of agro-industrial by-products also mitigate the problem of disposal of industrial wastes. Inclusion of non-conventional feeds reduces the quantum of concentrate in the diet for optimum animal performance.

Success Stories: Using Un-Conventional Resource of Desert Region as feed

Tumba (Citrullus colocynthis) Seed Cake - A Cheaper Feed Resource for Livestock Feeding in Arid Region

Tumba plant, an annual creeper, grows naturally in abundance in hot arid areas of the country with minimum possible water availability. It grows and multiply very fast during monsoon season and its fruits; Tumba - are available in the month of October-November. Mature fruits are golden yellow in colour, ball like, 10-12 cm in diameter. The taste of the fruit is very bitter. Tumba seed are rich in fat, having upto 16% oil. The Tumba (*Citrullus colocynthis*) seed cake (TSC) a byproduct of the oil extraction industry is nutritionally rich as it contains 16 to 22% CP. Presently, TSC is available in abundance in arid regions and is being used as a fuel for furnances in factories, and its thus wasted.

Feeding trials were conducted in Central Arid Zone Research Institute, Jodhpur since 1986 on cattle to identify and evaluate TSC as a source of livestock feed. The studies clearly indicates that in cattle the conventional concentrates can safely be replaced by TSC to the extent of 25%, as a regular ingredient, which constituted guar (*Cymposis tetragonaloba*) korma and oil cakes (mostly til (*Sesamum indicum*) and cotton (*Gossipuim spps.*) seed cake) and also pelleted cattle feeds.

It is observed that the TSC replacement did not affect the palatability and intake of feeds and fodder. There was no significant

(1>0.5) difference in milk yield pattern of control and treatment groups, in terms of quantity and quality. TSC feeding to heifers upto calving and onwards did not show any ill effect on different reproductive parameters.

It can, therefore, be inferred that a simple practice of inclusion of tumba seed cake (TSC) in animal feed will definitely lower the cost of animal feeding by 18% to 20% without having any adverse effect on the production, general health and reproductive performance. In addition, it would result in the utilization of locally available non-conventional cheaper protein source, abridging the gap between demand and supply of the scarce protein, thus, benefiting the marginal farmers appreciably (Mathur *et al.*, 1989, Mathur 1996).

Feeding Vilayati Babool (Prosopis juliflora) Pods Powder in small ruminants

Goats: Ten (10) goats in late lactation were divided into two group of 5 each i.e. group I control and group II treatment. Animals of each group were offered weighed quantity of roughage and concentrate on as such basis (consisting of 35% grinded bajara, 40% tumba seed cake and 25% groundnut cake). However, in treatment group 35% bajra in concentrate was replaced by *Prosopis juliflora* pod powder (PJPP) making ration near about isocaloric. Study showed that *Prosopis juliflora* pod powder can be used up to 35% in the concentrate of goats (Mathur *et al.,* 2004). No significant effect was observed on blood parameters and milk yield of goats in late lactation, during extremes of summer in arid zone(Mathur *et al.,* 2003).

Sheep: A feeding trial was conducted to evaluate the acceptability and palatability of ration comprised of *Prosopis juliflora* pod husk (PJPH) in Marwari sheep. The results of the study indicated that *P. juliflora* pod husk can be used up to 50 percent level in the concentrate along with tumba (*Citrullus colocynthis*) seed cake as low cost ration of the sheep without any adverse effect on animal health (Mathur *et al.*, 2002).

The acceptability and palatability of formulated concentrate mixture having *Prosopis juliflora* pods was high with no ill effect on health, increases milk and the cost of cattle production reduces significantly (Mathur *et al.,* 2009).

(a) (b)

Fig. 1 (a) Grazing and feeding of handmade Concentrate **(b)** Mixture having Villayati Babool (*Prosopis juliflora*) pods powder

Milk Yield of Lactating Marwari Goat Fed on Prosopis cineraria v/s Colophospermum mopane Leaves

Twelve (12) lactating Marwari goats were divided into 3 groups of 4 each namely T_1, T_2 and T_3. T1 and T_2 groups were stall fed for about five week while T_3 group was sent for grazing. Animals of T_1 and T_2 groups were offered balanced diet as per requirement comprising of concentrate feed and *C. ciliaris* as roughage. Additional T_1 group W.S offered with *P. cineraria* leaves while T2 group replaced by *C. mopane* leaves. It was observed that T_1 group the palatability of *P.cineraria* was 95% of the quantity offered while in T_2 group it was in reducing trend with 5% to 15% from first and fifth week of trial. The final body weights after V week of trial for T_1, T_2 and T_3 group was 31.95±1.099, 32.40 ±1.982; and 28.4 ±1.699, kg respectively. Animals of T_1 and T_2 group maintained their body weight while the grazing group showed in average group weight. T_1 and T_2 group consumed the concentrate offered equally, however *Cenchrus ciliaris* grass kutter consumption was higher in T_1 group (1.65 to 1.94 kg/day/group) as compared to T_2 group (1.41 to 1.72 kg/day/group). Water intake in groups T_2 and T_3 ranged from 7.40 to 10.62 and 6.14 to 9.0 liters/day. The total milk yield on an average / animal for a period of 45 days in T_1, T_2 and T_3 was 28.07., 22.53 and 19.73 litres respectively. The milk yield was 19.74% higher in *P.cineraria* fed group (T_1) than *C. mopane* fed group (T_2) and 29.71% more than T_3 group maintained exclusively on grazing. Thus it was observed that the palatability of dry *Colophospermum mopane* leaves was low and decreased with progress of feeding from 15 to 5% from 1st to 5th week and the traditional local *P. cineraria* leaves were better source of supplementation even in the dry form supporting the milk yield of goats (Mathur *et al.,* 2006). Although sole feeding of fresh C. *mopane* leaves is not possible in goats but it can very well be fed as a part of complete rations at 35-40% of whole ration (Patil *et al.,* 2007).

Lani (*Salsola baryosma*)

Salty shrub Lani (*Salsola baryosma*) of arid region available at vegetative stage can be a good source of fodder to animals. Shoots consists of mainly fleshy stem, since leaves are very minute, consisting not more than 5- 7% of the total biomass. The acceptability and palatability of fresh cut shoots showed that Lani is palatable (*Mathur et al., 2007*).

Lana (*Haloxylon salicornicum*) Seeds

Lana (*Haloxylon salicornicum*) salty weed available in the arid region consumed mostly by sheep and goats.Lana produces seeds having CP 18.60% found to be good source of protein in the concentrate part of lactating cows replacing about 25% of the conventional sesamum cake (Anonymous, 2007) and in camel, replacing concentrate mixture showed good palatability and digestibility (Mathur *et al.,* 2009).

Fig. 2. Lana (*Haloxylon salicornicum*) Seed Arid Feed Resource for Livestock

Hardwickia binata as a Supplement in Goats

H. binata is available in plenty in the arid region having rainfall above 300 mm. the study showed that the fresh leaves are palatable to goats and sole feeding supports the body weight growth of growing kids (Patil *et al.,* 2009).

Nutritional value of fresh leaves of Colophospermum mopane and Hardwickia binata as a supplement in goats

In a feeding trial of 45 days duration ten numbers of Marwari and Prabatsari growing bucks of 8-9 months age and the results showed that sole feeding of *C. mopane* leaves during 45 days of feeding to growing bucks was found to result in body weight loss of 3.36 Kg at the rate of 74.66 g/h/d where as the sole feeding of *H. binata* leaves resulted in body weight gain of 4.12 Kg at the rate of 91.65 g/h/d thus the fresh leaves of *C. mopane* and *H. binata* are the feeds of sub maintenance and productive type, respectively. Total DM Intake (g/h/d) was very less 515.58±47.25 for *C. mopane* with low fibre digestibility. The sole feeding of *C. mopane* leaves is not possible due to low acceptability and palatability.

Utilization of Seasonal Weeds of Arid Zone

The monsoonal weeds available in the rocky and sandy habitat of arid region have a distinct relative preference index for grazing sheep and goats and it was studied to be in the order of 1. Kanti (*Tribulus terristris*), 2. Kagio (*Tetrapogon tenellus*), 3. Santo (*Trianthema protulacastrum*), 4. Lolaru (*Digeria muricata*), 5. Bekario (*Indigofera cordifolia*), and 6. Gangan (*Grewia tanax*) for varieties tested in the region. (Patil *et al., 2005*).

Feeding of Mineral Mixtures and Common Salt

Feeding managers of animals should have mineral bricks and common salt, so that deficiency of minerals and common salt may not occur. Calcium, phosphorus, copper and magnesium was significantly low in livestock of villages in arid zone of western Rajasthan. Pica in cattle, buffaloes, camels and goats in this area is another problem related to mineral deficiencies (Mathur *et al.,* 2005, Mathur *et al.,* 2009). Milk fever in this area occurs mainly due to calcium deficiency only within one week post-parturient having sufficient level of magnesium.

The infertile cattle bears a history of under fed with malnutrition, and per rectum examination of genitalia revealed hypoplastic and/or smooth ovaries and persistent corpus luteum was very common, and they responds to mineral mixture–vitamin supplementation. There is need of area specific mineral mixture supplementation.

Coping Strategies for Minimizing Impact of Extreme Climatic Conditions on Production Performance of Grazing Sheep in Thar Desert

Thar Desert is characterized by high ambient temperature, low rainfall, high sun shine and wind velocity. An experiment was initiated in the month of February, 2012 at Chandan area of Jaisalmer district, maintaining sheep under continuous grazing system on fenced over mature sevan (*Lasiurus sindicus*) pasture as per the carrying capacity. The objective of study was to develop coping strategies for minimizing impact of extreme climatic conditions on livestock production of Thar Desert. Thirty six (36) growing sheep of Jaisalmeri breed consisting of 32 female and 4 male were divided into four groups, i.e., C1 (Control), and T1 (Treatment) having 6 animal each (5 female +1 male) /hectare as per the standard norm, while C2 (Control), T2 (Treatment) having double grazing pressure (over grazing) with 12 animals each (11 female+1male) /hectare . On 2.0 x 2.0 sq. meter quadrate basis the initial and final dry matter (D.M) percentage, D.M. yield in quintals, protein percentage were 86.25 ± 1.47 and 54.47 ± 1.84(%), 50.44 ± 3.0 and 19.43(qtls) ± 2.37, 2.62 ± 0.55 and 5.27 ± 0.29(%). Animals of control groups were provided balanced concentrate feed@ 200gms /animal, and health care as per requirements whereas animals of treatment group were not provided with such facilities. Live body weight changes over a period of 18 months with 3 wool shearing viz., September 2012, March 2013 and September 2013 were taken (Fig. 4).

Fig. 3. Sheep grazing on Sewan (*Lasurus sindicus*) pasture at CAZRI farm, Chandan, Jaisalmer

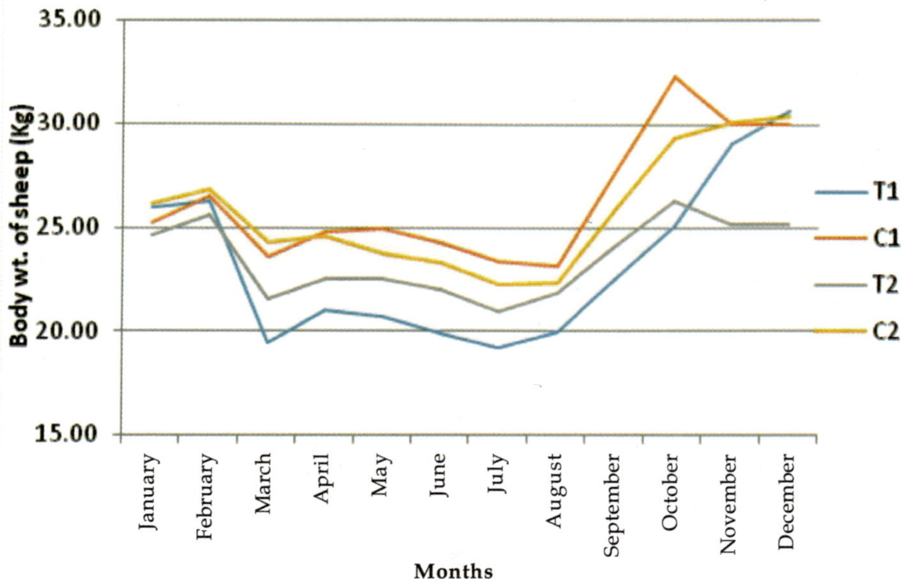

Fig 4. Body Weight Changes in Jaisalmeri Sheep on Lasiurus Sindicus Pasture

During the first shearing ,the control group viz., C1 and C2 gained average live body weight of 2.7 and 3.26 kg respectively, over their initial, while the treatment groups T1 and T2 gained/loss 0.77 and (-) 0.55 kg over the initial body weight. The average wool weight of first shearing were 731.83 ± 47.31, 702.27 ± 57.96, and 726.83 ± 61.26, 593.50 ± 30.24 gms. of C1, C2 and T1, T2 respectively. Average wool weight for second shearing were 735 ± 26.88, 675.91 ± 57.17, and 619.50 ± 39.09, 640.83 ± 43.66 and for third shearing 1025.33 ± 97.81,746.08 ± 39.96, and 988.33 ± 101.91, 599.33 ± 57.20 gms. of C1, C2 and T1, T2 respectively. The grazing time taken minutes/hour by sheep of treatment group was higher than the control group (Fig. 5). Water consumed was higher in summer months in both the control and treatment groups. Water consumption was significantly higher in the month of November 2012 than November 2013, may be due to changing ambient conditions during the period (Mathur *et al.*, 2013).

Thus, supplementation of concentrate and health management results into increase in live body weights and wool yield. The study shows that maintaining grazing sheep under extreme climate condition exclusively on fenced over mature Sevan (*Lasiurus sindicus*) pasture causes animal production loss, which can be mitigated by adopting scientific feeding and health management practices.

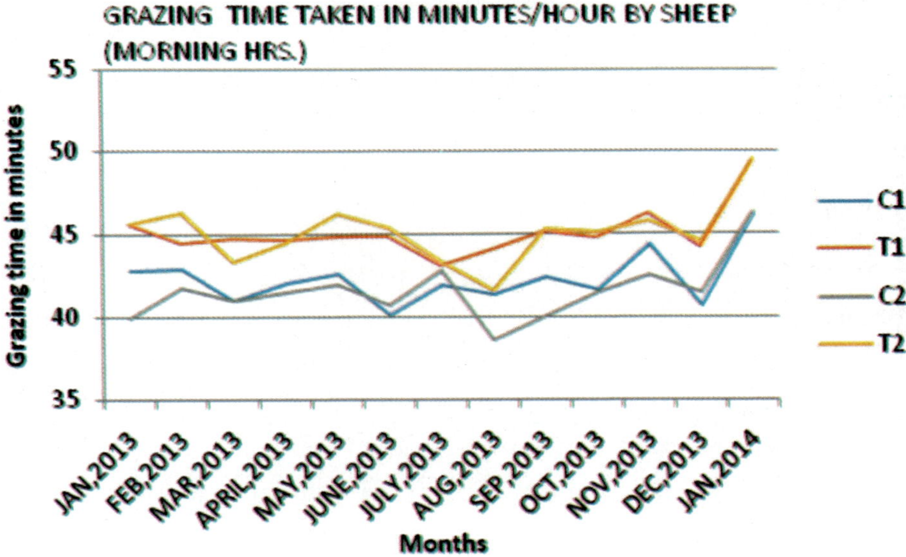

Fig. 5. Grazing time in minutes/hour by Sheep on *L.Sindicus* over matured pasture

Feeding management of livestock during drought can be summarized as follow :

1. Preparation of ration utilizing top feeds
2. Providing vitamin "A" doses
3. Cheaper and balanced concentrate utilizing local non-conventional feed resources
4. Addition of leguminous crop by products
5. Chopping of straw
6. Mineral mixture and common salt
7. Deworming

Health Management of Livestock

Broadly common diseases are divided in to four groups:

1. Viral diseases: Rinder pest, (Cattle plague), Foot & Mouth disease, Rabies and Sheep pox.
2. Bacterial diseases: Anthrax, Tuberculosis, Mastitis, Haemorrhagic Septicaemia, Black-quarter and Pneumonia.
3. Parasites and Parasitic diseases: Ectoparasites (ticks and mites), Endoparasites (Protozoan, Nematodes, Trematodes and Cestods). Deworming: Animal should be dewormed with broad spectrum anthelmintic(Albendazole/Fenbendazole/Closantel) thrice yearly. In case of ectoparasites: Spray/dusting should be done at regular intervals with ectoparasiticide (eg., deltamethrin). CLOSANTEL anthelmintic action is both endoparasiticide, and ectoparasiticide. These simple treatment will increase the availability of nutrients to livestock.
4. Miscellaneous diseases & Pathological conditions: Urinary calculi, Milk fever, Grass tetany, Cobalt deficiency, PICA, Vitamin A' Deficiency and Paraplegia.

Future Research and Development Needs

1. Agricultural production in arid region is basically based on animal husbandry, hence agriculture policy should be readjusted.

2. Establishment of fodder banks at district/town levels and veterinary first aid facilities with pedigreed males can ensure tremendous increase in livestock productivity.
3. Establishment of perennial components in pasture lands and planning of fodder trees will be helpful.
4. Sustainable integrated farming system models should be developed for different situations to balance livestock, environment and human needs.
5. Land use should be according to its capacity, with priority to agro forestry, pasture development and agro- horticulture.
6. Greater emphasis is to be given on livestock farming rather than crop farming in arid region.
7. Low- producing males should be castrated and selection of genetically superior animals within local breeds may be ensured for future breeding programmes.
8. Conservation of local biodiversity is necessary through need based policies and programmes.
9. Rainfed farming should be banned in non- irrigated areas receiving up to 150 mm precipitation annually. Instead, the farmers should be encouraged, educated and financed to grow endemic, nutritive and highly productive grasses.
10. Use locally available suitable alternate feed resources, e.g., agro industrial byproducts and value addition of top feeds, to bridge the gap between demand and supply of feed.

References

Administrative Bulletin 2008-09. Directorate of Animal Husbandry, Govt. of Rajasthan Jaipur, Rajasthan.

Annual Report 2008. Central Arid Zone Research Institute, Jodhpur

Anonymous 2007. Feeding Lana (*Haloxylon salicornicum*) and Saji (*Haloxylon-recurvum*) to Lactating Rathi Cattle. Annual Report, CAZRI 2007, pp. 101-102.

Balak Ram., 2003. Impact of human activities on land use changes in arid Rajasthan: Retrospect and prospects. In *Human Impact on Desert environment*,. (Eds. PratapNarain, Kathju, S., Amal Kar, Singh, M.P. and Praveen-Kumar), pp. 44-59 Scientific Publishers, Jodhpur, India.

Bohra, H.C. and Ghosh, P.K. 1977. Effect of restricted water intake during summer on the digestibility of cell wall constituents, nitrogen retention and water excretion in Marwari sheep. *Journal of Agricultural Science* (Camb.) 89: 605-608.

Census of Livestock 2012.Government of Rajasthan, Jaipur.

Chakrawarti Singh and Rollefson, I.K. 2005. Sheep pastoralism in Rajasthan-Still a viable livelihood option? *Workshop report*, LHPPS, Sadri, Dist.-Pali, Rajasthan.

Dutta, T., Taneja, V.K. and Singh, A., 1995. Effect of climatic variables on average daily milk production in crossbred cattle. *Indian Journal of Animal Sciences* 65(a) : 1004-1007.

Gahlot, A.K. 2005. Animal health management in drought prone area. Seminar on *Sustainable livestock production in arid region* March 22-23, 2005.College of Veterinary and Animal Science, Bikaner (Raj.).

Kaushish, S.K. 2006. Sheep breeds of India and Their production. In: *Strategies for Sustainable Livestock Production* (Eds. Kaushish, S.K., Bohra, H.C., Mathur, B.K., Patel, A.K. and A.C. Mathur). Kalyani Publishers Ludhiana. Pp. 323

Kaushish, S.K., Bohra, H.C., Mathur, B.K. and Patel, A.K. 1998. In *Fifty Years of Arid Zone Research in India* (Eds. Faroda, A.S. and Manjeet Singh) pp. 393-426.

Mathur, B. K. and Bohra, H.C. 1993. Nutritive value of *Prosopis juliflora* leaves and pods. Presented at the workshop on potential of *Prosopis* spp. for Arid and Semi-Arid Regions of India, Jodhpur. Jodhpur: *Central Arid Zone Research Institute*, November 22-23, 1993.

Mathur, B.K, Bhati, T.K and Tewari, J.C. 2006b. Comparative palatability of *Prosopis cineraria* v/s *Colophospermum mopane* leaves in lactating Marwari goat in arid region. In *National Symposium on "Livelihood Security and Diversified Farming Systems in Arid Region"*. January 14-16, 2006. Organized by Arid Zone Research Association of India, CAZRI, Jodhpur. No.3-2, pp.51-52.

Mathur, B.K., Patil, N.V., Mathur, A.C. and Patel, A.K. 2005. Impaction of Rumen of desert buck due to Polythene bags- a case study. *Livestock International* 10(5):17-20.

Mathur, B.K. 1996. Effect of replacement of cotton (*Gossypium* spp.) seed cake from Tumba (*Citrullus colocynthus*) seed cake and sun dried poultry droppings on the performance of sheep. Ph.D. thesis submitted to Rajasthan Agriculture University, Bikaner, Rajasthan.

Mathur, B.K. 2003. Livestock: human need for sustainability in arid environment. In *Human Impact on Desert Environment* (Eds. PratapNarain, S. Kathju, Amal Kar, M. P. Singh and Praveen- Kumar). Arid Zone Research Association of India and Scientific Publishers (India), Jodhpur, India. pp. 506-514.

Mathur, B.K., Abichandani, R.K. and Patel, A.K. 1999. Higher incidences of twinning in Marwari goats by supplementary feeding in hot arid region. In *Proceedings IX Animal Nutrition Conference*2-4th December 1999. Acharya N.G. Ranga Agricultural University, Rajendra Nagar, Hyderabad. Abstract No. 237, pp. 144.

Mathur, B.K., Bohra, H.C., Patel, A.K. and Kaushish, S.K. 1998. *In situ* Rumen degradability of Khejri (*Prosopis cineraria*) leaves in Marwari sheep. Presented at the Golden Jubilee Seminar on. "Sheep, Goat and Rabbit Production and Utilization for Maximizing Production Efficiency, pp., 27. *Indian Society for Sheep and Goat Production and Utilization* Jaipur: CSWRI, April 1998.

Mathur, B.K., Harsh, L.N., Mathur, A.C., Kushwaha, H.L., Bohra, H.C., Prajapat, Dinesh and Tiwari, J.C. 2009.Acceptability and Palatability of a Easily Processed, Balanced and Cheaper Concentrate Mixture Containing *Prosopis juliflora* Pods in Lactating Tharparkar Cattle. In *International Conference on Nurturing Arid zones for People and the Environment: Issues and Agenda for the 21st Century* Proceedings (Abstracts) of AZCONF 2009, 24-28 November, CAZRI, Jodhpur.pp.242.

Mathur, B.K., Mittal, J.P. and Prasad, S. 1989. Effect of tumba cake (*Citrullus colocynthis*) feeding on cattle production in arid region. *Indian Journal of Animal Sciences* 59: 1464-1465.

Mathur, B.K., Patel A.K., Mathur, A.C., Abichandani, K.K. and Kaushish, S.K. 2000. Wheat straw (untreated treated) consumption by tharparkar heifers during summer under stall fed conditions in *Third Biennial Conference of Animal Nutrition Association* on *Livestock feeding strategies in the new millennium* Nov. 7-9, 2000. CCSHAU, Hissar. No. 282.pp.206.

Mathur, B.K., Patel, A.K. and Kaushish, S.K. 2000. Utilization of tumba (*Citrullus colocynthis*) seed cake of desert for goat production 70 (4): 431-433.

Mathur, B.K., Patil, N.V., Mathur, A.C., Bohra, H.C., Bohra, R.C. and Sharma, K.L. 2009. Comparative Mineral Status of Jodhpur District Villagers' Cattle v/s Institute Farm Managed Tharparkar Cattle in Hot Arid Zone. *Proceedings of Animal Nutrition World Conference*14-17 Feb., New Delhi, India. pp.61.

Mathur, B.K., Singh J.P., Beniwal, R.K. and Singh, N.P. 2007. Utilization of salty shrub-Lani (*Salsola baryosma*) of arid region as drought feed for goats. FR-55. In *International Tropical Animal Nutrition Conference* October 4-7, 2007 NDRI, Karnal. pp.33.

Mathur B. K. 2008. Role of livestock in sustainability of farming systems in dry land and arid regions,pp16-24. In proceedings training programme on MTC on *Farming Systems Approach in Dryland and Arid Area* 16- 23 October, 2008.79p. For District level Government officers from all over India.At

Division of Agricultural Economics, Extension and Training, CAZRI, Jodhpur.

Mathur, B.K., Singh J.P., Sani, Nirmala, Pathak, K.M.L., Rathore,V.S., Kiraroo, B.D., Lukha, Arjun and Beniwal,R.K. 2009. Blood Biochemical & Mineral Profile of Camel Maintained on Different Feeding Regimen in Hot Arid Region of India. In: *Diversification of Animal Nutrition Research in the Changing Scenario volume II. Proceedings (Abstract Papers) of 13th Biennial conference of Animal Nutrition Society of India,*17-19 December 2009, *Banglore.*No.RNPE-134. Pp.80.

Mathur, B.K., Siyak, S.R. and Paharia, S. 2002. In corporation of *Prosopis juliflora* Pod husk as Ingredient of low cost ration for sheep in arid region. *Current Agriculture* 26 (1-2): 95-98.

Mathur, B.K., Siyak S.R. and Bohra, H. C. 2003. Feeding of Vilayati babool (*Prosopis juliflora*) pods powder to Marwari Goats in Arid Region. *Current Agriculture* 27(1-2):57-59.

Mathur, B.K., Sirohi, A.S., Mishra, A.K., Patel, A.K., Mathur, A.C., and Bohra, R.C. 2013. Performance of goats fed on Ardu (*Alianthu sexcelsa*) and Neem (*Azadirachta indica*) leaves in arid region. *Veterinary Practitioner*14 (2):48-350.

Mathur, B.K.2013.Juliflora Based Animal Feed. A Compendium of Agro Technologies. Agri-Tech Investors Meet,18-19 July 2013,New Delhi.pp.122-123. *National Agricultural Innovation Project* Krishi Anusandhan Bhawan–II.PusaCampus, New Delhi -110 012.127 p.

Mathur, B.K., Singh, J.P., Jat, S.R., Bhatt, R.K and Roy, M.M. 2013. Improving sheep productivity in the Thar Desert through strategic supplementation and health care.Abst.no.21,pp.58. In National Seminar on *Prospects in improving production, marketing and value addition of carpet wool* in collaboration with Indian Society for Sheep and Goat Production and Utilization at Central Sheep and Wool Research Institute 31st December, 2013, Arid Region Campus, Beechwal, Bikaner, Rajasthan, India-334006.

Mathur B.K.2014. Nutritional Constraints and Health Management of Livestock. Assistance to States for Control of animal Diseases (ASCAD) Seminar 2014 on *New Horizons: - Effective and Productive Livestock Development*19-20, February, pp.31-41.Animal Husbandry Department, Government of Rajasthan, Ajmer (Rajasthan) 42p.

Narain, P., Sharma, K.D., Rao, A.S., Singh, D.V., Mathur, B.K. and Ahuja, Usha Rani 2000. Strategy to combat Drought and Famine in the Indian Arid Zone. Central Arid Zone Research Institute, Jodhpur, 65 p.

Patidar, M., Mathur, B.K., Rajora, M.P. and Mathur, D. 2008. Effect of grass – legume strip cropping and fertility levels on yield and quality of fodder in silvipastoral system under hot arid condition. *Indian Journal of Agricultural Sciences* 78: 394-398.

Patil, N.V, Mathur, B.K, Bohra, H.C. and Patel, A.K. 2005. Relative preference index for arid forages for goat and sheep. National Seminar on *Conservation, Processing and Utilization of Monsoon Herbage for Augmenting Animal Production* December 17-18, 2005. Organized by Indian Society for Sheep and Goat Production and Utilization, at RRS, CSWRI, Bikaner. pp.190.

Patil, N.V., Mathur, B.K., Patel, A.K. and Bohra, R.C. 2011. Nutritional evaluation of *Colophospermum mopane* as fodder. *The Indian Veterinary Journal* 88 (1): 87-88.

Venkatateswarlu, J., Aantharam, K., Purohit, M.L., Khan, M.S., Bohra, H.C., Singh, K.C. and Mathur, B.K. 1992. Forage 2000 AD *The scenario for arid Rajasthan* CAZRI, Jodhpur pp. 32.

□□□

ABOUT THE EDITORS

Dr. R. K. Bhatt is basically Plant Physiologist having more than 30 years research experience in the fields of Plant Physiology,Natural Resource Management, Climate Change, Agroforestry and Seed Science & Technology. Contribution on plant processes under elevated CO_2, carbon sequestration and selection of photosynthetically efficient genotypes is noteworthy. He has published number of research papers in National and International Journals of high impact factors and chapters in books, proceedings, compendium and bulletins. On the basis of his significant contributions he has been awarded Outstanding Team Research Award of ICAR in 2002, Academy for Advancement of Agricultural Science Award of ISPP, New Delhi, presented by Honourable Dr. A. P. J. Abdul Kalam in 2003 and Dr. K. A. Shankar Narayan Memorial Award of CAZRI, Jodhpur in 2004. He was also awarded Fellow of Range Management Society of India (2007), National Institute of Ecology (2008) and Indian Society of Plant Physiology (2008). He has guided six students for their Ph.D. He visited UK, USA, Denmark, Sweden, Vietnam, China, Rome and Dubai.

Dr. Uday Burman, Principal Scientist (Plant Physiology) at ICAR-CAZRI, Jodhpur has extended his services on physiology of arid zone plants with emphasis on abiotic stresses for past 30 years. Born on 29 January 1959, he passed BSc (Botany, Hons.) in 1978 from Delhi University. He earned M.Sc. (Plant Physiology) from GBPUA&T Pantnagar and PhD (Plant Physiology) from H.A.U., Hisar. Dr. Burman also visited Nano Research Facility at Washington University at St. Louis (USA) and worked on 'Plant response to nano-particles'. Over the years he has also gained experience in intellectual property, technology management and its commercialization as Incharge Institute Technology Management Unit & as Member Consultancy Processing Cell. He has been Co-Principal Investigator in externally funded multidisciplinary projects from different agencies (AP Cess Fund, Indo-US, National Wasteland Development Board, NAIP on Nanotechnology and National Initiative on Climate Resilient Agriculture). He is life member of ISPP and AZRAI and has published more than 50 research papers, coauthored one bulletin, ten book chapters, four reviews and edited one book. He has reviewed papers from Journal of Plant Physiology, Journal of Nanomaterial, Journal of Agronomy and Crop Science, Indian Journal of Plant Physiology and Annals of Arid Zone besides serving as member (editorial board) of AAZ and ISPP.

Dr. D K Painuli, born on 28 September, 1954 in Rishikesh, Uttarakhand is a Soil Physicist by education, training and practice of nearly 40 years standing. He has been trained in various National and International Institutes of repute. After MSc from GBPUA & T, Pantnagar he completed PhD from CSSRI, Karnal. He has worked as Assistant Professor in BAU, Ranchi (1982-84); Postdoctoral Scientist/ Visting Scientist at IRRI, Philippines (1984-1987) and under ICTP - CNR at Istituto per La Chimica del Terreno, Pisa, Italy (1987-1989). He joined ICAR - IISS, Bhopal as Senior Scientist (1990 - 1997) and was Principal Scientist and Project Coordinator of the All India Coordinated Research Project on Tillage/ Soil Physical Constraints (1997-2004). For last more than 11 years he is working at CAZRI, Jodhpur and has headed Divisions of Extension and Training (2004-2007) and Integrated Land Use Management and Farming Systems (2008-2009). Besides Italy and Philippines Dr Painuli has visited IFS, Kandy, Sri Lanka and Michigan University, USA for professional activities. He has more than 100 publications to his credit including research articles, book chapters, technical bulletins, etc. He has been for a long period in the editorial board of Journal of Water Management. Dr Painuli's persistent interest is soil technology with focus on natural resource conservation; currently he is pursuing research on Bio-tillage.

Dr. D.V. Singh is presently working as Sr. Scientist (Agronomy) in Division of Natural Resources and Environment, Central Arid Zone Research Institute, Jodhpur. He obtained his Masters degree (Agronomy) in 1986 and Ph.D. degree in 1992 from HAU, Hisar. He joined CAZRI, Jodhpur as Scientist (Agronomy) in 1990. He has worked on long-term effects of organic and inorganic nutrient management in pearl millet - mustard sequence and nutrient management in clusterbean based cropping systems. He has worked on pearl millet based cropping systems involving crop sequences (pearl millet - green gram, pearl millet - clusterbean, pearl millet - mustard, pearl millet -isabgol) and intercropping with clusterbean. Irrigation and nutrient management studies in *Aloe vera* and seed treatment studies in cumin were also conducted. He has also worked on testing, validation/calibration of CERES-Millet, Mustard-GRO and INFOCROP models under arid conditions. He has about 18 research papers, 22 book chapters and three bulletins to his credit. Presently he is engaged in mid-season and end of season yield forecast of pearl millet, sorghum, cotton, wheat and mustard crops in arid Rajasthan and issuing bi-weekly weather based agro-advisories for farmers of five arid districts of Rajasthan.

Dr Ramavtar Sharma graduated from SKN, College of Agriculture, Jobner, Rajasthan and post-graduation and doctorate from IARI, New Delhi in Genetics. He started his career as Assistant Professor in Rajasthan Agricultural University, Bikaner in 1989 and selected Associate professor in 1996 and Professor in 2004. He did his postdoctoral training during 1995-1997, from NRCPB, New Delhi, with fellowship of DBT. Recently, he has joined Central Arid Zone Research Institute, Jodhpur as Principal Scientist. Dr Sharma has vast experience of working in cytogenetics, genetics, plant breeding and molecular biology. He has published more than 100 research papers and abstracts in national and international journals. He has guided 5 PhD students and 22 MSc students. He developed easy and cheap protocols for DNA isolation, direct PCR, biolistic gun mediated transformation and mutagenesis through microwave. He developed heat, drought and salt tolerant mutants in *Vigna aconitifolia* and fast growing thick tetraploid somaclones of *Aloe vera*. Additionally he has worked on molecular characterization, diversity analysis and tagging; and has also identified ESTs associated with heat tolerance in *Vigna aconitifolia*. Presently he is working on clusterbean improvement.

Dr. S.P.S. Tanwar, born on 20th December, 1974 is an Agronomist of repute having diverse experience of 15 years in research, teaching, extension and policy planning. He has completed his graduation from SKN college of Agriculture Jobner and Post-Graduation and doctorate from Maharana Pratap University of Agriculture & Technology, Udaipur. He also got trained in molecular biology at NRC on DNA Fingerprinting, NBPGR, New Delhi. He joined MPUAT as Assistant Professor and worked as Agronomist in AICRP on chickpea at Kota and AICRP on Wheat at Udaipur besides teaching at all levels (2005-2008). Lateron he got selected as Senior Scientist (Agronomy) in ICAR- Central Arid Zone Research Institute, RRS, Pali (2008-12) and presently working at CAZRI, Jodhpur campus. His research work comprises of resource conservation technologies, crop agronomy, climate change and integrated farming system. He has more than 80 publications to his credit including research articles, book chapters, books, Technical bulletins, pamphlets etc. He has also represented India in NPC- APO workshop on farming system in Colombo, Sri Lanka.